METHODS OF MATHEMATICAL PHYSICS, MECHANICS AND GENERAL PROPERTIES OF MATTER

METHODS OF MATHEMATICAL PHYSICS, MECHANICS AND GENERAL PROPERTIES OF MATTER

Chittaranjan Basu MSc

Head of the Department of Physics (Retd.)
Asutosh College, Kolkata

New Central Book Agency (P) Ltd

LONDON

HYDERABAD ERNAKULAM BHUBANESWAR

NEW DELHI KOLKATA PUNE GUWAHATI

REGD. OFFICE

8/1 Chintamoni Das Lane, Kolkata 700 009, India
email: ncbapvtltd@eth.net

OVERSEAS

NCBA (UK) Ltd, 149 Park Avenue North, Northampton, NN3 2HY, UK
email: ncbauk@yahoo.com

EXPORT

NCBA Exports Pvt. Ltd, 212 Shahpur Jat, New Delhi 110 049
email: ncbaexp@ncbapvtltd.com

BRANCHES

212 Shahpur Jat
New Delhi 110 049
email: ncbadel@ncbapvtltd.com

Shop Nos. 3 & 4, Vinayak Towers
681/B Budhwar Peth, Appa Balwant Chowk
Pune 411 002
email: ncbapune@ncbapvtltd.com

Shop No. 15, Bharati Towers, Ground Floor
Forest Park, **Bhubaneswar** 751 001
email: ncbabub@ncbapvtltd.com

House No. 3-1-315, 1st Floor
Nimboliadda, Kachiguda, **Hyderabad** 500 027
email: ncbahydb@ncbapvtltd.com

GSS Shopping Complex, 1st Floor
Opposite College Post Office, Convent Road
Ernakulam 682 035
email: ncbaernk@ncbapvtltd.com

Radhamayee Complex (Second Floor)
Opp. SBI, Panbazar, Jaswanta Road
Guwahati 781 001
email: ncbaguwahati@ncbapvtltd.com

METHODS OF MATHEMATICAL PHYSICS, MECHANICS
AND GENERAL PROPERTIES OF MATTER • Chittaranjan Basu

First Published: 1998
Second Edition: August 2010
Reprinted: 2014

PUBLISHER AND TYPESETTER
New Cental Book Agency (P) Ltd
8/1 Chintamoni Das Lane, Kolkata 700 009

PRINTER
New Central Book Agency (P) Ltd
Web-Offset Division, Dhulagarh, Sankrail, Howrah

COVER DESIGNER
Dulal Halder

PROOF CHECKING
Author and Prabhat Jas

ISBN 978 81 7381 317 7

Price ₹ 325.00

Dedicated to

ALL MY WELL-WISHERS

Preface to the Second Edition

It is highly gratifying for me to write down the preface for the Second Edition in which the book has been thoroughly revised and corrections have been incorporated as far as practicable.

Calcutta University Honours Questions up to 2000, have been answered.

I express my gratitude to all those well-wishers who have helped me in the circulation of this book.

My thanks are due to all personalities and employees of 'New Central Book Agency Private Limited' but for whose help this volume would not have been published.

All suggestions and criticisms for the improvement of the book will be cordially received.

Chittaranjan Basu

The 15th January 2010
49/1 Gurupada Halder Road
Kolkata 700 026

Preface to the First Edition

Written in accordance with the new C.U.[Hons.] syllabus this humble publication claims no originality. In fact, in writing this text all the standard books have been freely consulted. In some cases, I was so impressed that in few cases even their methods have been followed by me without proper permission and acknowledgement. I must apologise and express my gratitude to these renowned authors.

I do not claim this book to be of super standard. What I have attempted is to present the problems in such a manner that a student of average merit finds no difficulty in going through them. He may choose my writing as the course of first study and then consult higher volumes.

At this stage I must recall, with gratitude, the name of Prof. Jnanesh Patranabis, but for whose active cooperation and goodwill this book would not have been published.

My compliments are due to Prof. R. K. Ghosh and Prof. K. C. Maity for the interest they have taken for the publication.

I also feel indebted to Mr J. N. Sen, publisher of the book, who did not hesitate to publish the book with least acquaintance with me.

My graditude goes in full measure to the honourable and kind-hearted surgeon Prof. Asoke Banerjee, F.R.C.S. who was the first to encourage me in this writing.

I express my thanks to the workers and personalities of 'New Central Book Agency (P) Ltd' for the job they have done.

Finally, I express deep regret for the drawbacks of the book which has been written by me in a state of health not at all fit for the job.

Any attempt to point out my shortcomings will be highly appreciated.

Chittaranjan Basu

The 15th August 1997
49/1 Gurupada Halder Road
Kolkata 700 026

Acknowledgements

1.	Applied Mathematics for Engineers and Physicists	— *Pipes and Harvill*
2.	Theoretical Mechanics	— *Murray R. Spiegel*
3.	Classical Mechanics	— *Goldstein*
4.	Mechanics	— *Synge and Griffiths*
5.	Properties of matter	— *C. J. Smith*
6.	Properties of matter	— *Newman and Searle*
7.	Classical Mechanics	— *Gupta and Satya Prakash*
8.	Mathematical Physics	— *Rajput*
9.	Mathematical Physics	— *Gupta, Yadav and Malik*
10.	Mathematical Physics	— *P. K. Chattopadhyay*
11.	Introduction to Classical Mechanics	— *Takwale and Purnanik*
12.	Classical Mechanics of Particles and Rigid bodies	— *Kiran C. Gupta*
13.	Theoretical Physics	— *Joos*
14.	Theoretical Physics	— *Gray*
15.	Properties of Matter	— *Champion and Davy*
16.	Vector Analysis	— *Spiegel*

Contents

CHAPTER	*PAGES*

PART I

METHODS OF MATHEMATICAL PHYSICS AND MECHANICS

CHAPTER I

VECTOR ANALYSIS

Fundamental Concepts :

Definition of vectors in terms of transformation properties. Vector algebra ; Scalar and vector products, difference between polar and a real vector ; differentiation of vectors and integration of vectors. Gradient operator ; Divergence and curl of vectors—Applications to simple problems ; Gauss's Divergence theorem, Stokes' theorem and Green's theorem—Application in simple cases. Orthogonal curvilinear co-ordinate system—unit vectors in curvilinear co-ordinate systems (illustrators by spherical and cylindrical polar co-ordinate system).

Physical quantities are broadly classified into two types—Vectors and Scalars.

Scalar quantities are those which have magnitude only ; *e.g.* mass, density, speed, volume etc.

Vector quantities are those which have magnitude as well as direction ; e.g. displacement, velocity, force etc. A vector is represented by a directed segment $\overrightarrow{AB}$ of a straight line on which there are distinct initial and terminal points.

The length of the straight line represents on a certain chosen scale the magnitude of the vector and the arrow gives the direction of the vector. In this writing we shall represent vectors with arrows above letters such as $\vec{A}$, $\vec{a}$ etc. The magnitude or the modulus of the vectors is represented by the unarrowed letters such as A, a etc.

Tensers :

In many physical problems vector notation is not sufficiently general to express the relation between quantities involved. For example, in electrostatics, if a dielectric medium is isotropic, the relation between the electric displacement vector $\vec{D}$ and the electric intensity vector $\vec{E}$ is given by the vector equation $\vec{D} = K\vec{E}$, where the scalar K is the dielectric constant of the medium. If however the dielectric medium is not isotropic, the electric displacement vector $\vec{D}$ does not have the same direction as the vector $\vec{E}$, so that the scalar K must be replaced by a more general quantity to effect the required change in direction and magnitude. Such a quantity is called a tensor. A similar relation arises in elasticity. The stress vector $\vec{F}$ and the strain vector $\vec{X}$ are related in an isotropic medium as $\vec{F} = K\vec{X}$, where k is a scalar. If the medium is not isotropic, $\vec{F}$ and $\vec{X}$ will not have the same direction and the above equation takes the more general form in which K is a tensor.

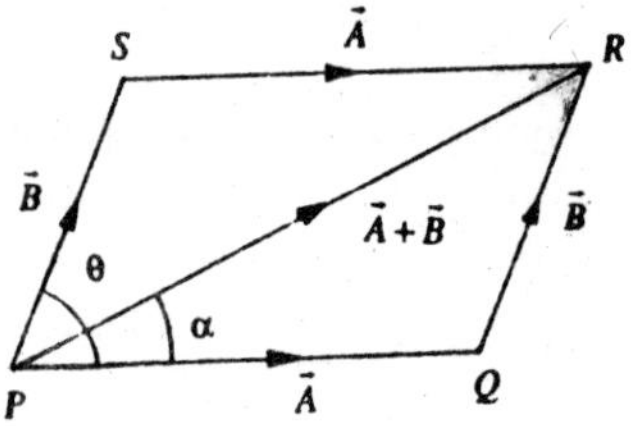

Fig. 1.1

Two vectors will be called identical when their magnitudes are same and directions are parallel.

Addition of Vectors :

Vector addition can be done by the law of parallelogram of vectors.

Thus, if $\vec{A}$ is represented by $\overrightarrow{PQ}$ and $\vec{B}$ by $\overrightarrow{PS}$, then $\vec{A} + \vec{B}$ will be represented vectorially by $\overrightarrow{PR}$. Thus $\vec{A} + \vec{B} = \overrightarrow{PQ} + \overrightarrow{QR} = \overrightarrow{PR}$.

In magnitudes $PR =$

$$= \sqrt{PQ^2 + QR^2 - 2PQ \cdot QR \cos (180° - \theta)}$$

$$= \sqrt{PQ^2 + QR^2 + 2PQ \cdot QR \cos \theta}$$

$$\text{or, } |\vec{A} + \vec{B}| = \sqrt{|\vec{A}|^2 + |\vec{B}|^2 + 2|\vec{A}||\vec{B}| \cos \theta}$$

$= \sqrt{A^2 + B^2 + 2AB \cos \theta}$, θ being the angle between the directions of $\vec{A}$ and $\vec{B}$. Again, if α be the angle between the directions of the resultant $\vec{A} + \vec{B}$ and the component vector $\vec{A}$, Then $\tan \alpha = \dfrac{B \sin \theta}{A + B \cos \theta}$.

Also $\vec{B} + \vec{A} = \overrightarrow{PS} + \overrightarrow{SR} = \overrightarrow{PR}$ (by triangle of vectors)

$\therefore \vec{A} + \vec{B} = \vec{B} + \vec{A}$ *i.e*, commutative law holds for the addition of two vectors.

Let us confine our attention to Fig. 1.2 which is self explanatory.

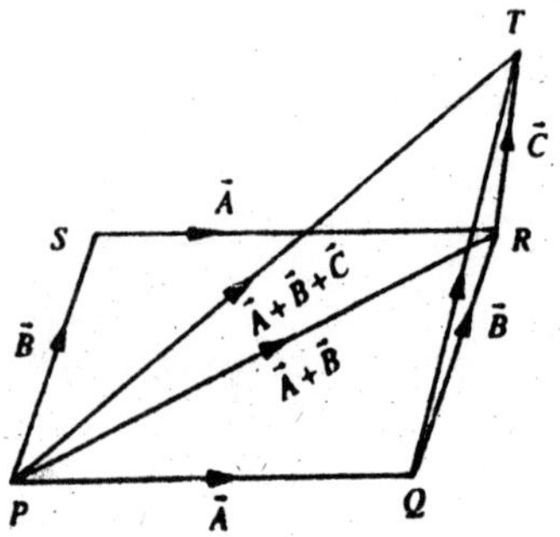

Fig. 1.2

Now, $(\vec{A} + \vec{B}) + \vec{C}$
$= (\vec{PQ} + \vec{QR}) + \vec{RT} = \vec{PR} + \vec{RT} = \vec{PT}$
Again $\vec{A} + (\vec{B} + \vec{C}) = \vec{PQ} + (\vec{QR} + \vec{RT}) = \vec{PQ} + \vec{QT} = \vec{PT}$
$\therefore\ (\vec{A} + \vec{B}) + \vec{C} = \vec{A} + (\vec{B} + \vec{C})$

i.e., associative law holds for addition of two vectors.

Also $n(\vec{A} + \vec{B}) = n\vec{A} + n\vec{B}$ (Fig. 1.3).

This is distributive law.

Let $\vec{PQ} = \vec{A}$
$\vec{PU} = n\vec{A}$
$\vec{QR} = \vec{B}$
and $\vec{UT} = n\vec{B}$
$\therefore\ \vec{PT} = n(\vec{A} + \vec{B})$

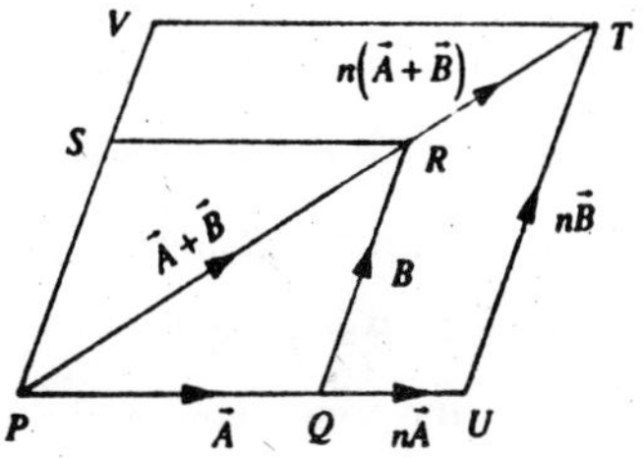

Fig. 1.3

Vector in terms of its components :

Let $\vec{OP} = \vec{A}$ and its components are A_x, A_y, A_z along X, Y and Z directions. Let $\hat{i}, \hat{j}, \hat{k}$ be unit vectors along OX, OY and OZ respectively.

[Unit vectors are vectors in the specified directions each having a unit magnitude.]

Now $\vec{A} = \vec{OP} = \vec{ON} + \vec{NP}$
$= \vec{OM} + \vec{MN} + \vec{NP}$

(PN is $\perp$ to XY plane)

$$= \hat{i}A_x + \hat{j}A_y + \hat{k}A_z \qquad (1.1)$$

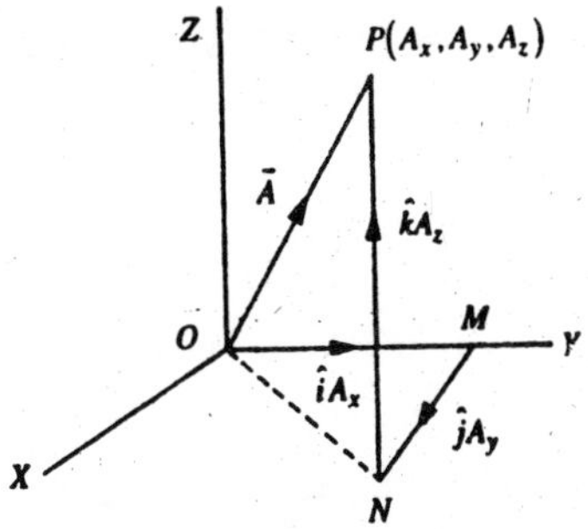

Fig. 1.4

$$|\vec{A}| = OP = \sqrt{(OM^2 + MN^2 + NP^2)} = \sqrt{A_x^2 + A_y^2 + A_z^2} \qquad (1.2)$$

Scalar or Dot Product of two Vectors :

Sometimes two vectors can be multiplied to give a scalar quantity.

e.g. : (Force) (Displacement) = Work.

This is called scalar or dot product of two vectors.

Mathematically, $\vec{A} \cdot \vec{B} = AB \cos\theta$ where θ is the angle between the directions of $\vec{A}$ and $\vec{B}$.

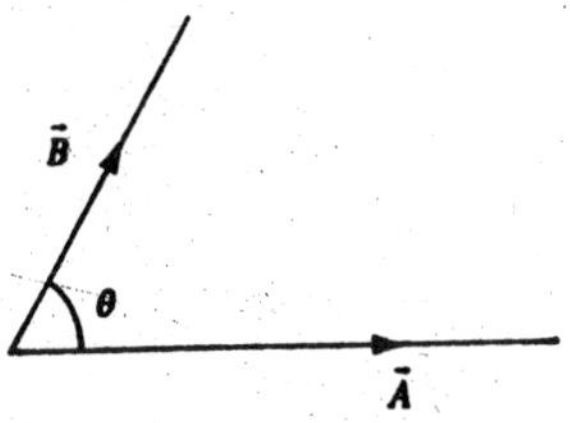

Fig. 1.5

If $\theta = 0$, $\vec{A} \cdot \vec{B} = AB$

If $\theta = 90°$, $\vec{A} \cdot \vec{B} = 0$

Coro. : $\vec{A} \cdot \vec{A} = AA = A^2$

Similarly $\hat{i} \cdot \hat{i} = \hat{k} \cdot \hat{k} = \hat{j} \cdot \hat{j} = 1$ and $\hat{i} \cdot \hat{j}$ etc. $= 0$.

Now $\vec{A} \cdot \vec{B} = (\hat{i} Ax + \hat{j} Ay + \hat{k} Az) \cdot (\hat{i} Bx + \hat{j} By + \hat{k} Bz)$

$$= AxBx + AyBy + AzBz \quad (1.3)$$

Again $\vec{A} \cdot \vec{B} = AB \cos\theta$ and $\vec{B} \cdot \vec{A} = BA \cos(-\theta) = AB \cos\theta$

$\therefore \quad \vec{A} \cdot \vec{B} = \vec{B} \cdot \vec{A}$.

Hence, the commutative law holds for scalar product of two vectors.

Now $\vec{A} \cdot (\vec{B} + \vec{C}) = \vec{A} \cdot \vec{B} + \vec{A} \cdot \vec{C}$.

Proof : L. H. S.

$$= (\hat{i}Ax + \hat{j}Ay + \hat{k}Az) \cdot [\,\hat{i}(Bx + Cx) + \hat{j}(By + Cy) + \hat{k}(Bz + Cz)\,]$$

$$= Ax(Bx + Cx) + Ay(By + Cy) + Az(Bz + Cz)$$

$$= (AxBx + AyBy + AzBz) + (AxCx + AyCy + AzCz)$$

$$= \vec{A} \cdot \vec{B} + \vec{A} \cdot \vec{C} = \text{R. H. S.}$$

Hence, the distributive law holds for scalar product of two vectors.

Vector or Cross Product of two Vectors :

Sometimes two vectors can be multiplied to give a vector quantity.

e.g. : Force (F) acting on a current-carrying conductor (element) placed in a magnetic field (H) is given by $F = i\,dl\,H \sin\theta$ (Fig. 1.5a), where dl = elementary length of the conductor, i = current in e.m.u. carried by the conductor.

Vector product of two vectors $\vec{A}$ and $\vec{B}$ is given by $\vec{A} \times \vec{B} = \hat{n} AB \sin\theta$, (1.4) (Fig. 1.6), where $\hat{n}$ is unit vector perpendicular to the plane containing $\vec{A}$ and $\vec{B}$, and coinciding with the direction of motion of a right-handed screw as it is rotated from $\vec{A}$ to $\vec{B}$.

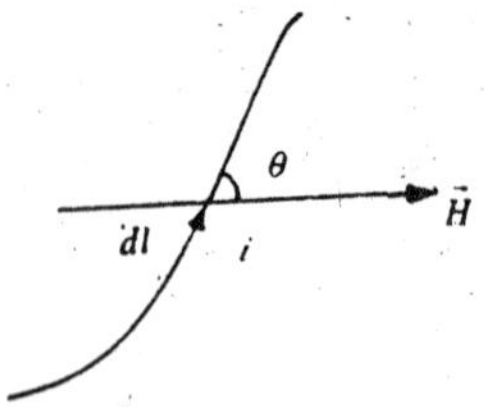

Fig. 1.5a

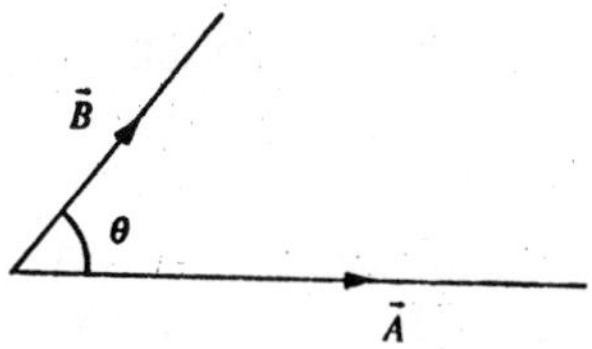

Fig. 1.6

Clearly, $\vec{B} \times \vec{A} = -\hat{n} AB \sin\theta$

$\therefore \vec{A} \times \vec{B} = -\vec{B} \times \vec{A}$ and the commutative law does not hold.

It is evident that $\hat{i} \times \hat{i} = \hat{j} \times \hat{j} = \hat{k} \times \hat{k} = 0$

and $\hat{i} \times \hat{j} = \hat{k}$; $\hat{j} \times \hat{i} = -\hat{k}$ etc.

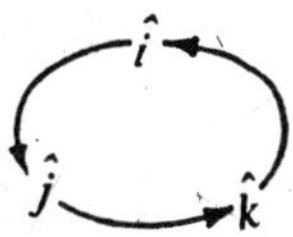

$\therefore \vec{A} \times \vec{B} = (\hat{i}Ax + \hat{j}Ay + \hat{k}Az) \times (\hat{i}Bx + \hat{j}By + \hat{k}Bz)$

$= \hat{k}AxBy - \hat{j}AxBz - \hat{k}AyBx + \hat{i}AyBz + \hat{j}(AzBx) - \hat{i}AzBy$

$= \hat{i}(AyBz - AzBy) + \hat{j}(AzBx - AxBz) + \hat{k}(AxBy - AyBx)$

$$= \begin{vmatrix} \hat{i} & \hat{j} & \hat{k} \\ Ax & Ay & Az \\ Bx & By & Bz \end{vmatrix} \qquad (1.4a)$$

It is to be noted that the magnitude of the vector product $|(\vec{A} \times \vec{B})| = AB \sin\theta$ represents the area of the parallelogram formed with $\vec{A}$ and $\vec{B}$ as coterminous edges.

Polar and Axial Vectors :

It is sometimes necessary to distinguish between the vectors which results from the vector product of two vectors and an ordinary vector. The form implies an area in a definite plane, the contour of the area being traced out in an anticlockwise direction as viewed from a point on the resulting vector. Both these properties are lacking completely in an ordinary vector where only two properties are magnitude and direction. Ordinary vectors are known as polar vectors and the vector multiplication of two polar vectors results in a vector known as the axial vector. The axial vector involves the idea of rotation about an axis perpendicular to the plane area it represents. The essential difference between a polar and an axial vector is that the scalar components of a polar vector change their signs when the positive directions of the co-ordinate axes are reversed, but those of the axial vector do not.

Scalar Tripple Product :

(Product of three vectors to give a scalar quantity)

$$\vec{A}.(\vec{B} \times \vec{C}) = \vec{B}.(\vec{C} \times \vec{A}) = \vec{C}.(\vec{A} \times \vec{B}) \qquad (1.5)$$

To prove that the above represents the volume of the parallelopiped formed with the vectors $\vec{A}$, $\vec{B}$ and $\vec{C}$ as co-terminous edges (Fig. 1.7). Let $\hat{n}$ be a unit vector perpendicular to the parallelogram l in the direction of $\vec{B} \times \vec{C}$.

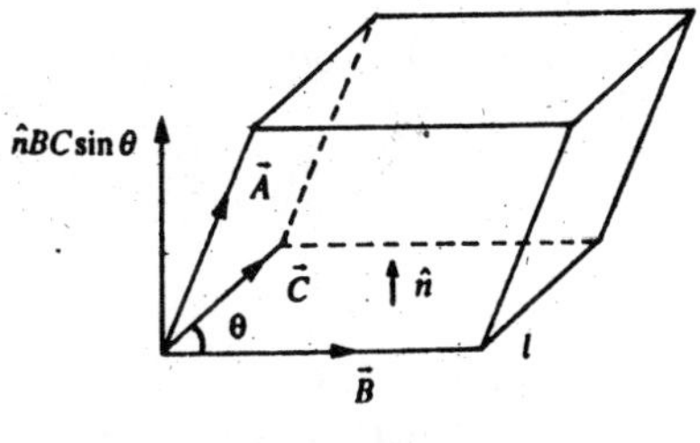

Fig. 1.7

Volume of the parallelopiped = $h\,(BC\sin\theta)$, where h = height of the parallelopiped and θ is the angle shown.

$$= (\vec{A}\,.\,\hat{n})\,BC\sin\theta = \vec{A}\,.\,(\vec{B}\times\vec{C})$$

In terms of Components,

$$\vec{A}\,.\,(\vec{B}\times\vec{C})$$

$$= (\hat{i}Ax + \hat{j}Ay + \hat{k}Az)\,.\,\{(\hat{i}Bx + \hat{j}By + \hat{k}Bz)\times(\hat{i}Cx + \hat{j}Cy + \hat{k}Cz)\}$$

$$= (\hat{i}Ax + \hat{j}Ay + \hat{k}Az)\,.\,\{\hat{i}(ByCz - BzCy) + \hat{j}(BzCx - BxCz) + \hat{k}(BxCy - ByCx)\}$$

$$= Ax\,(ByCz - BzCy) + Ay\,(BzCx - BxCz) + Az\,(BxCy - ByCx)$$

By expanding in terms of the components, the truth of equation (1.5) can be established.

Since $\vec{B}\times\vec{C} = -\vec{C}\times\vec{B}$

$$\vec{A}.\left(\vec{B}\times\vec{C}\right) = -\vec{A}.\left(\vec{C}\times\vec{B}\right).$$

In determinant notation

$$\vec{A}.\left(\vec{B}\times\vec{C}\right) = \begin{vmatrix} Ax & Ay & Az \\ Bx & By & Bz \\ Cx & Cy & Cz \end{vmatrix}$$

Hence, the truth is maintained so long as the order of $\vec{A}$, $\vec{B}$ and $\vec{C}$ is maintained.

Again by expanding in terms of the components it can be proved that

$$\vec{A}\,.\,(\vec{B}\times\vec{C}) = (\vec{A}\times\vec{B})\,.\,\vec{C}.$$

Thus the dot and the cross can be interchanged without affecting the result.

N.B. : The necessary and sufficient condition for the vector $\vec{A}$, $\vec{B}$ and $\vec{C}$ to be coplanner is that $\vec{A}\,.\,(\vec{B}\times\vec{C}) = 0$. Since $\vec{B}\times\vec{C}$ is a vector perpendicular to the plane containing $\vec{B}$ and $\vec{C}$, $\vec{A}$ must also be perpendicular to that plane, *i.e.*, it must lie in the plane of $\vec{B}$ and $\vec{C}$.

Also $\vec{A}\,.\,(\vec{A}\times\vec{C}) = (\vec{A}\times\vec{A})\,.\,\vec{C} = 0$.

Vector Tripple Product :

(Product of three vectors to give a vector.)

To prove that $\vec{A}\times(\vec{B}\times\vec{C}) = \vec{B}\,(\vec{A}\,.\,\vec{C}) - \vec{C}\,(\vec{A}\,.\,\vec{B})$ (1.6)

$$\text{L. H. S.} = (\hat{i}Ax + \hat{j}Ay + \hat{k}Az)\times\begin{vmatrix} \hat{i} & \hat{j} & \hat{k} \\ Bx & By & Bz \\ Cx & Cy & Cz \end{vmatrix}$$

$$= (\hat{i}Ax + \hat{j}Ay + \hat{k}Az)\times\{\hat{i}(ByCz - BzCy) + \hat{j}(BzCx - BxCz) + \hat{k}(BxCy - ByCx)\}$$

$$= \begin{vmatrix} \hat{i} & \hat{j} & \hat{k} \\ Ax & Ay & Az \\ ByCz - BzCy & BzCx - BxCz & BxCy - ByCx \end{vmatrix}$$

$$= \hat{i}\{ Ay(BxCy - ByCx) - Az(BzCx - BxCz)\}$$
$$+ \hat{j}\{ Az(ByCz - BzCy) - Ax(BxCy - ByCx)\}$$
$$+ \hat{k}\{ Ax(BzCx - BxCz) - Ay(ByCz - BzCx)\}$$

Again $\vec{B}(\vec{A}.\vec{C}) - \vec{C}(\vec{A}.\vec{B})$

$$(\hat{i}Bx + \hat{j}By + \hat{k}Az)(AxCx + AyCy + AzCz)$$
$$-(\hat{i}Cx + \hat{j}Cy + \hat{k}Cz)(AxBx + AyBy + AzBz)$$
$$= \hat{i}\{ Ay(BxCy - ByCx) - Az(BzCx - BxCz)\}$$
$$+ \hat{j}\{ Az(ByCz - BzCy) - Ax(BxCy - ByCx)\}$$
$$+ \hat{k}\{ Ax(BzCx - BxCz) - Ay(ByCz - BzCx)\}$$

Hence, the result.

$$(\vec{A} \times \vec{B}) \times \vec{C} = -\vec{C} \times (\vec{A} \times \vec{B})$$
$$= -\{\vec{A}(\vec{C}.\vec{B}) - \vec{B}(\vec{C}.\vec{A})\} = \vec{B}(\vec{C}.\vec{A}) - \vec{A}(\vec{C}.\vec{B})$$
$$\neq \vec{A} \times (\vec{B} \times \vec{C})$$

Hence, the associative law does not hold for the vector tripple product.

Vector Quadruple Product :

To prove that

$$\left.\begin{aligned} &(\vec{A} \times \vec{B}) \times (\vec{C} \times \vec{D}) \\ &= \vec{B}(\vec{A}.\vec{C} \times \vec{D}) - \vec{A}(\vec{B}.\vec{C} \times \vec{D}) \\ &= \vec{C}(\vec{A}.\vec{B} \times \vec{D}) - \vec{D}(\vec{A}.\vec{B} \times \vec{C}) \end{aligned}\right\} \qquad (1.7)$$

Let $\vec{Y} = \vec{A} \times \vec{B}$

$$\therefore \text{ L.H.S.} = \vec{Y} \times (\vec{C} \times \vec{D}) = \vec{C}(\vec{Y}.\vec{D}) - \vec{D}(\vec{Y}.\vec{C})$$
$$= \vec{C}(\vec{A} \times \vec{B}.\vec{D}) - \vec{D}(\vec{A} \times \vec{B}.\vec{C})$$
$$= \vec{C}(\vec{A}.\vec{B} \times \vec{D}) - \vec{D}(\vec{A}.\vec{B} \times \vec{C})$$

Again putting $\vec{Z} = \vec{C} \times \vec{D}$.

$$\left.\begin{aligned} \text{L. H. S.} &= (\vec{A} \times \vec{B}) \times \vec{Z} = -\vec{Z} \times (\vec{A} \times \vec{B}) \\ &= -\{\vec{A}(\vec{Z}.\vec{B}) - \vec{B}(\vec{Z}.\vec{A})\}, \\ &= \vec{B}(\vec{A}.\vec{C} \times \vec{D}) - \vec{A}(\vec{B}.\vec{C} \times \vec{D}) \end{aligned}\right\}$$

The Differential Operator $\vec{\nabla}$ (Del) :

Let us consider the function $u = f(x, y, z) = C$, where C is a constant. The function represents a surface in space. By changing C, a family of surfaces is obtained. If C becomes $C + \Delta C$, then $f(x, y, z) = C + \Delta C$ is a surface close to initial surface (Fig. 1.8).

In passing from one surface to another, the change in C is the same irrespective of the path chosen. But the rate of change will be maximum if we

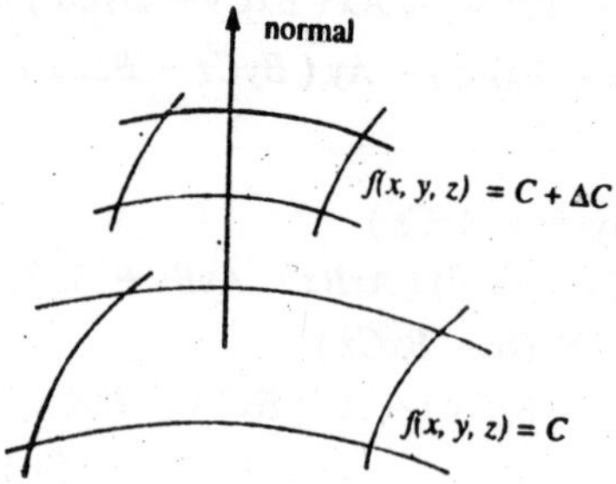

Fig. 1.8

proceed along the normal to the surface. If $\frac{d}{dn}$ represents differentiation along the normal, then $\frac{du}{dn}$ is the magnitude of a vector called the gradiant of u. If $\hat{n}$ is unit vector along the outward drawn normal, then

$$\hat{n}\frac{du}{dn} = \text{gradiant } u = \vec{\nabla} u \tag{1.8}$$

when $\vec{\nabla}$ is called del or nabla.

Let $\hat{n}\frac{\partial u}{\partial n} = \hat{i}a_x + \hat{j}a_y + \hat{k}a_z$

$$\therefore \quad \frac{\partial u}{\partial n}\hat{n}\,.\,\hat{i} = a_x = \frac{\partial u}{\partial n}\cos\alpha = \frac{\partial u}{\partial x}$$

Similarly, $a_y = \frac{\partial u}{\partial y}, \quad a_z = \frac{\partial u}{\partial z}$

$$\text{Consequently, } \vec{\nabla} u = \hat{n}\frac{du}{dn} = \hat{i}\frac{\partial u}{\partial x} + \hat{j}\frac{\partial u}{\partial y} + \hat{k}\frac{\partial u}{\partial z}$$

$$= \left(\hat{i}\frac{\partial}{\partial x} + \hat{j}\frac{\partial}{\partial y} + \hat{k}\frac{\partial}{\partial z}\right)u \tag{1.9}$$

So the differential vector operator $\vec{\nabla}$ is defined as

$$\vec{\nabla} \equiv \hat{i}\frac{\partial}{\partial x} + \hat{j}\frac{\partial}{\partial y} + \hat{k}\frac{\partial}{\partial z} \tag{1.10}$$

Problem : Find $\vec{\nabla} u$, where $u = x^2 + 2yz$

Ans. $\vec{\nabla} u = \left(\hat{i}\frac{\partial}{\partial x} + \hat{j}\frac{\partial}{\partial y} + \hat{k}\frac{\partial}{\partial z}\right)(x^2 + 2yz)$

$= 2x\hat{i} + 2z\hat{j} + 2y\hat{k}.$

Considering two functions $u = f(x, y, z)$ and $v = f_1(x, y, z)$, it can be shown that

$$\vec{\nabla}(u + v) = \vec{\nabla} u + \vec{\nabla} v \tag{1.11}$$

$$\vec{\nabla}(uv) = u\,\vec{\nabla} v + v\,\vec{\nabla} u \tag{1.12}$$

$$\text{Again } \vec{\nabla}\,.\,(\vec{\nabla} u) = \left(\hat{i}\frac{\partial}{\partial x} + \hat{j}\frac{\partial}{\partial y} + \hat{k}\frac{\partial}{\partial z}\right).\left(\hat{i}\frac{\partial u}{\partial x} + \hat{j}\frac{\partial u}{\partial y} + \hat{k}\frac{\partial u}{\partial z}\right)$$

$$= \frac{\partial^2 u}{\partial x^2} + \frac{\partial^2 u}{\partial y^2} + \frac{\partial^2 u}{\partial z^2}$$

or, $$\nabla^2 u = \left(\frac{\partial^2}{\partial x^2} + \frac{\partial^2}{\partial y^2} + \frac{\partial^2}{\partial z^2}\right) u$$

Thus $$\nabla^2 \equiv \frac{\partial^2}{\partial x^2} + \frac{\partial^2}{\partial y^2} + \frac{\partial^2}{\partial z^2} \qquad (1.13)$$

The *operator* ∇^2 is called the *Laplacian Operator.*

If $\vec{A}$ is a vector and u is a scalar,

$$\left.\begin{aligned} \text{then} \quad \vec{\nabla} \cdot \vec{A} &= \frac{\partial Ax}{\partial x} + \frac{\partial Ay}{\partial y} + \frac{\partial Az}{\partial z} \\ \text{and} \quad \vec{\nabla} u &= \hat{i}\frac{\partial u}{\partial x} + \hat{j}\frac{\partial u}{\partial y} + \hat{k}\frac{\partial u}{\partial z} \end{aligned}\right\} \qquad (1.14)$$

To prove that $\vec{\nabla} \cdot (u\vec{A}) = u(\vec{\nabla} \cdot \vec{A}) + \vec{\nabla} u \cdot \vec{A}$ (1.15)

$$\text{L.H.S.} = \left(\hat{i}\frac{\partial}{\partial x} + \hat{j}\frac{\partial}{\partial y} + \hat{k}\frac{\partial}{\partial z}\right) \cdot (u\hat{i}Ax + u\hat{j}Ay + u\hat{k}Az)$$

$$= \frac{\partial}{\partial x}(uAx) + \frac{\partial}{\partial y}(uAy) + \frac{\partial}{\partial z}(uAz)$$

$$= u\left(\frac{\partial Ax}{\partial x} + \frac{\partial Ay}{\partial y} + \frac{\partial Az}{\partial z}\right) + \frac{\partial u}{\partial x}Ax + \frac{\partial u}{\partial y}Ay + \frac{\partial u}{\partial z}Az$$

$$= u(\vec{\nabla} \cdot \vec{A}) + (\hat{i}Ax + \hat{j}Ay + \hat{k}Az) \cdot \left(\hat{i}\frac{\partial u}{\partial x} + \hat{j}\frac{\partial u}{\partial y} + \hat{k}\frac{\partial u}{\partial z}\right)$$

$$= u\vec{\nabla} \cdot \vec{A} + \vec{\nabla} u \cdot \vec{A}.$$

To prove that $\vec{\nabla} \times (u\vec{A}) = u(\vec{\nabla} \times \vec{A}) + \vec{\nabla} u \times \vec{A}$ (1.16)

$$\text{L.H.S.} = \left(\hat{i}\frac{\partial}{\partial x} + \hat{j}\frac{\partial}{\partial y} + \hat{k}\frac{\partial}{\partial z}\right) \times (\hat{i}uAx + \hat{j}uAy + \hat{k}uAz)$$

$$= \begin{vmatrix} \hat{i} & \hat{j} & \hat{k} \\ \frac{\partial}{\partial x} & \frac{\partial}{\partial y} & \frac{\partial}{\partial z} \\ uAx & uAy & uAz \end{vmatrix} = \hat{i}\left\{\frac{\partial}{\partial y}(uAz) - \frac{\partial}{\partial z}(uAy)\right\}$$

$$+ \hat{j}\left\{\frac{\partial}{\partial z}(uAx) - \frac{\partial}{\partial x}(uAz)\right\} + \hat{k}\left\{\frac{\partial}{\partial x}(uAy) - \frac{\partial}{\partial y}(uAx)\right\}$$

$$= \sum \hat{i}\left\{u\frac{\partial Az}{\partial y} + Az\frac{\partial u}{\partial y} - u\frac{\partial Ay}{\partial z} - Ay\frac{\partial u}{\partial z}\right\}$$

$$= u\sum \hat{i}\left(\frac{\partial Az}{\partial y} - \frac{\partial Ay}{\partial z}\right) + \sum i\left(\frac{\partial u}{\partial y}Az - \frac{\partial u}{\partial z}Ay\right)$$

$$= u(\vec{\nabla} \times \vec{A}) + \vec{\nabla} u \times \vec{A}.$$

To prove that $\nabla^2\left(\frac{1}{r}\right) = 0$

$$\nabla^2\left(\frac{1}{r}\right) = \left(\frac{\partial^2}{\partial x^2} + \frac{\partial^2}{\partial y^2} + \frac{\partial^2}{\partial z^2}\right)\left(\frac{1}{\sqrt{x^2+y^2+z^2}}\right)$$

Now $\frac{\partial}{\partial x}\left(\frac{1}{\sqrt{x^2+y^2+z^2}}\right) = \frac{1}{2}\frac{-2x}{(x^2+y^2+z^2)^{3/2}}$

$$= -x(x^2+y^2+z^2)^{-3/2}$$

$\therefore \quad \frac{\partial^2}{\partial x^2}\left(\frac{1}{\sqrt{x^2+y^2+z^2}}\right) = -(x^2+y^2+z^2)^{-3/2}$

$$+ x\frac{3}{2}\cdot 2x(x^2+y^2+z^2)^{-5/2}$$

$$= \frac{3x^2 - (x^2+y^2+z^2)}{(x^2+y^2+z^2)^{5/2}} = \frac{2x^2 - y^2 - z^2}{(x^2+y^2+z^2)^{5/2}}$$

Similarly, $\frac{\partial^2}{\partial y^2}\left(\frac{1}{\sqrt{x^2+y^2+z^2}}\right) = \frac{2y^2 - x^2 - z^2}{(x^2+y^2+z^2)^{5/2}}$

and $\frac{\partial^2}{\partial z^2}\left(\frac{1}{\sqrt{x^2+y^2+z^2}}\right) = \frac{2z^2 - x^2 - z^2}{(x^2+y^2+z^2)^{5/2}}$

Adding, $\left(\frac{\partial^2}{\partial x^2} + \frac{\partial^2}{\partial y^2} + \frac{\partial^2}{\partial z^2}\right)\left(\frac{1}{\sqrt{x^2+y^2+z^2}}\right) = 0.$

Divergence of a Vector Field :

Physically, the term divergence of a vector means the net outflow if some physical entity per unit volume associated with the vector. The physical entity may be electrical, magnetic or heat flux or outflow of the fluid flowing per unit volume per unit time.

Mathematically div. $\vec{A}$ over the surface of unit volume surrounding the point P (Fig. 1.9) is given by

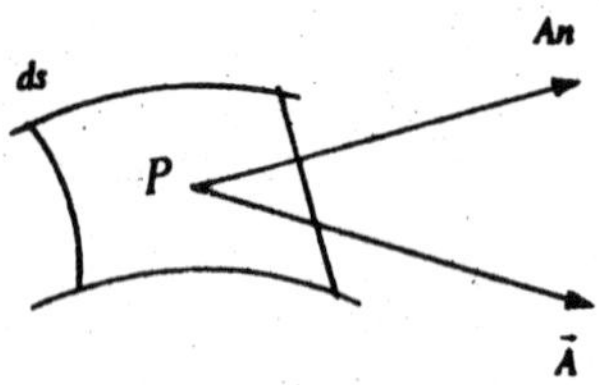

Fig. 1.9

$$\text{div}\,\vec{A} = \underset{v\to 0}{\text{Lt}} \iint \frac{An\,dS}{v}$$

$= \underset{v\to 0}{\text{Lt}} \iint \frac{\vec{A}\cdot\vec{dS}}{v}$; where dS is an elementary area near P and $\vec{dS}$ is a vector of magnitude dS whose direction is along the outward drawn normal to dS. The double integration involves integration over the whole surface surrounding P.

To find an analytical expression for div. $\vec{A}$, let us consider an elementary rectangular parallelopiped (Fig. 1.10) around the point $P(x, y, z)$ whose edges are of lengths δx, δy and δz and parallel to those of co-ordinate axes OX, OY and OZ respectively.

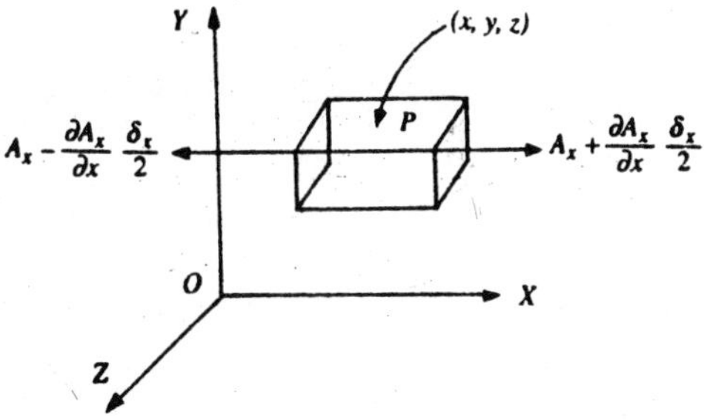

Fig. 1.10

The vector $\vec{A}$ at P can be resolved into three components Ax, Ay, Az parallel to the co-ordinate axes. The contribution of An dS over the surface to the right and perpendicular to OX is given by $\left(Ax + \frac{\partial Ax}{\partial x}\frac{\delta x}{2}\right)\delta y\delta z$. For the parallel surface to the left the contribution is $-\left(Ax - \frac{\partial Ax}{\partial x}\frac{\delta x}{2}\right)\delta y\delta z$. The total contribution over both the surfaces taken together is $\frac{\partial Ax}{\partial x}\delta x\delta y\delta z$.

Similar contributions from the surfaces perpendicular to OY and OZ are $\frac{\partial Ay}{\partial y}\delta x\delta y\delta z$ and $\frac{\partial Az}{\partial z}\delta x\delta y\delta z$ respectively.

The total contribution over the whole surface of the parallelopiped is $\left(\frac{\partial Ax}{\partial x} + \frac{\partial Ay}{\partial y} + \frac{\partial Az}{\partial z}\right)\delta x\delta y\delta z$.

$$\text{Hence, div. } \vec{A} = \frac{\left(\frac{\partial Ax}{\partial x} + \frac{\partial Ay}{\partial y} + \frac{\partial Az}{\partial z}\right)\delta x\delta y\delta z}{\delta x\delta y\delta z}$$

$$= \frac{\partial Ax}{\partial x} + \frac{\partial Ay}{\partial y} + \frac{\partial Az}{\partial z} = \vec{\nabla} . \vec{A}. \qquad (1.17)$$

If $\vec{A}$ can be expressed as the gradiant of a scalar *i.e.*, if $\vec{A} =$ grad. $u = \vec{\nabla} u$, Then div. $\vec{A} =$ div. grad. $u = \vec{\nabla} . \vec{\nabla} u$

$$= \left(\hat{i}\frac{\partial}{\partial x} + \hat{j}\frac{\partial}{\partial y} + \hat{k}\frac{\partial}{\partial z}\right) . \left(\hat{i}\frac{\partial u}{\partial x} + \hat{j}\frac{\partial u}{\partial y} + \hat{k}\frac{\partial u}{\partial z}\right)$$

$$= \frac{\partial^2 x}{\partial x^2} + \frac{\partial^2 u}{\partial y^2} + \frac{\partial^2 u}{\partial z^2} = \nabla^2 u. \qquad (1.18)$$

Again grad. div. $\vec{A} = \vec{\nabla}(\vec{\nabla} . \vec{A})$

$$= \left(\hat{i}\frac{\partial}{\partial x} + \hat{j}\frac{\partial}{\partial y} + \hat{k}\frac{\partial}{\partial z}\right)\left(\frac{\partial Ax}{\partial x} + \frac{\partial Ay}{\partial y} + \frac{\partial Az}{\partial z}\right)$$

$$= \hat{i}\left(\frac{\partial^2 Ax}{\partial x^2}+\frac{\partial^2 Ay}{\partial x\partial y}+\frac{\partial^2 Az}{\partial x\partial z}\right)+\hat{j}\left(\frac{\partial^2 Ax}{\partial y\partial x}+\frac{\partial^2 Ay}{\partial y^2}+\frac{\partial^2 Az}{\partial y\partial z}\right)$$
$$+\hat{k}\left(\frac{\partial^2 Ax}{\partial z\partial x}+\frac{\partial^2 Ay}{\partial z\partial y}+\frac{\partial^2 Az}{\partial z^2}\right) \quad (1.19)$$

The Curl of a Vector :

Let us consider a vector $\vec{A}$ to be defined everywhere in a given region containing the points P_1 and P_2 where the vector field exists and $\vec{dl}$ is an elementary path length vector between P_1 and P_2. Let us consider the scalar product $\int_{P_1}^{P_2} \vec{A}\,.\,\vec{dl} = \int_{P_1}^{P_2} A\,dl\cos\theta$, where θ is the angle between $\vec{A}$ and $\vec{dl}$, the line integral extends along the curve from P_1 to P_2.

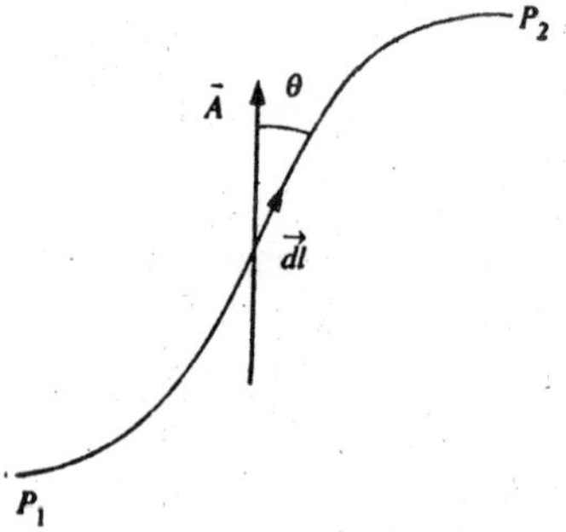

Fig. 1.11

For a closed path the integral sign is written as $\oint \vec{A}\,.\,\vec{dl}$. If L be a closed curve surrounding an area S about a point P associated with the vector $\vec{A}$, then the line integral of the vector $\vec{A}$ over the closed curve per unit area represent the component of the vector curl $\vec{A}$ in a direction perpendicular to the area enclosed by the curve L.

Expressed mathematically, $\text{Curl}_n\,\vec{A} = \underset{S\to 0}{\text{Lt}} \dfrac{\oint \vec{A}\,.\,\vec{dl}}{S}$.

Curl $\vec{A}$ is also called rot $\vec{A}$ and represents the rotational effect associated with the vector $\vec{A}$.

An analytical expression for curl $\vec{A}$, considering one component at a time can be defined in the following way.

Let $P\,(y, z)$ be a point in a vector field produced by $\vec{A}$. Let a rectangle $ABCD$ (Fig. 1.12) be drawn in the $y-z$ plane symmetrically around P, where AB is parallel to Y axis is of length δy and BC is parallel to Z axis and is of length δz.

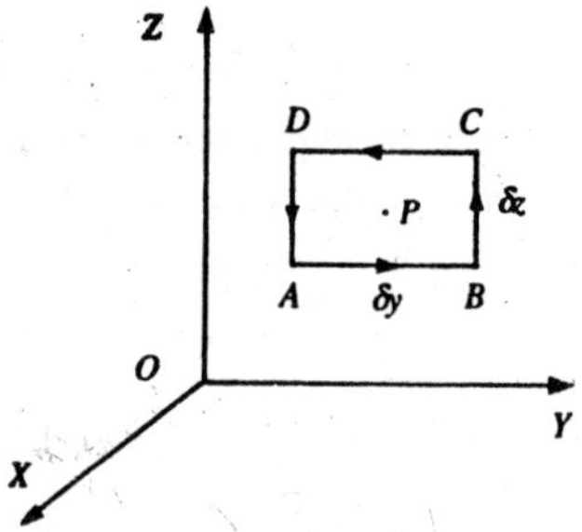

Fig. 1.12

Then $\int_{AB} \vec{A}\,.\,\vec{dl} = \left(Ay - \frac{\partial Ay}{\partial z}\frac{\delta z}{2}\right)\delta y$

and $\int_{CD} \vec{A}\,.\,\vec{dl} = -\left(Ay + \frac{\partial Ay}{\partial z}\frac{\delta z}{2}\right)\delta y$

Similarly $\int_{BC} \vec{A}.\vec{dl} = \left(Az + \frac{\partial Az}{\partial y}\frac{\delta y}{2}\right)\delta z$

and $\int_{DA} \vec{A}.\vec{dl} = -\left(Az - \frac{\partial Az}{\partial y}\frac{\delta y}{2}\right)\delta z$

Adding, $\int_{ABCD} \vec{A}.\vec{dl} = \left(\frac{\partial Az}{\partial y} - \frac{\partial Ay}{\partial z}\right)\delta y \delta z$

$$\therefore \text{Curl}_x \vec{A} = \underset{S \to 0}{\text{Lt}} \frac{\int_{ABCD} \vec{A}.\vec{dl}}{S} = \frac{\left(\frac{\partial Az}{\partial y} - \frac{\partial Ay}{\partial z}\right)\delta y \delta z}{\delta y \delta z} = \frac{\partial Az}{\partial y} - \frac{\partial Ay}{\partial z}.$$

In a like manner the y and the z components of curl $\vec{A}$ are

$$\text{curl}_y \vec{A} = \frac{\partial Ax}{\partial z} - \frac{\partial Az}{\partial x}.$$

$$\text{and curl}_z \vec{A} = \frac{\partial Ay}{\partial x} - \frac{\partial Ax}{\partial y}$$

$$\text{Hence, curl } \vec{A} = \hat{i}\left(\frac{\partial Az}{\partial y} - \frac{\partial Ay}{\partial z}\right) + \hat{j}\left(\frac{\partial Ax}{\partial z} - \frac{\partial Az}{\partial x}\right) + \hat{k}\left(\frac{\partial Ay}{\partial x} - \frac{\partial Ax}{\partial y}\right)$$

$$= \begin{vmatrix} \hat{i} & \hat{j} & \hat{k} \\ \frac{\partial}{\partial x} & \frac{\partial}{\partial y} & \frac{\partial}{\partial z} \\ Ax & Ay & Az \end{vmatrix} = \vec{\nabla} \times \vec{A} \tag{1.20}$$

To prove that curl curl $\vec{A}$ = grad. div. $\vec{A} = -\nabla^2 \vec{A}$

$$\text{or, } \vec{\nabla} \times (\vec{\nabla} \times \vec{A}) = \vec{\nabla}(\vec{\nabla}.\vec{A}) - \nabla^2 \vec{A} \tag{1.21}$$

$$\vec{\nabla} \times (\vec{\nabla} \times \vec{A}) = \vec{\nabla} \times \begin{vmatrix} \hat{i} & \hat{j} & \hat{k} \\ \frac{\partial}{\partial x} & \frac{\partial}{\partial y} & \frac{\partial}{\partial z} \\ Ax & Ay & Az \end{vmatrix}$$

$$= \vec{\nabla} \times \left\{\hat{i}\left(\frac{\partial Az}{\partial y} - \frac{\partial Ay}{\partial z}\right) + \hat{j}\left(\frac{\partial Ax}{\partial z} - \frac{\partial Az}{\partial x}\right) + \hat{k}\left(\frac{\partial Ay}{\partial x} - \frac{\partial Az}{\partial y}\right)\right\}$$

$$= \begin{vmatrix} \hat{i} & \hat{j} & \hat{k} \\ \frac{\partial}{\partial x} & \frac{\partial}{\partial y} & \frac{\partial}{\partial z} \\ \frac{\partial Az}{\partial y} - \frac{\partial Ay}{\partial z} & \frac{\partial Ax}{\partial z} - \frac{\partial Az}{\partial x} & \frac{\partial Ay}{\partial x} - \frac{\partial Ax}{\partial y} \end{vmatrix}$$

$$= \sum \hat{i}\left\{\frac{\partial}{\partial y}\left(\frac{\partial Ay}{\partial x} - \frac{\partial Ax}{\partial y}\right) - \frac{\partial}{\partial z}\left(\frac{\partial Ax}{\partial z} - \frac{\partial Az}{\partial x}\right)\right\}$$

$$= \sum \hat{i}\left\{-\frac{\partial^2 Ax}{\partial y^2}-\frac{\partial^2 Ax}{\partial z^2}\right\}+\sum i\left(\frac{\partial^2 Ay}{\partial y \partial x}+\frac{\partial^2 Az}{\partial z \partial x}\right)$$

$$= -\sum \hat{i}\left(\frac{\partial^2 Ax}{\partial x^2}+\frac{\partial^2 Ax}{\partial y^2}+\frac{\partial^2 Ax}{\partial z^2}\right)+\sum \hat{i}\left(\frac{\partial^2 Ax}{\partial x^2}+\frac{\partial^2 Ay}{\partial y \partial x}+\frac{\partial^2 Az}{\partial z \partial x}\right)$$

$$= -\left(\frac{\partial^2}{\partial x^2}+\frac{\partial^2}{\partial y^2}+\frac{\partial^2}{\partial z^2}\right)(\hat{i}Ax+\hat{j}Ay+\hat{k}Az)$$

$$+\hat{i}\frac{\partial}{\partial x}\left(\frac{\partial Ax}{\partial x}+\frac{\partial Ay}{\partial y}+\frac{\partial Az}{\partial z}\right)+\hat{j}\frac{\partial}{\partial y}\left(\frac{\partial Ax}{\partial x}+\frac{\partial Ay}{\partial y}+\frac{\partial Az}{\partial z}\right)$$

$$+\hat{k}\frac{\partial}{\partial z}\left(\frac{\partial Ax}{\partial x}+\frac{\partial Ay}{\partial y}+\frac{\partial Az}{\partial z}\right)$$

$$= -\nabla^2\vec{A}+\vec{\nabla}(\vec{\nabla}.\vec{A}) = \vec{\nabla}(\vec{\nabla}.\vec{A})-\nabla^2\vec{A}.$$

The result may also be obtained from the vector tripple product

$$\vec{\nabla}\times(\vec{\nabla}\times\vec{A}) = \vec{\nabla}(\vec{\nabla}.\vec{A})-(\vec{\nabla}.\vec{\nabla})\vec{A}$$

$$= \vec{\nabla}(\vec{\nabla}.\vec{A})-\nabla^2\vec{A}.$$

To prove that $\vec{\nabla}\times(\vec{\nabla}\varphi)=0$ (1.22)

$$\vec{\nabla}\times(\vec{\nabla}\varphi) = \vec{\nabla}\times\left(\hat{i}\frac{\partial\varphi}{\partial x}+\hat{j}\frac{\partial\varphi}{\partial y}+\hat{k}\frac{\partial\varphi}{\partial z}\right)$$

$$= \begin{vmatrix} \hat{i} & \hat{j} & \hat{k} \\ \frac{\partial}{\partial x} & \frac{\partial}{\partial y} & \frac{\partial}{\partial z} \\ \frac{\partial\varphi}{\partial x} & \frac{\partial\varphi}{\partial y} & \frac{\partial\varphi}{\partial z} \end{vmatrix} = \hat{i}\left\{\frac{\partial}{\partial y}\left(\frac{\partial\varphi}{\partial z}\right)-\frac{\partial}{\partial z}\left(\frac{\partial\varphi}{\partial y}\right)\right\}$$

$$+\hat{j}\left\{\frac{\partial}{\partial z}\left(\frac{\partial\varphi}{\partial x}\right)-\frac{\partial}{\partial x}\left(\frac{\partial\varphi}{\partial z}\right)\right\}$$

$$+\hat{k}\left\{\frac{\partial}{\partial x}\left(\frac{\partial\varphi}{\partial y}\right)-\frac{\partial}{\partial y}\left(\frac{\partial\varphi}{\partial x}\right)\right\}=0$$

Thus the vector which can be expressed as the gradient of a scalar is irrotational. *i.e.*, curl of the vector must be zero.

To prove that $\vec{\nabla}\times(\varphi\vec{A}) = \varphi\vec{\nabla}\times\vec{A}+\vec{\nabla}\varphi\times\vec{A}$ (1.23)

$$\vec{\nabla}\times(\varphi\vec{A}) = \vec{\nabla}\times\{\hat{i}\varphi A_x+\hat{j}\varphi A_y+\hat{k}\varphi A_z\}$$

$$= \hat{i}\left\{\frac{\partial}{\partial y}(\varphi Az)-\frac{\partial}{\partial z}(\varphi Ay)\right\}+\hat{j}\left\{\frac{\partial}{\partial z}(\varphi Ax)-\frac{\partial}{\partial x}(\varphi Az)\right\}$$

$$+\hat{k}\left\{\frac{\partial}{\partial x}(\varphi Ay)-\frac{\partial}{\partial y}(\varphi Ax)\right\}$$

$$= \sum i\left\{\varphi\left(\frac{\partial Az}{\partial y}-\frac{\partial Ay}{\partial z}\right)+Az\frac{\partial\varphi}{\partial y}-Ay\frac{\partial\varphi}{\partial z}\right\}$$

$$= \varphi(\vec{\nabla} \times \vec{A}) + \begin{vmatrix} \hat{i} & \hat{j} & \hat{k} \\ \frac{\partial \varphi}{\partial x} & \frac{\partial \varphi}{\partial y} & \frac{\partial \varphi}{\partial z} \\ A_x & A_y & A_z \end{vmatrix}$$

$$= \varphi(\vec{\nabla} \times \vec{A}) - (\vec{\nabla}\varphi) \times \vec{A}$$

To show that $\vec{\nabla}.(\vec{\nabla} \times \vec{A}) =$ div. curl $\vec{A} = 0$ (1.24)

$$\text{L.H.S.} = \vec{\nabla}.(\vec{\nabla} \times \vec{A})$$

$$= \left(\hat{i}\frac{\partial}{\partial x} + \hat{j}\frac{\partial}{\partial y} + \hat{k}\frac{\partial}{\partial z}\right) \cdot \begin{vmatrix} \hat{i} & \hat{j} & \hat{k} \\ \frac{\partial}{\partial x} & \frac{\partial}{\partial y} & \frac{\partial}{\partial z} \\ Ax & Ay & Az \end{vmatrix}$$

$$= \left(\hat{i}\frac{\partial}{\partial x} + \hat{j}\frac{\partial}{\partial y} + \hat{k}\frac{\partial}{\partial z}\right) \cdot \left\{\hat{i}\left(\frac{\partial Az}{\partial y} - \frac{\partial Ay}{\partial z}\right) + \hat{j}\left(\frac{\partial Ax}{\partial z} - \frac{\partial Az}{\partial x}\right) + \hat{k}\left(\frac{\partial Ay}{\partial x} - \frac{\partial Ax}{\partial y}\right)\right\} = 0$$

The vector field for which div. $A = 0$ is called a solenoidal field. Hence a vector field will be solenoidal if it can be expressed as the curl of another vector. The vector field for which curl $\vec{A} = 0$ is called irrotational.

To prove that $\vec{\nabla}.(\vec{A} \times \vec{B}) = \vec{\nabla} \times \vec{A}.\vec{B} - \vec{\nabla} \times \vec{B}.\vec{A}$ (1.25)

$$\text{L. H. S.} = \sum \hat{i}\frac{\partial}{\partial x} \cdot (\vec{A} \times \vec{B})$$

$$= \sum \hat{i}.\left\{\left(\frac{\partial \vec{A}}{\partial x} \times \vec{B}\right) + \vec{A} \times \frac{\partial \vec{B}}{\partial x}\right\}$$

$$= \sum i \times \frac{\partial \vec{A}}{\partial x} \cdot \vec{B} - \sum \hat{i} \times \frac{\partial \vec{B}}{\partial x} \cdot \vec{A}$$

$$= \vec{\nabla} \times \vec{A}.\vec{B} - \vec{\nabla} \times \vec{B}.\vec{A}$$

To prove that

$$\vec{\nabla} \times (\vec{A} \times \vec{B}) = (\vec{B}.\vec{\nabla})\vec{A} - (\vec{A}.\vec{\nabla})\vec{B} + \vec{A}(\vec{\nabla}.\vec{B}) - \vec{B}(\vec{\nabla}.\vec{A}) \quad (1.26)$$

$$\text{L. H. S.} = \hat{i} \times \frac{\partial}{\partial x}(\vec{A} \times \vec{B}) + \hat{j} \times \frac{\partial}{\partial y}(\vec{A} \times \vec{B}) + \hat{k} \times \frac{\partial}{\partial z}(\vec{A} \times \vec{B})$$

$$= \sum \hat{i} \times \left(\frac{\partial \vec{A}}{\partial x} \times \vec{B} + \vec{A} \times \frac{\partial \vec{B}}{\partial x}\right)$$

$$= \sum \hat{i} \times \left(\frac{\partial \vec{A}}{\partial x} \times \vec{B}\right) + \sum \hat{i} \times \left(\vec{A} \times \frac{\partial \vec{B}}{\partial x}\right)$$

$$= \sum (\hat{i}.\vec{B})\frac{\partial \vec{A}}{\partial x} - \sum \left(\hat{i}\cdot\frac{\partial \vec{A}}{\partial x}\right)\vec{B} + \sum \left(\hat{i}\cdot\frac{\partial \vec{B}}{\partial x}\right)\vec{A} - \sum (\hat{i}.\vec{A})\frac{\partial \vec{B}}{\partial x}$$

$$= \sum \left\{\left(\vec{B}.\hat{i}\frac{\partial}{\partial x}\right)\vec{A}\right\} - \vec{B}\sum \hat{i}\cdot\frac{\partial \vec{A}}{\partial x} + \vec{A}\sum \hat{i}\cdot\frac{\partial \vec{B}}{\partial x} - \sum \left(\vec{A}.\hat{i}\frac{\partial}{\partial x}\right)\vec{B}$$

$$= (\vec{B}.\vec{\nabla})\vec{A} - \vec{B}(\vec{\nabla}.\vec{A}) + \vec{A}(\vec{\nabla}.\vec{B}) - (\vec{A}.\vec{\nabla})\vec{B}$$

N. B. : $(\vec{B}.\vec{\nabla})\vec{A} \longrightarrow \vec{\nabla}$ operation to be performed on $\vec{A}$ only.

$A(\vec{\nabla}.\vec{B}) \longrightarrow \vec{\nabla}$ operation to be performed on $\vec{B}$ only.

Gauss's Divergence Theorem :

Statement : The volume integral of the divergence of a vector $\vec{A}$ in a vector field, taken through a given volume, is equal to the surface integral of the normal component of $\vec{A}$ over the surface enclosed by the volume. Expressed analytically,

$$\iiint (\text{div.}\vec{A})\,dv. = \iint \vec{A}.\vec{dS}.$$

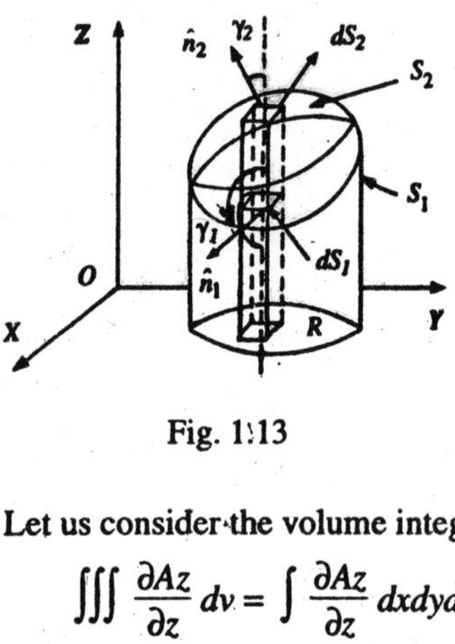

Fig. 1.13

Proof : Let S be a closed surface which is such that any line parallel to the co-ordinate axes cuts S in atmost two points..

Let the equation of the lower and the upper portions S_1 and S_2 (Fig. 1.13) be $z = f_1(x, y)$ and $z = f_2(x, y)$ respectively. The projection of the surface in the x-y plane be denoted by R. Let the vector $\vec{A}$ have components A_x, A_y and A_z in the three directions.

Let us consider the volume integral

$$\iiint \frac{\partial Az}{\partial z}\,dv = \int \frac{\partial Az}{\partial z}\,dxdydz \quad \ldots \quad (a)$$

$$= \iint_R \left\{\int_{Z=f_1(x,y)}^{Z=f_2(x,y)} \frac{\partial Az}{\partial z}\,dz\right\} dxdy = \iint_R [Az(x,y,z)]_{z=f_1}^{z=f_2}\,dxdy$$

$$= \iint_R [Az(x,y,f_2) - Az(x,y,f_1)]\,dxdy.$$

For the upper surface S_2, $dS_2 \cos\gamma_2 = dxdy = dS_2\,\hat{k}.\hat{n}_2$, where $\hat{n}_2$ is unit vector normal to dS_2, $\hat{k}$ is unit vector along Z and the angle γ_2 between them is acute. For the lower surface S_1, $-\ dS_1 \cos\gamma_1 = dxdy = -\ dS_1\ \hat{k}\cdot\hat{n}_1$, where $\hat{n}_1$ is unit vector normal to dS_1 and γ_1 is obtuse.

Then $$\iint_R A_z(x,y,f_2)\,dxdy = \iint_{S_2} Az\,\hat{k}\cdot\hat{n}_2\,dS_2$$

and $$\iint_R A_z(x,y,f_1)\,dxdy = -\iint_{S_1} A_z\,\hat{k}\cdot\hat{n}_1\,dS_1$$

$$\therefore\quad \iint_R A_z(x,y,f_2)\,dxdy - \iint_R A_z(x,y,f_1)\,dxdy$$

$$= \iint_{S_2} A_z\,\hat{k}\cdot\hat{n}_2\,dS_2 + \iint_{S_1} A_z\,\hat{k}\cdot\hat{n}_1\,dS_1$$

$$= \iint_S Az\,\hat{k}\cdot\hat{n}\,dS.$$

Hence (a) gives $$\iiint \frac{\partial Az}{\partial z}\,dv = \iint Az\,\hat{k}\cdot\hat{n}\,dS.$$

Similarly, by projecting S on other co-ordinate planes,

$$\iiint \frac{\partial Ax}{\partial x}\,dv = \iint Ax\,\hat{i}\cdot\hat{n}\,dS$$

and $$\iiint \frac{\partial Ay}{\partial y}\,dv = \iint Ay\,\hat{j}\cdot\hat{n}\,dS$$

Adding $$\iiint\left(\frac{\partial Ax}{\partial x}+\frac{\partial Ay}{\partial y}+\frac{\partial Az}{\partial z}\right)dv = \iint (A_x\hat{i}+A_y\hat{j}+A_z\hat{k})\cdot\hat{n}dS$$

or $$\iiint \vec{\nabla}\cdot\vec{A}\,dv = \iint \vec{A}\cdot\hat{n}\,dS$$

or, $$\iiint div.\vec{A}\,dv = \iint \vec{A}\cdot\vec{dS} \qquad (1.27)$$

Stokes' Theorem :

Statement : The integral over a surface of the normal component of the vector curl $\vec{A}$ is equal to the line integral of the tangential component of the vector $\vec{A}$ around the path enclosing the area.

Expressed analytically,

$$\oint \vec{A}\cdot\vec{dl} = \iint \text{curl}\,\vec{A}\cdot\vec{dS},$$

Proof : Let us first compute the line integral of a vector $\vec{A}$ around an infinitesimal rectangle of side δx and δy lying in the xy plane as shown in Fig. 1.14.

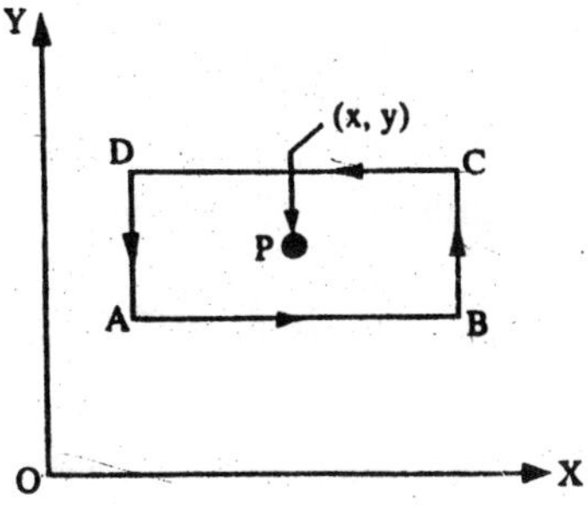

Fig. 1.14

Let $P(x, y)$ be the mid point of the rectangle at which the associated field vector is $\vec{A}$.

Then $$\int_{AB} \vec{A}.d\vec{l} = \left(Ax - \frac{\partial Ax}{\partial y}\frac{\delta y}{2}\right)\delta x$$

$$\int_{BC} \vec{A}.d\vec{l} = \left(Ay + \frac{\partial Ay}{\partial x}\frac{\delta x}{2}\right)\delta y$$

$$\int_{CD} \vec{A}.d\vec{l} = -\left(Ax + \frac{\partial Ax}{\partial y}\frac{\delta y}{2}\right)\delta x$$

$$\int_{DA} \vec{A}.d\vec{l} = -\left(Ay - \frac{\partial Ay}{\partial x}\frac{\delta x}{2}\right)\delta y$$

Adding the various contributions

$$\oint_{ABCDA} \vec{A}.d\vec{l} = \left(\frac{\partial Ay}{\partial x} - \frac{\partial Ax}{\partial y}\right)\delta x\,\delta y = (\vec{\nabla} \times \vec{A})_z\, dS_{xy} \quad \ldots \quad (a)$$

where $(\vec{\nabla} \times \vec{A})_z$ is the z component of the vector curl $\vec{A}$ and dS_{xy} is the area of the rectangle ABCD.

Let us now consider a closed curve in the xy plane as shown in Fig. 1.15.

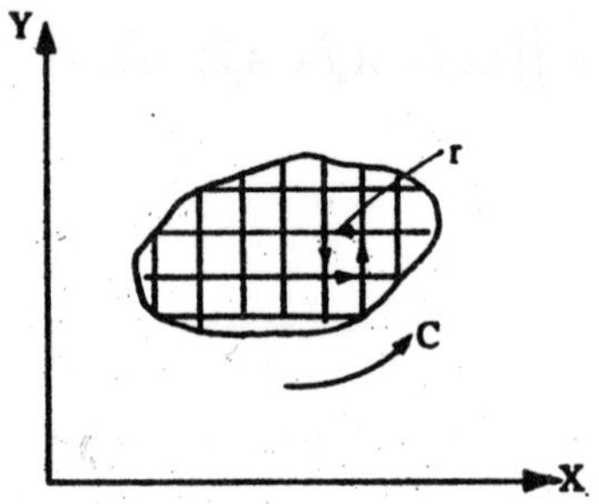

Fig. 1.15

Let us divide the space inside C by a network of lines forming a network of infinitesimal rectangles. If we take the sum of the line integrals around the various meshes, we obtain

$$\sum_{r=1}^{\infty} \oint_r \vec{A}.d\vec{l} = \sum_{r=1}^{\infty} (\vec{\nabla} \times \vec{A})_z\, dS_{xy} \quad \ldots \quad (b)$$

It is easily seen that the contributions to the line integrals of the adjoining meshes nutralise each other, because they are traversed in opposite directions. The contributions which remain are those along the boundary of the surface.

Hence, $$\sum_{r=1}^{\infty} \oint_r \vec{A}.d\vec{l} = \oint_c \vec{A}.d\vec{l} \quad \ldots \quad (c)$$

where the integral on the right is taken along the boundary curve C in the anticlockwise direction.

Now the sum of the right hand side of equation (b) reduces to $\sum (\vec{\nabla} \times \vec{A})_z \, dS_{xy} = \iint (\vec{\nabla} \times \vec{A})_z \, dS_{xy}$.

Hence, substituting in (b),

$$\oint_C \vec{A}.d\vec{l} = \iint \left(\vec{\nabla} \times \vec{A}\right)_z dS_{xy} \ldots \text{ (d)}$$

Let us consider a triangular surface as shown in Fig. 1.16. It is quite evident that

$$\oint_{ABC} \vec{A} \cdot \vec{dl}$$

$$= \oint_{OBC} \vec{A}.\vec{dl} + \oint_{OCA} \vec{A} \,.\, \vec{dl} + \oint_{OAB} \vec{A}.\vec{dl},$$

since the contributions along the lines OA, OB and OC cancel each other. However as a result of equation (d), we have

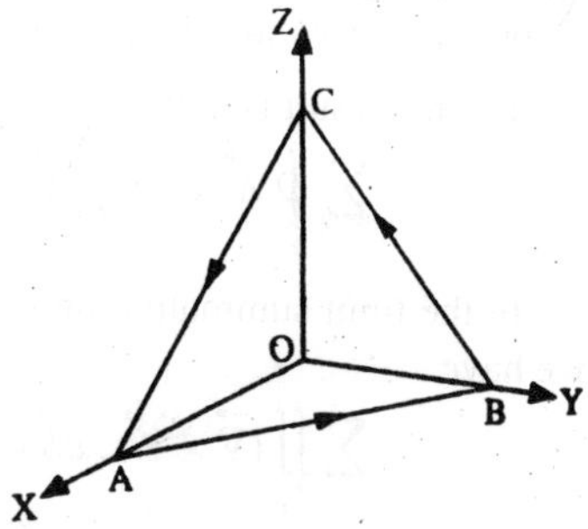

Fig. 1.16

$$\left.\begin{aligned} \int_{OBC} \vec{A} \cdot \vec{dl} &= \iint (\vec{\nabla} \times \vec{A})_x \, dydz \\ \int_{OCA} \vec{A} \cdot \vec{dl} &= \iint (\vec{\nabla} \times \vec{A})_y \, dzdx \\ \text{and } \int_{OAB} \vec{A} \cdot \vec{dl} &= \iint (\vec{\nabla} \times \vec{A})_z \, dxdy \end{aligned}\right\} \ldots \text{ (e)}$$

On addition taking the help of equation (d)

$$\int_{ABC} \vec{A} \cdot \vec{dl} = \iint_{OBC} (\vec{\nabla} \times \vec{A})_x \, dydz + \iint_{OCA} (\vec{\nabla} \times \vec{A})_y \, dzdx + \iint_{OAB} (\vec{\nabla} \times \vec{A})_z \, dxdy \quad \ldots \text{ (f)}$$

Next let us consider

$\vec{ds} = \hat{i}dydz + \hat{j}dzdx + \hat{k}dxdy$. Using this notation equation (f) becomes

$$\oint_{ABC} \vec{A} \,.\, \vec{dl} = \iint (\vec{\nabla} \times \vec{A}) \,.\, \vec{ds} \ \ldots \quad \text{(g)}$$

Now let us consider the open surface as shown in Fig. 1.17. We can regard the surface S of the open surface as being made up of an infinite number of elementary rectangular and triangular surfaces. If we denote these typical of them to be rth surface we have from equation (g)

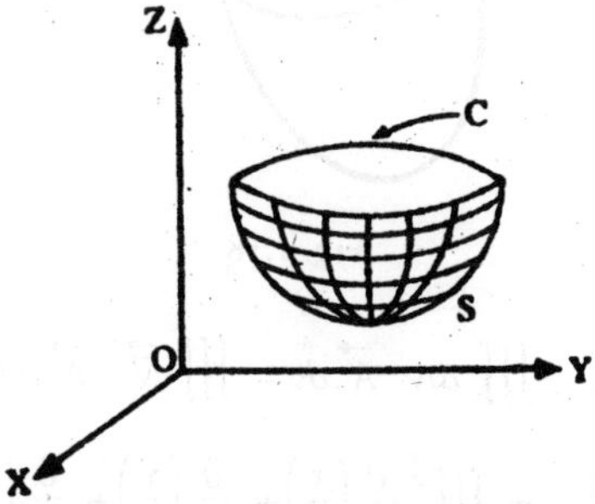

Fig. 1.17

$$\sum_{r=1}^{\infty} \oint_r \vec{A} \,.\, \vec{dl} = \sum \iint (\vec{\nabla} \times \vec{A}) \,.\, \vec{dS} \quad \ldots \text{ (h)}$$

Now the sum of the line integrals of the elementary rectangles or triangles reduces to the line integral about the boundary C of the open surface S, since the line integrals along the adjacent elementary surfaces are described in opposite senses and hence cancel.

The net result is that

$$\sum_r \oint \vec{A}\,.\,\vec{dl} = \oint_c \vec{A}\,.\,\vec{dl}$$

In the limit summation on the right side of (h) reduces to an integral, and we have

$$\sum_r \iint (\vec{\nabla} \times \vec{A})\,.\,\vec{dS} = \iint (\vec{\nabla} \times \vec{A})\,.\,\vec{dS}.$$

And consequently, equation (c) gives

$$\oint_c \vec{A}\,.\,\vec{dl} = \iint (\vec{\nabla} \times \vec{A})\,.\,\vec{dS}$$

EXERCISE

1. Show by the application of Gauss's divergen theorem and Stokes' theorem that the divergence of curl of a vector is zero.

Let there be a vector $\vec{A}$ which can be expressed as the curl of another vector $\vec{B}$ associated with a closed surface. By Gauss's divergence theorem

$$\iiint div.\vec{A}\,dv = \iint \vec{A}.\vec{ds} = \iint curl\,\vec{B}.dS$$

$$= \oint \vec{B}\,.\,\vec{dl} \text{ by Stokes' theorem}$$

$= 0$, as the closed surface has no boundary (Fig. 1.18).

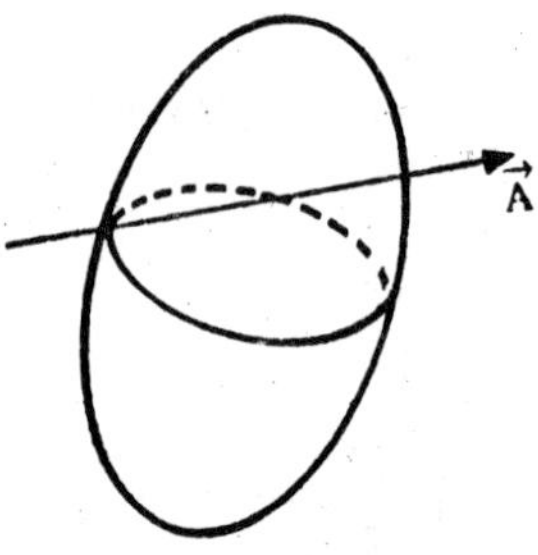

Fig. 1.18

2. Verify the Divergence theorem for the vector $\vec{A} = \hat{i}\frac{x}{r} + \hat{j}\frac{y}{r} + \hat{k}\frac{z}{r}$, the region of integration being the surface of a sphere of radius a with centre at the origin.

[C. U. 1981]

$$\iiint div.\,\vec{A}\,dv = \iiint \vec{\nabla}\,.\,\vec{A}\,dv = \iiint \left(\frac{\partial Ax}{\partial x} + \frac{\partial Ay}{\partial y} + \frac{\partial Az}{\partial z}\right) dv$$

$$= \iiint \left\{\frac{\partial}{\partial x}\left(\frac{x}{r}\right) + \frac{\partial}{\partial y}\left(\frac{y}{r}\right) + \frac{\partial}{\partial z}\left(\frac{z}{r}\right)\right\} dv \qquad r = \sqrt{x^2 + y^2 + z^2}$$

Now $\frac{\partial}{\partial x}\left(\frac{x}{r}\right) = \frac{r - x\frac{\partial r}{\partial x}}{r^2} = \left(r - \frac{x\,2x}{2\sqrt{x^2+y^2+z^2}}\Big/ r^2\right)$

$$= \frac{\frac{r^2 - x^2}{r}}{r^2} = \frac{y^2+z^2}{r^3}$$

$$\therefore \iiint div\,\vec{A}\,dv = \iiint \frac{2(x^2+y^2+z^2)}{r^3}dv$$

$$= \int_0^a \frac{2}{r}4\pi r^2 dr = 8\pi\int_0^a r dr = 4\pi a^2$$

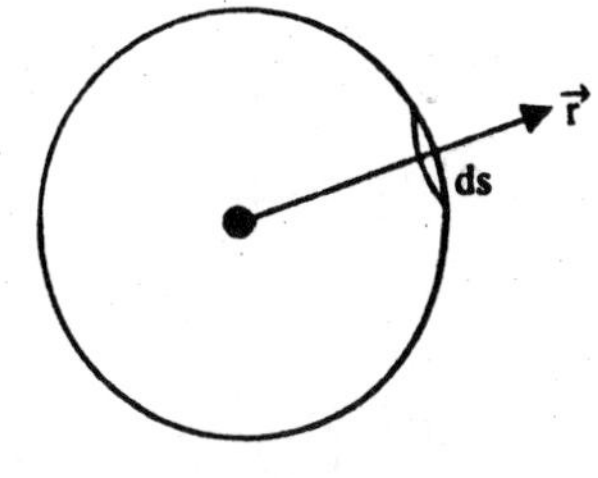

Fig. 1.19

And $\iint \vec{A}.\vec{dS} = \iint \frac{\vec{r}}{r}.\vec{dS}$

$$= \iint \frac{r\,\hat{n}.\vec{dS}}{r} = \iint dS = 4\pi a^2$$

Hence, the theorem is verified.

3. Gauss's divergence theorem is

$\iiint \vec{\nabla}.\vec{A}\,dv = \iint \vec{A}\;\hat{n}\,dS$. Suppose that $\vec{A} = r^3\,\hat{r}$. Evaluate the left side of the above equation over the volume of a sphere given by $x^2+y^2+z^2=a^2$. Evaluate also the R. H. S. over the surface of the same sphere and show that the divergence theorem is satisfied. **[C. U. 1984]**

$\vec{A} = r^3\hat{r} = r^2\vec{r} = (x^2+y^2+z^2)(\hat{i}x+\hat{j}y+\hat{k}z)$

$$\therefore \vec{\nabla}.\vec{A} = \left(\hat{i}\frac{\partial}{\partial x}+\hat{j}\frac{\partial}{\partial y}+\hat{k}\frac{\partial}{\partial z}\right).\{(x^2+y^2+z^2)(\hat{i}x+\hat{j}y+\hat{k}z)\}$$

$$= \frac{\partial}{\partial x}\{x(x^2+y^2+z^2)\} + \frac{\partial}{\partial y}\{y(x^2+y^2+z^2)\} + \frac{\partial}{\partial z}\{z(x^2+y^2+z^2)\}$$

$$= x^2+y^2+z^2+x.2x+x^2+y^2+z^2+y.2y+x^2+y^2+z^2+z.2z$$

$$= 3(x^2+y^2+z^2) + 2(x^2+y^2+z^2)$$

$$= 5(x^2+y^2+z^2) = 5r^2$$

$$\therefore \iiint \vec{\nabla}.\vec{A}\,dv = \int_0^a 4\pi r^2(5r^2)\,dr$$

$$= 20\pi\int_0^a r^4\,dr$$

$$= 4\pi a^5$$

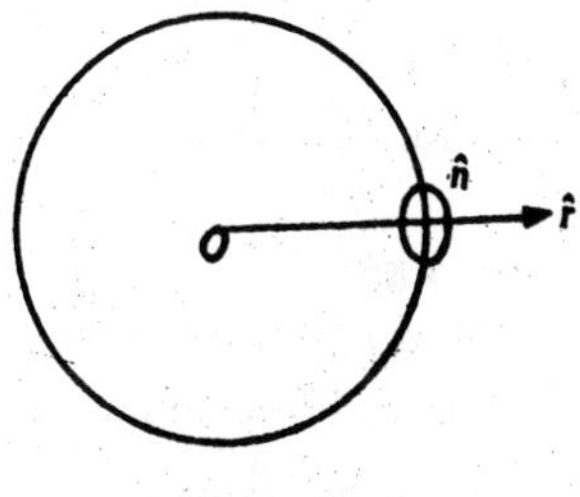

Fig. 1.20

Again $\iint \vec{A}.\hat{n}\,ds = \iint r^3\,\hat{r}.\hat{n}\,ds$

$$= \iint r^3\, ds = a^3\, 4\pi a^2 = 4\pi a^5.$$

Hence, the result

$\because \left|\vec{r}\right| = a$ on the surface.

4. Given $\vec{V} = \hat{i}x + \hat{j}y + \hat{k}z$, evalulate $\iint \vec{V}\,.\,\hat{n}\, d\sigma$ on the surface of a cylinder (top and bottom included) of radius R and length h, $\hat{n}$ is unit vector normal to the surface. The centre of the bottom surface is at origin of co-ordinates system and the axis of the cylinder is along Z axis.

[C. U. 1989]

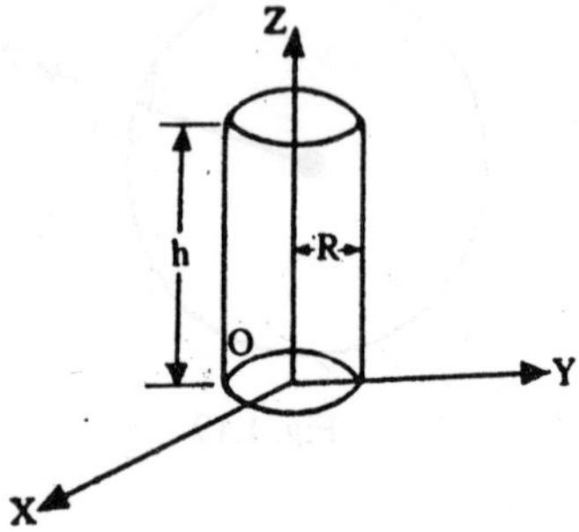

Fig. 1.21

By Gauss's divergence theorem

$$\iint \vec{V}\,.\,\hat{n}\, d\sigma$$

$$= \iiint div.\, \vec{V}\, dv$$

$$= \iiint (\vec{\nabla}\,.\,\vec{V})\, dv$$

$$= \int_{x=-R}^{R} \int_{-y_1}^{y_1} \int_{0}^{h} \left[\left\{\hat{i}\frac{\partial}{\partial x} + \hat{j}\frac{\partial}{\partial y} + \hat{k}\frac{\partial}{\partial z}\right\}.\left\{\hat{i}x + \hat{j}y + \hat{k}z\right\}\right] dx\, dy\, dz$$

$$= \int_{-R}^{R} \int_{-y_1}^{y_1} \int_{0}^{h} (1+1+1)\, dxdydz$$

$$= \int_{-R}^{R} \int_{0}^{h} \int_{-\sqrt{R^2-x^2}}^{\sqrt{R^2-x^2}} 3\, dxdzdy \qquad \because x^2 + y^2 = R^2$$

$$= \int_{-R}^{R} \int_{0}^{h} [3y]_{-\sqrt{R^2-x^2}}^{\sqrt{R^2-x^2}}\, dx\, dz = 6\int_{-R}^{R} \int_{0}^{h} \sqrt{R^2 - x^2}\, dxdz$$

$$= 6h \int_{-R}^{R} \sqrt{R^2 - x^2}\, dx = 6h \left\{\frac{x}{2}\sqrt{R^2 - x^2} + \frac{R^2}{2}\,.\,\sin^{-1}\frac{x}{R}\right\}_{-R}^{R}$$

$$= 6h \left\{\frac{R^2}{2}\,\pi/2 - \frac{R^2}{2}\,(-\pi/2)\right\} = \frac{6h\,\pi R^2}{2}$$

$$= 3\,\pi R^2 h\,.$$

Direct Method :

For the lower surface, $z = 0$

$$\therefore \iint \vec{V}\,.\,\hat{n}\, d\sigma$$

$$= \iiint (\hat{i}x + \hat{j}y) \cdot (-\hat{k}\, d\sigma) = 0.$$

For the upper surface

$$\iint \vec{V} \cdot \hat{n}\, d\sigma = \iint (\hat{i}x + \hat{j}y + \hat{k}h) \cdot \hat{k}\, d\sigma$$

$$= h \iint d\sigma = \pi R^2 h.$$

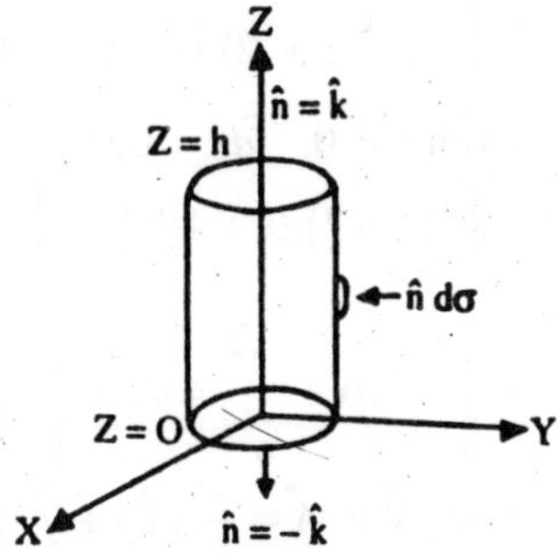

Fig. 1.22

For the curved surface $\iint \vec{V} \cdot \hat{n}\, d\sigma$

$$= \iint (\hat{i} R \sin\theta + \hat{j} R\cos\theta + \hat{k} z) \cdot \hat{n}\, R d\theta\, dz$$

$$= \iint (R\cos^2\theta + R\sin^2\theta)\, R\, d\theta\, dz$$

$\because \hat{n} \cdot \hat{i} = \cos\theta$

$\hat{n} \cdot \hat{j} = \sin\theta$

$$= \int_0^{2\pi} \int_0^{h} R^2 (\cos^2\theta + \sin^2\theta)\, d\theta\, dz$$

$$= 2\pi R^2 h.$$

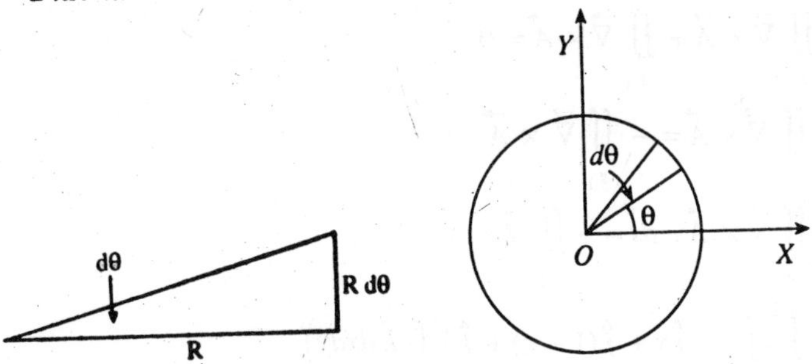

Fig. 1.23

Hence, for the whole surface

$$\iint \vec{V} \cdot \hat{n}\, d\sigma = \pi R^2 h + 2\pi R^2 h = 3\pi R^2 h.$$

5. Verify Stokes' theorem for the vector
$\vec{A} = \hat{i}(y - z + 2) + \hat{j}(yz + 4) - \hat{k}xz$,
the region of integration being the surface of the cube $x = 0$, $y = 0$, $z = 0$ and $x = 2$, $y = 2$, $z = 2$ above the xy plane.

It is evident that the boundary of the open surface is the square $OABC$ (Fig. 1.24).

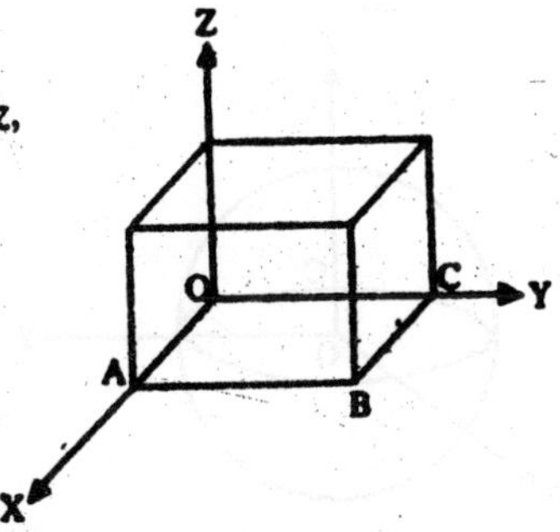

Fig. 1.24

$$\therefore \oint \vec{A} \cdot \vec{dl}$$

$$= \oint \{(y - z + 2)\hat{i} + (yz + 4)\hat{j} - xz\hat{k}\} . \quad (\hat{i}\,dx + \hat{j}\,dy + \hat{k}\,dz)$$

(with $z = 0$, $dz = 0$)

$$= \oint \{(y+2)dx + 4dy\} = \int_{OA} + \int_{AB} + \int_{BC} + \int_{CO}$$

$$= \int_0^2 2\,dx + \int_0^2 4\,dy + \int_2^0 4\,dx + \int_2^0 4\,dy = -4$$

Now $\vec{\nabla} \times \vec{A} = -\{\hat{i}y + \hat{j}(1 - z) + \hat{k}\}$

The surface S of the cube above xy plane together with the plane of the square $OA\ BC$ (S_1) constitute the closed surface S', say.

By Gauss's div. theorem

$$\iint_{s'} \vec{\nabla} \times \vec{A} = \iiint \vec{\nabla}\ (\vec{\nabla} \times \vec{A})\ dv = 0 \quad \because \text{div. curl } \vec{A} = 0$$

$$\therefore \iint_s \vec{\nabla} \times \vec{A} + \iint_{s_1} \vec{\nabla} \times \vec{A} = 0$$

$$\text{or, } \iint_s \vec{\nabla} \times \vec{A} = -\iint_{s_1} \vec{\nabla} \times \vec{A}$$

$$\therefore \iint_s \vec{\nabla} \times \vec{A} . \vec{ds} = -\iint_{s_1} \vec{\nabla} \times \vec{A} . \vec{ds}$$

$$= -\int_0^2 \int_0^2 - \{\hat{i}y + \hat{j}(1 - z) + \hat{k}\} . \left(-\hat{k}\,dxdy\right)$$

$$= -\int_0^2 \int_0^2 dydx = -4$$

Hence, the result.

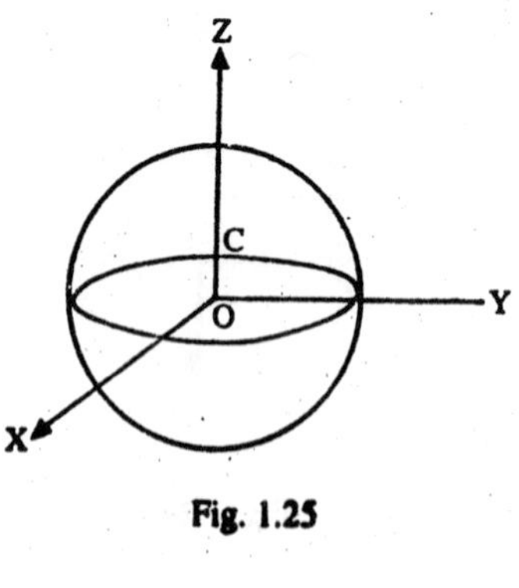

Fig. 1.25

6. Verify Stokes' theorem for $\vec{A} = (2x - y)\,\hat{i} - yz^2\hat{j} - y^2 z\hat{k}$, the region of integration being the upper half surface of the sphere $x^2 + y^2 + z^2 = 1$.

The boundary is the circle of unit radius in the xy plane (Fig. 1.25).

Putting $x = \cos\theta$, $y = \sin\theta$, $z = 0$

$$\oint_a \vec{A} . \vec{dl} = \oint_c \{(2x - y)\ dx - yz^2\,dy - y^2$$

$$= \int_0^{2\pi} (2\cos\theta - \sin\theta)(-\sin\theta\, d\theta) = \pi$$

$$\vec{\nabla} \times \vec{A} = \hat{k}$$

$$\iint_s \vec{\nabla} \times \vec{A} \cdot \hat{n}\, ds = \iint \hat{k} \cdot \hat{n}\, ds = \iint dxdy$$

$$= \int_{-1}^{+1} \left\{ \int_{-\sqrt{1-x^2}}^{\sqrt{1-x^2}} dydx = \pi \right.$$

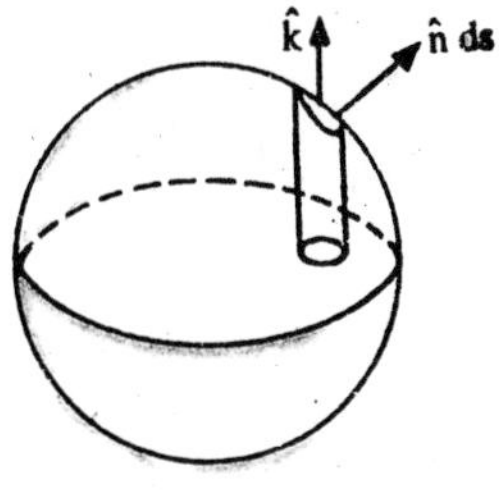

Fig. 1.26

Hence, the result.

N.B. : $\hat{k} \cdot \hat{n}\; dS = dxdy =$ projection of hemisphere on the xy plane.

Green's Theorem :

(A) Green's Theorem in a Plane :

Statement : If S is a closed region in the xy plane bounded by a simple closed curve C and if M and N are continuous functions of x and y having continuous derivatives in S, then

$$\oint_c Mdx + Ndy = \iint_S \left(\frac{\partial N}{\partial x} - \frac{\partial M}{\partial y} \right) dxdy,$$ where C is traversed in the anticlockwise (+ ve) sense.

Proof : Let the equations of the curves AEB and AFB be $y = y_1(x)$ and $y = y_2(x)$ respectively (Fig. 1.27). If S be the region bounded by C, we have

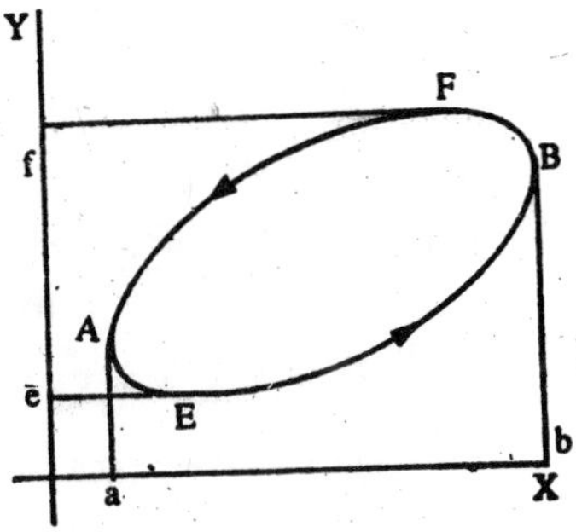

Fig. 1.27

$$\iint_S \frac{\partial M}{\partial y}\, dxdy$$

$$= \int_{x=a}^{b} \left\{ \int_{y=Y_1(x)}^{y=Y_2(x)} \frac{\partial M}{\partial y}\, dy \right\} dx$$

$$= \int_a^b [M(x, y)]_{y_1(x)}^{y_2(x)}\, dx$$

$$= \int_a^b \{M(x, y_2) - M(x, y_1)\}\, dx$$

$$= -\int_a^b M(x, y_1)\, dx - \int_b^a M(x, y_2)\, dx = -\oint_c Mdx$$

$$\therefore \oint_c Mdx = -\iint \frac{\partial M}{\partial y}\, dxdy.$$

Similarly, let the equations of curves *EAF* and *EBF* be $x = X_1(y)$ and $x = X_2'(y)$ respectively.

Then $$\iint_S \frac{\partial N}{\partial x}\, dx dy = \int_{y=e}^{f} \left\{ \int_{x=X_1(y)}^{X_2(y)} \frac{\partial N}{\partial x}\, dx \right\} dy$$

$$= \int_e^f \left\{ N(X_2, y) - N(X_1, y) \right\} dy$$

$$= \int_f^e N(X_1, y)\, dy + \int_e^f N(X_2, y)\, dy$$

$$= \oint_c N dy$$

Adding $$\oint (M dx + N dy) = \iint \left(\frac{\partial N}{\partial x} - \frac{\partial M}{\partial y} \right) dx dy \tag{1.28}$$

First and Second Forms of Green's Theorem :

Let us consider the vector $\vec{A} = u\, \vec{\nabla} v$, that is the vector field is the product f the scalar function *u* and the gradient of another scalar function *v*.

Now $$\vec{\nabla} . \vec{A} = \frac{\partial Ax}{\partial x} + \frac{\partial Ay}{\partial y} + \frac{\partial Az}{\partial z}$$

$$= \frac{\partial}{\partial x}\left(u \frac{\partial v}{\partial x} \right) + \frac{\partial}{\partial y}\left(u \frac{\partial v}{\partial y} \right) + \frac{\partial}{\partial z}\left(u \frac{\partial v}{\partial z} \right)$$

$$= u \left(\frac{\partial^2 v}{\partial x^2} + \frac{\partial^2 v}{\partial y^2} + \frac{\partial^2 v}{\partial z^2} \right) + \frac{\partial u}{\partial x}\frac{\partial v}{\partial x} + \frac{\partial u}{\partial y}\frac{\partial v}{\partial y} + \frac{\partial u}{\partial z}\frac{\partial v}{\partial z}$$

$$= u\, \nabla^2 v + \vec{\nabla} u . \vec{\nabla} v .$$

From Gauss's divergence theorem

$$\iiint (\vec{\nabla} . \vec{A})\, dV = \iint \vec{A} . \vec{dS}$$

or, $$\iiint (u \nabla^2 v + \vec{\nabla} u . \vec{\nabla} v)\, dV = \iint \left(u \vec{\nabla} v \right) . \vec{ds} \tag{1.29}$$

This transformation is referred to as the first form of Green's theorem.

Interchanging the functions *u* and *v* in equation (1.29) we get

$$\iiint \left(v \nabla^2 u + \vec{\nabla} v . \vec{\nabla} u \right) dV = \iint \left(v \vec{\nabla} u \right) . d\vec{S} \tag{1.30}$$

Subtracting (1.30) from (1.29)

$$\iiint (u\nabla^2 v - v\nabla^2 u)\, dV = \iint (u\vec{\nabla} v - v\vec{\nabla} u) \cdot d\vec{S} \tag{1.31}$$

This transformation is referred to as the second form of Green's theorem.

Orthogonal Curvilinear Co-ordinate System :

Many calculations in applied mathematics can be simplified by choosing instead of a cartesian co-ordinate system another kind of system that takes the advantage of relations of symmetry involved in the particular problem under consideration.

Let these new co-ordinates be denoted by u_1, u_2, u_3. These are defined by specifying the cartesian co-ordinates (x, y, z) as functions of u_1, u_2, u_3 as follows :

$$\left.\begin{aligned} x &= X(u_1, u_2, u_3) \\ y &= Y(u_1, u_2, u_3) \\ z &= Z(u_1, u_2, u_3) \end{aligned}\right\} \tag{1.32}$$

Solving terms of (x, y, z)

$$u_1 = u_1(x, y, z) \quad \text{etc.} \tag{1.33}$$

We shall confine ourselves to the case where the surfaces u_1 = const., u_2 = const. and u_3 = constant are orthogonal to one another (Fig. 1.28).

In this case the line element ds is given by

$$ds^2 = h_1^2\, du_1^2 + h_2^2\, du_2^2 + h_3^2\, du_3^2 \tag{1.34}$$

where h_1, h_2 and h_3 may be functions of u_1, u_2 and u_3. Set of (u_1, u_2, u_3) are called curvilinear co-ordiantes.

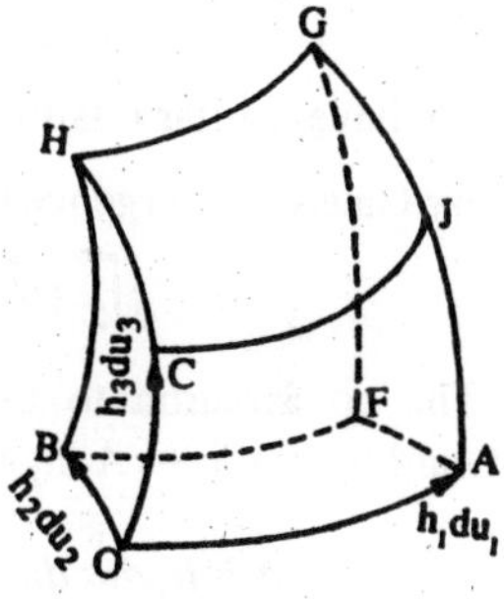

Fig. 1.28

Unit Vectors in Curvilinear Co-ordinates :

The position vector of any general point P is given by

$$\begin{aligned} \vec{r} &= \hat{i}\,x + \hat{j}\,y + \hat{k}\,z \\ &= \hat{i}\,X(u_1, u_2, u_3) + \hat{j}\,Y(u_1, u_2, u_3) + \hat{k}\,Z(u_1, u_2, u_3) \\ &= \vec{r}(u_1, u_2, u_3) \end{aligned}$$

A tangent vector to the curve u_1 at P having equation $u_2 =$ constant and $u_3 =$ constant is $\frac{\partial \vec{r}}{\partial u_1}$. A unit tangent vector in this direction is given by

$$\hat{e}_1 = \frac{\frac{\partial \vec{r}}{\partial u_1}}{\left|\frac{\partial \vec{r}}{\partial u_1}\right|} = \frac{\frac{\partial \vec{r}}{\partial u_1}}{h_1},$$

where $h_1 = \left|\frac{\partial \vec{r}}{\partial u_1}\right|$ from (1.34) writing $ds = dr$

Similarly if $\hat{e}_2$ and $\hat{e}_3$ are unit vectors tangent to u_2 and u_3 curves at P, then

$$\hat{e}_2 = \frac{\frac{\partial \vec{r}}{\partial u_2}}{h_2} \quad \text{and} \quad \hat{e}_3 = \frac{\frac{\partial \vec{r}}{\partial u_3}}{h_3} \tag{1.35}$$

The lengths of the edges of the parallelopiped are h_1du_1, h_2du_2 and h_3du_3.

Let $V(u_1, u_2, u_3)$ be a scalar function and $\vec{A}$ be a vector field with components A_1, A_2, A_3 in the three direction in which the co-ordinates u_1, u_2, u_3 increase. The u_1 component of gradient of V is given by

$$\vec{\nabla}_1 V = (\text{grad. } V)_1 = \underset{du_1 \to 0}{\text{Lt}} \frac{V_{(A)} - V_{(0)}}{h\, du_1} = \frac{1}{h_1}\frac{\partial V}{\partial u_1} \tag{1.36}$$

We have similar relations in the directions 2 and 3.

By Gauss's divergence theorem,

$$\iiint (\vec{\nabla} . \vec{A})\, dv = \iint \vec{A} . \vec{dS} \tag{1.37}$$

The contribution to the integral by the area $OBHC$, taken along the outward normal is $-\, A_1\, h_2\, h_3\, du_2\, du_3$, while through the area $AFGJ$ is

$$A_1h_2h_3\, du_2\, du_3 + \frac{\partial}{h_1\, \partial u_1} (A_1h_2h_3)\; h_1\, du_1\, du_2\, du_3$$

Similar expressions for the other two pairs of surfaces, we have

$$\text{div.} \vec{A} = \underset{v \to 0}{\text{Lt}} \iiint (\vec{\nabla} . \vec{A})\, dv = \underset{v \to 0}{\text{Lt}} \iint \vec{A} . \vec{dS}$$

$$= \iint \vec{A} . \vec{dS} \tag{1.38}$$

Thus $$\vec{\nabla} . \vec{A} = \frac{1}{h_1h_2h_3}\left\{\frac{\partial}{\partial u_1}(A_1h_2\, h_3) + \frac{\partial}{\partial u_2}(A_2h_3h_1) + \frac{\partial}{\partial u_3}(A_3h_1h_2)\right\} \tag{1.39}$$

If $\vec{A} = \vec{\nabla} V$

$$\nabla^2 V = \frac{1}{h_1 h_2 h_3}\left\{\frac{\partial}{\partial u_1}\left(\frac{h_2 h_3}{h_1}\frac{\partial V}{\partial u_1}\right) + \frac{\partial}{\partial u_2}\left(\frac{h_1 h_3}{h_2}\frac{\partial V}{\partial u_2}\right) + \frac{\partial}{\partial u_3}\left(\frac{h_1 h_2}{h_3}\frac{\partial V}{\partial u_3}\right)\right\} \quad (1.40)$$

Spherical Polar Co-ordinates :

In spherical polar co-ordinates (Fig. 1.29)

$$\left.\begin{aligned} x &= r\sin\theta\cos\varphi \\ y &= r\sin\theta\sin\varphi \\ z &= r\cos\theta \end{aligned}\right\} \quad (1.41)$$

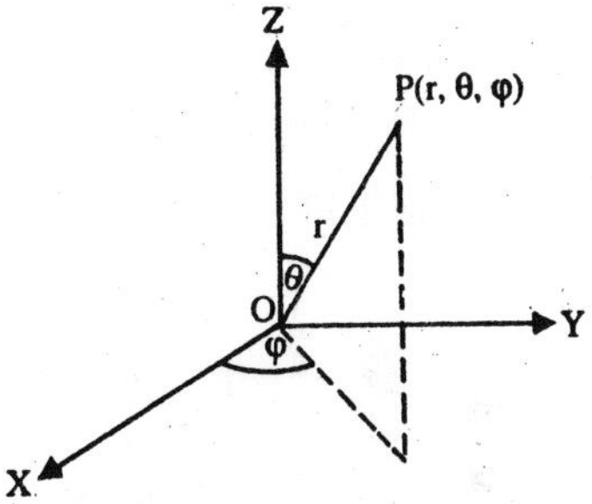

Fig. 1.29

$$ds^2 = dr^2 + r^2 d\theta^2 + r^2 \sin^2\theta d\varphi^2$$

Hence, we have

$$\left.\begin{aligned} u_1 &= r, \quad h_1 = 1 \\ u_2 &= \theta, \quad h_2 = r \\ u_3 &= \varphi, \quad h_3 = r\sin\theta \end{aligned}\right\} \quad (1.42)$$

The unit vectors are

$$\hat{e}_1 = \frac{\frac{\partial\vec{r}}{\partial u_1}}{h_1} = \hat{r} = \hat{i}_r \qquad \therefore \ \vec{r} = \hat{r}\, r$$

$$\hat{e}_2 = \frac{\frac{\partial\vec{r}}{\partial u_2}}{h_2} = \frac{\frac{\partial\vec{r}}{\partial\theta}}{r} = \frac{\partial\hat{r}}{\partial\theta} = \hat{i}_\theta$$

$$\hat{e}_3 = \frac{\frac{\partial\vec{r}}{\partial u_3}}{h_3} = \frac{\frac{\partial\vec{r}}{\partial\varphi}}{r\sin\theta} = \frac{\partial\hat{r}}{\partial\varphi\sin\theta} = \hat{i}\varphi$$

Also $\nabla^2 V = \frac{1}{r^2\sin\theta}\left\{\frac{\partial}{\partial r}\left(r^2\sin\theta\frac{\partial V}{\partial r}\right)\right\} + \frac{1}{r^2\sin\theta}\left\{\frac{\partial}{\partial\theta}\left(\frac{r\sin\theta}{r}\frac{\partial V}{\partial\theta}\right)\right.$

$$+ \left.\frac{1}{r^2\sin\theta}\frac{\partial}{\partial\varphi}\left(\frac{r}{r\sin\theta}\frac{\partial V}{r\varphi}\right)\right\} \text{ from (1.41)}$$

$$= \frac{1}{r^2}\frac{\partial}{\partial r}\left(r^2\frac{\partial V}{\partial r}\right) + \frac{1}{r^2\sin\theta}\frac{\partial}{\partial\theta}\left(\sin\theta\frac{\partial V}{r\theta}\right) + \frac{1}{r^2\sin^2\theta}\frac{\partial^2 V}{\partial\phi^2} \quad (1.43)$$

Cylindrical Co-ordinates :

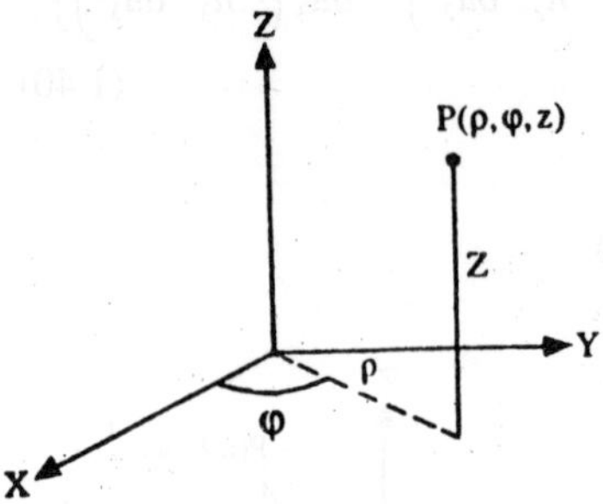

Fig. 1.30

In cylindrical co-ordinates

$$\left.\begin{aligned} x &= \rho\cos\varphi \\ y &= \rho\sin\varphi \\ z &= z \end{aligned}\right\} \quad (1.44)$$

$$ds^2 = dp^2 + p^2\, d\varphi^2 + dz^2$$

We have, therefore,

$$\left.\begin{aligned} u_1 &= \rho,\ h_1 = 1 \\ u_2 &= \varphi,\ h_2 = \rho \\ u_3 &= z,\ h_3 = 1 \end{aligned}\right\} \quad (1.45)$$

The unit vectors are

$$\hat{e}_1 = \frac{\dfrac{\partial \vec{r}}{\partial u_1}}{h_1} = \frac{\partial \vec{r}}{\partial \rho} = \hat{i}_\rho$$

$$\hat{e}_2 = \frac{\dfrac{\partial \vec{r}}{\partial u_2}}{h_2} = \frac{\dfrac{\partial \vec{r}}{\partial \phi}}{\rho} = \hat{i}_\varphi$$

$$\hat{e}_3 = \frac{\dfrac{\partial \vec{r}}{\partial u_3}}{h_3} = \frac{\partial \vec{r}}{\partial z} = \hat{i}_z$$

Also $\nabla^2 V = \dfrac{1}{r}\dfrac{\partial}{\partial r}\left(r\dfrac{\partial V}{\partial r}\right) + \dfrac{1}{r}\dfrac{\partial}{\partial \theta}\left(\dfrac{1}{r}\dfrac{\partial V}{\partial \theta}\right) + \dfrac{1}{r}\dfrac{\partial}{\partial z}\left(r\dfrac{\partial V}{\partial z}\right)$

$$= \frac{1}{r}\frac{\partial}{\partial r}\left(\frac{r\,\partial V}{\partial r}\right) + \frac{1}{r^2}\frac{\partial^2 V}{\partial \theta^2} + \frac{\partial^2 V}{\partial z^2} \quad (1.46)$$

Writing $\rho = r$ and $\varphi = \theta$ for simplicity.

EXERCISE (C.U.)

If $A\rho$, $A\varphi$ and Az be the components of a vector $\vec{A}$ in cylindrical co-ordinates, prove that

$$\frac{d\vec{A}}{dt} = \left\{\frac{dA_\rho}{dt} + \left(A_\varphi \frac{d\varphi}{dt}\right)\right\}\hat{\rho} + \left(\frac{dA\varphi}{dt} + A\rho\frac{d\varphi}{dt}\right)\hat{\varphi} + \frac{dA_z}{dt}\hat{z}$$

$$\vec{r} = \hat{i}x + \hat{j}y + \hat{k}z$$

$$= \hat{i}\rho\cos\varphi + \hat{j}\rho\sin\varphi + \hat{k}z \quad \ldots (a)$$

$$\hat{\rho} = \frac{\frac{\partial \vec{r}}{\partial \rho}}{\left|\frac{\partial \vec{r}}{\partial \rho}\right|} = \frac{\hat{i}\cos\varphi + \hat{j}\sin\varphi}{\sqrt{\cos^2\varphi + \sin^2\varphi}} = \hat{i}\cos\varphi + \hat{j}\sin\varphi \quad \ldots \text{(b)}$$

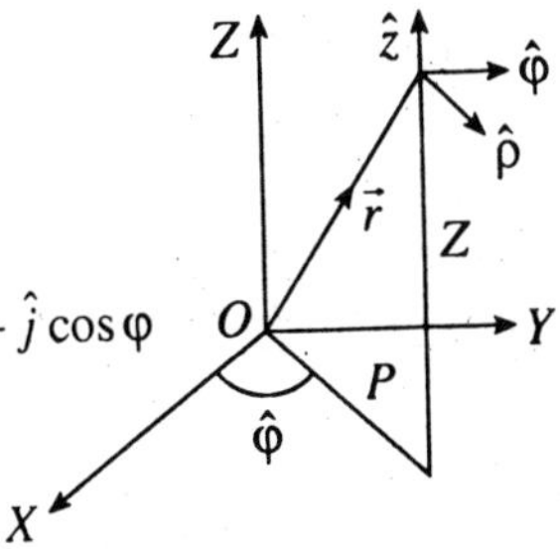

$$\vec{\varphi} = \frac{\frac{\partial \vec{r}}{\partial \varphi}}{\left|\frac{\partial \vec{r}}{\partial \varphi}\right|} = \frac{-\hat{i}\rho\sin\varphi + \hat{j}\rho\cos\varphi}{\sqrt{\rho^2\sin^2\varphi + \rho^2\cos^2\varphi}} = -\hat{i}\sin\varphi + \hat{j}\cos\varphi \quad \ldots \text{(c)}$$

$$\hat{z} = \frac{\frac{\partial \vec{r}}{\partial z}}{\left|\frac{\partial \vec{r}}{\partial z}\right|} = \hat{k} \quad \ldots \text{(d)}$$

$$\dot{\hat{\rho}} = -\hat{i}\sin\varphi\,\dot{\varphi} + \hat{j}\cos\varphi\,\dot{\varphi} = \dot{\varphi}(-\hat{i}\sin\varphi + \hat{j}\cos\varphi)$$
$$= \dot{\varphi}\,\hat{\varphi}$$

$$\dot{\hat{\varphi}} = -\hat{i}\cos\varphi\,\dot{\varphi} - \hat{j}\sin\varphi\,\dot{\varphi} = \dot{\phi}\left(-\hat{i}\cos\varphi - \hat{j}\sin\varphi\right) = \dot{\varphi}\,\hat{\rho}$$

Again $\vec{A} = \hat{\rho}A\rho + \hat{\varphi}A\varphi + \hat{z}Az \quad \ldots \text{(e)}$

$$\frac{d\vec{A}}{dt} = \dot{\hat{\rho}}A\rho + \hat{\rho}\frac{dA\rho}{dt} + \dot{\hat{\varphi}}A\varphi + \hat{\varphi}\frac{dA\varphi}{dt} + \hat{z}\frac{dAz}{dt}$$
$$= \dot{\varphi}\hat{\varphi}A\rho + \hat{\rho}\frac{dA\rho}{dt} + \dot{\varphi}\hat{\rho}A\varphi + \hat{\varphi}\frac{dA\varphi}{dt} + \hat{z}\frac{dAz}{dt}$$
$$= \hat{\rho}\left(\frac{dA\rho}{dt} - A\varphi\frac{d\varphi}{dt}\right) + \hat{\varphi}\left(\frac{dA\varphi}{dt} + \dot{\varphi}A\rho\right) + \hat{z}\frac{Az}{dt}$$

Vector Components in Spherical Polar Co-ordinates :

As in the previous example [equation (a)]

$$\vec{A} = \hat{r}\,Ar + \hat{\theta}\,A_\theta + \hat{\varphi}\,A\varphi$$

Now $\vec{r} = \hat{i}x + \hat{j}y + \hat{k}z$

$$= \hat{i}r\sin\theta\cos\varphi + \hat{j}r\sin\theta\sin\varphi + \hat{k}r\cos\theta$$

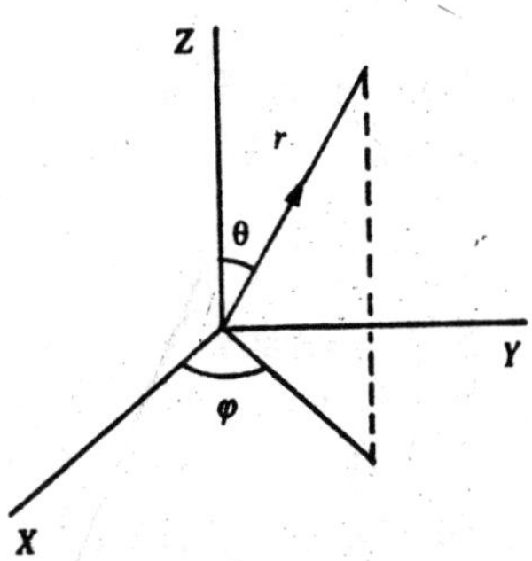

Fig. 1.32

$$\hat{\theta} = \frac{\dfrac{\partial \vec{r}}{\partial \theta}}{\left|\dfrac{\partial \vec{r}}{\partial \theta}\right|} = \frac{\hat{i}\, r\cos\theta\cos\varphi + \hat{j}\, r\cos\theta\sin\varphi - \hat{k}\, r\sin\theta}{\sqrt{r^2\cos^2\theta\left(\cos^2\varphi + \sin^2\dot{\phi}\right) + r^2\sin^2\theta}}$$

$$= \hat{i}\cos\theta\cos\varphi + \hat{j}\cos\theta\sin\varphi - \hat{k}\sin\theta$$

$$\hat{\varphi} = \frac{\dfrac{\partial \vec{r}}{\partial \varphi}}{\left|\dfrac{\partial \vec{r}}{\partial \varphi}\right|} = \frac{\hat{i}\, r\sin\theta(-\sin\varphi) + \hat{j}\, r\sin\theta\cos\varphi}{\sqrt{r^2\sin^2\varphi\left(\cos^2\varphi + \sin^2\varphi\right)}} = -\hat{i}\sin\varphi + \hat{j}\cos\varphi$$

Hence, $\vec{A} = A_r\hat{r} + A_\theta\left\{\hat{i}\cos\theta\cos\varphi + \hat{j}\cos\theta\sin\varphi - \hat{k}\sin\theta\right\}$

$$+ A_\varphi\left(-\hat{i}\sin\varphi + \hat{j}\cos\varphi\right)$$

EXERCISE

1. Given three vectors $\vec{A}$, $\vec{B}$ and $\vec{C}$. Construct the vectors.

(*a*) $3\vec{A} - 2\vec{B} + \dfrac{\vec{C}}{3}$, (*b*) $4\vec{A} + 3\vec{B} - 2\vec{C}$.

2. A triangle has vertices at the points A (2, 1, – 1), B (– 1, 3, 2), C (1, – 2, 1). Find the length of the median to the side AB. $\left[\textbf{Ans.} \dfrac{\sqrt{60}}{2}\right]$

3. If $\varphi(x, y, z) = 3x^2y - y^2z^3$, find $\vec{\nabla}\varphi$ at the point (1, – 2, – 1).

[**Ans.** $-12\hat{i} - 9\hat{j} - 16\hat{k}$]

4. Define Dot and Cross products of two vectors. Show that the area of the triangle ABC formed by the vectors $\vec{AB}$ and $\vec{AC}$ is given by

$$\Delta\, ABC = \frac{1}{2}\left|\vec{AB} \times \vec{AC}\right|.$$

5. Prove that $\vec{\nabla}.\ (\vec{A} + \vec{B}) = \vec{\nabla}.\ \vec{A} + \vec{\nabla}.\ \vec{B}$.

6. If $\vec{A} = 2\hat{i}x^2 - 2\hat{j}y^2z^2 + \hat{k}xy^2z$. Find $\vec{\nabla}.\ \vec{A}$ at (1, – 1, 1)

[**Ans.** – 3]

7. If $\vec{r} = \hat{i}x + \hat{j}y + \hat{k}z$, show that $\vec{\nabla}r = \hat{r}$.

8. Prove that $\vec{\nabla}(r^n) = nr^{n-2}\,\vec{r}$.

9. What is the value of the constant C, so that the vector $\hat{i}(x + 3y) + \hat{j}(y - 2z) + \hat{k}(n + cz)$ is solenonidal. [**Ans.** $C = -2$]

10. If $\vec{A} = \dfrac{\vec{r}}{r}$, find $\vec{\nabla}(\vec{\nabla}.\vec{A})$. [**Ans.** $-2r^2\,\vec{r}$]

11. If $\vec{A}$ is a constant vector, prove that $\vec{\nabla}.\ (\vec{A} \times \vec{r}) = 0$.

12. Define gradiant of a scalar and divergence of a vector. What do you mean by curl of a vector ?

13. Prove that curl $(\vec{A} + \vec{B}) =$ curl $\vec{A} +$ curl $\vec{B}$.

14. If $\vec{A} = (\vec{a} \times \vec{r})\, r^n$, show that

(*i*) div $\vec{A} = 0$, (*ii*) curl $\vec{A} = (n+2)\, r^n \vec{a} - n r^{n-2} (\vec{a}.\vec{r})\, \vec{r}$.

15. Find a unit vector perpendicular to the plane of the vectors $\vec{A} = 3\hat{i} - 2\hat{j} - 4\hat{k}$ and $\vec{B} = \hat{i} + \hat{j} - 2\hat{k}$. $\left[\textbf{Ans.} \pm \left(\dfrac{2\hat{j} + \hat{k}}{\sqrt{5}}\right)\right]$

16. State and prove the divergence theorem and verify it for the vector $\vec{A} = 4x\hat{i} - 2y^2\hat{j} + z^2\hat{k}$ taken over the surface $x^2 + y^2 = 4$, $z = 0$, $z = 3$.

[**Ans.** 84π]

17. Verify the divergence theorem for this vector $\vec{A} = 4xz\hat{i} - y^2\hat{j} + yz\hat{k}$ and a cube bounded by the plane $x = 0, 1$, $y = 0, 1$, and $z = 0, 1$.

18. If $\vec{R}(t)$ is the vector from the origin to the instanteneous position of particle moving along a curve C, show that $\vec{R} \times d\vec{R}$ is equal to twice the area of curve defined by the two vectors $\vec{R}(t)$, $\vec{R}(t+dt) = \vec{R} + d\vec{R}$ and the arc of C which they intercept.

(*a*) Hence, show that

(*i*) If $\vec{R} \times d\vec{R} = 0$, the vector $\vec{R}$ has a constant direction.

(*ii*) If $\vec{R} \times \dfrac{d^2\vec{R}}{dt^2} = 0$, the particle moves so that the radius vector sweeps out equal areas in equal times.

19. A scalar funtion $\phi(x, y, z) = xy^2 + yz^3$. Find the component of grad. φ at the point $(2, -1, 1)$ in the direction of the vector $\hat{i} - 2\hat{j} + 2\hat{k}$.

20. Verify Stokes' Theorem for the vector $V(x + y,\ x - z,\ y + z)$ taken over the triangle ABC cut from the plane $3x + 2y + z = 0$ by the co-ordinate planes.

21. A plane central field $\vec{A}$ is defined by $\vec{A} = \hat{r}\, F(r)$. Determine $F(r)$, so that the field may be irritational and solenoidal.

22. Show that if $\vec{r}$ be the position vector of any point of a closed surface S, then $\iint_S \vec{r}.\, d\vec{S} = 3V$, where V is the volume enclosed by S.

23. If $\vec{A} = 2xz^2\hat{i} - yz\hat{j} + 3xz^3\hat{k}$ and $\varphi = x^2yz$, find $\nabla + (\varphi \vec{A})$

[**Ans.** $5\hat{i} - 3\hat{j} - 4\hat{k}$]

24. Show that the vector $r^n\vec{r}$ is an irrotational vector for any value of n, but is solenoidal only if $n = -3$.

25. if $\vec{F} = 2xz\hat{i} + y^2\hat{j} + yz\hat{k}$, evaluate $\iint \vec{F}.\hat{n}dS$ where S is surface of the cube bounded by $x = 0, x = 1, y = 0, y = 1, z = 0, z = 1$. $\left[\textbf{Ans. } \frac{5}{2}\right]$

26. Show that $\iint_S \vec{F}.\hat{n}\, dS = 64$ where $\vec{F} = z\hat{i} + x\hat{j} - 3y^2z\hat{k}$ and S is the surface of the cylinder $x^2 + y^2 = 16$ included in the first octant between $z = 0$ and $z = 4$.

27. Verify Green's theorem in a plane for $\oint_C \left\{(x^2 + xy)\, dx + x\, dy\right\}$ where C is the curve enclosing the region bounded by $y = x^2$ and $y = x$. $\left[\textbf{Ans. } \frac{1}{12}\right]$

28. Verify Stokes' theorem for the vector field defined by $\vec{F} = (x^2 - y^2)\hat{i} - 2xy\hat{j}$ in the rectangular region in the xy plane bounded by the straight lines $x = 0, x = a\,; y = 0, y = b$.

29. Verify that the angle between the planes $\vec{r} = 2\hat{i} + 3\hat{j} + \hat{k} = 7$ and $\vec{r} = 3\hat{i} - 2\hat{j} + 5\hat{k} = 5$ is $\cos^{-1} = \dfrac{5}{\sqrt{14}\sqrt{35}}$.

ADDITIONAL EXERCISE

1. Find grad r^n, where $\vec{r}$ is the position vector of a point w. r. t. to the origin and show that the differential $d\varphi = \vec{dr}\,.\,\text{grad}\,\varphi$.

We know that $\varphi(x, y, z) = r^n = (x^2 + y^2 + z^2)^{n/2}$

$$\text{Now } \frac{\partial\varphi}{\partial x} = (x^2 + y^2 + z^2)^{n/2 - 1}.\, 2x = 2x\,(r^2)^{n/2 - 1}$$

$$= 2x\, r^{n-2}$$

$$\text{Similarly } \frac{\partial\varphi}{\partial y} = 2y\, r^{n-2} \text{ and } \frac{\partial\varphi}{\partial z} = 2z\, r^{n-2}$$

$$\text{Hence grad } \varphi = \nabla\varphi = \left(\hat{i}\frac{\partial\varphi}{\partial x} + \hat{j}\frac{\partial\varphi}{\partial y} + \hat{k}\frac{\partial\varphi}{\partial r}\right)$$

$$= 2r^{n-2}(\hat{i}x + \hat{j}y + \hat{k}z) = 2r^{n-2}\vec{r}$$

Again differential $d\varphi = \frac{\partial \varphi}{\partial x} dx + \frac{\partial \varphi}{\partial y} dy + \frac{\partial \varphi}{\partial z} dz$

$$= (\hat{i}\, dx + \hat{j}\, dy + \hat{k}\, dz) \cdot \left(\hat{i} \frac{\partial \varphi}{\partial x} + \hat{j} \frac{\partial \varphi}{\partial y} + \hat{k} \frac{\partial \varphi}{\partial z} \right)$$

$$= \vec{dr} \cdot \text{grad } \varphi .$$

2. Prove that $\text{div } (\varphi \vec{A}) = \varphi \text{ div } \vec{A} + \vec{A} \cdot \text{grad } \varphi$

$$\text{div } (\varphi \vec{A}) = \sum \hat{i} \frac{\partial}{\partial x} (\varphi Ax)$$

$$= \sum \hat{i} \left(\varphi \frac{\partial A_x}{\partial x} + \frac{\partial \varphi}{\partial x} A_x \right)$$

$$= \sum \hat{i} \varphi \frac{\partial A_x}{\partial x} + \sum \hat{i} \frac{\partial \varphi}{\partial x} A_x$$

$$= \varphi \sum \hat{i} \frac{\partial A_x}{\partial x} + \left(\sum \hat{i} \frac{\partial \varphi}{\partial x} \right) A_x$$

$$= \varphi \text{ div } \vec{A} + \text{grad } \varphi \cdot \vec{A}$$

3. Evaluate $\iint \vec{A} \cdot \hat{n}\, dS$ where $\vec{A} = z\hat{i} + x\hat{j} - 3y^2 z\hat{k}$ and S is the surface of a cylinder $x^2 + y^2 = 16$ included in the first octant between $z = 0$ and $z = 5$.

Let us project S on the $x - z$ plane

$$\iint \vec{A} \cdot \hat{n}\, ds = \iint \vec{A} \cdot \hat{n} \frac{dx\, dz}{|\hat{n} \cdot \hat{j}|}$$

A normal to $x^2 + y^2 = 16$ is $\vec{\nabla}(x^2 + y^2) = 2x\hat{i} + 2y\hat{j}$ and the unit normal to S is

$$\hat{n} = \frac{2x\hat{i} + 2y\hat{j}}{\sqrt{(2x)^2 + (2y)^2}} = \frac{x\hat{i} + y\hat{j}}{4}$$

Since $x^2 + y^2 = 16$ on S

$$\therefore \quad \vec{A} \cdot \hat{n} = (z\hat{i} + x\hat{j} - 3y^2 z\hat{k}) \cdot \frac{x\hat{i} + y\hat{j}}{4} = \frac{xz + xy}{4}$$

and $\hat{n} \cdot \hat{j} = \frac{x\hat{i} + y\hat{j}}{4} \cdot \hat{j} = \frac{y}{4}$.

So the required surface integral equals

$$\iint \frac{x\hat{i} + y\hat{j}}{y} dxdz = \int_{z=0}^{5} \int_{x=0}^{4} \left(\frac{xz}{\sqrt{16 - x^2}} + x \right) dzdx$$

$$= \int_{z=0}^{5} (4z + 8)\, dz = 90.$$

4. A particle moves so that its position vector is given by $\vec{r} = \hat{i}\cos\omega t + \hat{j}\sin\omega t$, where ω is a constant. Show that (*i*) its velocity, $\vec{v}$ is perpendicular to $\vec{r}$, (*ii*) its acceleration is directed towards the origin and has magnitude proportional to the distance from the origin, (*iii*) $\vec{r} \times \vec{v}$ is a constant vector.

(i) $\vec{v} = \dfrac{d\vec{r}}{dt} = -\hat{i}\,\omega\sin\omega t + \hat{j}\,\omega\cos\omega t$

Hence, $\vec{r}.\vec{v} = \left(\hat{i}\cos\omega t + \hat{j}\sin\omega t\right).\left(-\hat{i}\,\omega\sin\omega t + \hat{j}\omega\cos\omega t\right)$

$= 0$; so $\vec{r}$ is perpendicular to $\vec{v}$.

(*ii*) acceleration $\vec{f} = \dfrac{d\vec{v}}{dt} = -\hat{i}\,\omega^2\cos\omega t - \hat{j}\,\omega^2\sin\omega t$

$= -\omega^2(\hat{i}\cos\omega t + \hat{j}\sin\omega t) = -\omega^2\vec{r}$.

Thus the magnitude of the acceleration is proportional to $|\vec{r}|$ and it is directed towards the origin.

(*iii*) $\vec{r} \times \vec{v} = \begin{vmatrix} \hat{i} & \hat{j} & \hat{k} \\ \cos\omega t & \sin\omega t & 0 \\ -\omega\sin\omega t & \omega\cos\omega t & 0 \end{vmatrix}$

$= \omega(\cos^2\omega t + \sin^2\omega t)\,\hat{k} = \omega\hat{k} =$ constant vector.

5. If (r, θ) represent the polar co-ordinates of the position of a particle and $\hat{r}$ and $\hat{\theta}$ are unit vectors in direction of $\vec{r}$ and perpendicular to it respectively (in the direction of θ increasing), then show that

(*i*) $\hat{r} = \hat{i}\cos\theta + \hat{j}\sin\theta$,

$\hat{\theta} = -\hat{i}\sin\theta + \hat{j}\cos\theta$

(*ii*) $\hat{i} = \hat{r}\cos\theta - \hat{\theta}\sin\theta$,

$\hat{j} = \hat{r}\sin\theta + \hat{\theta}\cos\theta$

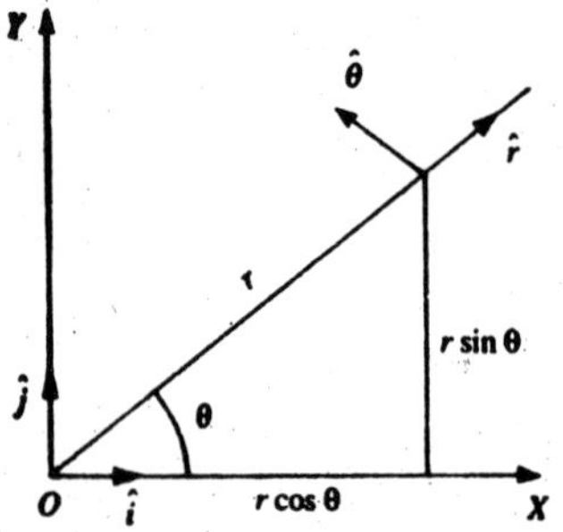

Fig. 1.33

(*i*) If $\vec{r}$ be the instanteneous position vector of the particle, then unit vector in the direction of r increasing is given by

$$\hat{r} = \frac{\dfrac{\partial\vec{r}}{\partial r}}{\left|\dfrac{\partial\vec{r}}{\partial r}\right|}$$

Since $\vec{r} = \hat{i}x + \hat{j}y = \hat{i}r\cos\theta + \hat{j}r\sin\theta$, we have

$$\frac{\partial \vec{r}}{\partial r} = \hat{i}\cos\theta + \hat{j}\sin\theta \text{ and } \left|\frac{\partial \vec{r}}{\partial r}\right| = 1$$

Hence, $\hat{r} = \hat{i}\cos\theta + \hat{j}\sin\theta$ ... (A)

$$\text{Similarly, } \hat{\theta} = \frac{\dfrac{\partial \vec{r}}{\partial \theta}}{\left|\dfrac{\partial \vec{r}}{\partial \theta}\right|} = \frac{-\hat{i}r\sin\theta + \hat{j}r\cos\theta}{r}$$

$$= -\hat{i}\sin\theta + \hat{j}\cos\theta \qquad \text{... (B)}$$

Solving (A) and (B) $\hat{i} = \hat{r}\cos\theta - \hat{\theta}\sin\theta$

and $\hat{j} = \hat{r}\sin\theta + \hat{\theta}\cos\theta$

6. Find the volume of the parallelopiped whose edges are given by $\vec{A} = 2\hat{i} + 3\hat{j} - \hat{k}$, $\vec{B} = \hat{i} - 2\hat{j} + 2\hat{k}$, $\vec{C} = 3\hat{i} - \hat{j} - 2\hat{k}$

Volume $= \vec{A} \cdot \vec{B} \times \vec{C} = 31.$

7. Show that $\vec{\nabla} \times (r^2 \vec{r}) = 0$

where $\vec{r} = \hat{i}x + \hat{j}y + \hat{k}z$ and $r = |\vec{r}|$

$$\text{R. H. S.} = \left(\hat{i}\frac{\partial}{\partial x} + \hat{j}\frac{\partial}{\partial y} + \hat{k}\frac{\partial}{\partial z}\right) \times \left\{(x^2 + y^2 + z^2)(\hat{i}x + \hat{j}y + \hat{k}z)\right\}$$

$$= \begin{vmatrix} \hat{i} & \hat{y} & \hat{k} \\ \dfrac{\partial}{\partial x} & \dfrac{\partial}{\partial y} & \dfrac{\partial}{\partial z} \\ x(x^2 + y^2 + z^2) & y(x^2 + y^2 + z^2) & z(x^2 + y^2 + z^2) \end{vmatrix}$$

$$= \hat{i}\left\{\frac{d}{\partial y}(x^2 + y^2 + z^2)z - \frac{\partial}{\partial z}(x^2 + y^2 + z^2)y\right\} + \cdots\cdots$$

$$= \hat{i}\{2yz - 2yz\} + \cdots\cdots = 0.$$

8. If $\vec{A} = 3xz^2\,\hat{i} - yz\hat{j} + (x + 2z)\,\hat{k}$, find curl curl $\vec{A}$

Now curl curl $\vec{A} = \vec{\nabla}(\vec{\nabla} \cdot \vec{A}) - \nabla^2 \vec{A}$

$$\left(\hat{i}\frac{\partial}{\partial x} + \hat{j}\frac{\partial}{\partial y} + \hat{k}\frac{\partial}{\partial z}\right)\left(\hat{i}\frac{\partial}{\partial x} + \hat{j}\frac{\partial}{\partial y} + \hat{k}\frac{\partial}{\partial z}\right) \cdot [\hat{i}3xz^2 - \hat{j}yz + \hat{k}(x + 2z)]$$

$$- \left(\frac{\partial^2}{\partial x^2} + \frac{\partial^2}{\partial y^2} + \frac{\partial^2}{\partial z^2}\right)\{\hat{i}3xz^2 - \hat{j}yz + \hat{k}(x + 2z)\}$$

$$= \left(\hat{i}\frac{\partial}{\partial x} + \hat{j}\frac{\partial}{\partial y} + \hat{k}\frac{\partial}{\partial z}\right)\left(3z^2 - z + 2\right) - 6\hat{i}x$$

$$= -6\hat{i}x + (6z - 1)\hat{k}$$

(*i*) If $\vec{A} = \left(2xy + z^3\right)\hat{i} + \left(x^2 + 2y\right)\hat{j} + \left(3xz^2 - 2\right)\hat{k}$, show that $\int_C \vec{A}.d\vec{r}$ is independent of the path C joining the points $(1, -1, 1)$ and $(2, 1, 2)$.

$$\vec{\nabla} \times \vec{A} = \begin{vmatrix} \hat{i} & \hat{j} & \hat{k} \\ \dfrac{\partial}{\partial x} & \dfrac{\partial}{\partial y} & \dfrac{\partial}{\partial z} \\ 2xy + z^3 & x^2 + 2y & 3xz^2 - 2 \end{vmatrix} = 0$$

Then $\vec{A}$ can be expressed as the gradiant of a scalar. Let it be φ; i.e. $\vec{A} = \vec{\nabla}\varphi$

$$= \hat{i}\frac{\partial\varphi}{\partial x} + \hat{j}\frac{\partial\varphi}{\partial y} + \hat{k}\frac{\partial\varphi}{\partial z}$$

Then $\dfrac{\partial\varphi}{\partial x} = 2xy + z^3$; $\dfrac{\partial\varphi}{\partial y} = x^2 + 2y$; $\dfrac{\partial\varphi}{\partial z} = 3xz^2 - 2$

Hence, $\vec{A}.\,d\vec{r} = \left(\hat{i}\dfrac{\partial\varphi}{\partial x} + \hat{j}\dfrac{\partial\varphi}{\partial y} + \hat{k}\dfrac{\partial\varphi}{\partial z}\right)\cdot(\hat{i}\,dx + \hat{j}dy + \hat{k}dz)$

$$= \frac{\partial\varphi}{\partial x}dx + \frac{\partial\varphi}{\partial y}dy + \frac{\partial\varphi}{\partial z}dz = d\varphi$$

$$= (2xy + z^3)\,dx + (x^2 + 2y)\,dy + (3xz^2 - 2)\,dz$$

$$= (2xy + z^3)\,dx + x^2dy + 3xz^2dz + 2ydy - 2dz$$

$$= d\,(x^2y + xz^3) + d\,(y^2) + d\,(-2z)$$

$$= d\,(x^2y + xz^3 + y^2 - 2z)$$

Hence, the integral is independent of the path and its value is

$$\int_{1,-1,1}^{2,1,2} \vec{A}.\,d\vec{r} = \int_{1,-1,1}^{2,1,2} d\,(x^2y + xz^3 + y^2 - 2z)$$

$$= \left[x^2y + xz^3 + y^2 - 2z\right]_{1,-1,1}^{2,1,2} = 18$$

9. Given $\vec{B} = \hat{i}(x^2 + 2yz) + \hat{j}\,2yz - \hat{k}\,(z^2 + 2xz)$, find a vector $\vec{A}$, such that $\vec{\nabla} \times \vec{A} = \vec{B}$.

Let us find the divergence of $\vec{B}$.

$$\text{div.}\ \vec{B} = \vec{\nabla}\cdot\vec{B} = \frac{\partial}{\partial x}(x^2 + 2yz) + \frac{\partial}{\partial y}(2yz) + \frac{\partial}{\partial z}\{-(z^2 + 2xz)\}$$

$$= 2x + 2z - 2z - 2x = 0$$

It is evident that the vector $\vec{B}$ must be solenoidal. Vector $\vec{A}$ can not be determined uniquely and to find its y and z component, let us assume that the x component of the vector is zero, as a particular case. Thus since

$$\vec{\nabla}\times\vec{A} = \vec{B}$$

$$\text{curl}_y\ \vec{A} = -\frac{\partial Az}{\partial x} = 2yz$$

and $$\text{curl}_z\ \vec{A} = \frac{\partial A_y}{\partial x} = -\left(z^2 + 2zx\right)$$

which an integration yields

$$A_z = -2yzx + C_1(y, z)$$

$A_y = -(z^2x + zx^2) + C_2(y_1 z)$, C_1 and C_2 being constants of integration depending on y and z.

x-component of $\vec{B}$ is given by

$$B_x = x^2 + 2yz = \frac{\partial A_z}{\partial y} - \frac{\partial A_y}{\partial z}$$

$$\text{Hence, } B_x = -2zx + \frac{\partial C_1}{\partial y} + 2zx + x^2 - \frac{\partial C_2}{\partial z}$$

Equating the two values we have

$$x^2 + 2yz = \frac{\partial C_1}{\partial y} + x^2 - \frac{\partial C_2}{\partial z}.$$

This equality will be satisfied only if we choose

$$C_1 = 0, \quad C_2 = -yz^2$$

$$\text{or,}\quad C_1 = y^2 z,\ C_2 = 0 \text{ or, } C_1 = \frac{1}{2}y^2z,\ C_2 = \frac{1}{2}yz^2.$$

Hence, with one of the choice vector $\vec{A}$ can be expressed as

$$\vec{A} = -\frac{\hat{i}}{2}(z^2x + zx^2) - \hat{k}(2xyz - y^2z).$$

CHAPTER II

ORDINARY DIFFERENTIAL EQUATIONS

2.1. First order linear differential equation

Exact differential equations :

A differential equation of the form $Mdx + Ndy = 0$, where M and N are functions of x and y, is said to be exact, where there is a function u of x and y such that $Mdx + Ndy = du$ i.e., when $Mdx + Ndy$ becomes a perfect differential or $\frac{\partial M}{\partial y} = \frac{\partial N}{\partial x}$.

The method of finding the solution of those equations is as follows. Let us first integrate the terms in Mdx as if y is constant, then integrate the terms in Ndy taking x as constant and then equate this sum to a constant.

Illustration :

(1) Solve :

$$xdx + ydy + (x^2 + y^2)\, dy = 0$$

or, $$\frac{d\,(x^2 + y^2)}{2} = -\,(x^2 + y^2)\, dy$$

or, $$\frac{1}{2} \int \frac{d\,(x^2 + y^2)}{x^2 + y^2} = -\int dy$$

or, $$\log\,(x^2 + y^2) = -\,2y + \log c$$

or, $$x^2 + y^2 = ce^{-2y}$$

(2) Solve :

$$xdy - ydx + a\,(x^2 + y^2)\, dx = 0$$

or, $$\frac{xdy - ydx}{x^2} = -\,a\left(1 + \frac{y^2}{x^2}\right) dx$$

or, $$\int \frac{d\left(\frac{y}{x}\right)}{1 + \frac{y^2}{x^2}} = -\,a \int dx$$

or, $\tan^{-1}\left(\frac{y}{x}\right) = -ax + c$

(3) Solve :

$$xdx + ydy + \frac{xdy - ydx}{x^2 + y^2} = 0$$

or, $\int \frac{d(x^2 + y^2)}{2} + \int \frac{d\left(\frac{y}{x}\right)}{1 + \frac{y^2}{x^2}} = 0$

or, $\frac{x^2 + y^2}{2} + \tan^{-1}\frac{y}{x} = c$

(4) Solve :

$$y(1 + xy)\, dx + x(1 - xy)\, dy = 0$$

or, $y\,dx + xy^2\,dx + x(1 - xy)\,dy = 0$

or, $y\,dx + xy^2\,dx + x\,dy - x^2y\,dy = 0$

or, $d(xy) + xy(y\,dx - x\,dy) = 0$

or, $\frac{d(xy)}{x^2y^2} + \frac{y\,dx - x\,dy}{y^2\left(\frac{x}{y}\right)} = 0$

or, $\int \frac{d(xy)}{x^2y^2} + \int \frac{d\left(\frac{x}{y}\right)}{\frac{x}{y}}$

or, $-\frac{1}{xy} + \log\left(\frac{x}{y}\right) = c$

An equation of the form

$\frac{dy}{dx} + Py = Q$, where P and Q are functions of x or constant is called a linear differential equation of the first order. To integrate, each side is to be multiplied by a factor $e^{\int P dx}$ known as the integrating factor.

Illustration :

(5) Solve :

$$\cos^2 x \frac{dy}{dx} + y = \tan x$$

or, $\frac{dy}{dx} + y\sec^2 x = \sec^2 x \tan x$

I.F. is $e^{\int \sec^2 x\, dx} = e^{\tan x}$

$$\therefore \int dy\, e^{\tan x} + \int y \sec^2 x\, e^{\tan x}\, dx = \int \sec^2 x \tan x\, e^{\tan x} dx$$

or, $y e^{\tan x} = \tan x\, e^{\tan x} - \int \sec^2 x e^{\tan x}\, dx$

$$= \tan x\, e^{\tan x} - e^{\tan x} + c$$

or, $y = \tan x - 1 + c e^{-\tan x}$

(6) Solve :

$$\frac{dy}{dx} + xy = x$$

I.F. $= e^{\int x\, dx} = e^{x^2/2}$

Hence $e^{x^2/2}\, dy + xy\, e^{x^2/2}\, dx = x\, e^{x^2/2}\, dx$

Integrating

$$y e^{x^2/2} = \int x e^{\frac{x^2}{2}}\, dx = e^{x^2/2} + c$$

or, $y = 1 + c\, e^{-x^2/2}$

(7) Solve :

$$(1 - x^2) \frac{dy}{dx} - xy = 1$$

or, $\dfrac{dy}{dx} - \dfrac{x}{1 - x^2}\, y = \dfrac{1}{1 - x^2}$

I.F. $= e^{-\int \frac{x}{1-x^2} dx} = e^{\log \frac{(1-x^2)}{2}} = e^{\log \sqrt{1 - x^2}}$

$$= \sqrt{1 - x^2}$$

Multiplying by the I.F. and integrating

$$\int \sqrt{1 - x^2}\, dy - \int \frac{y\, x\, dx}{(1 - x^2)^{1/2}} = \int \frac{dx}{\sqrt{1 - x^2}}$$

or, $y \sqrt{1 - x^2} = \sin^{-1} x + c$

Equations of the form

$\dfrac{dy}{dx} + Py = Qy^n$ can be solved by :

(8) Solve :

$$\frac{dy}{dx} + x \sin 2y = x^3 \cos^2 y$$

Dividing by $\cos^2 y$

$$\sec^2 y \frac{dy}{dx} + 2x \tan y = x^3$$

Putting $\tan y = z$

$$\frac{dz}{dx} + 2xz = x^3$$

$$\text{I.F.} = e^{\int 2x dx} = e^{x^2}$$

$$\therefore \quad e^{x^2} dz + 2xze^{x^2} dx = x^3 e^{x^2} dx$$

Integrating

$$ze^{x^2} = \frac{x^2}{2} \int 2x\, e^{x^2} dx - \int x e^{x^2} dx$$

$$= \frac{x^2}{2} e^{x^2} - e^{x^2}/2 + c$$

or, $\tan y = \dfrac{x^2 - 1}{2} + c e^{-x^2}$

(9) Solve :

$$\frac{xdy}{dx} + y = y^2 \log x$$

Dividing by $y^2 x$

$$\frac{1}{y^2} \frac{dy}{dx} + \frac{1}{xy} = \frac{\log x}{x}$$

Putting $\dfrac{1}{y} = z$

$$-\frac{1}{y^2} \frac{dy}{dz} = \frac{dz}{dx}$$

$$\therefore \quad \frac{dz}{dx} - \frac{z}{x} = -\frac{\log x}{x}$$

$$\text{I.F.} = e^{\int -\frac{1}{x} dx} = e^{-\log x} = e^{\log \frac{1}{x}} = \frac{1}{x}$$

Multiplying by I.F.

$$\frac{dz}{x} - \frac{z}{x^2} dx = -\frac{\log x}{x^2} dx$$

Integrating

$$\frac{z}{x} = \int \log x \left(-\frac{1}{x^2}\right) dx$$

$$= \log x \left(\frac{1}{x}\right) - \int \frac{dx}{x^2}$$

$$= \frac{\log x}{x} + \frac{1}{x} + c$$

or, $\frac{1}{xy} = \frac{\log x}{x} + \frac{1}{x} + c$

or, $y\,(cx + 1 + \log x) = 1$

(10) Solve :

$$\frac{dy}{dx} + \frac{y}{x} \log y = \frac{y}{x^2} (\log y)^2$$

Dividing by $y (\log y)^2$

$$\frac{1}{y (\log y)^2} \frac{dy}{dx} + \frac{1}{x \log y} = \frac{1}{x^2}$$

Put $\frac{1}{\log y} = z$

$$-\frac{1}{y (\log y)^2} \frac{dy}{dx} = \frac{dz}{dx}$$

$$\therefore \quad \frac{dz}{dx} - \frac{z}{x} = -\frac{1}{x^2}$$

I.F. is $e^{\int -\frac{1}{x} dx} = e^{-\log x} = e^{\log \frac{1}{x}} = \frac{1}{x}$

Multiplying by the I.F.

$$\frac{dz}{x} - \frac{z dx}{x^2} = -\frac{dx}{x^3}$$

Integrating

$$\frac{z}{x} = \frac{1}{2x^2} + c$$

or, $zx = \frac{1}{2} + cx^2$

or, $\frac{x}{\log y} = \left(\frac{1}{2} + cx^2\right)$

$$\therefore \quad x = \log y \left(\frac{1}{2} + cx^2\right)$$

Second order linear differential equation with constant coefficient (Homogeneous and inhomogeneous both) :

When the right hand member is zero, the typical equation is

$$\frac{d^2y}{dx^2} + P_1 \frac{dy}{dx} + P_2 y = 0, \quad \ldots \text{(a)}$$

where P_1 and P_2 are constants.

The method of trial solution is as follows.

Let $y = e^{mx}$ be a trial solution of equation.

(a) Finding $\frac{dy}{dx}, \frac{d^2y}{dx^2}$ and substituting in (a) $(m^2 + P_1 m + P_2)\ e^{mx} = 0$

Since $e^{mx} \neq 0$, we have

$$m^2 + P_1 m + P_2 = 0. \quad \ldots \text{(b)}$$

This is known as the auxiliary equation. If m_1 and m_2 are the roots of equation,

(b) the two particular solutions are $y = Ae^{m_1 x}$ and $Be^{m_2 x}$.

The general solution is $y = Ae^{m_1 x} + Be^{m_2 x}$. ... (c)

N. B. : It must be remembered that the general solution of a second order differential equation must contain two arbitrary constants.

If $m_1 = m_2 = \alpha$ say

Then $y = (A + B)\ e^{\alpha x} = ce^{\alpha x}$ is not a general solution, since it involves only one arbitrary constant.

In such a case let $y = c^{\alpha x} v$,

where $v = f(x)$, be the solution.

Since the auxiliary equation has two equal roots α, $P_1 = -2\alpha$ and $P_2 = \alpha^2$ i.e., the differential equation is of the form

$$\frac{d^2y}{dx^2} - 2\alpha \frac{dy}{dx} + \alpha^2 y = 0 \quad \ldots \text{(d)}$$

Substituting for y, $\frac{dy}{dx}$ and $\frac{d^2y}{dx^2}$

$$c^{\alpha x} \frac{d^2v}{dx^2} = 0$$

$$\because\ e^{\alpha x} \neq 0,\ \frac{d^2v}{dx^2} = 0$$

Hence $\frac{dv}{dx} = \text{constant} = c_1$

and $v = c_1 x + c_2$

$\therefore$ the solution is $y = (c_1 x + c_2)\ e^{\alpha x}$ ... (e)

If the roots of the auxiliary equation are complex i.e., $m_1 = \alpha + i\beta$ and $m_2 = \alpha - i\beta$

$$y = e^{\alpha x}(Ae^{i\beta x} + Be^{-i\beta x})$$

$$= e^{\alpha x}[A(\cos\beta x + i\sin\beta x) + B(\cos\beta x - i\sin\beta x)]$$

$$= ce^{\alpha x}\cos(\beta x + \gamma) \qquad \ldots \text{(f)}$$

Illustration :

(1) Solve the equation :

$$\frac{d^2y}{dx^2} - 7\frac{dy}{dx} + 12y = 0$$

let $y = e^{mx}$ be a trial solution.

Auxiliary equation is $m^2 - 7m + 12 = 0$

or, $(m - 3)(m - 4) = 0$

$\therefore m = 3, 4$

Hence the solution is

$$y = Ae^{3x} + Be^{4x}$$

(2) Solve :

$\dfrac{d^2y}{dx^2} + 2\dfrac{dy}{dx} + y = 0$; $y = e^{mx}$ is the trial solution.

Auxiliary equation is $m^2 + 2m + 1 = 0$

or, $(m + 1)^2 = 0 \quad \therefore m = -1, -1$

Hence the solution is $y = (Ax + B)e^{-x}$

(3) Solve :

$$\frac{d^2y}{dx^2} - 4\frac{dy}{dx} + 13y = 0$$

$y = e^{mx}$ is the trial solution.

Auxiliary equation is

$$m^2 - 4m + 13 = 0$$

$$\therefore m = 2 \pm 3i$$

Solution is $y = e^{2x}\{(A\cos 3x + i\sin 3x) + B(\cos 3x - i\sin 3x)\}$

$$= e^{2x}\{\cos 3x(A + B) + \sin 3x(A - B)i\}$$

$$= e^{2x}(C\cos 3x + D\sin 3x)$$

When the right hand member is a function of x :

The typical differential equation is

$$\frac{d^2y}{dx^2} + P_1 \frac{dy}{dx} + P_2 y = X, \text{ where } X \text{ represents the function of } x. \quad \ldots (a)$$

The solution consists of two parts :

(1) Complementary function (C.F.) $\varphi(x)$ obtained by solving

$$\frac{d^2y}{dx^2} + P_1 \frac{dy}{dx} + P_2 y = 0$$

(2) Particular Integral $\psi(x)$

The general solution will be $y = \varphi(x) + \psi(x)$

For finding out the P.I., let us introduce the operator symbols

$$D \equiv \frac{d}{dx}, \quad D^2 \equiv \frac{d^2}{dx^2}$$

and $\frac{1}{D} \equiv \int (\quad) \, dx$

$$\frac{1}{D^2} \equiv \int \left\{ \int (\quad) \, dx \right\} dx \text{ etc.}$$

In operator symbols equation (a) reduces to $(D^2 + P_1 D + P_2)\, y = X$

or, $f(D)\; y = X$

$$\therefore \; y = \frac{X}{f(D)}$$

Methods of finding P.I.

(A) When $X = x^m$, $m =$ Constant

$$\therefore \text{ (P.I.). } y = \frac{X}{f(D)} = \frac{x^m}{f(D)} = f(D)^{-1} \, x^m$$

Expanding $f(D^{-1})$ and operating on x^m, the P.I. can be found out.

(B) When $X = e^{ax} V$, where V is a function of x or constant.

Let V_1 be another function of x such that

$$f(D + a)\; V_1 = V \quad \text{i.e., } V_1 = \frac{V}{f(D + a)}$$

Now $f(D)\; e^{ax} V_1 = e^{ax} f(D + a)\; V_1$

$$= e^{ax} V$$

$$\therefore \; \frac{e^{ax} V}{f(D)} = e^{ax} V_1 = e^{ax} \frac{V}{f(D + a)}$$

Lemma :

$$D\left\{e^{ax}V_1\right\} = ae^{ax}V_1 + e^{ax}DV_1$$

$$= e^{ax}(D+a)V_1 \text{ etc.}$$

N. B. : It is to be noted that the function to be operated is to be written on the R.H. side of the operator. When written on the L.H. side means the operation is not to be done on this function.

In a similar way,

$$\frac{e^{ax}K}{f(D)} = e^{ax}\frac{K}{f(D+a)} = Ke^{ax}\frac{1}{f(D+a)}$$

(C) when $X = e^{ax}$

If $f(a) \neq 0$, $f(D)\,e^{ax} = f(a)\,e^{ax}$

$$\therefore \frac{e^{ax}}{f(D)} = \frac{e^{ax}}{f(a)}$$

If $f(a) = 0$, $(D-a)$ is a factor of $f(D)$

$\therefore f(D) = (D-a)\,\varphi(D)$ if $\varphi(a) \neq 0$

or, $(D-a)^2$ if $\varphi(a) = 0$

In the first case, $\dfrac{e^{ax}}{f(D)} = \dfrac{e^{ax}}{(D-a)\,\varphi(D)}$

$$= \frac{e^{ax}}{\varphi(a)(D-a)} = \frac{1}{\varphi(a)}e^{ax}\frac{1}{(D-a+a)}$$

$$= \frac{e^{ax}}{\varphi(a)}\frac{1}{D} = \frac{e^{ax}x}{\varphi(a)}$$

In the second case,

$$\frac{e^{ax}}{f(D)} = \frac{e^{ax}}{(D-a)^2} = e^{ax}\frac{1}{D^2} = e^{ax}\frac{x^2}{2}$$

(D) When $X = \sin(ax+b)$ or $\cos(ax+b)$

Since $\varphi(D^2)\sin(ax+b) = \varphi(-a^2)\sin(ax+b)$

$$\frac{\sin(ax+b)}{\varphi(D^2)} = \frac{\sin(ax+b)}{\varphi(-a^2)}$$

Similarly $\dfrac{\cos(ax+b)}{\varphi(D^2)} = \dfrac{\cos(ax+b)}{\varphi(-a^2)}$ provided $\varphi(-a^2) \neq 0$

If $\varphi(-a^2) = 0$, then procedure will as shown in the illustration

Illustration :

(1) Solve :

$$\frac{d^2y}{dx^2} + 4y = 2x + 3$$

For C.F., $\frac{d^2y}{dx^2} + 4y = 0$

Let $y = e^{mx}$ be on trial solution.

Auxiliary equation is $m^2 + 4 = 0$

$$\therefore \; m = \pm\, 2i$$

Hence the C.F. is $y = (A \cos 2x + B \sin 2x)$

For P.I. we have

$$(D^2 + 4)\; y = 2x + 3$$

$$y = \frac{2x + 3}{D^2 + 4} = \frac{1}{4}\left(1 + \frac{D^2}{4}\right)^{-1} (2x + 3)$$

$$= \frac{1}{4}\left(1 - \frac{D^2}{4} + \ldots\ldots\right)(2x + 3) = \frac{2x + 3}{4}$$

The general solution is

$$y = A \cos 2x + B \sin 2x + \frac{2x + 3}{4}$$

(2) Solve :

$$\frac{d^2y}{dx^2} - y = e^{2x}$$

For C.F. we have $\frac{d^2y}{dx^2} - y = 0$

Auxiliary equation is $m^2 - 1 = 0$ or $m^2 = 1$

$$\therefore \; m = \pm\, 1$$

$\therefore$ C.F. is $y = Ae^x + Be^{-x}$

For P.I. we have using operator symbols $(D^2 - 1)\, y = e^{2x}$

$$\therefore \; y = \frac{e^{2x}}{(D^2 - 1)} = e^{2x} \cdot \frac{1}{(2^2 - 1)}$$

$$= \frac{e^{2x}}{3}$$

Hence the general solution is

$$y = A e^{x} + B e^{-x} + e^{2x}/3$$

(3) Solve :

$$(D^2 + 2D + 2)\, y = xe^{-x}$$

For C.F. $(D^2 + 2D + 2)\, y = 0$

A.E. is $m^2 + 2m + 2 = 0$

or, $(m + 1)^2 = -1$

or, $m + 1 = \pm i$

$\therefore\ m = -1 \pm i$

Hence the C.F. is $y = e^{-x} (A \cos x + B \sin x)$

For P.I. $(D^2 + 2D + 2)\ y = xe^{-x}$

$$\therefore\ y = \frac{xe^{-x}}{D^2 + 2D + 2} = e^{-x} \frac{x}{(D-1)^2 + 2(D-1) + 2}$$

$$= e^{-x} \frac{x}{D^2 - 2D + 1 + 2D - 2 + 2} = e^{-x} (1 + D^2)^{-1}\, x$$

$$= e^{-x} (1 - D^2)\, x = xe^{-x}$$

$\therefore$ General solution is

$$y = e^{-x} (A \cos x + B \sin x + x)$$

(4) Solve :

$$\frac{d^2y}{dx^2} + y = \cos^2 x$$

For C.F. $\dfrac{d^2y}{dx^2} + y = 0$

A.E. is $m^2 + 1 = 0$ or, $m^2 = -1 \ \therefore\ m = \pm i$

$\therefore$ C.F. is $y = (A \cos x + B \sin x)$

For P.I.

$$(D^2 + 1)\, y = \cos^2 x = \frac{1 + \cos 2x}{2}$$

$$\therefore\ y = \frac{1 + \cos 2x}{2(1 + D^2)} = \frac{1}{2}\left(\frac{1}{1 + D^2} + \frac{\cos 2x}{1 + D^2}\right)$$

$$= \frac{1}{2}\left[(1 - D^2)\, 1 + \frac{\cos 2x}{1 - 4}\right] = \frac{1}{2} - \frac{\cos 2x}{6}$$

Hence the general solution is

$$y = A\cos x + B\sin x - \frac{\cos 2x}{6} + \frac{1}{2}$$

(5) Solve :

$$\frac{d^2y}{dx^2} + y = \sin x$$

C.F. is $y = A\cos x + B\sin x$ as in previous example.

For P.I. $(D^2 + 1)\, y = \sin x$

$$\therefore\ y = \frac{\sin x}{1 + D^2}$$

$$\text{Hence } iy = \frac{i\sin x}{D^2 + 1}$$

$$\text{Let } y^1 = \frac{\cos x}{D^2 + 1}$$

$$\text{Adding } y^1 + iy = \frac{\cos x + i\sin x}{D^2 + 1} = \frac{e^{ix}}{D^2 + 1}$$

$$= e^{ix}\frac{1}{(D + i)^2 + 1} = e^{ix}\frac{1}{D^2 + 2iD + i^2 + 1}$$

$$= e^{ix}\frac{1}{D^2 + 2iD} = e^{ix}\frac{1}{2iD\left(1 + \frac{D}{2i}\right)}$$

$$= e^{ix}\frac{(1 - D/2i)}{2iD}\,.\,1 = e^{ix}\,.\frac{1}{2iD} = e^{ix}\frac{x}{2i}$$

$$= \frac{x(\cos x + i\sin x)}{2i}$$

Equating the imaginary parts

$$iy = \frac{x\cos x}{2i} = -\frac{ix\cos x}{2}$$

$\therefore$ Complete solution is

$$y = A\cos x + B\sin x - \frac{x\cos x}{2}$$

(6) Solve :

$$(D^2 - 9)\, y = e^{3x}\cos x$$

For C.F. Auxiliary equation is $m^2 - 9 = 0$

$$\text{or,} \quad m = \pm 3$$

Hence the C.F. is $y = Ae^{3x} + Be^{-3x}$

For P.I.

$$(D^2 - 9)\, y = e^{3x} \cos x$$

$$\therefore \; y = \frac{e^{3x} \cos x}{D^2 - 9} = e^{3x} \frac{\cos x}{(D+3)^2 - 9}$$

$$= e^{3x} \frac{\cos x}{D^2 + 6D} = e^{3x} \frac{(D^2 - 6D)}{D^4 - 36D^2} \cos x$$

$$= \frac{e^{3x}(-\cos x + 6 \sin x)}{(-1)^2 + 36} = \frac{e^{3x}(6 \sin x - \cos x)}{37}$$

complete solution is $y = Ae^{3x} + Be^{-3x} + \dfrac{e^{3x}}{37}(6 \sin x - \cos x)$

EXERCISE

1. A particle is said to excute S.H.M. when it moves on a straight line, with acceleration always directed towards a fined point on the line and is proportional to the displacement from it in any position. If it starts from rest at a distance a from the fixed point, find its velocity in any position, and the time for that position. Deduce that the motion is oscillatory and find the periodic time.

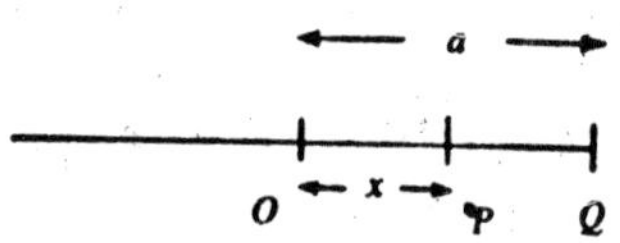

Fig. 2.1

Let x be the displacement of the particle from the fixed point O any instant of tune 't'.

From definition $\dfrac{d^2x}{dt^2} \propto x$

$$\text{or,} \quad \frac{d^2x}{dt^2} = -\mu x \qquad \ldots \quad (a)$$

Let $y = e^{mt}$ be a trial solution of equn. (a). The Auxiliary equation is

$$m^2 + \mu = 0 \text{ or } m = \pm i\sqrt{\mu}$$

Hence the solution is

$$x = Ae^{\sqrt{\mu}t} + Be^{-\sqrt{\mu}t}$$

$$= C\cos\sqrt{\mu}t + D\sin\sqrt{\mu}t \qquad \ldots \quad (b)$$

$$\therefore \frac{dx}{dt} = -\sqrt{\mu}\ C \sin \sqrt{\mu}\,t + \sqrt{\mu}\ D \cos \sqrt{\mu}\,t \qquad \ldots \text{(c)}$$

Initial condition is at $t = 0$, $x = a$, $\frac{dx}{dt} = 0$

Hence from (b) and (c)

$$c = a, \text{ and } D = 0$$

So the displacement equation is

$$x = a \cos \sqrt{\mu}\,t. \qquad \ldots \text{(d)}$$

Motion is oscillatory

From (d), instanteneous velocity

$$|(v)| = \left(\frac{dx}{dt}\right) = a\sqrt{\mu} \sin \sqrt{\mu}\,t$$

$$= a\sqrt{\mu}\sqrt{1 - \frac{x^2}{a^2}} = \sqrt{\mu}\sqrt{a^2 - x^2} \qquad \ldots\ldots \text{(e)}$$

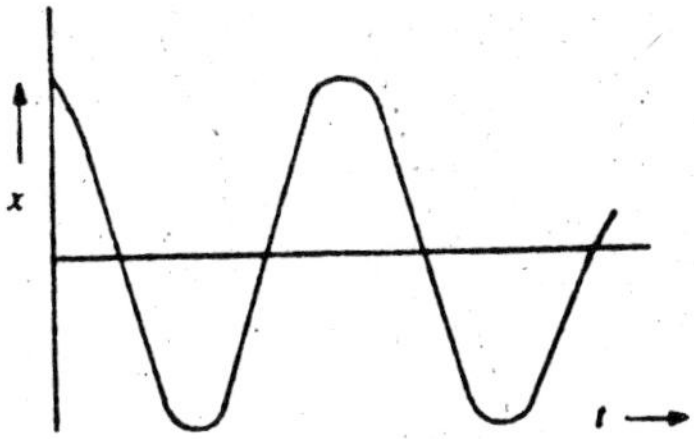

Fig. 2.2

From (d) $t = \frac{1}{\sqrt{\mu}} \cos^{-1} \frac{x}{a}$

If t' be the time from Q to O

$$\cos \sqrt{\mu}\,t' = 0 = \cos \pi/2$$

$$\text{or, } \quad t' = \frac{\pi}{2\sqrt{\mu}}$$

Hence the periodic time $T = 4t' = \frac{2\pi}{\sqrt{\mu}}$

2. A particle executes S.H.M. being acted upon by a restoring force portional to the displacement from a fixed point of the path and directed towards the fixed point and a force of resistance $\propto$ velocity. To investigate the motion, let m be the mass of the particle. The differential equation of motion of the particle is

$$\frac{md^2x}{dt^2} = -\mu x - \alpha \frac{dx}{dt} \qquad \ldots \text{(a)}$$

$$\text{or, } \quad \frac{d^2x}{dt^2} + 2b\frac{dx}{dt} + n^2x = 0 \qquad \ldots \text{(b)}$$

where $2b = \frac{\alpha}{m}$ and $n^2 = \frac{\mu}{m}$

Let $x = e^{m't}$ be a trial solution

A.E. is $m'^2 + 2bm' + n^2 = 0$

$$\therefore\ m' = \frac{-2b \pm \sqrt{4b^2 - 4n^2}}{2} = -b \pm \sqrt{b^2 - n^2}$$

Hence the general solution is

$$x = e^{-bt}\left(Ae^{\sqrt{b^2 - n^2}\,t} + Be^{-\sqrt{b^2 - n^2}\,t}\right) \quad \ldots \text{(c)}$$

Three cases may arise :

Case I : If $b > n$, $\sqrt{b^2 - n^2}$ is real and x decays rather slowly with time. The motion is called heavily damped or dead beat. Variation of x with t is shown graphical below.

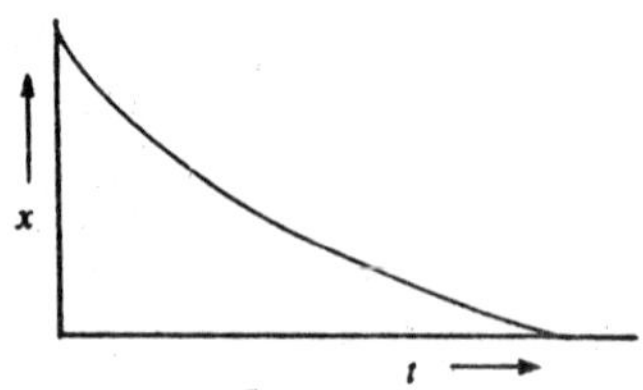

Fig. 2.3

Case II : If $b < n$, $\sqrt{b^2 - n^2} = i\sqrt{n^2 - b^2} = in'$ say

$$x = e^{-bt}\left(Ae^{-in't} + Be^{-in't}\right)$$

$$= e^{-bt}\, c \cos(n't + \alpha)$$

The motion is oscillatory with amplitude exponentially decaying. Variation of x with t is shown graphically below.

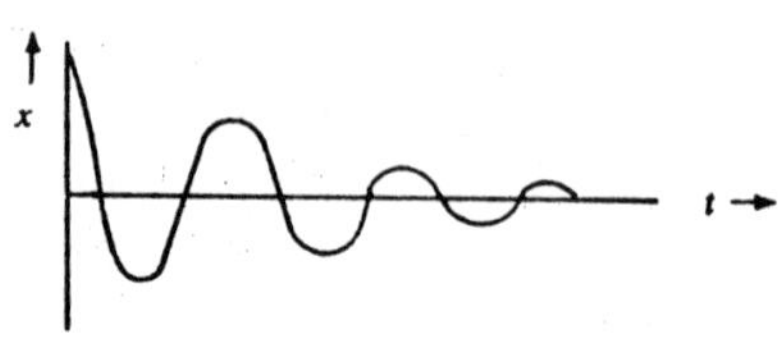

Fig. 2.4

Case III : If $b \to n$, the roots of the A.E. will be coincident. Hence the solution of it will be $x = e^{-bt}(A + Bt)$.

x decays with t at a rather faster rate. The motion is said to be critically damped. Variation of x with t is shown graphically in Fig. 2.5.

3. A particle falls towards the earth starting from rest at a height h above the surface. If the attraction varies inversely as the square of the distance from the centre, find the velocity in reaching the earth's surface. a = radius of the earth, g = acceleration due to gravity on the earth's surface.

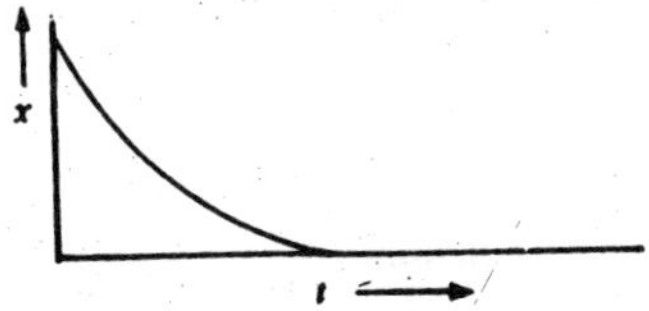

Fig. 2.5

Differential equation of motion of the particle is

$$\frac{d^2x}{dt^2} = -\frac{\mu}{x^2} \qquad \ldots \text{(A)}$$

Multiplying both sides by $2\frac{dx}{dt}$ and integrating w. r. t. 't'.

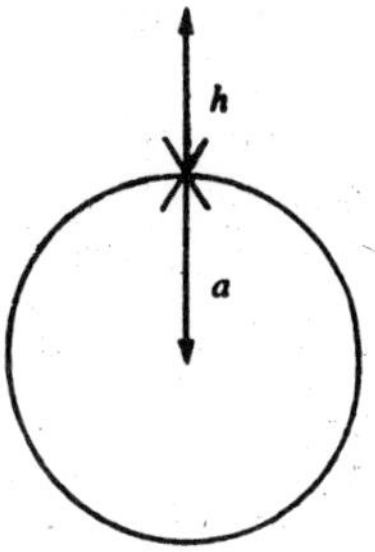

Fig. 2.6

$$\int 2\frac{dx}{dt}\frac{d^2x}{dt^2}\,dt = -2\mu\int\frac{1}{x^2}\frac{dx}{dt}\,dt$$

or, $$\left(\frac{dx}{dt}\right)^2 = \frac{2\mu}{x} + A'$$

$$\frac{dx}{dt} = 0, \text{ when } x = a + h$$

$$\therefore\ A' = \frac{-2\mu}{a+h}$$

Substituting for A'

$$\left(\frac{dx}{dt}\right)^2 = 2\mu\left\{\frac{1}{x} - \frac{1}{a+h}\right\}$$

On the surface of the earth $x = a$

$$\therefore\ \left(\frac{dx}{dt}\right)^2 = 2\mu\left\{\frac{1}{a} - \frac{1}{a+h}\right\} = \frac{2\mu h}{a(a+h)}$$

Again $$\left(\frac{d^2x}{dt^2}\right)_{x=a} = \frac{\mu}{a^2} = g \quad \therefore\ \mu = ga^2$$

Hence $$\left(\frac{dx}{dt}\right)^2 = \frac{2ga^2h}{a(a+h)} \quad \therefore\ \frac{dx}{dt} = \sqrt{\frac{2gah}{a+h}}$$

4. Radium decomposes at a rate portional to the amount present. If 5% of the amount disappears in 50 years, how much will remain at the end of 100 years ?

Let N_o = amount present initially,

N = amount present at any time.

given $\frac{dN}{dt} \alpha N$ ... (A)

or, $\frac{dN}{dt} = -\lambda N \quad \lambda = \text{constant}$.

or, $\frac{dN}{N} = -\lambda dt$

Integrating, $\log_e^N = -\lambda t + A'$

$N = N_0$, at $t = 0 \quad \therefore A' = \log_e N_0$

Substituting for A'

$$\log_e \left(\frac{N}{N_0}\right) = -\lambda t$$

$$\therefore N = N_0 e^{-\lambda t} \qquad \text{... (B)}$$

$$\frac{N}{N_0} = \frac{95}{100} \quad \text{when } t = 50$$

$$\therefore \frac{95}{100} = e^{-50\lambda}$$

$$\text{or, } 50\lambda = \log_e \left(\frac{100}{95}\right) \qquad \text{... (C)}$$

If N' be the amount present after 100 years

$$\log_e \left(\frac{N_0}{N'}\right) = 100\lambda \qquad \text{... (D)}$$

From (D) and (C)

$$\log_e \left(\frac{N_0}{N'}\right) = \frac{100\lambda}{50\lambda} \log_e \frac{100}{95}$$

$$= \log_e \frac{400}{361}$$

$$\therefore \frac{N'}{N_0} = 361/400 \quad \text{or} \quad N' = \frac{361}{400} N_0$$

5. A particle falls in a vertical line undergravity (supposed constant) and the air resistance to the motion is proportional to the velocity. Show that the velocity can not exceed a particular limit. **(C.U.H. '99) [C. U.]**

Let m be the mass of the particle. Differential equation of its motion is

$$m\frac{dv}{dt} = mg - mkv$$

or, $\frac{dv}{dt} = g - kv$

or, $\int \frac{dv}{g - kv} = \int dt$

or, $\frac{\log_e(g - kv)}{-k} = t + A'$

or, $\log_e(g - kv) = -kt + A$

At $t = 0$, $v = 0$

$\therefore \log g = A$

Substituting for A,

$$\log_e\left(\frac{g - kv}{g}\right) = -kt$$

or, $1 - \frac{kv}{g} = e^{-kt}$

or, $\frac{kv}{g} = 1 - e^{-kt}$

When $t \to \infty$, $e^{-kt} \to 0$

$\therefore V_{max} = g/k.$

6. A particle is thrown vertically upwards with a velocity v_0. Assuming the air resistance to be proportional to the velocity of the particle, show that the maximum height attained by it will be given by

$$x_m = \frac{v_0}{k} + g/k^2 \log\frac{g}{g + kv_0},$$

$k =$ resisting force constant,
$g =$ acceleration due to gravity.

[C. U. (Hons.)]

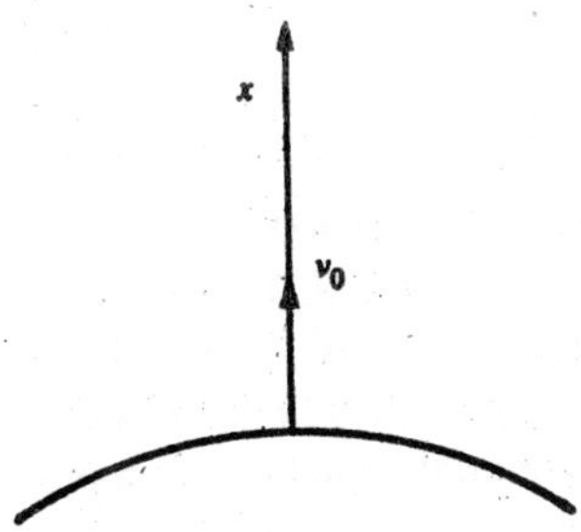

Fig. 2.7

Let m be the mass of the particle.

Differential equation of its motion is

$$m\frac{d^2x}{dt^2} = -mg - mk\frac{dx}{dt}$$

or, $\frac{d^2x}{dt^2} + k\frac{dx}{dt} + g = 0$

or, $\frac{dv}{dt} + kv + g = 0$

or, $\int \frac{dv}{kv+g} = -\int dt$

or, $\frac{1}{k} \log (kv+g) = -t + A$

At $t = 0$, $v = v_0$

$\therefore A = \frac{\log (kv_0 + g)}{k}$

Substituting for A

$$\log \left(\frac{Kv+g}{kv_0+g} \right) = -kt$$

or, $\frac{kv+g}{kv_0+g} = e^{-kt}$

or, $k\frac{dx}{dt} + g = (kv_0+g)\, e^{-kt}$

or, $\int dx = \int \frac{kv_0+g}{k} e^{-kt}\, dt - \int \frac{g}{k} dt$

or, $x = \frac{kv_0+g}{-k^2} e^{-kt} - \frac{gt}{k} + B$

At $t = 0$, $x = 0$ $\therefore B = \frac{kv_0+g}{k^2}$

$\therefore x = \frac{kv_0+g}{-k^2} e^{-kt} - \frac{g}{k} t + \frac{kv_0+g}{k^2}$

when x is max, $\frac{dx}{dt} = v = 0$ $\therefore e^{-kt'} = \frac{g}{kv_0+g}$

$$\therefore x_m = \frac{kv_0+g}{-k^2} \frac{g}{kv_0+g} - \frac{g}{k}\left(-\frac{1}{k}\right) \log \left(\frac{g}{kv_0+g}\right) + \frac{kv_0+g}{k^2}$$

$$= \frac{v_0}{k} + \frac{g}{k^2} \log_e \left(\frac{g}{kv_o+g}\right)$$

7. The natural length of a spring is 10 cm. When a mass 'm' gm is suspended from the lower end of the spring held vertical, the equilibrium length becomes 12 cm. The mass is now given a blow such that it starts moving vertically downwards with a velocity of 3 cm/sec. Establish the differential equation of motion. Find the frequency and amplitude of the motion. **[C. U. (Hons.) 1984]**

The sequence of facts is shown in Fig. 2.8(i), (ii) and (iii).

Let k be the spring constant of the spring.

From the condition given $mg = k(12 - 10) = 2k$

$$\therefore\ k = mg/2$$

Differential equation of motion of the spring is

$$m\frac{d^2x}{dt^2} = k(2 - x) - mg = -kx$$

$$= -\frac{mg}{2}x$$

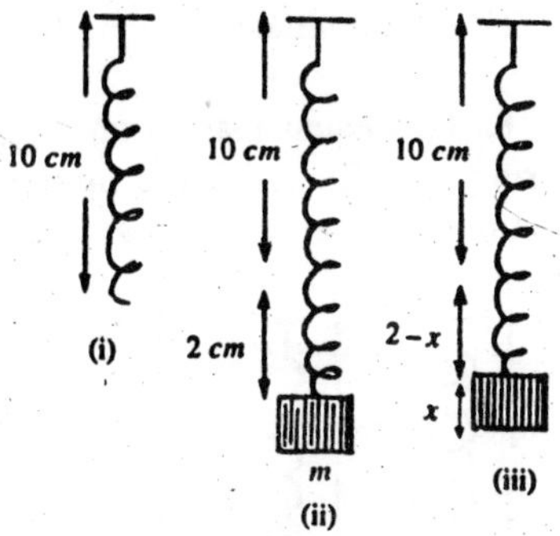

Fig. 2.8

or, $\dfrac{d^2x}{dt^2} = -\dfrac{g}{2}x$

Proceeding in the usual manners

$$\left(\frac{dx}{dt}\right)^2 = -\int \frac{g}{2}x\ 2\frac{dx}{dt}\,dt = -\frac{gx^2}{2} + A$$

Initial condition is—when $x = 0$, $\dfrac{dx}{dt} = 3$

$\therefore\ A = 9$

Substituting

$$\left(\frac{dx}{dt}\right)^2 = 9 - \frac{gx^2}{2}$$

or, $\dfrac{dx}{dt} = \sqrt{9 - \dfrac{gx^2}{2}}$

or, $\displaystyle\int \frac{dx}{\sqrt{9 - \dfrac{gx^2}{2}}} = \int dt$

or, $\displaystyle\int \frac{dx}{\sqrt{\dfrac{g}{2}}\sqrt{\dfrac{18}{g} - x^2}} = \int dt$

or, $\sin^{-1}\left(\dfrac{x}{\sqrt{\dfrac{18}{g}}}\right) = \sqrt{\dfrac{g}{2}}\,t + \delta$

At $t = 0$, $x = 0$ $\quad\therefore\ \delta = 0$

$\therefore\ x = \sqrt{\dfrac{18}{g}}\sin\sqrt{\dfrac{g}{2}}\,t.$

Hence amplitude $= \sqrt{\frac{18}{g}}$, frequency $= \frac{\sqrt{g/2}}{2\pi}$

8. Two equal masses are suspended at the end of a weightless elastic spring of stiffness k. One mass falls off. Solve the differential equation of motion of the remaining mass. (C.U.H.)

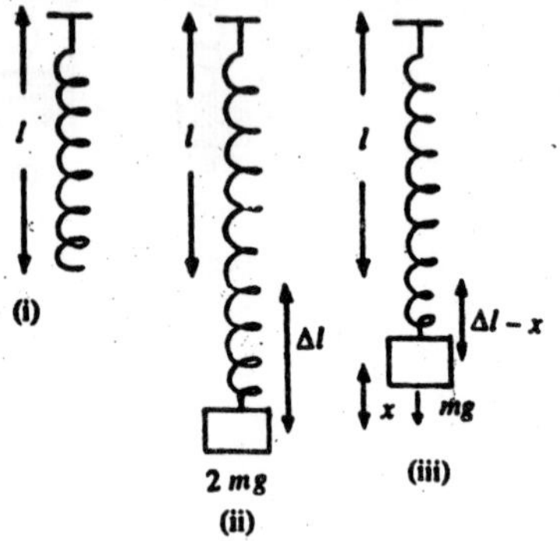

Fig. 2.9

The sequence of events is shown in Fig. 2.9(i), (ii) and (iii).

Initially $k \Delta l = 2mg$.

When one mass falls off,

$$F = k(\Delta l - x) - mg.$$

$$= 2mg - kx - mg = mg - kx$$

or, $$m\frac{d^2x}{dt^2} = mg - kx$$

or, $$\frac{d^2x}{dt^2} = g - \frac{k}{m}x$$

Proceeding as usual

$$\left(\frac{dx}{dt}\right)^2 = \int \left(g - \frac{k}{m}x\right) 2\frac{dx}{dt}\,dt$$

$$= 2gx - \frac{k}{m}x^2 + A$$

When $x = 0$, $\frac{dx}{dt} = 0$ $\therefore A = 0$

Hence $$\int \frac{dx}{\sqrt{2gx - \frac{k}{m}x^2}} = \int dt$$

or, $$\frac{1}{\sqrt{k/m}} \int \frac{dx}{\sqrt{\frac{2gm}{k}x - x^2}} = \int dt$$

or, $$\int \frac{dx}{\sqrt{\frac{g^2m^2}{k^2} - \left(x - \frac{gm}{k}\right)^2}} = \sqrt{\frac{k}{m}} \int dt$$

or, $$\sin^{-1}\left(\frac{x - gm/k}{\frac{gm}{k}}\right) = \sqrt{\frac{k}{m}}\,t + B$$

At $t = 0$, $x = 0$ $\quad \therefore \; B = \sin^{-1}(-1) = -\pi/2$

Substituting for B

$$\frac{x - gm/k}{gm/k} = \sin\left(-\pi/2 + \sqrt{\frac{k}{m}}\,t\right)$$

$$= -\sin\left(-\sqrt{\frac{k}{m}}\,t + \pi/2\right) = -\cos\sqrt{\frac{k}{m}}\,t$$

$$\therefore \; x = \frac{gm}{k}\left(1 - \cos\sqrt{\frac{k}{m}}\,t\right)$$

9. A uniform chain of length a lies on a smooth table so that a length b sags. Find the time taken by the chain to slip down. (C.U.H.)

At any instant let a length x of the chain hang vertically down. If ρ be the mass per unit length if the chain we have

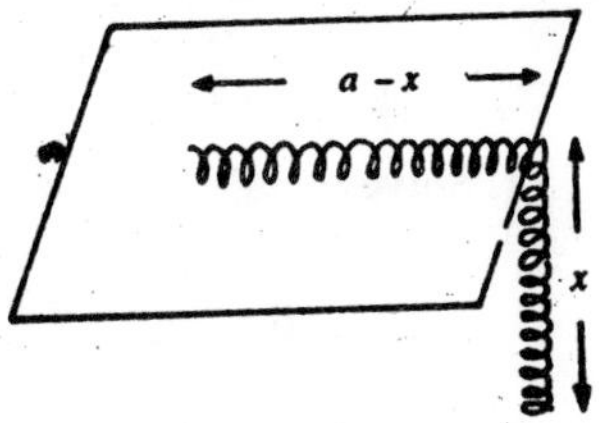

Fig. 2.10

$$a\rho \frac{d^2x}{dt^2} = x\rho g$$

or, $$\frac{d^2x}{dt^2} = \frac{g}{a}x$$

or, $$\int \frac{d^2x}{dt^2}\, 2\frac{dx}{dt}\, dt = \frac{g}{a}\int 2x\frac{dx}{dt}\, dt$$

or, $$\left(\frac{dx}{dt}\right)^2 = \frac{g}{a}x^2 + c$$

When $x = b$, $\dfrac{dx}{dt} = 0$ $\quad \therefore \; c = -\dfrac{g}{a}b^2$

Substituting, $$\left(\frac{dx}{dt}\right)^2 = \frac{g}{a}(x^2 - b^2)$$

or, $$\int_b^a \frac{dx}{\sqrt{x^2 - b^2}}\, dx = \sqrt{\frac{g}{a}}\int_0^{t_0} dt$$

or, $$\left[\log_e\left(x + \sqrt{x^2 - b^2}\right)\right]_b^a = \sqrt{\frac{g}{a}}\; t_0$$

$$\therefore \; t_0 = \sqrt{\frac{a}{g}}\,\log_e\left(a + \frac{\sqrt{a^2 - b^2}}{b}\right)$$

10. Solve :

$$y\frac{dy^2}{dx^2} = \left(\frac{dy}{dx}\right)^2$$

$$\text{or,} \quad \int \frac{\frac{d^2y}{dx^2}}{\frac{dy}{dx}}\, dx = \int \frac{\frac{dy}{dx}}{y}\, dx$$

$$\text{or,} \quad \log\left(\frac{dy}{dx}\right) = \log y + \log c$$

$$\text{or,} \quad \log\left(\frac{\frac{dy}{dx}}{y}\right) = \log c$$

$$\text{or,} \quad \int \frac{\frac{dy}{dx}\, dx}{y} = \int c\, dx$$

$$\text{or,} \quad \log y = cx + c'$$

11. Solve :

$$\frac{\partial u}{\partial x} = 4\frac{\partial u}{\partial y} \quad \text{given } u(o, y) = 8e^{-3y}$$

Let $u = u_1 u_2$ be the solution where

$u_1 =$ function of x only

$u_2 =$ " " y "

Substituting in the given differential equation

$$u_2 u'_1 = 4 u_1 u'_2$$

or, $\dfrac{u'_1}{u_1} = \dfrac{4u'_2}{u_2} = k =$ constant, since the R.H. side is a function of x alone, while the L.H. side is a function of y alone.

Now $\dfrac{u'_1}{u_1} = k$

Integrating with respect to x

$$\log u_1 = kx + k_1$$

$$\text{or,} \quad u_1 = c^{kx + k_1}$$

Similarly $u_2 = e^{ky/4 + k_2}$

Hence $u = u_1 u_2 = ce^{k\left(x + \frac{y}{4}\right)}$

$\because u(o, y) = 8e^{-3y}$

$$c = 8 \text{ and } \frac{k}{4} = -3 \text{ or, } k = -12$$

$$\therefore\ u = 8e^{-12}\left(x + \frac{y}{4}\right)$$

2.2. Second order linear differential equations with variable coefficients. Frobeneous' method of power series solutions for Legendra and Hermite differential equation

Let us study ordinary, homogeneous, linear, second order differential equations. The general form of the equation will be of the form

$$\frac{d^2y}{dx^2} + P(x)\frac{dy}{dx} + Q(x)\ y = 0 \qquad \ldots\ldots \quad (1)$$

where $P(x)$ and $Q(x)$ are known functions of x.

A second order linear differential equation has two linearly independent solutions. Two solutions y_1 and y_2 are said to be linearly independent if the condition

$$\alpha_1 y_1 + \alpha_2 y_2 = 0$$

where α_1 and α_2 are constants, can be satisfied only by the choice $\alpha_1 = 0$, $\alpha_2 = 0$.

If y_1 and y_2 are two linearly independent solutions of (1), the general solution can be expressed as

$$y = c_1 y_1 + c_2 y_2 \qquad \ldots\ldots \quad (2)$$

where, c_1 and c_2 are arbitrary contants. The point x_0 is said to be an ordinary point of the differential equation (1), if both $P(x)$ and $Q(x)$ are analytic at $x = x_0$.

If either $P(x)$ or $Q(x)$ or both fail to be analytic at $x = x_0$, then x_0 is called a singular point of the differential equation.

Series integration method of the solution of Linear differential equation (Frobenius' method) :

Method of series integration of the solution of a differential equation enables to represent one variable in a series of ascending or descending powers of another variable.

Let us consider the second order homogeneous linear differential equation

$$\frac{d^2y}{dx^2} + P(x)\frac{dy}{dx} + Q(x)\,y = 0 \qquad \ldots\ldots \quad (1)$$

and suppose that this equation has no solution which is expressible as a finite linear combination of known elementary functions. Let us assume, however, that it does have a solution which can be expressed in the form an infinite series. Specially we assume that it has a solution expressible in the form

$$c_o + c_1 (x - x_0) + c_2 (x - x_0)^2 + \ldots\ldots$$

$$= \sum_{n=0}^{\infty} C_n (x - x_0)^n \qquad \ldots\ldots \quad (2)$$

where c_0, c_1, c_2 are constants. An expression of this form is called a power series in $(x - x_0)$. Thus we assume that the differential equation (1) has a solution called the power series solution given by equation (2). To determine the coefficients $c_1, c_2, c_3 \ldots\ldots$, we assume that the series (2) converges in an interval $|x - x_0| < R$ about x_0. Then it may be differentiated term by term in this interval twice in succession to obtain.

$$\frac{dy}{dx} = c_1 + 2c_2 (x - x_0) + 3c_3 (x - x_0)^2 + \ldots\ldots$$

$$= \sum_{n=1}^{\infty} nc_n (x - x_0)^{n-1} \qquad \ldots\ldots \quad (3)$$

In a similar manner

$$\frac{d^2y}{dx^2} = \sum_{n=2}^{\infty} n(n-1) c_n (x - x_0)^{n-2} \qquad \ldots\ldots \quad (4)$$

Substituting for y, $\frac{dy}{dx}$ and $\frac{d^2y}{dx^2}$ in (1), we have

$$a_0 \sum_{n=2}^{\infty} n(n-1) c_n (x - x_0)^{n-2} + a_1 \sum_{n=1}^{\infty} n\, c_n (x - x_0)^{n-1}$$

$$+ a_2 \sum_{n=0}^{\infty} c_n (x - x_0)^n = 0 \qquad \ldots\ldots \quad (5)$$

or, $k_0 + k_1 (x - x_0) + k_2 (x - x_0)^2 = 0 \qquad \ldots\ldots \quad (6)$

where k_0, k_1, k_3 are functions of the coefficients c_n. In order that equation (6) is valid for all x_0, the coefficient of every power of x must be zero.

So $K_0 = K_1 = K_2 = \ldots\ldots = 0$.

From these equations, the coefficients c_0, c_1, c_2 etc. can be calculated and the required solution obtained.

Application :

(A) Solution of Legendra differential equation (To be done with Laplace's equation).

(B) Solution of Hermite differential equation.

Hermite differential equation is

$$\frac{d^2 H_n(x)}{dx^2} - 2x \frac{dH_n(x)}{dx} + 2n\, H_n(x) = 0 \qquad \ldots\ldots (1)$$

where n is an integer and $H_n(x)$ are Hermite Polynomials of degree n. This ferential equation occur in the problem of one dimensional harmonic cillator in quantum mechanics. The Polynominals are standardised by the condition that the coefficient of x^n is 2^n. The recurrence relations for $H_n(x)$ can be deduced from this differential equation (1). Differentiating both sides w.r.t. x, we get

$$\frac{d^2 H'_n}{dx^2} - 2x \frac{dH'_n}{dx} + 2(n-1)\, H'_n(x) = 0 \qquad \ldots\ldots (2)$$

where $H'_n = \frac{dH_n}{dx}$ is a polynomial of degree $(n-1)$

Hence it must be proportional to $H_{n-1}(x)$

$$\text{i.e. } H'_n(x) = C\, H_{n-1}(x) \qquad \ldots\ldots (3)$$

The constant c is determined by comparing the coefficient of x^{n-1} from both sides of (3)

$$2^n . n = c\ 2^{n-1}$$

$$\text{or, } c = 2n$$

$$\therefore\ H'_n(x) = 2n\, H_{n-1}(x) \qquad \ldots\ldots (4)$$

Again from the differential equation (1)

$$\frac{d}{dx} H'_n - 2x\, H'_n + 2n\, H_n = 0$$

Using the relation (4)

$$\frac{d}{dx}(2n\, H_{n-1}) = 2x\ 2n\ H_{n-1} - 2n\, H_n$$

or, $H'_{n-1} = 2x\, H_{n-1} - H_n$

In the L.H.S. using equation (4) we get

$$2(n-1)\ H_{n-2} = 2x\, H_{n-1} - H_n$$

Changing n to $(n+1)$ on both sides

$$H_{n+1}(x) = 2x\, H_n(x) - 2n\, H_{n-1}(x) \qquad \ldots\ldots (5)$$

The same recurrence relation (4) and (5) can be obtained using the generating functions

$$g(x_1 t) = e^{2xt - t^2} = \sum_{n=0}^{\infty} H_n(x)\, t^n \quad \ldots \quad (6)$$

To verify the above relation let us expand $g(x_1 t)$ by Taylor's series around $t = 0$

$$e^{2xt - t^2} = \sum_{n=0}^{\infty} f_n(x) \frac{t^n}{\angle n}$$

Where $f_n(x) = \frac{\partial^x}{\partial x^n} (e^{2xt - t^2})_{t=0}$

$$= e^{x^2} \frac{\partial^x}{\partial x^n} e^{-(x-t)^2}{}_{t=0}$$

Since, $\frac{\partial}{\partial t} f(x - t) = -\frac{\partial}{\partial x} f(x - t)$

$$f(x) = e^{x^2} (-1)^n \frac{\partial^n}{\partial x^n} [e^{-(x-t)^2}]_{t=0}$$

or, $$f_n(x) = (-1)^n e^{x^2} \frac{d^n}{dx^n} (e^{-x^2}) \quad \ldots \quad (7)$$

It is easy to see that $f_n(x)$ are polynomials of degree n, the coefficient of x^n being 2^n. Next we prove that $f^n(x)$ satisfy the Hermite differential equation.

Let $F = c^{-x^2}$, then $\frac{dF}{dx} = -2xF$.

Differentiating w.r.t. x, $(n + 1)$ times, we have

$$\frac{d^2}{dx^2}\left[\frac{d^n}{dx^n}(e^{-x^2})\right] = -2x \frac{d}{dx}\left[\frac{d^n}{dx^n}(e^{-x^2})\right] - 2(n+1)\frac{d^n}{dx^n}(e^{-x^2}) \quad \ldots \quad (8)$$

Substituting $\frac{d^n}{dx^n}(e^{-x^2}) = (-1)^n e^{-x} f_n(x)$ in (8)

We have

$$\frac{d^2 f_n}{dx^2} - 2x \frac{df_n}{dx} + 2nf_n = 0 \quad \ldots \quad (9)$$

Hence $f_n(x) = H_n(x)$

or, $$H_n(x) = (-1)^n e^{x^2} \frac{d^n}{dx^n} (e^{-x^2}) \quad \ldots \quad (10)$$

Equation (10) is Rodrignes' relation for Hermite polynomials.

EXERCISE

1. Solve :

$$\frac{dy}{dx} - x + y = 0$$

Putting $y - x = v$

$$\frac{dy}{dx} - 1 = \frac{dv}{dx}$$

Substituting above

$$1 + \frac{dv}{dx} = -v$$

or, $\dfrac{dv}{dx} = -v - 1$

or, $\dfrac{dv}{-v-1} = dx$

Integrating

$$\log(v+1) = -x + \log c$$

or, $\dfrac{v+1}{c} = e^{-x}$

or, $y - x + 1 = ce^{-x}$ is the general solution.

2. Solve :

$$\frac{d^2x}{dt^2} + 2\frac{dx}{dt} - 3x = 2\cos t - 4\sin t$$

For C.F. we have

$$\frac{d^2x}{dt^2} + 2\frac{dx}{dt} - 3x = 0$$

Auxiliary equation is

$$m^2 + 2m - 3 = 0$$

$$\text{or,} \quad (m+3)(m-1) = 0$$

$$\therefore m = -3, 1$$

Hence the complementary function is

$$x = c_1 e^{t} + c_2 e^{-3t}$$

For P.I. we have in operator symbols,

$$(D^2 + 2D - 3)x = 2\cos t - 4\sin t$$

$$\therefore \quad x = \frac{2\cos t - 4\sin t}{(D^2 + 2D - 3)}$$

$$= \frac{[(D^2 - 3) - 2D]\ (2\cos t - 4\sin t)}{(D^2 - 3)^2 - 4D^2}$$

$$= \frac{-2\cos t - 6\cos t + 4\sin t + 4\sin t + 12\sin t + 8\cos t}{(-1-3)^2 + 4}$$

$$= \frac{20\sin t}{20} = \sin t$$

Hence the general solution is

$$x = c_1 e^t + c_2 e^{-3t} + \sin t$$

3. Solve :

$$(3x^2 + y\cos x)\ dx + (\sin x - 4y^3)\ dy = 0$$

$$3x^2\ dx + d\,(y\sin x) - 4y^3\ dy = 0$$

Integrating $x^3 + y\sin x - y^4 = c$ which is the required solution.

4. A particle on x-axis weighing 8 lbs, distance 12 ft. apart from a fixed point O on the axis is given a velocity 16 ft/sec., is a direction perpendicular to x-axis. The particle is subjected to a force F along x-axis ($-\ ve$) whose magnitude is inversely proportional to the cube of the x-co-ordinate. Taking the constant of proportionality to be 10, find the distance of the particle from O as a function of the time t.

For x motion $8\frac{d^2x}{dt^2} = \frac{-10}{x^3}$

or, $\int 2\frac{dx}{dt}\frac{d^2x}{dt^2}\,dt = \frac{10}{-8}\int \frac{1}{x^3}\,2\frac{dx}{dt}\,dt$

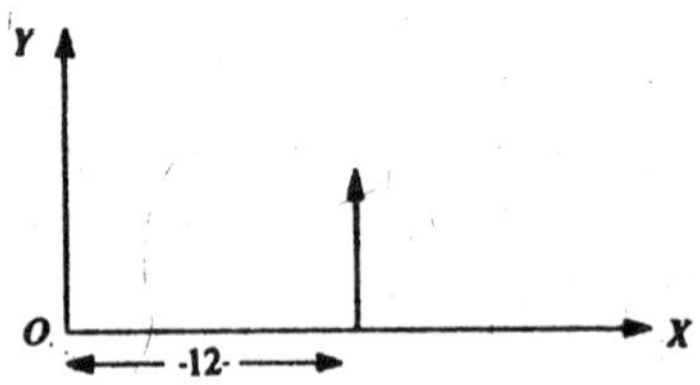

Fig. 2.11

or, $\left(\frac{dx}{dt}\right)^2 = \frac{5}{4x^2} + c_1$

when $x = 12$, $\frac{dx}{dt} = 0$

$\therefore\ c_1 = -\frac{5}{4 \times 12^2}$

Substituting above,

$$\left(\frac{dx}{dt}\right)^2 = \frac{5}{4}\left(\frac{1}{x^2} - \frac{1}{12^2}\right)$$

or, $\frac{dx}{dt} = \frac{\sqrt{5}}{2} \frac{\sqrt{12^2 - x^2}}{12x}$

or, $\int \frac{xdx}{\sqrt{12^2 - x^2}} = \int \frac{\sqrt{5}}{24}\, dt$

or, $-\sqrt{12^2 - x^2} = \frac{\sqrt{5}}{24} t + c_2$

At $t = 0$, $x = 12$ $\therefore c_2 = 0$

Hence $12^2 - x^2 = \frac{5}{576} t^2$

or, $x^2 = 144 - \cdot 0086\, t^2$

For y motion

$$y = 16\, t$$

$$\text{or, } y^2 = 256\, t^2$$

Hence $r^2 = x^2 + y^2 = 144 + 256\, t^2 - \cdot 0086\, t^2$

$= 144 - 255.9914\, t^2$

or, $r = (144 - 255.9914\, t^2)^{1/2}$

5. Solve :

$$x^3 \frac{d^3y}{dx^3} - \frac{x^2\, d^2y}{dx^2} + 2x \frac{dy}{dx} - 2y = x^2$$

Let us put $x = e^z$, $\therefore z = \log_e x$

Thus $\frac{dy}{dx} = \frac{dy}{dz} \frac{dz}{dx} = \frac{1}{x} \frac{dy}{dz}$ and

$\frac{d^3y}{dx^3} = \frac{1}{x^3} \left(\frac{d^3y}{dx^3} - 3 \frac{d^2y}{dx^2} + 2 \frac{dy}{dx} \right)$

Writing $\delta = \frac{d}{dx}$, we have the transformed equation as

$\left[\delta(\delta - 1)(\delta - 2) - \delta(\delta - 1) + 2\delta - 2\right] y = e^{2z}$

Where $x^2 \frac{d^2y}{dx^2} = \delta(\delta - 1) y$ and $x^3 \frac{d^3y}{dx^3} = \delta(\delta - 1)(\delta - 2) y$

and $x \frac{dy}{dx} = \delta y$

or, $(\delta-1)^2(\delta-2)y=e^{2z}$

The roots of the auxiliary equation are 1, 1 and 2

Hence C.F. is $y=(c_1+c_2z)e^z+c_3e^{+2z}$

P.I. is $\dfrac{e^{2z}}{(\delta-1)^2(\delta-2)}=e^{2z}(z-1)$

The general solution is

$$y=(c_1+c_2z)c^z+c_3e^{+2z}+e^{2z}(z-1)$$

$$\text{i.e.,}\quad y=(c_1+c_2\log x)x+c_3x^{+2}+x^2(\log_e x-1)$$

6. A particle moves in a plane such that

$$\frac{dx}{dt}=-50y\,;\quad \frac{dy}{dt}=18x$$

Given that it passes through the point (2, 0), find the equation of its path.

Dividing $\dfrac{dy}{dx}=-\dfrac{18x}{50y}$

or, $50y\,dy+18x\,dx=0$

Integrating $25y^2+9x^2=$ constant

Since it passes through (2, 0)

$$0+9\times 4=\text{constant}$$

Hence $25y^2+9x^2=36$ is the equation of the path.

CHAPTER III
EQUATIONS

Solutions by the method of separation of variables; Laplace's equation and its solution in Cartesian, Spherical Polar and Cylindrical co-ordinate system. Wave equation and its plane and spherical wave equation.

Laplace's Equation

The partial differential equation

$$\nabla^2 V = 0 \tag{3.1}$$

is known as Laplace's Equation.

The above equation is satisfied by the following functions :

(i) The gravitational potential in regions not occupied by attracting matter.

(ii) The eletrostatic potential in a uniform dielectric.

(iii) The magnetic potential in free space.

(iv) The electric potential in the theory of steady flow of electric currents in solid conductors.

(v) The temperature in the theory of equilibrium of solids.

(vi) The velocity potential at points of a homogeneous liquid moving irrotationally.

Solution of Laplace's Equation by the method of seperation of variables.

(a) **In Cartesian Co-ordinates :**

The equation $\nabla^2 V = 0$ when expressed in cartetsian co-ordinates is

$$\frac{\partial^2 V}{\partial x^2} + \frac{\partial^2 V}{\partial y^2} + \frac{\partial^2 V}{\partial z^2} = 0 \tag{3.2}$$

Let us put $V = X(x)\ Y(y)\ Z(z)$ as the solution of equation (3.2), where $X(x)$ is a function of x alone etc.

Now $\dfrac{\partial^2 V}{\partial x^2} = \dfrac{\partial^2 X(x)}{\partial x^2}\ Y(y)\ Z(z)$ etc.

$= X'YZ$ say,

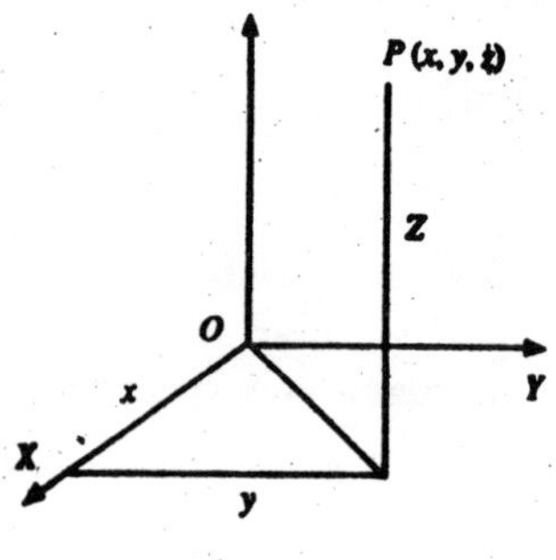

Fig. 3.1

substituting (in) (3.2) and dividing by $X(x)\ Y(y)\ Z(z)$

$$\frac{X''}{X} + \frac{Y''}{Y} + \frac{Z''}{Z} = 0 \qquad (3.3)$$

Since x, y, z are independent of each other, each of the above forms $\frac{X''}{X}$ etc. are seperately equal to a constant if equation (3.3) is to be true. It is also evident that the sum of there constants must vanish. Writing these constants as K_1^2, K_2^2, K_3^2 respectively,

$$\frac{X''}{X} = K_1^2 \quad \text{i.e.,} \quad \frac{\partial^2 X}{\partial x^2} = K_1^2 X$$

Solving $X = c_1 e^{k_1 x} + c_2 e^{-k_1 x}$

Similarly $Y = c_3 e^{k_2 y} + c_4 e^{-k_2 y}$ and

$$Z = c_5 e^{k_3 z} + c_6 e^{-k_3 z} \text{ with}$$

$$k_1^2 + k_2^2 + k_3^2 = 0 \qquad (3.4)$$

The general solution will be

$$V = (c_1 e^{k_1 x} + c_2 e^{-k_1 x})(c_3 e^{k_2 y} + c_4 e^{-k_2 y})(c_5 e^{k_3 z} + c_6 e^{-k_3 z})$$

A more general solution will be (3.5)

$$V = \sum_{k_1, k_2, k_3} C_{k_1 k_2 k_3}\, e^{k_1 x + k_2 y + k_3 z} \qquad (3.6)$$

with $k_1^2 + k_2^2 + k_3^2 = 0$.

(b) In Cylindrical Co-ordinates (ρ, φ, z) :

Writing $\rho = r$, and $\varphi = \theta$ (Fig. 3.2)

Laplace's equation in three dimensional cylindrical co-ordinates is

$$\nabla^2 V = \frac{1}{r}\frac{\partial}{\partial r}\left(r\frac{\partial V}{\partial r}\right) + \frac{1}{r^2}\frac{\partial^2 V}{\partial \theta^2} + \frac{\partial^2 V}{\partial z^2} = 0 \qquad (3.7)$$

Let us assume a solution of the form $V = F_1(\theta)\ F_2(r)\ F_3(z)$ (3.8),

where F_1 is a function of θ alone etc.

Substituting in (3.7)

$$\frac{F_1(\theta)\ F_3(z)}{r}\ \frac{\partial}{\partial r}\left\{r\ \frac{\partial F_2(r)}{\partial r}\right\}$$

$$+\ \frac{F_2(r)\ F_3(z)}{r^2}\ \frac{d^2F_1}{d\theta^2} + F_1(\theta)\ F_2(r)\ \frac{d^2 F_3}{dz^2}$$

or,

$$\frac{1}{r\,F_2}\left\{\frac{dF_2}{dr} + r\frac{d^2 F_2}{dr^2}\right\}$$

$$+\ \frac{1}{r^2 F_1}\ \frac{d^2F_1}{d\theta^2} + \frac{1}{F_3}\ \frac{d^3 F_3}{dz^2} = 0$$

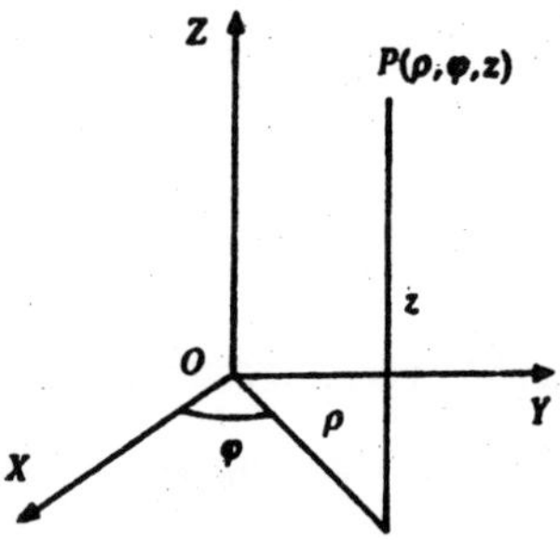

Fig. 3.2

or, $$\frac{1}{F_3}\frac{d^2F_3}{dz^2} = -\frac{1}{F_2}\frac{d^2F_2}{dr^2} - \frac{1}{rF_2}\frac{dF_2}{dr} - \frac{1}{r^2F_1}\frac{d^2F_1}{d\theta^2} \tag{3.9}$$

By Hypothesis F_3 is function of z only and by (3.9) a change of z does not change the right-hand side. Hence the left hand side reduces to a constant.

i.e., $$\frac{1}{F_3}\ \frac{d^2F_3}{dz^2} = k^2 \text{ say} \tag{3.10}$$

Hence $F_3 = c_1\, e^{kz} + c_2\, e^{-kz}$ (3.11)

From (3.9) and (3.10) we have

$$\frac{r^2}{F_2}\ \frac{d^2 F_2}{dr^2} + \frac{r}{F_2}\ \frac{dF_2}{dr} + k^2r^2 = -\ \frac{1}{F_1}\ \frac{d^2F_1}{d\theta^2} \tag{3.12}$$

By the same reasoning as above each member of (3.12) is a constant. Let it be m^2. Hence we have

$$\frac{d^2 F_1}{d\theta^2} = m^2 F_1 \tag{3.13}$$

Solution is $F_1 = c_3\ \cos m\theta + c_4\ \sin m\theta$ (3.14)

and $$r^2\ \frac{dF_2}{dr^2} + r\frac{dF_2}{dr} + (k^2 r^2 - m^2)\,F_2 = 0 \tag{3.15}$$

To solve let us put $kr = x$

Equation (3.15) reduces to

$$x^2\ \frac{d^2 F_2}{dx^2} + x\ \frac{dF_2}{dx} + (x^2 - m^2)\ F_2 = 0 \tag{3.16}$$

This is Bessel's differential equation. Its solution is

$$F_2 = c_5 J_m(x) + c_6 J_{-m}(x)$$

$$= c_5 J_m(kr) + c_6 J_{-m}(kr) \quad (3.17)$$

if m is fractional, or $F_2 = c_5 J_m(kr) + c_6 Y_m(kr)$ (3.18)

if m is integral $J_m(x)$ is called the Bessel function of mth order and 1st Kind and $Y_m(x)$ is Bessel function of mth order and 2nd kind.

Any one of the above values of F_1, F_2 and F_3 substituted in (3.8), gives a solution of Laplace's equation. If we let K be a fixed constant and if we require V to be a single valued function of θ, then m must take only integral values and we have the solution

$$V = \sum_{m=0}^{\infty} \left\{ e^{kz}(A_m \cos m\theta + B_m \sin m\theta) + e^{-kz}(C_m \cos m\theta + D_m \sin m\theta) \right\} J_m(kr) \quad (3.19)$$

The part $Y_m(kr)$ is dropped, because it $\rightarrow \infty$ when $r \rightarrow 0$

N. B. :

$$J_m(x) = \sum_{s=0}^{\infty} \frac{(-1)^s}{\pi(s)\,\pi(m+s)} \left(\frac{x}{2}\right)^{m+2s}$$

$$\text{and } J_{-m}(x) = \sum_{s=0}^{\infty} \frac{(-1)^s}{\pi(s)\,\pi(s-m)} \left(\frac{x}{2}\right)^{2s-m}$$

where $\pi(n) = \Gamma(n+1)$ when n is fractional

and $\pi(n) = \underline{|n}$ when n is integer

when n is not an integer $Y_n(x)$ is defined as

$$Y_n(x) = \frac{\cos n\pi\, J_n(x) - J_{-n}(x)}{\sin n\pi} \quad \text{and when } n \text{ is an integer}$$

$$Y_n(x) = \underset{r \to n}{\text{Lt}} \frac{J_r(x) \cos r\pi - J_{-r}(x)}{\sin r\pi}$$

(c) In Spherical Polar Co-ordinates :

Laplace's equation in spherical polar co-ordinates is

$$\nabla^2 V = \frac{1}{r^2}\frac{\partial}{\partial r}\left(r^2 \frac{\partial V}{\partial r}\right) + \frac{1}{r^2 \sin\theta}\frac{\partial}{\partial\theta}\left(\sin\theta \frac{\partial V}{\partial\theta}\right) + \frac{1}{r^2 \sin^2\theta}\left(\frac{\partial^2 V}{\partial\varphi^2}\right) = 0 \quad (3.20)$$

Let us have a solution of **equation** (3.20) in the form $V = R\,\Theta\,\Phi = RS$

where R is a function of r only, Θ is a function of θ only and Φ is a function of φ only. The function $S(\theta, \varphi) = \Theta\,\Phi$ is called a surface harmonic. The function Θ, when φ is constant is called a zonal surface harmonic.

Substituting $V = RS$ in (3.20)

$$S\frac{\partial}{\partial r}\left(r^2\frac{\partial R}{\partial r}\right) + \frac{R}{\sin\theta}\frac{\partial}{\partial\theta}\left(\sin\theta\frac{\partial S}{\partial\theta}\right) + \frac{R}{\sin^2\theta}\frac{\partial^2 S}{\partial\varphi^2} = 0$$

Dividing by RS

$$\frac{1}{R}\frac{\partial}{\partial r}\left(r^2\frac{\partial R}{\partial r}\right) + \frac{1}{S\sin\theta}\frac{\partial}{\partial\theta}\left(\sin\theta\frac{\partial S}{\partial\theta}\right) + \frac{1}{S\sin^2\theta}\frac{\partial^2 S}{\partial\varphi^2} = 0 \quad (3.22)$$

The first term is a function of r only, and the others involve only the angles. Equation (3.22) can be satisfied of

$$\frac{1}{R}\frac{\partial}{\partial r}\left(r^2\frac{\partial R}{\partial r}\right) = K \quad (3.23)$$

and $$\frac{1}{S\sin\theta}\frac{\partial}{\partial\theta}\left(\sin\theta\frac{\partial S}{\partial\theta}\right) + \frac{1}{S\sin^2\theta}\frac{\partial^2 S}{\partial\varphi^2} = -k \quad (3.24)$$

On simplification by putting $k = n(n+1)$ equation (3.23) takes the form

$$r^2\frac{\partial^2 R}{\partial r^2} + 2r\frac{\partial R}{\partial r} - n(n+1)R = 0 \quad (3.25)$$

Let $R = r^\alpha$ be a solution of (3.25).

Substituting in (3.25)

$$\alpha^2 - \alpha + 2\alpha - n(n+1) = 0$$

or, $\alpha(\alpha+1) - n(n+1) = 0$, so that

$$\alpha = n \text{ or } \alpha = -(n+1)$$

Hence $R = Ar^n + Br^{-(n+1)}$ ..... (3.26) is a solution of (3.23)

Multiplying (3.24) by S, we have

$$\frac{1}{\sin\theta}\frac{\partial}{\partial\theta}\left(\sin\theta\frac{\partial S}{\partial\theta}\right) + \frac{1}{\sin^2\theta}\frac{\partial^2 S}{\partial\varphi^2} + n(n+1)S = 0 \quad (3.26)$$

Putting $\dfrac{\partial^2 S}{\partial\varphi^2} = -m^2 S$, so that

$$S = ce^{\pm im\varphi} \quad (3.27)$$

We have $$\frac{1}{\sin\theta}\frac{\partial}{\partial\theta}\left(\sin\theta\frac{\partial S}{\partial\theta}\right) - \frac{m^2 S}{\sin^2\theta} + n(n+1)S = 0$$

Putting $\cos\theta = x$, such that $-\sin\theta = \dfrac{dx}{d\theta}$

$$\frac{1}{\sin\theta}\frac{\partial}{\partial x}\left(\sin\theta\frac{\partial S}{\partial\theta}\right)(-\sin\theta) + \left\{n(n+1) - \frac{m^2}{1-x^2}\right\}S = 0$$

$$\text{or, } -\left\{\cos\theta\frac{\partial\theta}{\partial x}\frac{\partial S}{\partial\theta} + \sin\theta\frac{\partial}{\partial x}\left(\frac{\partial S}{\partial\theta}\right)\right\} + \left\{n(n+1) - \frac{m^2}{1-n^2}\right\}S = 0$$

$$\text{or, } -\cos\theta\frac{\partial S}{\partial x} - \sin\theta\frac{\partial}{\partial x}\left(\frac{\partial S}{\partial x}\frac{\partial x}{\partial\theta}\right) + \left\{n(n+1) - \frac{m^2}{1-x^2}\right\}S = 0$$

$$\text{or, } -\cos\theta\frac{\partial S}{\partial x} - \sin\theta\frac{\partial^2 S}{\partial x^2}\frac{\partial x}{\partial x} - \sin\theta\frac{\partial S}{\partial x}\frac{\partial}{\partial x}(-\sin\theta)$$
$$+ \left\{n(n+1) - \frac{m^2}{1-x^2}\right\}S = 0$$

$$\text{or, } -x\frac{\partial S}{\partial x} + \sin^2\theta\frac{\partial^2 S}{\partial x^2} - \sin\theta\frac{\partial S}{\partial x}(-\cos\theta)\frac{\partial\theta}{\partial x}$$
$$+ \left\{n(n+1) - \frac{m^2}{1-x^2}\right\}S = 0$$

$$\text{or, } -x\frac{\partial S}{\partial x} + (1-x^2)\frac{\partial^2 S}{\partial x^2} - x\frac{\partial S}{\partial x} + \left\{n(n+1) - \frac{m^2}{1-x^2}\right\}S = 0$$

$$\text{or, } (1-x^2)\frac{\partial^2 S}{\partial x^2} - 2x\frac{\partial S}{\partial x} + \left\{n(n+1) - \frac{m^2}{1-x^2}\right\}S = 0 \quad (3.28)$$

To find the solution of equation on (3.28)

Let first solve $(1-x^2)\dfrac{\partial^2 S}{\partial x^2} - 2x\dfrac{\partial S}{\partial x} + n(n+1)\,S = 0$ (3.29)

which is known as **Legendre differential equation of degree *n*.**

Let us assume an infinite series solution of Legendre differential equation in the form

$$S = x^m \sum_{r=0}^{\infty} a_r x^r = \sum_{r=0}^{\infty} a_r x^{m+r} \quad (3.30)$$

If (3.30) is to be a solution of (3.29), it is necessary that when the former is substituted in the latter, the co-efficient of every power of x must vanish. Equating the co-efficient of $x^{m'+r-2}$ to zero, we have

$$(m'+r)(m'+r-1)\,a_r - (m'+r-2)(m'+r-3)\,a_{r-2} - 2(m'+r-2)\,a_{r-2}$$
$$+ n(n+1)\,a_{r-2} = 0$$

$$\text{or, } (m'+r)(m'+r-1)\,a_r + (n-m-r+2)(n+m'+r-1)\,a_{r-2} = 0 \quad (3.31)$$

Since the series begin with a_0,

$a_{-1} = 0$ and $a_{-2} = 0$

Hence putting $r = 0$ in (3.31)

$$m'(m' - 1)\, a_0 = 0 \tag{3.32}$$

Putting $r = 1$

$$(m' + 1)\, m'\, a_1 = 0 \tag{3.33}$$

Equation (3.32) gives $m' = 0$; $m' = 1$ $\because$ a_0 is arbitrary

Equation (3.33) gives $m' = 0$, $m' = -1$ $\because$ a_1 is arbitrary. Taking $m' = 0$ and substituting in (3.31)

$$a_r = -\, a_{r-2}\, \frac{(n - r + 2)\,(n + r - 1)}{r\,(r - 1)} \tag{3.34}$$

The above relation will enable us to reduce the even order terms to a_0 and odd order terms to a_1

Hence $S = a_0 \left\{1 - \frac{n\,(n+1)}{\underline{|2}}\, x^2 + \frac{n\,(n-2)\,(n+1)\,(n+3)}{\underline{|4}}\, x^4 + \ldots\ldots\right\}$

$$+\, a_1 \left\{x - \frac{(n-1)\,(n+2)}{\underline{|3}}\, x^3 + \frac{(n-1)\,(n-3)\,(n+2)\,(n+4)}{\underline{|5}}\, x^5 + \ldots\right\} \tag{3.35}$$

As a_0 and a_1 are arbitrary (3.35) is the general solution of Legendre differential equation and is represented by $P_n\,(x)$.

If we now give the arbitrary co-efficients the values that the polynomial becomes unity when x becomes unity, we have

$P_0(x) = 1$; $\quad P_1(x) = x$; $\quad P_2\,(x) = \frac{1}{2}\,(3x^2 - 1)$

$P_3(x) = \frac{1}{2}\,(5x^3 - 3x)$; $\quad P_4\,(x) = \frac{1}{8}\,(35x^4 - 30x^2 + 3)$ etc.

N. B. : $P_2(x) = a_0 \left\{1 - \frac{n\,(n+1)}{\underline{|2}}\, x^2\right\}$

Putting $x = 1$, $\quad P_2(1) = a_0\,\{1 - 3\} = -\,2a_0 = 1$

$\therefore\ a_0 = -\,\frac{1}{2}$

$$\therefore\ P_2(x) = -\frac{1}{2}\left\{1 - \frac{2.3}{2}x^2\right\} = \frac{1}{2}(3x^2 - 1)$$

$$P_3 x = a_1\left\{x - \frac{2.5}{\underline{|3}}x^3\right\}$$

Putting $x = 1,\ P_3(1) = a_1\left(1 - \frac{5}{3}\right) = a_1\left(-\frac{2}{3}\right) = 1$

$$\therefore\ a_1 = -\frac{3}{2}$$

$$\therefore\ P_3(x) = \frac{1}{2}(5x^3 - 3x) \text{ etc.}$$

$P_n(x)$ are called Legendre Polynomials. The general formula for arriving at them is due to Rodrigue and is $P_n(x) = \frac{1}{2^n \underline{|n}}\frac{d^n}{dx^n}(x^2 - 1)^n$.

It can be shown that if

$$(1 - x^2)^{m/2}\frac{d^m}{dx^m}\{P_n(x)\} \text{ be written as } P_n^{\ m}(x),$$

then $S = P_n^{\ m}(x)$ will be a particular solution of equation (3.28).

$P_n^{\ m}(x)$ is called associated Legendre function or associated spherical harmonic.

Combining we conclude that a solution of Laplace's Equation in spherical polar co-ordinates has the form

$$V = \left\{Ar^n + Br^{-(n+1)}\right\} P_n^{\ m}(\cos\theta)\ e^{\pm im\varphi} \qquad (3.35a)$$

and general solution will be a superposition of any number of such functions.

Wave Equation and its Solution

(a) In three dimensions the wave equation is

$$\nabla^2 \xi = \frac{1}{c^2}\frac{\partial^2\xi}{\partial t^2} \qquad (3.36)$$

c represents the velocity of the wave.

In cartesian co-ordinates the above equation reduce to

$$\nabla^2 \xi = \frac{\partial^2 \xi}{\partial x^2} + \frac{\partial^2 \xi}{\partial y^2} + \frac{\partial^2\xi}{\partial z^2} = \frac{1}{c^2}\frac{\partial^2 \xi}{\partial t^2} \qquad (3.37)$$

Let us consider the solution of the above equation, subject to the boundary conditions :

(i) $\frac{\partial \xi}{\partial x} = 0$, when $x = 0$, and $x = a$

(ii) $\frac{\partial \xi}{\partial y} = 0$, when $y = 0$, and $y = a$

(iii) $\frac{\partial \xi}{\partial z} = 0$, when $z = 0$, and $z = a$

(iv) $\xi \neq 0$ at $t = 0$, but $\xi = 0$

Let us assume the solution in the form

$$\xi(x, y, z, t) = X(x)\ Y(y)\ Z(z)\ T(t) \tag{3.38}$$

where X is a function of x alone etc.

From (3.38) we have

$$\frac{\partial^2 \xi}{\partial x^2} = YZT \frac{\partial^2 X}{\partial x^2} \ ; \ \frac{\partial^2 \xi}{\partial y^2} = XZT \frac{\partial^2 Y}{\partial y^2}$$

$$\frac{\partial^2 \xi}{\partial z^2} = XYT \frac{\partial^2 Z}{\partial z^2} \text{ and } \frac{\partial^2 \xi}{\partial t^2} = XYZ \frac{\partial^2 T}{\partial t^2}$$

Substituting these values in (3.37) and dividing by $XYZT$,

$$\frac{1}{X}\frac{\partial^2 X}{\partial x^2} + \frac{1}{Y}\frac{\partial^2 Y}{\partial y^2} + \frac{1}{Z}\frac{\partial^2 Z}{\partial z^2} = \frac{1}{c^2 T}\frac{\partial^2 T}{\partial t^2} \tag{3.39}$$

Since x, y, z, t are independent of each other, each of the above terms must be seperately equal to a constant. Writing these constants as $-k_1^2$, $-k_2^2$, $-k_3^2$ and $-k^2$ respectively.

$$k_1^3 + k_2^2 + k_3^2 = k^2 \tag{3.40}$$

Now $\frac{1}{X}\frac{d^2X}{dx^2} = -k_1^2$

$$\frac{1}{Y}\frac{d^2Y}{dy^2} = -K_2^2$$

$$\frac{1}{Z}\frac{d^2Z}{dz^2} = -K_3^2$$

$$\frac{1}{c^2T}\frac{d^2T}{dt^2} = -K^2$$

giving solutions

$$X = A_1 \cos k_1 x + B_1 \sin k_1 x$$

$$Y = A_2 \cos k_2 y + B_2 \sin k_2 y$$

$$Z = A_3 \cos k_3 z + B_3 \sin k_3 z$$

$$T = A_4 \cos kct + B_4 \sin kct$$

Hence the solution of equation (3.37) is

$$\xi\,(x, y, z, t) = (A_1 \cos k_1 x + B_1 \sin k_1 x)\,(A_2 \cos k_2 y + B_2 \sin k_2 y)$$

$$\times\,(A_3 \cos k_3 z + B_3 \sin k_3 z)\,(A_4 \sin k\,ct + B_4 \sin k\,ct) \qquad (3.41)$$

which may also be written in the form

$$\xi\,(x, y, z, t) = Ae^{\pm\, i(k_1 x + k_2 y + k_3 z + k\,ct)}$$

For real values k_1, k_2, k_3 the solution represents a plane wave moving in the direction of the vector $\vec{K}\,(k_1,\ k_2,\ k_3)$

Since $\dfrac{\partial \xi}{\partial x} = 0$, when $x = 0$ $\quad \therefore\ B_1 = 0$

" $\dfrac{\partial \xi}{\partial y} = 0$, when $y = 0$ $\quad \therefore\ B_2 = 0$

" $\dfrac{\partial \xi}{\partial z} = 0$, when $z = 0$ $\quad \therefore\ B_3 = 0$.

Also $\xi \neq 0$ at $t = 0$, but $\dot{\xi} = 0$ at $t = 0$

$$\therefore\ B_4 = 0$$

Hence we have

$$\xi\ (x,\ y,\ z,\ t) = A_1 \cos k_1 x\ A_2 \cos k_2 y\ A_3 \cos k_3 z\ A_4 \cos kct \qquad (3.42)$$

$$\therefore\ \frac{\partial \xi}{\partial x} = -\ Ak_1 \sin k_1 x \cos k_2 y \cos k_3 z \cos kct$$

where $A = A_1\ A_2\ A_3\ A_4$

$$\because\ \frac{\partial \xi}{\partial x} = 0, \text{ at } x = a$$

$\sin\ k_1 a\ \cos k_2 y \cos k_3 z \cos kct = 0$

i.e., $\sin k_1 a = 0$ or, $k_1 a = n_1 \pi$, $n_1 = 1, 2, 3$ .....

$$\therefore\ k_1 = \frac{n_1 \pi}{a}$$

Similarly $k_2 = \dfrac{n_2 \pi}{a}$, $\quad k_3 = \dfrac{n_3 \pi}{a}$

and $k = \sqrt{k_1^2 + k_2^2 + k_3^2} = \dfrac{\pi}{a}\sqrt{n_1^2 + n_2^2 + n_3^2}$

From (3.42) $\xi(x, y, z, t) = \sum_{n_1=1}^{\infty} \sum_{n_2=1}^{\infty} \sum_{n_3=1}^{\infty} A_{n_1 n_2 n_3} \cos \frac{n_1 \pi x}{a} \cos \frac{n_2 \pi y}{a}$

$$\cos \frac{n_3 \pi z}{a} \cos \sqrt{n_1^2 + n_2^2 + n_3^2} \frac{\pi ct}{a} \tag{3.43}$$

(b) **In Spherical Polar Co-ordinates :**

In spherical polar co-ordinates (r, θ, φ), the wave equation has the form

$$\frac{\partial^2 \xi}{\partial r^2} + \frac{2}{r}\frac{\partial \xi}{\partial r} + \frac{1}{r^2}\frac{\partial^2 \xi}{\partial \theta^2} + \frac{\cot\theta}{r^2}\frac{\partial \xi}{\partial \theta} + \frac{1}{r^2 \sin^2\theta}\frac{\partial^2 \xi}{\partial \varphi^2} = \frac{1}{c^2}\frac{\partial^2 \xi}{\partial t^2} \tag{3.43a}$$

Let us assume a solution

$$\xi(r, \theta, \varphi, t) = R(r)\Theta(\theta)\,\Phi(\varphi)\,T(t) \tag{3.44}$$

On substituting in equation 3.43(a) and dividing by $R\,\Theta\,\Phi\,T$

$$\frac{1}{R}\frac{\partial^2 R}{\partial r^2} + \frac{2}{Rr}\frac{\partial R}{\partial r} + \frac{1}{r^2\Theta}\frac{\partial^2 \Theta}{\partial \theta^2} + \frac{\cot\theta}{r^2\Theta}\frac{\partial \Theta}{\partial \theta} + \frac{1}{r^2\sin^2\theta\,\Phi}\frac{\partial^2 \Phi}{\partial \varphi^2} = \frac{1}{c^2 T}\frac{\partial^2 T}{\partial t^2} \tag{3.45}$$

Since r, θ, φ and t are independent of each other, each of the above differentials must be separately equal to a constant.

Let us write

$$\left.\begin{aligned} \frac{1}{c^2 T}\frac{d^2 T}{dt^2} &= -\lambda^2 \\ \frac{1}{\Phi}\frac{d^2 \Phi}{d\varphi^2} &= -k^2 \end{aligned}\right\} \tag{3.46}$$

and $$\frac{1}{\Theta}\frac{d^2\Theta}{d\theta^2} + \frac{\cot\theta}{\Theta}\frac{d\Theta}{d\theta} = \frac{k^2}{\sin^2\theta} - n(n+1) \tag{3.47}$$

Equation (3.45) now becomes

$$\frac{1}{R}\frac{d^2 R}{dr^2} + \frac{2}{Rr}\frac{dR}{dr} + \frac{k^2}{r^2\sin^2\theta} - \frac{n(n+1)}{r^2} - \frac{k^2}{r^2\sin^2\theta} = -\lambda^2$$

or, $$\frac{1}{R}\frac{d^2 R}{dr^2} + \frac{2}{Rr}\frac{dR}{dr} = \frac{n(n+1)}{r^2} - \lambda^2 \tag{3.47a}$$

Thus the variables are seperated

Solutions of equation (3.46) are

$$T = e^{\pm i\lambda ct}$$

$$\Phi = e^{\pm ik\varphi}$$

Equation (3.47) may be written as

$$\frac{d^2\Theta}{d\theta^2} + \cot\theta \frac{d\Theta}{d\theta} + \left\{n(n+1) - \frac{k^2}{\sin^2\theta}\right\}\Theta = 0$$

$$\text{or, } \frac{1}{\sin\theta}\left\{\sin\theta \frac{d^2\Theta}{d\theta^2} + \cos\theta \frac{d\Theta}{d\theta}\right\} + \left\{n(n+1) - \frac{k^2}{\sin^2\theta}\right\}\Theta = 0$$

$$\text{or, } \frac{1}{\sin\theta}\frac{d}{d\theta}\left(\sin\theta \frac{d\Theta}{d\theta}\right) + \left\{n(n+1) - \frac{k^2}{\sin^2\theta}\right\}\Theta = 0 \qquad (3.48)$$

If we put $\cos\theta = x$

$$\frac{dx}{d\theta} = -\sin\theta$$

$$\text{or, } -\frac{1}{\sin\theta}\frac{d}{d\theta} = \frac{d}{dx}$$

Equation (3.48) gives

$$\frac{d}{dx}\left\{\sin^2\theta \frac{d\Theta}{dx}\right\} + \left\{n(n+1) - \frac{k^2}{\sin^2\theta}\right\}\Theta = 0$$

$$\text{or, } \frac{d}{dx}\left\{(1-x^2)\frac{d\Theta}{dx}\right\} + \left\{n(n+1) - \frac{k^2}{1-x^2}\right\}\Theta = 0$$

This is associated Legendre equation and its solution is

$$\Theta = A_n P_n{}^k(x) + B_n Q_n{}^k(x) = A_n P_n{}^k(\cos\theta) + B_n Q_n{}^k(\cos\theta)$$

If Θ has to be finite when $x \to 0$, B_n must be zero and hence the above equation can be written as $\Theta = A_n P_n{}^k(\cos\theta)$. $P_n{}^k(x)$ are associated Legendre function.

Lastly equation (3.47a) may be written as

$$\frac{1}{R}\frac{d^2R}{dr^2} + \frac{2}{Rr}\frac{dR}{dr} + \lambda^2 - \frac{n(n+1)}{r^2} = 0$$

$$\text{or, } \frac{d^2R}{dr^2} + \frac{2}{r}\frac{dR}{dr} + \left\{\lambda^2 - \frac{n(n+1)}{r^2}\right\}R = 0$$

If we put $R = r^{-\frac{1}{2}} S(r)$, the above will reduce to

$$\frac{d^2S}{dr^2} + \frac{1}{r}\frac{dS}{dr} + \left\{\lambda^2 - \frac{\left(n+\frac{1}{2}\right)^2}{r^2}\right\}S = 0$$

Putting $\lambda r = x$, the above takes the form

$$\frac{d^2S}{dx^2} + \frac{1}{x}\frac{dS}{dx} + \left\{1 - \frac{\left(n+\frac{1}{2}\right)^2}{\lambda^2}\right\} S = 0 \tag{3.49}$$

This is Bessel's differential equation. Its solution is $S = C\,J_m(x) + D\,J_{-m}(x)$, where $m = n + \frac{1}{2}$. $J_m(x)$ is called Bessel's function of m the order and first kind. Hence S, i.e., R is determined. Substituting the values of R, Θ, Φ and T in equation (3.44), we get the solution of wave equation in spherical polar co-ordinates.

(c) **In Cylindrical Co-ordinates :**

In cylindrical co-ordinates (r, θ, z) the wave equation is

$$\frac{\partial^2\xi}{\partial r^2} + \frac{1}{r}\frac{\partial\xi}{\partial r} + \frac{1}{r^2}\frac{\partial^2\xi}{\partial\theta^2} + \frac{\partial^2\xi}{\partial z^2} = \frac{1}{c^2}\frac{\partial^2\xi}{dt^2} \tag{3.50}$$

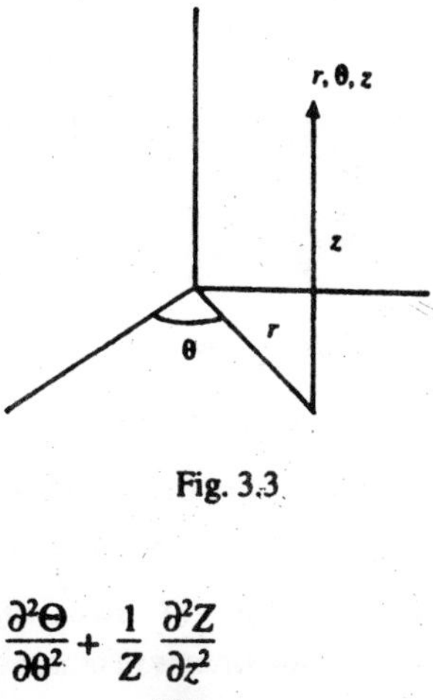

Fig. 3.3

Let us assume a solution of the form

$$\xi(r, \theta, z, t) = R(r)\ \Theta(\theta)\ Z(z)\ T(t) \tag{3.51}$$

Substituting in (3.50) and dividing both sides by $R\,\Theta\,ZT$ we have

$$\frac{1}{R}\left(\frac{\partial^2R}{\partial r^2} + \frac{1}{r}\frac{\partial R}{\partial r}\right) + \frac{1}{r^2\Theta}\frac{\partial^2\Theta}{\partial\theta^2} + \frac{1}{Z}\frac{\partial^2Z}{\partial z^2} = \frac{1}{c^2T}\frac{\partial^2T}{\partial t^2} \tag{3.52}$$

in which the variables are separated. Since r, θ, z and t are independent of each other, each differential will be seperately equal to a constant.

Let us put

$$\left.\begin{aligned} \frac{1}{\Theta}\frac{d^2\Theta}{d\theta^2} &= -k^2 \\ \frac{1}{z}\frac{d^2Z}{dz^2} &= -k_1{}^2 \\ \frac{1}{T}\frac{d^2T}{dt^2} &= -c^2\lambda^2 \end{aligned}\right\} \tag{3.53}$$

Equation (3.52) became

$$\frac{1}{R}\left\{\frac{d^2R}{dr^2}+\frac{1}{r}\frac{dR}{dr}\right\}-\frac{k^2}{r^2}=k_1^2-\lambda^2=-\alpha^2 \text{ say}$$

$$\text{or, } \frac{d^2R}{dr^2}+\frac{1}{r}\frac{dR}{dr}+\left(\alpha^2-\frac{k^2}{r^2}\right)R=0 \tag{3.54}$$

Putting $(\alpha r) = x$, the above equation takes the form

$$\frac{d^2R}{dx^2}+\frac{1}{x}\frac{dR}{dx}+\left(1-\frac{k^2}{x^2}\right)R=0 \tag{3.55}$$

The solutions are

$$\Theta = A_1 \cos k\theta + B_1 \sin k\theta$$

$$Z = A_2 \cos k_1 z + B_2 \sin k_1 z$$

$$T = A_3 \cos \lambda ct + B_3 \sin \lambda ct$$

$$R = A_4 J_k(\alpha r) + B_4 Y_k(\alpha r)$$

$$= A_4 J_k(\alpha r)$$

$\because Y_x(\alpha r) \to \infty$, when $r \to o$ $\therefore B_4 = 0$.

The complete solution is

$$\xi(r, \theta, z, t) = (A_1 \cos k\theta + B_1 \sin k\theta)(A_2 \cos k_1 z + B_2 \sin k_1 z) \times (A_3 \cos \lambda ct + B_3 \sin \lambda ct)\, A_4 J_k(\alpha r) \tag{3.56}$$

(d) **Solution of two dimensional wave equation in Cartesian Co-ordinates** (Transverse vibration of rectangular membrane) :

A theoretical membrane is a perfectly flexible thin lamina of solid matter of uniform thickness stretched in all directions by a tension so great that it remains sensibly unaltered during vibrations. The tension Γ and the density (ρ) are supposed to be constants.

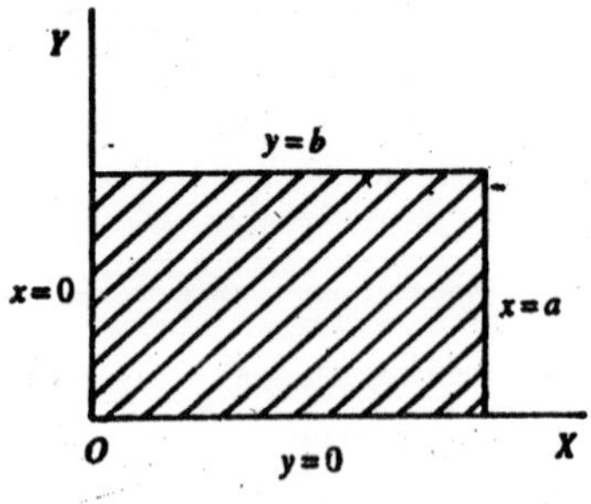

Fig. 3.4

The differential equation of free vibration of such a membrane is given by

$$\frac{1}{c^2}\frac{\partial^2\xi}{\partial t^2}=\frac{\partial^2\xi}{\partial x^2}+\frac{\partial^2\xi}{\partial y^2} \tag{3.57}$$

where $c^2 = \dfrac{\Gamma}{\rho}$

The boundary of the membrane is the rectangle formed by the co-ordinate axes and the lines $x = a$ and $y = b$ (Fig. 3.4).

Since for every point on the boundary $\xi = 0$, we have

$$\xi(0, y, t) = \xi(a, y, t) = \xi(x, 0, t) = \xi(x, b, t) = 0 \tag{3.58}$$

Initial displacement and velocity are given by

$$\xi(x, y, 0) = F(x, y) \tag{3.59}$$

$$\text{and } \xi(x, y, 0) = \left[\frac{\partial}{\partial t}\xi(x, y, t)\right]_{t=o} = \varphi(x, y) \tag{3.60}$$

As a solution of equation (3.57), let us assume

$$\xi(x, y, t) = X(x)\; Y(y)\; T(t) \tag{3.61}$$

Substituting in (3.57) and dividing by XYT.

$$\frac{1}{X}\frac{\partial^2 X}{\partial x^2} + \frac{1}{Y}\frac{\partial^2 Y}{\partial y^2} = \frac{1}{Tc^2}\frac{\partial^2 T}{\partial t^2} \tag{3.62}$$

Since x, y, t are independent of each other, we may write by choosing proper constants of seperation.

$$\frac{1}{X}\frac{\partial^2 X}{\partial x^2} = -\lambda_1^2, \quad \frac{1}{Y}\frac{\partial^2 Y}{\partial y^2} = -\lambda_2{}^2$$

$$\text{and } \frac{1}{c^2 T}\frac{\partial^2 T}{\partial t^2} = -\lambda^2,$$

so that $\lambda_1^2 + \lambda_2^2 = \lambda^2$

Solving each of the above, we have

$$\left.\begin{aligned} X &= A_1 \cos\lambda_1 x + B_1 \sin\lambda_1 x \\ Y &= A_2 \cos\lambda_2 y + B_2 \cos\lambda_2 y \\ T &= A_3 \cos\lambda ct + B_3 \sin\lambda ct \end{aligned}\right\} \tag{3.63}$$

From the condition (3.58), it is evident that

$$X(0) = X(a) = 0\,;\; Y(0) = Y(b) = 0$$

Hence $A_1 = A_2 = 0$

And $B_1 \sin\lambda_1 a = 0 = \sin m\pi$

$$\therefore\; \lambda_1 = \frac{m\pi}{a}$$

Similarly $B_2 \sin\lambda_2 b = 0 = \sin n\pi$

$$\therefore\; \lambda_2 = \frac{n\pi}{b}.$$

where m and n are integers.

Thus the solution of equation (3.57) is of the form

$$\xi = (A \cos \lambda_{mn} t + B \sin \lambda_{mn} t) \sin \frac{m\pi x}{a} \times \sin \frac{n\pi y}{b} \tag{3.64}$$

where A and B are constants and λ_{mn} is given by

$$\lambda_{mn}^2 = \left(\frac{m\pi}{a}\right)^2 + \left(\frac{n\pi}{b}\right)^2 \tag{3.65}$$

m, n are integers

Such values are known as Eigen values or characteristic values of vibrating membrane.

(e) **Two dimensional wave equation in cylindrical co-ordinates** (Transverse vibration of circular membrane) :

In cylindrical co-ordinates the two dimensional wave equation is

$$\frac{\partial^2 \xi}{\partial t^2} = c^2 \left\{\frac{\partial^2 \xi}{\partial r^2} + \frac{1}{r}\frac{\partial \xi}{\partial r} + \frac{1}{r^2}\frac{\partial^2 \xi}{\partial \theta^2}\right\} \tag{3.66}$$

Let us assume a solution of the form

$$\xi(r, \theta, t) = R(r)\ \Theta(\theta)\ T(t) \tag{3.67}$$

where R, Θ and T are respectively the functions of r, θ and t alone.

Substituting in (3.66) and dividing by $R\,\Theta\,T$

$$\frac{1}{c^2 T}\frac{\partial^2 T}{\partial t^2} = \frac{1}{R}\frac{\partial^2 R}{\partial r^2} + \frac{1}{Rr}\frac{\partial R}{\partial r} + \frac{1}{r^2 \Theta}\frac{\partial^2 \Theta}{\partial \theta^2} \tag{3.68}$$

Since the L.H. side is a function of t alone while the R.H. side involve (r, θ) which are quite independent of each other, the above differential equation will be true if each part is separately equal to a constant $= -\lambda^2$ say.

Thus $\dfrac{d^2 T}{dt^2} = -\lambda^2 c^2 T$ giving a solution

$$T = P \cos \lambda ct + Q \sin \lambda ct \tag{3.69}$$

To make the R.H. possible let us put

$\dfrac{1}{\Theta}\dfrac{d^2\Theta}{d\theta^2} = -n^2$ giving a solution

$$\Theta = K e^{\pm in\theta} \tag{3.70}$$

Since ξ is periodic in θ of period 2π, n must be an integer.

Finally the differential equation (3.68) reduces to

$$\frac{1}{R}\frac{d^2R}{dr^2}+\frac{1}{rR}\frac{dR}{dr}-\frac{n^2}{r^2}=-\lambda^2$$

or, $$\frac{d^2R}{dr^2}+\frac{1}{r}\frac{dR}{dr}+\left(\lambda^2-\frac{n^2}{r^2}\right)R=0 \qquad (3.71)$$

This is Bessel's differential equation which gives a solution.

$$R(r)=AJ_n(\lambda r)+BY_n(\lambda r) \qquad (3.72)$$

J_n is Bessel function of nth order and 1st kind.

Y_n is Bessel function of nth order and 2nd kind.

Since $Y_n(\lambda r)\to\infty$, as $r\to 0$; $B=0$

Hence the general solution of equation (3.66) is

$$\xi(r,\theta,t)=\sum_{m=1}^{\infty}\sum_{n=0}^{\infty}\left\{P_{mn}\cos\lambda_{mn}ct+Q_{mn}\sin\lambda_{mn}ct\right\}\times e^{\pm in\theta}J_n(\lambda_{mn}r) \qquad (3.73)$$

At $r=a$, $\xi=0$ for all θ and t

$\therefore\ J_n(\lambda a)=0$

when $n=0$, the equation giving the admissible values of λ is given by

$J_o(\lambda a)=0$, which yields $\lambda a=2.404$, 5.520, 8.624 etc. Hence the fundamental angular frequency is given by

$$\omega_1=\frac{2.404}{a},\ a=\text{radius of the membrane.}$$

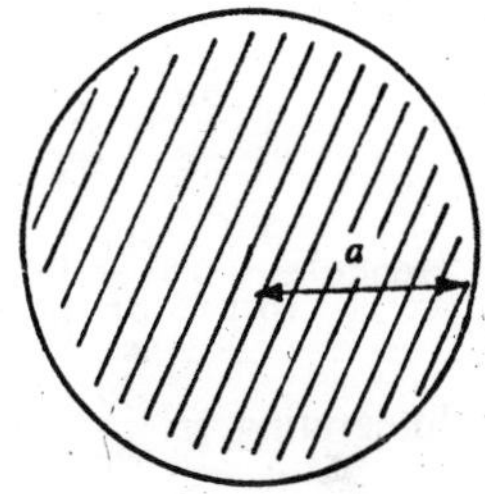

Fig. 3.5

EXERCISE

1. Write down the three dimensional equation of heat flow through a medium of thermal conductivity K. Show that under steady stable conditions, this reduces to the Laplace's equation $\nabla^2u=0$. When $u(r)$ denotes temperature, using spherical polar co-ordinates verify that this has the solution of the form

$u(r) = \dfrac{Q}{4\pi kr} + B$, where Q is the rate at which heat flows out isothermally from a point source at $r = 0$ and B is constant. If the temperatures on two concentric spheres of radius r_1 and r_2 with the point source at the centre be u_1 and u_2 respectively, show that

$$Q = \frac{4\pi k\,(u_1 - u_2)\,r_1 r_2}{r_2 - r_1}.$$

Ans. Fourier equation for heat flow in three dimensions is $k\nabla^2 u = \rho S \dfrac{\partial u}{\partial t}$

K = Thermal conductivity, ρ = density ; S = sp. heat

or, $h\nabla^2 u = \dfrac{\partial u}{\partial t}$ where $h = k/\rho s$

In the steady state $\dfrac{\partial u}{\partial t} = 0$

$\therefore\ \nabla^2 u = 0$ which is Laplace's equation.

In spherical polar co-ordinates

$$\frac{\partial}{\partial r}\left(r^2\frac{\partial u}{\partial r}\right) + \frac{1}{\sin\theta}\frac{\partial}{\partial\theta}\left(\sin\theta\frac{\partial u}{\partial\theta}\right) + \frac{1}{\sin^2\theta}\frac{\partial^2 u}{\partial\varphi^2} = 0$$

Since the Case is spherically symmetrical

$$\frac{\partial u}{\partial\theta} = \frac{\partial u}{\partial\varphi} = 0$$

Hence $\dfrac{\partial}{\partial r}\left(r^2\dfrac{\partial u}{\partial r}\right) = 0$

Integrating, $r^2\dfrac{\partial u}{\partial r}$ = Constant.

But $Q = -k \cdot 4\pi r^2 \dfrac{\partial u}{\partial r}$ from the theory of heat conduction

$$\therefore\ \frac{r^2\partial u}{\partial r} = -\frac{Q}{4\pi k}$$

or, $$\int du = -\frac{Q}{4\pi k}\int\frac{dr}{r^2}$$

or, $$u(r) = \frac{Q}{4\pi kr} + B$$

$$\therefore\ u_1 = \frac{Q}{4\pi kr_1} + B$$

$$u_2 = \frac{Q}{4\pi k r_2} + B$$

$$\therefore \ u_1 - u_2 = \frac{Q}{4\pi k}\left(\frac{1}{r_1} - \frac{1}{r_2}\right)$$

2. Expression for Capacitance using Laplace's equation :

Parallel Plate Capaciter :

Let V_1 and V_2 be the potentials of the Conducting parallel plates P and Q ($V_1 > V_2$). X-axis is taken at right angles to the plates with the origin O on the plate P. The plates are assumed to be large enough to avoid edge effects. The potential V at a point between the plates will depend on the x co-ordinate alone. Hence Laplace's equation takes the form

$$\frac{\partial^2 V}{\partial x^2} = 0 \qquad \ldots \quad (1)$$

since V is independent of y and z.

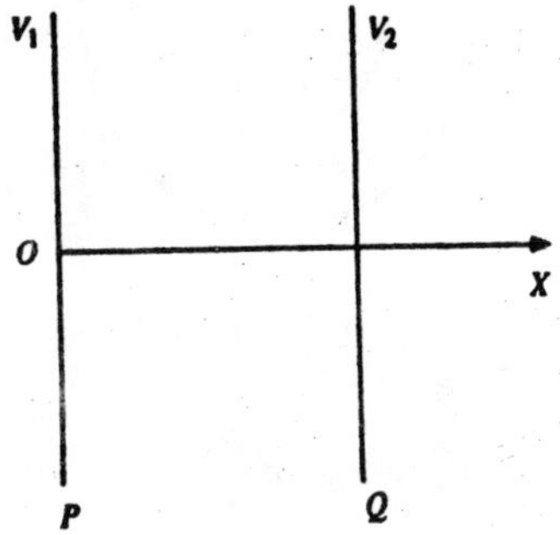

Fig. 3.6

Integrating $w.\ r.\ t.\ x$, $\frac{\partial V}{\partial x} = A \qquad \ldots \quad (2)$

and $V = Ax + B \qquad \ldots \quad (3)$

If t be the distance between plates, then $V = V_1$ at $x = 0$ and $V = V_2$ at $x = t$.

Substituting in (3)

$$V_1 = B \text{ and } V_2 = At + V_1 \text{ or, } A = \frac{V_2 - V_1}{t}$$

Hence $V = V_1 - \frac{x}{t}(V_1 - V_2)$

If σ be the surface density of charge on the positive plate, by Coulomb's theorem

$$\frac{4\pi\sigma}{\epsilon} = 4\pi Q/A\epsilon = -\left(\frac{\partial V}{\partial x}\right)_{x=0}$$

$$\therefore \quad 4\pi \frac{Q}{\epsilon A} = -A = -\frac{V_2 - V_1}{t}$$

$$\therefore \quad \frac{Q}{V_1 - V_2} = C = \frac{\epsilon A}{4\pi t}, \ \epsilon \to \text{relative permittivity of the medians.}$$

3. Spherical Condenser :

A Spherical Condenser consists of two concentric spherical conductors. The inner one is of radius a and is given a charge Q. The outer one is of radius b and is earthed. If V be the potential at a point between the two sphers and at a distance r from the centre, we have by Laplace's equation $\nabla^2 V = 0$. Using spherical polar co-ordinates we have

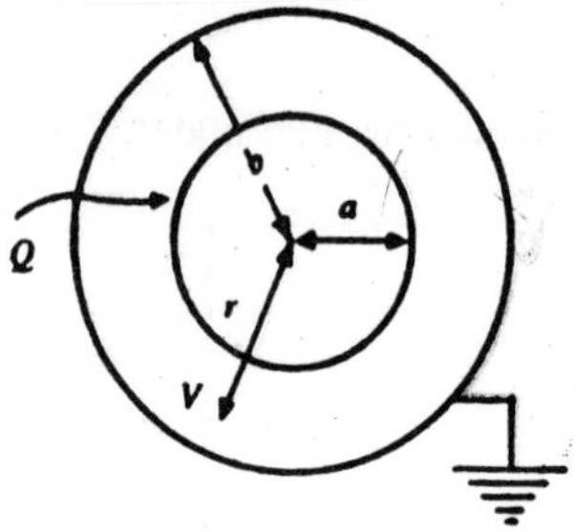

Fig. 3.7

$$\frac{1}{r^2}\frac{\partial}{\partial r}\left(r^2 \frac{\partial V}{\partial r}\right) = 0,$$

$$\text{since } \frac{\partial V}{\partial \theta} = \frac{\partial V}{\partial \varphi} = 0$$

Integrating, $r^2 \frac{\partial V}{\partial r} = A$ or, $\frac{\partial V}{\partial r} = \frac{A}{r^2}$

Integrating again $V = -\frac{A}{r} + B$

At $r = b$, $V = 0$ $\therefore$ $0 = -\frac{A}{b} + B$

At $r = a$, $V = V_a$ $\therefore$ $V_a = -\frac{A}{a} + B$

Solving $A\left(\frac{1}{b} - \frac{1}{a}\right) = V_a = A\left(\frac{a-b}{ab}\right)$

$\therefore$ $A = V_a \frac{ab}{a-b}$ and $B = \frac{V_a a}{a-b}$

Substituting for A and B

$$V = \frac{V_a ab}{(b-a)r} - \frac{V_a a}{b-a}.$$

If σ be the surface density of charge on the positive plate, we have by Coulomb's theorem.

$$\frac{4\pi\sigma}{\epsilon} = \frac{4\pi Q}{\epsilon A} = -\left(\frac{\partial V}{dr}\right)_{r=a}$$

$$= V_a \frac{ab}{b-a}\frac{1}{a^2} = V_a \frac{b}{a(b-a)}$$

or, $\dfrac{4\pi Q}{\in 4\pi a^2} = V_a \dfrac{b}{a(b-a)}$ $\in \rightarrow$ relative permittivity of the mediun

Hence $C = \dfrac{Q}{V_a} = \in \dfrac{V_a a^2 b}{a(b-a)V_a} = \in \dfrac{ab}{b-a}$.

4. Cylindrical Condenser :

A Cylindrical Condenser consists of two co-axial cylindrical conductors. The inner one is of radius a and is charged with a charge Q per unit length. The outer one is of radius b and is earthed. Let V be the potential at a point between the two cylinders distance r from the common axis. From Laplace's Equation $\nabla^2 V = 0$. Using cylindrical co-ordinates

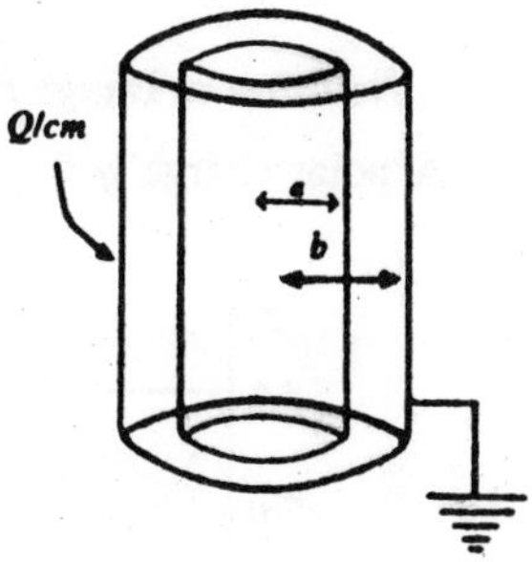

Fig. 3.8

$\dfrac{1}{r}\dfrac{\partial}{\partial r}\left(r\dfrac{\partial V}{\partial r}\right) = 0$ since V is independent of θ and z.

Integrating $r\dfrac{\partial V}{\partial r} = A$

or, $\dfrac{\partial V}{\partial r} = \dfrac{A}{r}$

Integrating again

$$V = A \log_e r + B$$

$$V = Va \text{ when } r = a \quad \therefore\ Va = A \log_e a + B$$

$$V = 0 \text{ when } r = b \quad \therefore\ 0 = A \log_e b + B$$

Solving $A = \dfrac{Va}{\log_e\left(\dfrac{a}{b}\right)}$

$$B = Va - Va \log_e {}^{a}/_{\log_e\left(\frac{a}{b}\right)} = Va \frac{\log_e b}{\log_e\left(\dfrac{a}{b}\right)}$$

Hence $V = Va \left\{\dfrac{\log_e r}{\log_e \dfrac{a}{b}} + \dfrac{\log_e b}{\log_e\left(\dfrac{a}{b}\right)}\right\}$

From Coulomb's theorem

$$\frac{4\pi Q}{\varepsilon 2\pi a} = -\left(\frac{\partial V}{\partial r}\right)_{r=a} = -\frac{V_a}{a \log_e\left(\frac{a}{b}\right)}$$

$$\text{or,} \quad Va = \frac{2Q}{\varepsilon} \log_e\left(\frac{b}{a}\right)$$

$$\therefore \ C = \frac{Q}{V_a} = \frac{\varepsilon}{2 \log_e\left(\frac{b}{a}\right)}$$

5. Solution of Image Problem by Laplace's Equation :

A point charge q is placed at a distance a from the centre of an earth connected conducting sphere of radius a_o ($a > a_o$). $P(r, \theta, \varphi)$ is a point outside the sphere. The potential V at P is due to the charge q plus the induced charge on the sphere. Thus $V = \frac{q}{s} + v'$ ... (1)

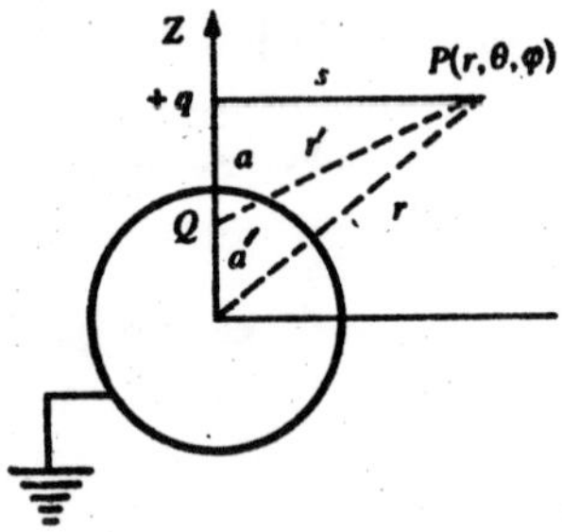

Fig. 3.9

where s is the distance shown and V' is the potential due to the induced charge on the sphere. Assuming that V' will satisfy Laplace's equation we have

$$V' = \Sigma\,(A_n r^n + B_n r^{-n-1})\, P_n(\cos\theta) \quad \ldots (2)$$

Because of axial symmetry V' will be independent of φ.

$\because\ V' = O$ at $r \to \infty$, $A_n \to O$

$$\text{Hence } V' = \Sigma \frac{B_n}{r^{n+1}} P_n(\cos\theta) = \Sigma \frac{B_n}{r^{n+1}} P_n(x)$$

Where $x = \cos\theta$.

As the conductor is earth connected

$V = 0$, when $r = a_o$

$$\text{Hence} \left(\frac{q}{s}\right)_{\text{on the surface}} + \Sigma \frac{B_n}{r^{n+1}} P_n(x) = 0 \quad \ldots (3)$$

$$\text{Now } s^2 = a^2 + r^2 - 2ar\cos\theta = a^2\left\{1 + \frac{r^2}{a^2} - \frac{2r}{a}\cos\theta\right\}$$

$$= a^2\{1 - 2xy + y^2\} \text{ putting } x = \cos\theta \text{ and } y = \frac{r}{a}$$

$$\therefore \frac{1}{s} = \frac{1}{a}(1 - 2xy + y^2)^{-\frac{1}{2}} = \frac{1}{a}\left\{1 + y^2\left(1 - \frac{2x}{y}\right)\right\}^{-\frac{1}{2}}$$

$$= \frac{1}{a}\left\{1 - \frac{1}{2}y^2\left(1 - \frac{2x}{y}\right) - \frac{1}{2}\left(-\frac{3}{2}\right)\frac{1}{\underline{|2}}\, y^4\left(1 - \frac{2x}{y}\right)^2 - \frac{1}{2}\left(-\frac{3}{2}\right)\left(-\frac{5}{2}\right)\frac{1}{\underline{|3}}\frac{y^6}{1}\left(1 - \frac{2x}{y}\right)^3 + \ldots\ldots\right\}$$

$$= \frac{1}{a}\left\{1 - \frac{y^2}{2} + xy + \frac{3}{8}y^4\left(1 - 4\frac{x}{y} + 4\frac{x^2}{y^2}\right) - \frac{5y^6}{16}\left(1 - \frac{6x}{y} + \frac{12x^2}{y^2} - \frac{8x^3}{y^3}\right) + \ldots\right.$$

$$= \frac{1}{a}\left\{y^o P_o(x) + y P_1(x) + y^2 P_2(x) + y^3 P_3(x) + \ldots\ldots\right\}$$

$$= \frac{1}{a}\Sigma y^n P_n(x) = \frac{1}{a}\Sigma\left(\frac{r}{a}\right)^n P_n(\cos\theta)$$

Condition (3) gives $\frac{q}{a}\Sigma\left(\frac{a_o}{a}\right)^n P_n(\cos\theta) + \Sigma\frac{B_n}{a_o{}^{n+1}} P_n(\cos\theta) = 0$

or, $\Sigma\left\{\frac{q}{a}\left(\frac{a_o}{a}\right)^n + \frac{B_n}{a_o{}^{n+1}}\right\} P_n(\cos\theta) = 0$

Since this is true for all values of n.

$$B_n = \frac{q\, a_o^{2n+1}}{a^{2n+1}}$$

$$\therefore V' = \Sigma - q\frac{a_o^{2n+1}}{a^{n+1}} \cdot \frac{1}{r^{n+1}} P_n(x)$$

$$= -\frac{qa_o}{ra}\Sigma\left(\frac{a_o^2}{ar}\right)^n P_n(x) \qquad \ldots (4)$$

It is evident from equation (4), that the potential due to the induced charge may be accounted for by a suitable charge at a suitable point. If we consider the point $Q(0, 0, a')$, such that $QP = s'$,

$$s'^2 = a'^2 + r^2 - 2a'r\cos\theta = r^2\left(1 - \frac{2a'\cos\theta}{r} + \frac{a'^2}{r^2}\right)$$

$$\therefore \frac{1}{s'} = \frac{1}{r}\Sigma\left(\frac{a'}{r}\right)^n P_n(\cos\theta)$$

If q' be the required charge at Q, then

$$V' = \frac{q'}{s'} = \frac{q'}{r} \Sigma \left(\frac{a'}{r}\right)^n P_n(x) = \frac{qa_o}{ar} \Sigma \left(\frac{a_o^2}{ar}\right)^n P_n(x) \qquad \ldots \quad (5)$$

This is an identity and true for all values of n.

Putting $n = 0$, $\frac{q'}{r} = \frac{qa_o}{ar}$ $\quad \therefore \; q' = \frac{qa_o}{a}$

Putting $n = 1$ $\frac{q'}{r} \frac{a'}{r} x = -\frac{qa_o}{ar} \frac{a_o^2}{ar} x$

$$\therefore \; a' = -\frac{qa_o{}^3}{a^2 r^2} \times \left(\frac{ar^2}{-qa_o}\right) = \frac{a_o^2}{a}.$$

The above gives both the magnitude and position of the image charge.

6. Heat Conduction in one dimension :

The Fourier differential equation for heat conduction in one dimension is

$$\frac{\partial u}{\partial t} = c^2 \frac{\partial^2 u}{\partial x^2} \qquad \ldots \quad (1)$$

where u is the temperature of a point at a distance x at time t and $c^2 = \frac{k}{\rho s}$,

$k =$ Thermal conductivity, $\rho =$ density

$s =$ sp. heat of the material.

Let us assume a solution of the form

$u(x, t) = X(x)\; T(t)$, where X is a function of x alone and T is a function of t only.

Substituting in (1) $XT' = c^2 TX''$

or, $$\frac{X''}{X} = \frac{1}{c^2} \frac{T'}{T} \qquad \ldots \quad (2)$$

clearly left hand side of (2) is a function of x only and the right hand side that of t alone. Since x and t are independent of each, equation (2) is valid if each side is equal to a constant, k say

Thus $$\frac{d^2X}{dx^2} - kx = 0 \qquad \ldots \quad (3)$$

$$\frac{dT}{dt} - kc^2 T = 0 \qquad \ldots \quad (4)$$

Solving (3) and (4) we get

(i) When k is + ve = p^2, say

$$X = c_1 e^{px} + c_2 e^{-px}, \quad T = c_3 e^{c^2 p^2 t}$$

(ii) When k is − ve = $-p^2$ say

$$X = (c_4 \cos px + c_5 \sin px), \ T = c_6 e^{-c^2 p^2 t}$$

(iii) When k is zero

$$X = c_7 x + c_8, \quad T = c_9$$

The various possible solutions are

$$u = (c_1 e^{px} + c_2 e^{-px}) \ c_3 e^{c^2 p^2 t}, \quad \ldots (5)$$

$$u = (c_4 \cos px + c_5 \sin px) \ c_6 e^{-c^2 p^2 t} \quad \ldots (6)$$

$$u = (c_7 x + c_8) \ c_9 \quad \ldots (7)$$

As we are dealing with the problem of heat conduction, it must be a transient solution i.e., u must decrease with the increase of t. Accordingly, the solution is given by (6) i.e.

$$u = (c_1 \cos px + c_2 \sin px) \ e^{-c^2 p^2 t} \quad \ldots (8)$$

is the only suitable solution of (1).

Application : An insulated rod of length l has its ends A and B maintained at 0°C and 100°C respectively until steady state conditions prevail. If B is suddenly reduced to 0°C and maintained at 0°C, find the temperature at a distance x from A at time t.

Under steady state condition $\frac{\partial u}{\partial t} = 0$.

Hence equation (1) redues to $\frac{\partial^2 u}{\partial x^2} = 0 \quad \ldots (9)$

The general solution is $u = Ax + B \quad \ldots (10)$

Since $u = 0$ at $x = 0$, $B = 0$

Again since $u = 100$ at $x = l$, $A = \frac{100}{l}$

$$\therefore \ u = \frac{100x}{l} \quad \ldots (11)$$

The boundary conditions for subsequent flow are

$$u(0, t) = 0 \text{ for all } t\text{'s} \quad \ldots (12)$$

$$u(l, t) = 0 \text{ for all } t\text{'s} \quad \ldots (13)$$

Taking equation (8) as the solution

$$\text{i.e.,}\quad u(x,t) = (c_1 \cos px + c_2 \sin px)\, e^{-c^2p^2t} \qquad \ldots \quad (14)$$

From (12), $0 = c_1 e^{-c^2p^2t}$ for all t's

$$\therefore\ c_1 = 0$$

Again from (13)

$$0 = c_2 \sin pl\, e^{-c^2p^2t} \text{ for all } t\text{'s}$$

$$\therefore\ \sin pl = 0 \text{ or, } pl = n\pi \quad \because c_2 \neq 0$$

$$\text{or,}\quad p = \frac{n\pi}{l}, \text{ where } n \text{ is an integer}$$

Hence the solution is $u(x,t) = A_n \sin \dfrac{n\pi x}{l}\, e^{\frac{-c^2n^2\pi^2}{l^2}t}$.

Adding all such solutions, the most general solution is

$$u(x,t) = \sum_{n=1}^{\infty} A_n \sin \frac{n\pi x}{l} \frac{e^{-c^2n^2\pi^2t}}{l^2} \qquad \ldots \quad (15)$$

Putting $t = 0$, $u(x,0) = \sum\limits_{n=1}^{\infty} A_n \sin \dfrac{n\pi x}{l}$ $\qquad \ldots \quad (16)$

Since the solutions given by (11) and (16) are identical

$$\frac{100x}{l} = \sum_{n=1}^{\infty} A_n \sin \frac{n\pi x}{l}$$

Hence $A_n \dfrac{l}{2} = \dfrac{100}{l} \displaystyle\int_0^l x \sin \dfrac{n\pi x}{l}\, dx$

$$\text{or, } A_x = \frac{200}{l^2} \left\{ \left[-\frac{lx}{n\pi} \cos \frac{n\pi x}{l} \right]_0^l + \frac{l}{n\pi} \int_0^l \cos \frac{n\pi x}{l}\, dx \right\}$$

$$= -\frac{200}{n\pi l} \{ l \cos n\pi \} = -\frac{200}{n\pi}(-1)^n = \frac{200}{n\pi}(-1)^{n+1}$$

Hence (15) gives

$$u(x,t) = \frac{200}{\pi} \sum \frac{(-1)^{n+1}}{n} \sin \frac{n\pi x}{l}\, e^{\frac{-c^2n^2\pi^2}{l^2}t}$$

EXERCISE

1. Apply Laplace's equation to the case of a charged conducting sphere of radius R for which a boundary condition is $V = v_0$ for $r = R$. Show that for all interior points $r < R$, $V = v_0$ and at all external points $r > R$, $V = \dfrac{v_0 R}{r}$.

2. A homogeneous spherical shell is made up of a material of diffusivity h. The internal and external radii of the shell are a and b respectively. If the inner shell is kept at a temperature θ_α, and the outer shell is kept at a constant temperature θ_b, show that the temperature of the shell is given by

$$\theta = (\theta_a - \theta_b)\,\frac{ab}{b-a}\cdot\frac{1}{r} + \frac{\theta_b\, b - \theta_a\, a}{b-a}.$$

3. Find the potential at any point of a thin plane sheet of infinite extent where all points of given line which lie to be left of the origin are kept at zero potential and all points which lie to right of this origin are kept a unit potential. $\left[\textbf{Ans. } V = \dfrac{1}{2\pi}\log_e (x^2 + y^2)\right]$

4. Show that the only solution of two dimensional Laplace's Equation depending only on $r = (x^2 + y^2)^{\frac{1}{2}}$ is $\Psi = c \log r + k$.

5. Verify that $u = \dfrac{c}{r}$ satisfies Laplace's Equation in polar co-ordinates.

6. A tightly stretched string with fixed ends $x = 0$ and $x = l$ is initially in a position given by $y = y_o \sin^3\left(\dfrac{\pi x}{l}\right)$ If it is released from this position find its displacement $y(x, t)$.

$$\left[\textbf{Ans. } y(x,t) = \frac{3y_0}{4}\sin\frac{\pi x}{l}\cos\frac{\pi ct}{l} - \frac{y_0}{4}\sin\frac{3\pi x}{l}\cos\frac{3\pi ct}{l}\right]$$

7. A long rectangular plate of width l cm with insulated surfaces has its temperature φ equal to zero on both the long sides and one of the short sides so that

$\theta(0, y) = 0, \quad \theta(a, y) = 0, \quad \theta(x, \infty) = 0, \quad \theta(x, 0) = kx$

Show that the steady state temperature within the plate is given by

$$\theta(x, y) = \frac{2ak}{\pi}\sum_{n=1}^{\infty}\frac{(-1)^n}{n}\, e^{-n\pi y/a}\sin\frac{n\pi x}{a}.$$

8. Show that when the values of a potential function on the boundary of a circle of radius a are given by

$V(r, \theta) = F(\theta)$, the potential at any interior point is given by

$$V(r,\theta) = \frac{1}{\pi}\int_{-\pi}^{\pi}\left\{\frac{1}{2} + \sum_{n=1}^{\infty}\left(\frac{r}{a}\right)^n \cos n(\theta - u)\right\} F(u)\, du$$

CHAPTER IV
FOURIER SERIES

[Definitions of linear independence and completeness of set of functions. Fourier's theorem (Proof not required)—Evaluation of Fourier co-efficients; Analysis of simple waveforms with Fourier series.]

The concept of linear dependence of a set of functions, say $f_n(x)$ can be developed as follows. If for a given set of functions $f_n(x)$ a set of constants c_n can be found that satisfy the relation

$$\sum_{i=1}^{n} c_i f_i(x) = 0 \text{ not all } c_i = 0 \tag{4.1}$$

then the set functions is said to be linearly dependent. If on the other hand, the only set of constants that will satisfy (4.1) is $c_i = 0$ for all i, then the set of functions are said to be linearly independent. In terms of only two functions $f(x)$ and $g(x)$ this implies that if $\dfrac{f(x)}{g(x)} = \text{constant}$ (4.2)

then $f(x)$ and $g(x)$ are linearly dependent. Conversely if $\dfrac{f(x)}{g(x)} =$ function of x then the two functions are linearly independent.

An easy method of testing the linear dependence of a set of functions is the Wronskian. The development of the test function proceeds as follows. Equation (4.1) is differentiated a total of $n-1$ times yielding the following system of homogeneous equations for the n unknown coefficients c_1

$$\begin{aligned} &c_1 f_1(x) + c_2 f_2(x) + \ldots\ldots + c_n f_n(x) = 0 \\ &c_1 f_1'(x) + c_2 f_2'(x) + \ldots\ldots + c_n f_n'(x) = 0 \\ &\ldots\ldots\ldots\ldots\ldots\ldots\ldots\ldots\ldots \\ &c_1 f_1^{(n-1)}(x) + c_2 f_2^{(n-1)}(x) + \ldots + c_n f_n^{n-1}(x) = 0 \end{aligned} \tag{4.3}$$

using matrix notation the system may be written as follows :

$$[W(x)]\{c\} = 0 \tag{4.3(a)}$$

$$\text{where } [w(x)] = \begin{bmatrix} f_1 & f_2 & \ldots\ldots & f_n \\ f_1' & f_2' & \ldots\ldots & f_n' \\ \ldots & \ldots & \ldots & \ldots \\ f_1^{n-1} & f_2^{n-1} & \ldots & f_n^{n-1} \end{bmatrix} \tag{4.4}$$

$$\text{and } \{c\} = \begin{Bmatrix} c_1 \\ c_2 \\ \cdot \\ \cdot \\ \vdots \\ c_n \end{Bmatrix}$$

It can be proved that a nontrivial solution for {c} exists only when $[w] = 0$. It may thus be stated that the set of functions $f_n(x)$ are linearly dependent whenever $[w] = 0$. The determinant $[w]$ is called the Wrönskian of the functions $f_n(x)$.

Completeness of Functions : A set of functions is said to be complete if an arbitrary function $f(x)$, satisfying the same boundary conditions as the functions of the set, can be expanded as follows :

$$f(x) = \sum_{n=1}^{\infty} a_n S_n(x) \tag{4.5}$$

a_n being constant coefficients. Completeness and orthogonality are the two important properties of eigen functions.

Fourier's Theorem : According to Fourier, if $f(x)$ is a periodic function of x, where the value of x extends from 0 to 2π or from $-\pi$ to π, then $f(x)$ can be expressed as an infinite series of the form

$$f(x) = a_0 + a_1 \cos x + a_2 \cos 2x + \ldots\ldots + a_n \cos nx$$
$$+ b_1 \sin x + b_2 \sin 2x + \ldots\ldots + b_n \sin nx \tag{4.5(a)}$$

where $a_0, a_1, a_2 \ldots\ldots a_n$,

$b_1, b_2 \ldots\ldots b_n$ are coefficients to be determined.

$f(x)$ can be expressed as the sum of such a series of sine and cosine terms provided the following conditions due to Dirichlet are observed :

(i) The function must be single valued.

(ii) In any range of 2π it must have at most a finite number of discontinuities, all of which discontinuities must themselves be finite.

(iii) There must be a finite number of maxima and minima in the range of 2π.

The various terms in the series are known as the harmonic terms. In general the nth harmonic will be represented by a term of the form $a_n \cos nx + b_n \sin nx$ or $a_n \cos nwt + b_n \sin nwt$ of frequency n times the frequency of the fundamental.

To valuate the coefficients of such a series, use is made of the following integrals :

$$\int_{-\pi}^{+\pi} \cos nx\, dx = \int_{-\pi}^{+\pi} \sin nx\, dx = 0 \tag{4.6}$$

$$\int_{-\pi}^{+\pi} \cos mn \cos nx\, dx = \int_{-\pi}^{+\pi} \sin mx \sin nx\, dx = 0 \text{ if } m \neq n$$

$$= \pi \text{ if } m = n \tag{4.7}$$

$$\int_{-\pi}^{+\pi} \sin mx \cos nx\, dx = 0 \tag{4.8}$$

where m and n are integers not equal to zero. The same results apply for integration over any range of 2π i.e. $\int_{\alpha}^{\alpha+2\pi}$ in general.

To find a_0 in equation (4.5(a)) let us integrate both sides from $-\pi$ to π

$$\therefore \int_{-\pi}^{\pi} f(x)\, dx = \int_{-\pi}^{\pi} a_0\, dx = 2\pi a_0$$

$$\therefore a_0 = \frac{1}{2\pi} \int_{-\pi}^{+\pi} f(x)\, dx \tag{4.9}$$

To find a_n multiply equation (4.5(a)) through by cos nx and integrate from $-\pi$ to π

$$\therefore \int_{-\pi}^{\pi} f(x) \cos nx\, dx = \int_{-\pi}^{\pi} a_0 \cos nx\, dx + \int_{-\pi}^{\pi} a_1 \cos x \cos nx + \ldots..$$

$$+ \int_{-\pi}^{\pi} b_1 \sin x \cos nx\, dx + \ldots..$$

It is evident from what has been stated above that all the terms on the R. H. S. will be zero except $\int_{-\pi}^{\pi} a_n \cos^2 nx\, dx = a_n \pi$

$$\therefore a_n = \frac{1}{\pi} \int_{-\pi}^{\pi} f(x) \cos nx\, dx \tag{4.10}$$

In a similar manner

$$b_n = \frac{1}{\pi} \int_{-\pi}^{\pi} f(x) \sin nx\, dx \tag{4.11}$$

It is to be noted that the general equation (4.10) **does not give the equation** (4.9) for a_0 on putting $n = 0$, but gives

$$a_0 = \frac{1}{\pi} \int_{-\pi}^{\pi} f(x)\, dx.$$

To ensure consistancy here, it is a simple matter, therefore, to put $\frac{a_0}{2}$ as the first term of the series. Thus the Fourier series is

$$f(x) = \frac{a_0}{2} + \sum_{n=1}^{\infty} (a_n \cos nx + b_n \sin nx) \tag{4.12}$$

where a_n and b_n are given by the equations (4.10) and (4.11) respectively.

Analysis of simple waveforms by Fourier's series :

(a) Saw-tooth curve : In this case the function $y = f(x)$ varies linearly with x over a range $-\pi$ to π or 0 to 2π. The variation can be represented by $y = kx$, where k is a constant (Fig. 4.1). The equation is that of a straight line. At the extreme ends of this range, however, there are sudden discontinuties. Hence the given function is not periodic over the whole range and as such can not be expanded in a Fourier series for all values of x. However, $y = kx$ can be expanded in a series within a range of 2π. Hence harmonic analysis of the waveform is possible in spite of the fact that $f(x)$ is not periodic. Let us try to evaluate the Fourier coefficients of the series in range $-\pi$ to $+\pi$.

Fig. 4.1

Thus using equation (4.12)

$$kx = \frac{1}{2} a_0 + \sum_{n=1}^{\infty} (a_n \cos nx + b_n \sin nx)$$

where $a_n = \frac{1}{\pi} \int_{-\pi}^{\pi} f(x) \cos nx\, dx = \frac{1}{\pi} \int_{-\pi}^{\pi} kx \cos nx\, dx$

$$= \frac{k}{\pi} \left[\frac{x \sin nx}{n} \right]_{-\pi}^{\pi} - \frac{k}{\pi} \int_{-\pi}^{\pi} \frac{\sin nx}{n}\, dx = 0 \text{ for all values of } x$$

$$b_n = \frac{1}{\pi} \int_{-\pi}^{\pi} kx \sin nx\, dx = \frac{k}{\pi} \left[-\frac{x \cos nx}{n} \right]_{-\pi}^{\pi} + \frac{k}{\pi} \int_{-\pi}^{\pi} \frac{\cos nx}{n}\, dx$$

$$= \frac{k}{\pi}\left\{-\frac{\pi \cos n\pi}{\pi} + \frac{\pi \cos(-n\pi)}{\pi}\right\} + 0$$

$$= -\frac{2k}{\pi}\cos n\pi$$

If n is even $b_n = \frac{-2k}{n}\cos n\pi = \frac{-2k}{\pi}$

If n is odd $b_n = \frac{2k}{\pi}$

Hence $f(x) = kx = 2k\sin x - \frac{2k}{2}\sin 2x + \frac{2k}{3}\sin 3x - \frac{2k}{4}\sin 4x + \ldots$

$$= 2k\left\{\sin x - \frac{\sin 2x}{2} + \frac{\sin 3x}{3} - \frac{\sin 4x}{4} + \ldots\right\} \tag{4.13}$$

Fig. 4.2 illustrate the harmonic components in the above equation in the range $-\pi$ to π.

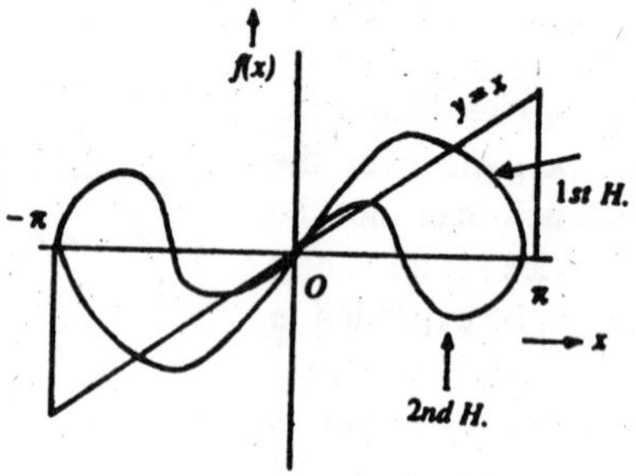

Fig. 4.2

(b) Find the Fourier series for the periodic function $f(x)$ with the period 2π in the interval $-\pi < x < \pi$. Such that

$$f(x) = \begin{cases} -k \text{ for } -\pi < x < 0 \\ k \text{ for } 0 < x < \pi \end{cases}$$

and deduce that $\pi/4 = 1 - \frac{1}{3} + \frac{1}{5} - \frac{1}{7} + \ldots$

The Fourier series $f(x)$ is given by

$$f(x) = \frac{a_0}{2} + \sum_{n=1}^{\infty}(a_n \cos nx + b_n \sin nx)$$

$$a_n = \frac{1}{\pi}\int_{-\pi}^{0} -k\cos nx\,dx + \frac{1}{\pi}\int_{0}^{\pi} k\cos nx\,dx = 0$$

$$\therefore\ a_0 = 0.$$

$$b_n = \frac{1}{\pi}\int_{-\pi}^{0} -k \sin nx\, dx + \frac{1}{\pi}\int_{0}^{\pi} k \sin nx\, dx$$

$$= \frac{2k}{n\pi}\{1 - \cos n\pi\} \quad \begin{cases} = 0 \text{ for } n \text{ even} \\ = \dfrac{4k}{n\pi} \text{ for } n \text{ odd} \end{cases}$$

Hence $$f(x) = \frac{4k}{\pi}\left\{\sin x + \frac{\sin 3x}{3} + \frac{\sin 5x}{5} + \ldots\ldots\right\} \tag{4.14}$$

Putting $x = \pi/2$ in equation (4.14)

$$f(\pi/2) = k = \frac{4k}{\pi}\left\{1 - \frac{1}{3} + \frac{1}{5} - \frac{1}{7} + \ldots\ldots\right\}$$

or, $$\frac{\pi}{4} = 1 - \frac{1}{3} + \frac{1}{5} - \frac{1}{7} + \ldots\ldots$$

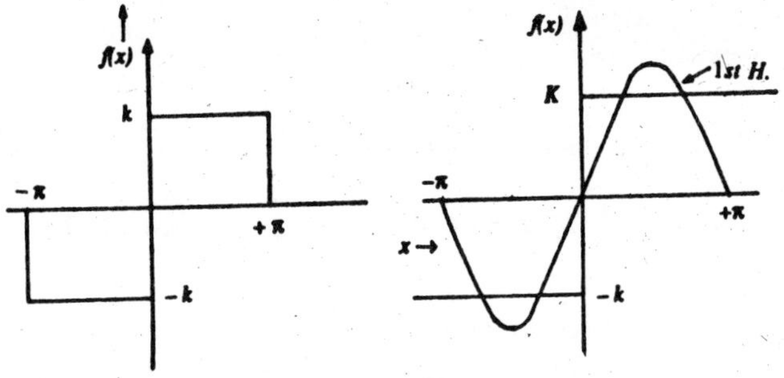

Fig. 4.3

(c) **Waveform of the half wave rectifier :** The output current of a half wave rectifier can be represented as $i = i_0 \sin \omega t$ in the first half cycle from $\omega t = 0$ to π and during the subsequent half cycle from π to 2π, $i = 0$

writing $\omega t = \theta$ for x, we have the Fourier series as

$$i = \frac{1}{2}a_0 + \sum_{n=1}^{\infty}(a_n \cos n\theta + b_n \sin n\theta) \tag{4.15}$$

From equation (4.10)

$$a_n = \frac{1}{\pi}\int_0^{2\pi} i_0 \sin\theta \cos n\theta\, d\theta.$$

$$= \frac{i_0}{2\pi}\int_0^{\pi}\{\sin(n+1)\theta - \sin(n-1)\theta\}\, d\theta, \quad \because i = 0 \text{ from } \theta = \pi \text{ to } 2\pi$$

Putting $n = 0$

$$a_0 = \frac{i_0}{\pi} \int_0^\pi \sin\theta \, d\theta = \frac{i_0}{\pi} \left[-\cos\theta \right]_0^\pi = \frac{2 i_0}{\pi} \tag{4.16}$$

Also $$a_n = \frac{i_0}{2\pi} \left[\frac{-\cos(n+1)\theta}{n+1} + \frac{\cos(n-1)\theta}{n-1} \right]_0^\pi$$

$$= \frac{i_0}{2\pi} \left\{ \frac{1-\cos(n+1)\pi}{n+1} + \frac{\cos(n-1)\pi - 1}{n-1} \right\} \tag{4.17}$$

when n is odd $(n+1)$ and $(n-1)$ are both even

$\therefore \; a_n = 0.$

This is however inapplicable for $n = 1$, because in that case the 2nd term of equation (4.17) becomes indeterminate.

However, when $n = 1$

$$a_1 = \frac{i_0}{\pi} \int_0^\pi \sin\theta \cos\theta \, d\theta = \frac{i_0}{2\pi} \int_0^\pi \sin 2\theta \, d\theta = 0$$

If n is even $\cos(n+1)\pi = \cos(n-1)\pi = 0$

$$\therefore \; a_n = \frac{i_0}{2\pi} \left\{ \frac{1}{n+1} - \frac{1}{n-1} \right\} = \frac{-i_0}{2(n-1)(n+1)\pi} \tag{4.18}$$

Again, $$b_n = \frac{1}{\pi} \int_0^{2\pi} i_0 \sin\theta \sin n\theta \, d\theta = \frac{i_0}{\pi} \int_0^\pi \sin\theta \sin n\theta \, d\theta$$

$$= \frac{i_0}{2\pi} \int_0^\pi \{\cos(n-1)\theta - \cos(n+1)\theta\} \, d\theta$$

$$= \frac{i_0}{2\pi} \left[\frac{\sin(n-1)\theta}{n-1} - \frac{\sin(n+1)\theta}{n+1} \right]_0^\pi \tag{4.19}$$

For $n \neq 1$, the above expression is zero.

For $n = 1$, the first term becomes indeterminate and the expression is invalid.

However b_1 is given by

$$b_1 = \frac{i_0}{\pi} \int_0^\pi \sin^2\theta \, d\theta = \frac{i_0}{\pi} \int_0^\pi \left(\frac{1-\cos 2\theta}{2} \right) d\theta$$

$$= \frac{i_0}{2\pi} \left[\theta - \frac{\sin 2\theta}{2} \right]_0^\pi = \frac{i_0}{2}$$

Hence the Forier series is

$$i = \frac{a_0}{2} + \sum_{n=1}^{\infty} (a_n \cos n\theta + b_n \sin n\theta)$$

$$= \frac{i_0}{\pi} + \frac{i_0 \sin\theta}{2} - \frac{i_0 \cos 2\theta}{1.3\pi} - \frac{i_0 \cos 4\theta}{3.5\pi} - \frac{i_0 \cos 6\theta}{5.7\pi}$$

$$= \frac{i_0}{\pi} \left\{1 + \frac{\pi \sin \omega t}{2} - \frac{\cos 2\,\omega t}{1.3} - \frac{\cos 4\,\omega t}{3.5} - \frac{\cos 6\,\omega t}{5.7} \cdots\cdots \right\} \quad (4.20)$$

The resultant is equivalent to a direct current $\frac{i_0}{\pi}$ together with alternating currents having frequencies $\frac{\omega}{2\pi}, \frac{2\omega}{2\pi}, \frac{4\omega}{2\pi}$, etc. and of amplitudes $\frac{i_0}{2\pi}, -\frac{i_0}{3\pi}, -\frac{i_0}{15\pi}, -\frac{i_0}{35\pi}$ etc.

(d) **Waveform of the full wave rectifier :** The output current of a full wave rectifier can be represented as

$$f(t) = i = i_0 \sin \omega t \qquad 0 < \omega t < \pi$$

$$= -\, i_0 \sin \omega t \qquad -\pi < \omega t < 0$$

Writing $\omega t = \theta$. The Fourier series can be written as

$$f(t) = \frac{a_0}{2} + \sum_{n=1}^{\infty} (a_n \sin n\theta + b_n \cos n\theta)$$

From equation (4.10)

$$a_n = \frac{1}{\pi} \int_0^{\pi} i_0 \sin \theta \cos n\theta \, d\theta + \frac{1}{\pi} \int_{-\pi}^{0} - i_0 \sin \theta \cos n\theta \, d\theta$$

$$= \frac{2}{\pi} \int_0^{\pi} i_0 \sin \theta \cos n\theta \, d\theta = \frac{i_0}{\pi} \int_0^{\pi} \{ \sin (n+1)\theta - \sin (n-1)\theta \} \, d\theta$$

Putting $n = 0$

$$a_0 = \frac{i_0}{\pi} \int_0^{\pi} 2 \sin \theta \, d\theta = \frac{2i_0}{\pi} [-\cos \theta] = \frac{4i_0}{\pi} \quad (4.21)$$

Now $a_n = \frac{i_0}{\pi} \left[-\frac{\cos (n+1)\theta}{n+1} + \frac{\cos (n-1)\theta}{n-1} \right]_0^{\pi}$

$$= \frac{i_0}{\pi} \left\{ \frac{2}{n+1} - \frac{2}{n-1} \right\} \text{ for } n \text{ even}$$

$= 0$ for n odd

Simplifying $a_n = -\dfrac{4}{\pi(n^2-1)}$ when n is even (4.22)

and $a_n = 0$ when n is odd.

In a similar manner.

$$b_n = \frac{1}{\pi}\int_0^{\pi} i_0 \sin\theta \sin n\theta \, d\theta + \frac{1}{\pi}\int_{-\pi}^{0} - i_0 \sin\theta \sin n\theta \, d\theta = 0.$$

Thus $$f(t) = \frac{2i_0}{\pi} - \frac{4i_0}{\pi}\left\{\frac{\cos 2\omega t}{3} + \frac{\cos 4\omega t}{15} + \ldots\ldots\right\} \quad (4.23)$$

It is to be noted that the lowest angular frequency present is 2ω. The higher frequency components fall off as $(n^2 - 1)$ showing that the full wave rectifier does a fairly good job of producing direct current.

(e) Express $f(x) = x$ as a Fourier series in the interval $-\pi < x < \pi$.

[C. U. 95]

Let $$f(x) = \frac{a_0}{2} + \sum_{n=1}^{\infty} (a_n \cos nx + b_n \sin nx) \, dx$$

From equation (4.10) $$a_n = \frac{1}{\pi}\int_{-\pi}^{\pi} x \cos nx \, dx$$

$$= \frac{1}{\pi}\left\{\left[x\frac{\sin nx}{n}\right]_{-\pi}^{\pi} - \int_{-\pi}^{\pi}\frac{\sin nx}{n}dx\right\} = \frac{1}{\pi}\left[\frac{\cos nx}{n^2}\right]_{-\pi}^{\pi} = 0$$

$$a_0 = \frac{1}{\pi}\int_{-\pi}^{\pi} x \, dx = \frac{1}{\pi}\left[\frac{x^2}{2}\right]_{-\pi}^{\pi} = 0$$

$$b_n = \frac{1}{\pi}\int_{-\pi}^{\pi} x \sin nx \, dx = \frac{1}{\pi}\left\{\left[-x\frac{\cos nx}{n}\right]_{-\pi}^{\pi} + \int_{-\pi}^{\pi}\frac{\cos nx}{n}dx\right\}$$

$$= \frac{1}{\pi}\left\{\frac{-\pi\cos n\pi - \pi\cos n\pi}{n}\right\} + 0 = -\frac{2\cos n\pi}{n}$$

Hence $$f(x) = \sum_{n=1}^{\infty} -\frac{2\cos n\pi}{n}\sin nx$$

$$= 2\left\{\frac{1}{1}\sin x - \frac{1}{2}\sin 2x + \frac{1}{3}\sin 3x + \ldots\ldots\right\}.$$

(f) Show that if A is a constant, then in the region $0 < \theta < \pi$

$$A = \frac{4A}{\pi}\left(\sin\theta + \frac{\sin 3\theta}{3} + \frac{\sin 5\theta}{5} + \ldots\ldots\right)$$ [C. U. 1996]

$f(\theta) = A$ for $0 \le \theta < \pi$.

Since we are to apply Fourier's theorem for a complete period, let us assume for convenience that $f(\theta) = -A$ for $-\pi < \theta < 0$.

The Fourier series $f(\theta)$ is given by

$$f(\theta) = \frac{a_0}{2} + \sum_{n=1}^{\infty}(a_n \cos n\theta + b_n \sin n\theta)$$

$$a_0 = \frac{1}{\pi}\left\{-A\int_{-\pi}^{0} d\theta + A\int_{0}^{\pi} d\theta\right\} = 0$$

$$a_n = \frac{1}{\pi}\left\{\int_{-\pi}^{0} -A\cos n\theta\, d\theta + \int_{0}^{\pi} A\cos n\theta\, d\theta\right\} = 0$$

$$b_n = \frac{1}{\pi}\left\{\int_{-\pi}^{0} -A\sin n\theta\, d\theta + \int_{0}^{\pi} A\sin n\theta\, d\theta\right\}$$

$$= \frac{1}{\pi}\left\{\left[\frac{A\cos n\theta}{n}\right]_{-\pi}^{0} - \left[\frac{A\cos n\theta}{n}\right]_{0}^{\pi}\right\}$$

$$= \frac{1}{\pi}\left\{\frac{A - A\cos n\pi}{n} - \frac{A\cos n\pi + A}{n}\right\}$$

$= 0$ where n is even

$= \dfrac{4A}{\pi n}$ when n is odd.

Thus $$f(\theta) = A = \frac{4A}{\pi}\left\{\sin\theta + \frac{\sin 3\theta}{3} + \frac{\sin 5\theta}{5} + \ldots\ldots\right\}$$

ADDITIONAL EXERCISE

1. Assuming $f(x) = x^4$, for $-\pi < x < \pi$,

$$\text{prove that } \frac{\pi^2}{12} = \sum_{n=1}^{\infty} \frac{(-1)^{n+1}}{\pi^2}.$$

2. Express $f(x)$ in Fourier series in the interval $(-2, 2)$ if $f(x) = 0$ for $-2 < x < 0$ and $f(x) = 1$ for $0 < x < 2$.

In this case $a_0 = \frac{1}{4}\int_{-2}^{2} f(x)\, dx = \frac{1}{4}\left\{\int_{-2}^{0} 0\, dx + \int_{0}^{2} 1\, dx\right\} = \frac{1}{2}$

$$a_n = \frac{1}{2}\left\{\int_{-2}^{2} f(x) \cos \frac{n\pi x}{2} \cdot dx\right\}$$

$$= \frac{1}{2}\left\{\int_{-2}^{0} 0 \times \cos \frac{n\pi x}{2}\, dx + \int_{0}^{2} 1 \times \cos \frac{n\pi x}{2}\, dx\right\}$$

$$= \frac{1}{n\pi}\left[\sin \frac{n\pi x}{2}\right]_0^2 = 0$$

Similarly $b_n = \frac{1}{2}\int_{-2}^{2} f(x) \sin \frac{n\pi x}{2}\, dx$

$$= \frac{1}{2}\left\{\int_{-2}^{0} 0 \times \sin \frac{n\pi x}{2}\, dx + \int_{0}^{2} 1 \times \sin \frac{n\pi x}{2}\, dx\right\}$$

$$= \frac{1}{n\pi}\{1 - \cos n\pi\} = 0 \text{ if } n \text{ is even}$$

$$= \frac{2}{n\pi} \text{ if } n \text{ is odd}$$

Hence $f(x) = \frac{1}{2} + \frac{2}{\pi}\left\{\sin \frac{\pi x}{2} + \frac{1}{3} \sin \frac{3\pi x}{2} + \ldots\ldots\right\}$

3. A displacement curve is defined as

$$y = \frac{4at}{T} \text{ when } 0 \le t \le \frac{T}{4}$$

$$y = 2a - \frac{4at}{T} \text{ when } \frac{T}{4} < t < \frac{3T}{4}$$

$$y = \frac{4at}{T} - 4a \text{ when } \frac{3T}{4} \le t \le T.$$

Find the amplitude of the fundamental and the first two harmonics.

Let $y = \frac{a_0}{2} + \sum_n \left(a_n \cos n\omega t + b_n \sin n\omega t\right)$

Now

$$\frac{a_0}{2} = \frac{1}{T}\int_0^T y\,dt = \frac{1}{T}\left\{\int_0^{T/4} \frac{4at}{T}\,dt + \int_{T/4}^{3T/4}\left(2a - \frac{4at}{T}\right)dt + \int_{\frac{3T}{4}}^{T}\left(\frac{4at}{T} - 4a\right)dt\right\}$$

$$= \frac{1}{T}\left\{\frac{4a}{T}\frac{T^2}{32} + 2a\left(\frac{3T}{4} - \frac{T}{4}\right) - \frac{4a}{T}\left(\frac{9T^2}{32} - \frac{T^2}{32}\right)\right.$$

$$\left. + \frac{4a}{T}\left(\frac{T^2}{2} - \frac{9T^2}{32}\right) - 4a\left(T - \frac{3T}{4}\right)\right\}$$

$$= \frac{1}{T}\left\{\frac{aT}{8} + 2a \cdot \frac{T}{2} - \frac{4a}{T}\frac{8T^2}{32} + \frac{4a}{T}\frac{7T^2}{32} - 4a \cdot \frac{T}{4}\right\}$$

$$= \left\{\frac{a}{8} - a + \frac{7}{8}a\right\} = 0$$

$$a_n = \frac{1}{T}\int_0^T y\cos n\omega t\,dt = \frac{1}{T}\left\{\int_0^{T/4}\frac{4at}{T}\cos n\omega t\,dt\right.$$

$$\left. + \int_{T/4}^{3T/4}\left(2a - \frac{4at}{T}\right)\cos n\omega t\,dt + \int_{3T/4}^{T}\left(\frac{4at}{T} - 4a\right)\cos n\omega t\,dt\right\}$$

Similarly $b_n = \frac{1}{T}\int_0^T y \sin n\omega t\,dt$

$$\left[\textbf{Ans. } \frac{8a}{\pi^2}, \frac{8a}{9\pi^2}; \frac{8a}{25\pi^2}\right]$$

4. A note on Fourier transform :

As usual we have

$$f(x) = \frac{a_0}{2} + \sum_{n=1}^{\infty}\left(a_n \cos\frac{n\pi x}{l} + b_n \sin\frac{n\pi x}{l}\right)$$

where $a_n = \frac{1}{l}\int_{-l}^{l} f(x)\cos\frac{n\pi x}{l}\,dx$ and

$$b_n = \frac{1}{l}\int_{-l}^{l} f(x)\sin\frac{n\pi x}{l}\,dx.$$

Substituting for a_n and b_n in the above expansion.

$$f(x) = \frac{1}{2l}\int_{-l}^{l} f(x')\,dx' + \frac{1}{l}\sum_{n=1}^{\infty}\cos\frac{n\pi x}{l}\int_{-l}^{l} f(x')\cos\frac{n\pi x'}{l}\,dx'$$

$$+ \frac{1}{l}\sum_{n=1}^{\infty}\sin\frac{n\pi x}{l}\int_{-l}^{l} f(x')\sin\frac{n\pi x'}{l}\,dx'$$

$$= \frac{1}{2l}\int_{-l}^{l} f(x')\,dx' + \frac{1}{l}\sum_{n=1}^{\infty}\int_{-l}^{l} f(x')\cos\frac{n\pi}{l}(x - x')\,dx' \quad \ldots \text{(i)}$$

Writing $\frac{n\pi}{l} = k$, so that $\delta k = \frac{\pi}{l}\,\delta n = \frac{\pi}{l}$,

$$\text{if } \left[\int_{-\infty}^{\infty} f(x')\,dx'\right] \text{ is finite.}$$

Then as $l \to \infty$

$$f(x) = \frac{1}{\pi}\int_{0}^{\infty} dk \int_{-\infty}^{\infty} f(x')\cos k(x - x')\,dx'$$

$$= \frac{1}{2\pi}\int_{-\pi}^{\pi} dk \int_{-\infty}^{\infty} f(x')\cos k(x - x')\,dx' \quad \ldots \text{(ii)}$$

Equation (ii) is known as Fourier Integral. This equation will be valid if Dirichlet conditions and $\left[\int_{-\infty}^{\infty} f(x)\,dx\right] < \infty$

Again as $K(x - x')$ is an odd function of K,

$$\frac{1}{2\pi}\int_{-\infty}^{\infty} dk \int_{-\infty}^{\infty} f(x')\sin k(x - x')\,dx' = 0 \quad \ldots \text{(iii)}$$

From (ii) and (iii) we have

$$f(x) = \frac{1}{2\pi}\int_{-\infty}^{\infty} dk \int_{-\infty}^{\infty} f(x')\,e^{ik(n - x')}\,dx'$$

$$= \frac{1}{\sqrt{2\pi}}\int_{-\infty}^{\infty} dk\, e^{ikn}\left\{\frac{1}{\sqrt{2\pi}}\int_{-\infty}^{\infty} f(x')\,e^{-ikx'}\,dx'\right.$$

$$= \frac{1}{\sqrt{2\pi}}\int_{-\infty}^{\infty} g(k)\,e^{ikx}\,dk \quad \ldots \text{(iv)}$$

where $$g(k) = \frac{1}{\sqrt{2\pi}} \int_{-\infty}^{\infty} f(x)\, e^{ikx}\, dx \qquad \ldots \quad \text{(v)}$$

and is called the Fourier transform of $f(x)$.

In symbols the above can be written as

$$g(k) = F\{f(x)\} \qquad \ldots \quad \text{(vi)}$$

By inverse transformation we have

$$f(x) = F^{-1}\{g(k)\} \qquad \ldots \quad \text{(vii)}$$

Example :

Let us find the Fourier transform of the function

$f(x) = \{1 - x^2\}$ for $|x| < 1$

$\quad 0 \quad$ „ $|x| > 1$

$$g(x) = \frac{1}{\sqrt{2\pi}} \int_{-1}^{1} (1 - x^2)\, e^{-ikx}\, dx$$

$$= \frac{1}{\sqrt{2\pi}} \left\{ \left[(1 - x^2) \frac{e^{ikx}}{ik} \right]_{-1}^{1} + \frac{2}{ik} \int_{-1}^{1} x e^{ikx}\, dx \right\}$$

$$= \frac{1}{\sqrt{2\pi}} \left\{ -\frac{4}{k^3} (k \cos k - \sin k) \right\}$$

CHAPTER V

MATRIX

Addition, Subtraction, Multiplication—non-commutivity, Hermitian and Unitary matrices, Diagonalization, Eigen value and Eigen vectors for real symmetric matrix.

5.1. A matrix is an arrangement of numbers in the form shown below

$$[A] = \begin{bmatrix} a_{11} & a_{12} \cdots\cdots & a_{1n} \\ a_{21} & a_{22} \cdots\cdots & a_{2n} \\ \cdots & \cdots\cdots & \cdots \\ \cdots & \cdots\cdots & \cdots \\ a_{n1} & a_{n2} \cdots\cdots & a_{nn} \end{bmatrix} \quad (5.1)$$

Each number, which can be either real or complex is called an element. The element a_{ij} belongs to the *i*th row and *j*th column. If the matrix has *m* rows and *n* columns it is called an $m \times n$ matrix. A matrix with $n = 1$ consists of a single column and is called a column matrix. On the other hand a matrix with $m = 1$ has a single row and is called a row matrix.

Bisides the square arrangement like (5.1) we may have also rectangular arrangement of matrices of *m* rows and *n* colums. Where there is only one row i.e., $m = 1$, the matrix will be termed a vector of the first kind or a prime. For example we have $[A] = [a_{11} \quad a_{12} \cdots\cdots a_{1n}]$. (5.2)

If there is a single column of *n* elements, the matrix will be termed a vector of the second kind or a point.

$$\text{thus } [B] = \begin{bmatrix} b_{11} \\ b_{21} \\ \cdots \\ \cdots \\ b_{n1} \end{bmatrix} \quad (5.3)$$

5.2. Transposition of Matrices

The matrix [A′] = [aji] obtained by a complete interchange of columns and rows of [A] is called the transposed matrix of [A]. The *i*th row of [A] is the same as *i*th column of [A′].

For rectangular matrices the transpose of

$$\begin{bmatrix} a_{11} & a_{12} & a_{13} \\ a_{21} & a_{22} & a_{23} \end{bmatrix} \text{ is} = \begin{bmatrix} a_{11} & a_{21} \\ a_{12} & a_{22} \\ a_{13} & a_{23} \end{bmatrix} \quad (5.4)$$

5.3. Different types of Matrices

Square Matrix : If the number of rows and columns of a matrix are equal to m, then such a matrix is said to be a square matrix of order m or simply a square of order m.

Diagonal Matrix : If all the elements other than those in the principal diagonal are zero, then the matrix is called a diagonal matrix

$$\begin{bmatrix} a_{11} & 0 \ldots\ldots 0 \\ 0 & a_{22} \quad 0 \quad 0 \\ \ldots\ldots\ldots\ldots \\ 0 & 0 \ldots\ldots a_{nn} \end{bmatrix}$$

Unit Matrix : The unit matrix of order n is defined to be diagonal matrix of order n which has all its diagonal elements unity. For example—

$$\begin{bmatrix} 1 & .0 \ldots\ldots 0 \\ 0 & 1 \quad 0 \ldots \\ \ldots\ldots\ldots\ldots \\ 0 & 0 \quad 0 \ldots 1 \end{bmatrix}$$

Null Matrix : A null matrix has all its elements zero. It is represented by [0].

Symmetrical Matrix : If a matrix is such that $a_{ij} = a_{ji}$ for all i and j the matrix is said to be symmetrical. In this case the matrix is identical with its transposed one i.e. [A] = [A′].

Skew-symmetrical Matrix : If $a_{ij} = -a_{ji}$ for all unequal i and j but the elements a_{ij} are not all zero, the matrix is termed a skew matrix. On the other hand if $a_{ij} = -a_{ji}$ for all i and j so that all $a_{ii} = 0$, the matrix is called a skew symmetric matrix. It is to be noted that both symmetrical and skew matrices are necessarily square.

5.4. Equality of Matrices

Two matrices [A] and [B] of the same order are defined to be equal if and only if their corresponding elements are identical, in other words

$$a_{ij} = b_{ij} \text{ for all } i \text{ and } j.$$

5.5. Addition of Matrices (Addition and Subtraction)

If [A] and [B] are matrices of the same order, then the sum of [A] and [B] is defined to be a matrix [C], the typical element of which is

$c_{ij} = a_{ij} + b_{ij}$. In other words

$$[C] = [A] + [B] \tag{5.5}$$

Provided $c_{ij} = a_{ij} + b_{ij}$ (5.6)

In a similar manner

$$[D] = [A] - [B] \text{ provided} \tag{5.7}$$

$$d_{ij} = a_{ij} - b_{ij} \tag{5.8}$$

If $[A] = \begin{bmatrix} a_{11} & a_{12} & a_{13} \\ a_{21} & a_{22} & a_{23} \end{bmatrix}$ and $[B] = \begin{bmatrix} b_{11} & b_{12} & b_{13} \\ b_{21} & b_{22} & b_{23} \end{bmatrix}$

then $[C] = \begin{bmatrix} a_{11} + b_{11} & a_{12} + b_{12} & a_{13} + b_{13} \\ a_{21} + b_{21} & a_{22} + b_{22} & a_{23} + b_{23} \end{bmatrix}$

Similarly $[D] = \begin{bmatrix} a_{11} - b_{11} & a_{12} - b_{12} & a_{13} - b_{13} \\ a_{21} - b_{21} & a_{22} - b_{22} & a_{23} - b_{23} \end{bmatrix}$

Properties of matrix additon :

(a) Matrix addition obeys the commutative law :

i.e. $[A] + [B] = [B] + [A]$

Let $[A] = [a_{ij}]_{m \times n}$ and $[B] = [b_{ij}]_{m \times n}$

Then $[A] + [B] = [a_{ij} + b_{ij}]_{m \times n}$ and $[B] + \{A] = [b_{ij} + a_{ij}]_{m \times n}$

Hence the result.

(b) Matrix addition obeys the associative law :

i.e. $[A] + ([B] + [C]) = ([A] + [B]) + [C]$ with previous notation.

$$([A] + [B]) + [C] = \{(a_{ij})_{m \times n} + (b_{ij})_{m \times n}\} + (c_{ij})_{m \times n}$$

$$= (a_{ij})_{m \times n} + \{(b_{ij})_{m \times n} + (c_{ij})_{m \times n}\}$$

Hence $([A] + [B]) + [C] = [A] + ([B] + [C])$

(c) Identity exists in matrix addition :

i.e. $[A] + [0] = [0] + [A] = [A]$, where $[0]$ is the null matrix comparable to $[A]$.

Since $a_{ij} + 0 = 0 + a_{ij}$, the above result follows.

Let us consider the matrix equation

$[A] + [B] = 0$ where $[B]$ is comparable to $[A]$.

This has got a unique solution

$[B] = (-a_{ij})_{m \times n}$ i.e. [B] is a matrix; each element of which is the negative of the corresponding element of [A]. The matrix [B] is called the additive inverse of [A] or the negative of [A] and is represented by the symbol − [A].

(d) The following laws also hold good

$$c\{[A + B]\} = c[A] + c[B]$$

$$(a + b)[A] = a[A] + b[A]$$

$$a(b[A]) = (ab)[A]$$

5.6. Matrix Multiplication

The idea of matrix multiplication is somewhat different from ordinary scalar multiplication. Two matrices can only be multiplied when the number of columns of the first is equal to the number rows of the second. Matrices that satisfy this condition are termed conformable matrices.

The product of a matrix [A] by a matrix [B] is defined by the equation

$$[A]\,[B] = [C] \tag{5.9}$$

$$\text{where } c_{ij} = \sum_{k=1}^{k=p} a_{ik}\, b_{kj} \tag{5.10}$$

and the order of the matrices [A], [B] and [C] are $(m \times p)$, $(p \times n)$ and $(m \times n)$ respectively.

To have a clear idea, let us consider the following example.

$$\text{let } [A] = \begin{bmatrix} a_{11} & a_{12} & a_{13} \\ a_{21} & a_{22} & a_{23} \\ a_{31} & a_{32} & a_{33} \end{bmatrix} \text{ and } [B] = \begin{bmatrix} b_{11} & b_{12} \\ b_{21} & b_{22} \\ b_{31} & b_{32} \end{bmatrix}$$

Applying the given definition we have

$$[A]\,[B] = \begin{bmatrix} a_{11} b_{11} + a_{12} b_{21} + a_{13} b_{31} & a_{11} b_{12} + a_{12} b_{22} + a_{13} b_{32} \\ a_{21} b_{11} + a_{22} b_{21} + a_{23} b_{31} & a_{21} b_{12} + a_{22} b_{22} + a_{23} b_{32} \\ a_{31} b_{11} + a_{32} b_{21} + a_{33} b_{31} & a_{31} b_{12} + a_{32} b_{22} + a_{33} b_{32} \end{bmatrix} \tag{5.11}$$

If the matrices are square and each of order n, then the corresponding relation [A] [B] = [C] is true for the determinants of the matrices [A], [B] and [C].

Since the matrices are regarded equal only when they are element for element identical, it follows that, since a row by column rule will, in general, give different elements than a column by row rule, the product [B] [A], when it exists, is usually different from [A] [B]. Hence it is necessary to distinguish between the premultiplication, as when [B] is premultiplied by [A] to give the result [A] [B], and postmultiplication, as when [B] is postmultiplied by [A] to give the result [B] [A]. If we have the equality [A] [B] = [B] [A], (5.12) the matrices [A] and [B] are said to commute or to be permutable. The unit matrix (U), it may be noted, commutes with any square matrix of the same order. That is, we have for all such [A],

$$[A]\,[U] = [U]\,[A] = [A] \tag{5.13}$$

Properties of matrix multiplication :

(a) Multiplication of matrices is associative :

This means that [A] . ([B] [C]) = ([A] [B]) . [C]

where [A], [B], [C] are $m \times n$, $n \times p$ and $p \times q$ matrices respectively.

Proof : Let [A] = (a_{ij}) ; [B] = (b_{jk}) and [C] = (c_{kl})

where $i = 1, 2, \ldots\ldots. m$, $j = 1, 2, \ldots\ldots n$,

$k = 1, 2 \ldots\ldots p$ and $l = 1, 2, \ldots\ldots q$.

Let us obtain the (i, l)th element of each of the two products, [A] ([B] [C]) and ([A] [B]) [C]

By the product [B] [C] of $(n \times p)$ matrix [B] = b_{ik} and $(p \times q)$ matrix [C] = c_{kl} is meant $n \times q$ matrix [D] = d_{jl}, where d_{jl} = (j, l)th element of the product [B] [C] = $\sum_{k=l}^{p} b_{jk}\, c_{kl}$.

For different values of j, this will give the different elements in the lth column of [B] [C]. Now [A] is an $m \times n$ matrix and [D] is $n \times q$ matrix.

Hence [A] ([B] [C]) = [A] [D] will be $m \times q$ matrix. Let [A] ([B] [C]) = (w_{il}), where w_{il} is the (i, l)th element of [A] ([B] [C]). Then w_{il} is the sum of the products of the corresponding elements in the ith row of [A] and the lth column of [D] i.e.

$$[B]\,[C] = \sum_{j=l}^{n} a_{ij}\, d_{il} = \sum_{j=l}^{n} [a_{ij} \cdot \sum_{k=l}^{p} (b_{ik}\, c_{kl})] = \sum_{k=1}^{p} \left[\sum_{j=l}^{n} (a_{ij}\, c_{jk}) \right] b_{kl} \tag{5.14}$$

Again [A] [B] represents an $m \times p$

matrix [E] = (c_{ik}), where $c_{ik} = \sum_{j=l}^{n} a_{ij}\, b_{ik}$

Hence ([A] [B]) [C] denotes the $m \times q$ matrix [E] [C]

$$= [F] = (f_{il}), \text{ where } (f_{il}) = \sum_{k=1}^{p} e_{ik}\, c_{kl}$$

$$= \sum_{k=1}^{p} \left(\sum_{j=1}^{n} a_{ij}\, b_{jk} \right) c_{kl} \tag{5.15}$$

From (5.14) and (5.15), it is clear that

$$[A] ([B] [C]) = ([A] [B]) [C].$$

(b) Except in special cases matrix multiplication is not commutative :

The reason for multiplication of two matrices not commutative is that (i, j)th element of [A] [B] is different from the (i, j)th element of [B] [A] even if the orders of [A] [B] and [B] [A] are same.

For example, let [A] and [B] both be the matrices of order $n \times n$ so that [A] [B] and [B] [A] are of the same order $n \times n$. Then

$$(AB)_{ij} = \sum_{k} a_{ik}\, b_{kj}$$

$$\text{and } (BA)_{ij} = \sum_{k} b_{ik}\, a_{kj}$$

in general $a_{ik}\, b_{kj} \neq b_{ik}\, a_{kj}$

and hence [A] [B] $\neq$ [B] [A].

If the orders of [A] [B] and [B] [A] are different, the question of [A] [B] and [B] [A] being equal does not arise.

(c) Matrix multiplication follows the law of distribution :

Let us consider a matrix of [A] of order $m \times n$ and matrices [B] and [C] each of order $n \times p$, then if [A] ([B] + [C]) = [D]

$$\text{We have } (D)_{ij} = \sum_{k=1}^{n} [A]_{ik} ([B] + [C])_{kj}$$

Let $[A] = (a_{ij}), [B] = (b_{jk}), [C] = (c_{jk})$

where $1 \leq i \leq m,\ 1 \leq k \leq n,\ 1 \leq j \leq p$

$$\text{then } D_{ij} = \sum_{k=1}^{n} a_{ix} (b_{kj} + c_{kj})$$

$$= \sum_{k=1}^{n} a_{ik}\, b_{kj} + \sum_{k=1}^{n} a_{ik}\, c_{kj}$$

$$= [AB]_{ij} + [AC]_{ij}$$

$$= (AB + AC)_{ij}$$

or, $\{[A]([B]+[C])\}_{ij} = [AB]_{ij} + [AC]_{ij}$, for every i and j

Hence $[A]([B]+[C]) = [AB] + [AC]$ **(5.16)**

which is the law of distribution.

(d) **Direct product of Matrices :**

There is another way of multiplying the matrices, known as the direct product or tensor product defined as follows :

If [A] is a $(m \times m)$ matrix and [B] is a $(n \times n)$ matrix, then the direct product is

$$[A] \otimes [B] = [C] \quad (5.16(a))$$

where [C] is a $(mn \times mn)$ matrix with elements

$$C_{ik,\, jl} = [A]_{ij}\, [B]_{kl} \quad (5.17)$$

As an example let us consider the matrices

$$[A] = \begin{bmatrix} a_{11} & a_{12} \\ a_{21} & a_{22} \end{bmatrix} \text{ and } [B] = \begin{bmatrix} b_{11} & b_{12} \\ b_{21} & b_{22} \end{bmatrix}$$

each of order 2×2. The direct product of [A] and [B] is given by

$$[A] \times [B] = \begin{bmatrix} a_{11}B & a_{12}B \\ a_{21}B & a_{22}B \end{bmatrix}$$

$$= \begin{bmatrix} a_{11}b_{11} & a_{11}b_{12} & a_{12}b_{11} & a_{12}b_{12} \\ a_{11}b_{21} & a_{11}b_{22} & a_{12}b_{21} & a_{12}b_{22} \\ a_{21}b_{11} & a_{21}b_{12} & a_{22}b_{11} & a_{22}b_{12} \\ a_{21}b_{21} & a_{21}b_{22} & a_{22}b_{21} & a_{22}b_{22} \end{bmatrix}$$

5.7. Conjugate of a Matrix

If [A] is a matrix having conjugate numbers as elements, the matrix obtained from [A] by replacing each element by its conjugate is called the conjugate of [A] and is denoted by $[\bar{A}]$, e.g.

$$\text{if if } [A] = \begin{bmatrix} a+ib & c+id & e+if \\ \alpha - i\beta & \Upsilon - i\delta & \varphi + i\Psi \end{bmatrix}$$

$$\text{then } [\bar{A}] = \begin{bmatrix} a-ib & c-id & e-if \\ \alpha + i\beta & \Upsilon + i\delta & \varphi - i\Psi \end{bmatrix}$$

It is evident that the conjugate of the conjugate of a matrix coincides with itself.

Conjugate of the transpose of a matrix [A] is called the conjugate transpose of [A] and is denoted by $[A^+]$ or $[A^\theta]$. Thus

$$[A^{+}] = [\bar{A}']$$

If $[A] = (a_{ij})$ of order $m \times n$

Then $[A'] = (a'_{ij})$ of order $n \times m$

and $[\bar{A}'] = [A^{+}] = (\bar{a}'_{ij})$ of order $n \times n$

where $a'_{ij} = a_{ji}$ and $\bar{a}'_{ij} = \bar{a}_{ji} =$ Complex conjugate of a_{ij}

5.8. Hermitian Matrices

A square matrix $[A] = (a_{ij})$ such that $[\bar{A}'] = [A]$

or, $[A'] = [\bar{A}]$ or $\bar{a}_{ij} = a_{ji}$ i.e. whose conjugate transpose coincides with the matrix itself is called Hermitian. The diagonal elements are obviously real, e.g. $\begin{pmatrix} a & b+ic \\ b-ic & d \end{pmatrix}$ in a Hermitian matrix.

A Hermitian matrix of real elements is nothing but a real symmetric matrix.

A square matrix $[A] = (a_{ij})$ such that

$[\bar{A}'] = -[A]$ or $[A'] = -[\bar{A}]$ i.e. $a_{ij} = -a_{ji}$ is called skew Hermitian. If the diagonal elements are zero or pure imaginary

i.e. $\begin{pmatrix} 0 & 1+i \\ -1+i & 0 \end{pmatrix}$ or $\begin{pmatrix} 3c & 2+5i \\ -2+5i & 4i \end{pmatrix}$

the matrices are skew symmetric.

Examples :

1. If $[A^{\theta}]$ and $[B^{\theta}]$ be the transpose conjugate of [A] and [B] respectively, then

(i) $\left[(A^{\theta})^{\theta}\right] = [A]$

(ii) $\left[(A+B)^{\theta}\right] = A^{\theta} + B^{\theta}$, A and B being conformable.

(iii) $\left[(AB)^{\theta}\right] = [B^{\theta}]\,[A^{\theta}]$, [A] and [B] being conformable to multiplication.

Proof :

(i) Let $[A] = (a_{ij})_{m \times n}$

Then $[A^{\theta}] = [\bar{A}'] = [(\bar{A})']$ ($\because$ the conjugate of the transpose is the same as the transpose of the conjugate.)

Also $\left[(A^{\theta})^{\theta}\right] = [\,\{(\bar{\bar{A}})'\}'] = [(\bar{\bar{A}})] = [A]$

$\because$ the conjugate of (conjugate of [A]) = [A]

(ii) $\left[(A+B)^{\theta}\right] = [(\overline{A+B})'] = [(\overline{A'+B'})] = [(\overline{A}')] + [(\overline{B}')]$

$$= [A^{\theta}] + [B^{\theta}]$$

(iii) $\left[(AB)^{\theta}\right] = [\{(\overline{AB})'\}] = [(\overline{B'A'})] = [\overline{B}'][\overline{A}'] = B^{\theta}A^{\theta}$

2. If [A] be any square matrix, prove that

$[A] + [A^{\theta}]$, $[AA^{\theta}]$ and $[A^{\theta}A]$ are Hermitian and $[A - A^{\theta}]$ is skew Hermitian.

Proof :

(i) $\left[(A+A^{\theta})^{\theta}\right] = [A^{\theta}] + \left[(A^{\theta})^{\theta}\right]$

Again $\left[(A+A^{\theta})^{\theta}\right] = A^{\theta} + A = A + A^{\theta}$ by previous example

This shows that the conjugate of the transpose of the matrix $\left[A+A^{\theta}\right]$ is equal to the matrix $\left[A+A^{\theta}\right]$ itself. Hence the matrix $A + A^{\theta}$ is Hermitian

(ii) $\left[(AA^{\theta})^{\theta}\right] = \left[(A^{\theta})^{\theta}\right]\left[A^{\theta}\right] = AA^{\theta}$

This shows that the matrix AA^{θ} is Hermitian

(iii) Again $\left[(A-A^{\theta})^{\theta}\right] = \left[A^{\theta}\right] - \left[(A^{\theta})^{\theta}\right] = \left[A^{\theta} - A\right] = -\left[A - A^{\theta}\right]$

This shows that the matrix $\left[A - A^{\theta}\right]$ is skew Hermitian.

5.9. Reciprocal or Inverse of a Matrix

If [A] is a non-singular square matrix and [B] is another matrix such that

[A] [B] = [B] [A] = [I], I being unit matrix, then [B] is called the inverse of [A].

If is clear that if [A] [B] and [B] [A] are both defined and equal, both of them must be square matrices of the same order. Inverse of the matrix [A] is denoted by $[A^{-1}]$, and thus we have

$$[A][A^{-1}] = [A^{-1}][A] = [I].$$

Hence the product of a matrix and its inverse is the unit matrix in which ever order the multiplication is carried out. If the determinant of [A] vanishes, [A] has no inverse and is said to be singular. Thus we may state a theorem for the existence of the inverse. The necessary and sufficient condition for the existence of inverse of a square matrix [A] is that $|A| \neq 0$. (5.18)

5.10. Unitary Matrix

A square matrix is said to be unitary if its inverse is equal to its conjugate transpose i.e.

$$\left[A^{+}\right] = \left[A^{-1}\right] \tag{5.19}$$

$$\text{or, } [\bar{A}'] = \left[A^{-1}\right]$$

$$\text{or, } [A'] = \left[(\bar{A}^{-1})\right]$$

Hence for a matrix to be unitary

$$[A]\left[A^{+}\right] = \left[A^{+}\right][A] = [I] \tag{5.20}$$

Taking the determinants of both sides, we have

$$|A\,A^{+}| = |I|$$

$$\text{or, } |A|\;|A^{+}| = 1$$

$$\text{or, } |A|\;|\bar{A}| = 1 \tag{5.21}$$

which shows that the modulus of the determinant of a unitary matrix is unity. If a unitary matrix is also Hermitian, then

$$\left[A^{+}\right] = [A] = \left[A^{-1}\right]$$

$$\text{or, } [A\,A] = [I] = \left[A^{2}\right]$$

Taking an example :

$$\text{If } [A] = \begin{bmatrix} \cos\theta & i\sin\theta \\ i\sin\theta & \cos\theta \end{bmatrix}$$

$$\text{then } [A^{+}] = \begin{bmatrix} \cos\theta & -i\sin\theta \\ -i\sin\theta & \cos\theta \end{bmatrix}$$

$$\text{and} \left[AA^{+}\right] = \begin{bmatrix} 1 & 0 \\ 0 & 1 \end{bmatrix} = [I]$$

which shows that [A] is unitary matrix.

In a similar manner it can be shown that the matrix $\begin{bmatrix} \frac{1}{\sqrt{2}} & \frac{1}{2} \\ -\frac{1}{\sqrt{2}} & -\frac{1}{\sqrt{2}} \end{bmatrix}$ is a unitary matrix.

It can be proved that every multiplication of two unitary matrices is a unitary matrix. Let us consider two unitary matrices [A] and [B] of order $n \times n$,

$$\text{then} \left[A^{+}\right] = [A] \text{ and } \left[B^{+}\right] = [B]$$

$$\text{and} \left[(AB)^{+}\right][AB] = \left[B^{+}\right]\left[A^{+}\right][A][B] = \left[B^{+}\right][B] = [I]$$

$$\text{Similarly } [AB]\left[(AB)^{+}\right] = [I]$$

Thus shows that [AB] is a unitary matrix.

In a similar manner it can be shown that [BA] is a unitary matrix.

A real unitary matrix is said to be an orthognal matrix. If [A] be real equation (5.20) gives

$$[A]\,[A'] = [A']\,[A] = [I] \tag{5.21a}$$

since for real [A]

$[A^+] = [A']$

Hence $[A'] = [A^{-1}]$ (5.22)

Thus for a matrix to be orthogonal, its inverse coincides with its transpose. Considering the determinants on both sides by (5.21a) we get

$$|A\,A'| = |A|\;|A'| = |A|^2 = |I| = 1$$

$$\text{or, } |A| = \pm 1 \tag{5.23}$$

Thus determinant of an orthogonal matrix is 1 or − 1. If $|A| = 1$, then [A] is said to be proper orthogonal matrix. If $|A| = -1$, it is called improper orthogonal matrix.

5.11. Eigen values and Eigen vectors of a Matrix

Let [A] be a square matrix ($n \times n$) and [X] a column matrix with n elements. Then [A] [X] is also a column matrix with n rows. Generally [A] [X] will be different from [X]. For certain special choices of [X], however, [A] [X] and [X] will be proportional

$$\text{i.e. } [A]\,[X] = \lambda\,[X] \tag{5.24}$$

In such a case [X] is called an eigen vector of [A] belonging to the eigen value λ.

To have a precise idea regarding [X] let us rewrite equation (5.24) in an explicit form as shown below :

$$\begin{bmatrix} a_{11} & a_{12} & \dots & a_{1n} \\ a_{21} & a_{22} & \dots & a_{2n} \\ \dots & \dots & \dots & \dots \\ a_{n1} & a_{n2} & \dots & a_{nn} \end{bmatrix} \begin{bmatrix} x_1 \\ x_2 \\ \cdot \\ \cdot \\ \cdot \\ x_n \end{bmatrix} = \lambda \begin{bmatrix} x_1 \\ x_2 \\ \cdot \\ \cdot \\ \cdot \\ x_n \end{bmatrix} \tag{5.25}$$

After matrix multiplication and rearrangement we get the following set of linear homogeneous equations in $x_1, x_2, \dots x_n$

$$\left.\begin{array}{l}(a_{11}-\lambda)x_1+a_{12}x_2+\ldots\ldots+a_{1n}x_n=0\\ a_{21}x_1+(a_{22}-\lambda)x_2+\ldots\ldots+a_{2n}x_n=0\\ \ldots\ldots\ldots\ldots\ldots\ldots\ldots\ldots\ldots\ldots\\ a_{n1}x_1+a_{n2}x_2+\ldots\ldots+(a_{nn}-\lambda)x_n=0\end{array}\right\} \quad (5.26)$$

This set of equation will have a nontrivial solution, if and only if the determinant

$$\begin{vmatrix} a_{11}-\lambda & a_{12} & \cdots & a_{1n} \\ a_{21} & a_{22}-\lambda & \cdots & a_{2n} \\ \cdots & \cdots & \cdots & \cdots \\ a_{n1} & a_{n2} & \cdots & a_{nn}-\lambda \end{vmatrix} = 0 \quad (5.27)$$

The above equation can also be written as det $(A - \lambda X) = 0$ and is known as the characteristic equation of the matrix [A]. It is an nth degree algebraic equation in the unknown λ and will have n roots. For each value of λ, equation (5.24) can now be used to determine X. The process of finding the eigen values and eigen vectors of a matrix is known as the matrix diagonalisation.

5.12. Diagonalisation of Matrices

Let us consider a matrix [A] having $\lambda_1, \lambda_2, \lambda_3 \ldots\ldots \lambda_n$ as its n eigen values and $X_1, X_2, \ldots\ldots X_n$ the corresponding eigen vectors.

Let the column vector

$$X_1 = \begin{bmatrix} x_{11} \\ x_{12} \\ \cdot \\ \cdot \\ \cdot \\ x_{1n} \end{bmatrix} \quad (5.28)$$

Let the matrix [E] be such that the column vectors are the n-eigen vectors

$$\text{i.e. } [E] = \begin{bmatrix} x_{11} & x_{12} & \cdots & x_{1n} \\ x_{21} & x_{22} & \cdots & x_{2n} \\ \cdots & \cdots & \cdots & \cdots \\ x_{n1} & x_{n2} & \cdots & x_{nn} \end{bmatrix} \quad (5.29)$$

Let [F] be a diagonal matrix such that

$$[F] = \begin{bmatrix} \lambda_1 & 0 & 0 \\ 0 & \lambda_2 & 0 \\ \cdot\cdot & \cdot\cdot & \cdot\cdot \\ 0 & 0 & \lambda_n \end{bmatrix} \quad (5.30)$$

Now [E] [F] = $\begin{bmatrix} \lambda_1 x_{11} & \lambda_2 x_{12} & \ldots\ldots & \lambda_\nu x_{1n} \\ \lambda_1 x_{21} & \lambda_2 x_{22} & \ldots\ldots & \lambda_\nu x_{2n} \\ \ldots\ldots & \ldots\ldots & \ldots\ldots & \ldots\ldots \\ \lambda_1 x_{n1} & \lambda_2 x_{n2} & \ldots\ldots & \lambda_\nu x_{nn} \end{bmatrix} = (\lambda_\varphi x_{ij})$ (5:31)

$$= (AX_1, AX_2, \ldots AX_n)$$

$$= A(X_1, X_2, \ldots X_n) = [A]\,[E]$$

In case [E] is a non-singular matrix, then premultiplication of $[E^{-1}]$ gives

$$[E^{-1}]\,[A]\,[E] = [F].$$

Thus premultiplication of [A] by $[E^{-1}]$ and postmultiplication by [E], gives the diagonal matrix whose elements are the eigen values. The above operation is called the diagonalisation of the matrix [A].

EXERCISE

1. Find the Eigen values and Eigen vectors of the matrix $[A] = \begin{bmatrix} 0 & i \\ -i & 0 \end{bmatrix}$.

The characteristic equation is

$$\begin{bmatrix} -\lambda & i \\ -i & -\lambda \end{bmatrix} = 0 \qquad \ldots \text{(i)}$$

or, $\lambda^2 - 1 = 0 \quad \therefore \lambda = \pm 1$

The Eigen values are ± 1.

Using the Eigen value equation

$$[A]\,[X] = \lambda\,[X]$$

Taking $\lambda = +1$ we have

$$\begin{bmatrix} 0 & i \\ -i & 0 \end{bmatrix} \begin{bmatrix} x_1 \\ x_2 \end{bmatrix} = +1 \begin{bmatrix} x_1 \\ x_2 \end{bmatrix}$$

or, $ix_2 = x_1$ and $-ix_1 = x_2$ ... (ii)

The above equations only give the ratio of x_1 and x_2 and fail to determine their values. We can choose either x_1 or x_2. Thus an eigen vector of [A] belonging to the eigen value + 1 can be written as $\begin{bmatrix} 1 \\ -i \end{bmatrix}$ where the value of x_1 has been taken.

Similarly a second solution can be obtained for $\lambda = -1$,

2. Find the Eigen values of the following matrix :

$$\begin{bmatrix} 1 & 0 & 1 \\ 1 & 1 & 0 \\ 1 & 0 & 1 \end{bmatrix}$$ [**Ans.** 0, 1, 2]

3. $$\begin{bmatrix} 1 & -1 & -1 \\ -1 & 1 & -1 \\ -1 & -1 & 1 \end{bmatrix}$$ [**Ans.** – 1, 2, 2]

4. $$\begin{bmatrix} 1 & 0 & 0 \\ 0 & 1 & 1 \\ 0 & 1 & 1 \end{bmatrix}$$ [**Ans.** 0·1, 2]

5. If [A] be Hermitian matrix such that

$$[A]^2 = 0, \text{ show that } [A] = 0.$$

6. If [A] and [B] are Hermitian matrices, show that

([A] [B] + [B] [A]) and ([A] [B] – [B] [A]) are also Hermitian.

7. Find the reciprocal of the matrix

$[A] = \begin{bmatrix} 0 & 1 & 1 \\ 1 & 0 & 1 \\ 1 & 1 & 0 \end{bmatrix}$ and also show that the transform of the matrix

$$[B] = \begin{bmatrix} b+c & c-a & b-a \\ c-b & c+a & a-b \\ b-c & a-c & a+b \end{bmatrix} \text{ by } [A] \text{ i.e.}$$

$[A]\,[B]\,[A^{-1}]$ is a diagonal matrix.

8. For two matrices [A] and [B],

Prove that transpose [A] [B] = transpose [B] [A].

9. Find the inverse of the matrix $\begin{bmatrix} \cos\theta & \sin\theta \\ -\sin\theta & \cos\theta \end{bmatrix}$

10. Are the four matrices

$$c_1 = \begin{bmatrix} 0 \\ 1 \\ 1 \\ 1 \end{bmatrix}, c_2 = \begin{bmatrix} 1 \\ 0 \\ 1 \\ 1 \end{bmatrix}, c_3 = \begin{bmatrix} 1 \\ 1 \\ 0 \\ 1 \end{bmatrix} \text{ and } c_4 = \begin{bmatrix} 1 \\ 1 \\ 1 \\ 0 \end{bmatrix}$$

linearly independent ?

11. For any matrix [A] prove that [A] [A′] and [A′] [A] are both symmetric.

12. Add the matrices

$$\begin{bmatrix} 7 & 8 & 7 \\ 9 & 0 & 3 \\ 6 & 5 & 4 \end{bmatrix} \text{ and } \begin{bmatrix} 6 & 7 & 0 \\ 0 & 7 & 9 \\ 3 & 2 & 1 \end{bmatrix}$$

13. Show that the matrix $\frac{1}{2}\begin{bmatrix} 1 & 1 & 1 & 1 \\ 1 & -1 & 1 & -1 \\ 1 & 1 & -1 & -1 \\ 1 & -1 & -1 & 1 \end{bmatrix}$ is orthogonal.

14. Show that the inverse of unitary matrix is unitary.

5.13. Solution of Linear Equation

Matrices have important applications in the study of solution of linear differential equation.

Let us consider m linear equations in n unknown as

$$\left.\begin{array}{l} a_{11}x_1 + a_{12}x_2 + \ldots\ldots + a_{1n}x_n = b_1 \\ a_{21}x_1 + a_{22}x + \ldots\ldots + a_{2n}x_n = b_2 \\ \ldots\ldots \\ a_{m1}x_1 + a_{m2}x_2 + \ldots\ldots + a_{mn}x_n = b_n \end{array}\right\} \qquad (5.31(a))$$

In the above equations m is not necessarily equal to n. If one of the m constants b_i is different from zero, the system is said to be non-homogeneous. If all the b's are zero, the system is said to be homogeneous.

The $m \times n$ matrix

$$[A] = \begin{bmatrix} a_{11} & a_{12} & \ldots\ldots & a_{1n} \\ a_{21} & a_{22} & \ldots\ldots & a_{2n} \\ \ldots & \ldots & \ldots\ldots & \ldots \\ a_{m1} & a_{m2} & \ldots\ldots & a_{mn} \end{bmatrix} \qquad (5.32)$$

the elements of which are coefficients of linear equations (5.31a) is called the coefficient matrix of the system.

If we define two column matrices [X] and [B] of orders $n \times 1$ and $m \times 1$ respectively, as follows

$$[X] = \begin{bmatrix} x_1 \\ x_2 \\ \cdot \\ \cdot \\ x_n \end{bmatrix} \text{ and } [B] = \begin{bmatrix} b_1 \\ b_2 \\ \cdot \\ \cdot \\ b_m \end{bmatrix} \qquad (5.33)$$

Equation (5.31a) can now be written as

$$[A]\,[X] = [B]. \tag{5.34}$$

In a similar way, the system of m-homogeneous equations given by equation (5.31a) with all the b's equal to zero can be written as

$$[A]\,[X] = [0], \tag{5.35}$$

where [0] is the column null matrix given as

$$[0] = \begin{bmatrix} 0 \\ 0 \\ . \\ . \\ . \\ . \\ 0 \end{bmatrix} \tag{5.36}$$

EXERCISE

1. Consider the following transformation in three dimensions. **[C. U. 1996]**

$$x' = x\cos\theta + y\sin\theta$$
$$y' = -x\sin\theta + y\cos\theta$$
$$z' = z$$

(i) Write down that transformation matrix.

Ans.

$$[A] = \begin{bmatrix} \cos\theta & \sin\theta & 0 \\ -\sin\theta & \cos\theta & 0 \\ 0 & 0 & 1 \end{bmatrix}$$

(ii) Show that $[A(\theta_1)] + [A(\theta_2)] = [A(\theta_1 + \theta_2)]$

$$[A(\theta_1)] + [A(\theta_2)] = \begin{bmatrix} \cos\theta_1 & \sin\theta_1 & 0 \\ -\sin\theta_1 & \cos\theta_1 & 0 \\ 0 & 0 & 1 \end{bmatrix} + \begin{bmatrix} \cos\theta_2 & \sin\theta_2 & 0 \\ -\sin\theta_2 & \cos\theta_2 & 0 \\ 0 & 0 & 1 \end{bmatrix}$$

$$= \begin{bmatrix} \cos\theta_1\cos\theta_2 - \sin\theta_1\sin\theta_2 + 0 & \cos\theta_1\sin\theta_2 + \sin\theta_1\cos\theta_2 + 0 & 0+0+0 \\ -\sin\theta_1\cos\theta_2 - \cos\theta_1\sin\theta_2 + 0 & -\sin\theta_1\sin\theta_2 + \cos\theta_1\cos\theta_2 + 0 & 0+0+0 \\ 0+0+0 & 0+0+0 & 0+0+1 \end{bmatrix}$$

$$= \begin{bmatrix} \cos(\theta_1+\theta_2) & \sin(\theta_1+\theta_2) & 0 \\ -\sin(\theta_1+\theta_2) & \cos(\theta_1+\theta_2) & 0 \\ 0 & 0 & 1 \end{bmatrix} = [A(\theta_1+\theta_2)]$$

(iii) Is [A (θ)] unitary ?

Ans.

$$A' = \begin{bmatrix} \cos\theta & -\sin\theta & 0 \\ \sin\theta & \cos\theta & 0 \\ 0 & 0 & 1 \end{bmatrix}$$

$$[A]\,[A'] = \begin{bmatrix} \cos\theta & \sin\theta & 0 \\ -\sin\theta & \cos\theta & 0 \\ 0 & 0 & 1 \end{bmatrix} \begin{bmatrix} \cos\theta & -\sin\theta & 0 \\ \sin\theta & \cos\theta & 0 \\ 0 & 0 & 1 \end{bmatrix}$$

$$= \begin{bmatrix} \cos^2\theta + \sin^2\theta + 0 & -\cos\theta\sin\theta + \sin\theta\cos\theta + 0 & 0+0+0 \\ -\sin\theta\cos\theta + \cos\theta\sin\theta + 0 & \sin^2\theta + \cos^2\theta + 0 & 0+0+0 \\ 0+0+0 & 0+0+0 & 0+0+1 \end{bmatrix}$$

$$= \begin{bmatrix} 1 & 0 & 0 \\ 0 & 1 & 0 \\ 0 & 0 & 1 \end{bmatrix} = [I]$$

Hence [A] is unitary.

(iv) What are the eigen values of A(θ) ?

Ans. The characteristic equation is

$$\begin{bmatrix} \cos\theta - \lambda & \sin\theta & 0 \\ -\sin\theta & \cos\theta - \lambda & 0 \\ 0 & 0 & 1-\lambda \end{bmatrix} = 0$$

or, $(\cos\theta - \lambda)\left\{(\cos\theta - \lambda)(1-\lambda)\right\} + \sin\theta\,(1-\lambda)\sin\theta = 0$

or, $(\cos\theta - \lambda)^2(1-\lambda) + (1-\lambda)\sin^2\theta = 0$

or, $(1-\lambda)\left\{\sin^2\theta + \cos^2\theta - 2\cos\theta\,\lambda + \lambda^2\right\} = 0$

or, $(1-\lambda)\left\{1 - 2\cos\theta\,\lambda + \lambda^2\right\} = 0$

$$\therefore\ \lambda = 1 \text{ or } \lambda = \frac{2\cos\theta \pm \sqrt{4\cos^2\theta - 4}}{2} = (\cos\theta \pm \sin\theta)$$

ADDITIONAL EXERCISE

1. Explain—(i) square matrix, (ii) unit matrix, (iii) diagonal matrix and (iv) row matrix.

2. Show that a matrix multiplication is associative.

3. If [B] is a matrix of order $m \times n$ and [C] is a matrix of order $n \times p$ than [B] [C] is a matrix of order $m \times p$.

4. Find the eigen values and eigen vectors of the Hermitian matrix

$$[A] = \begin{bmatrix} 0 & 1 & 1 \\ 1 & 0 & 1 \\ 1 & 1 & 0 \end{bmatrix}$$

Ans. The characteristic equation is

$$\begin{vmatrix} -\lambda & 1 & 1 \\ 1 & -\lambda & 1 \\ 1 & 1 & -\lambda \end{vmatrix} = 0$$

Simplifying,

$$(2-\lambda)\begin{vmatrix} 1 & 1 & 1 \\ 1 & -\lambda & 1 \\ 1 & 1 & -\lambda \end{vmatrix} = 0$$

$$\text{or, } (2-\lambda)\begin{vmatrix} 1 & 1 & 1 \\ 0 & -\lambda-1 & 0 \\ 0 & 0 & -\lambda-1 \end{vmatrix} = 0$$

or, $(2 - \lambda)\ (\lambda + 1)^2 = 0$

or, $\lambda = 2, -1, -1$ (eigen values)

To determine the eigen vector

$$X_1 = \begin{bmatrix} x_1 \\ y_1 \\ z_1 \end{bmatrix} \text{ belonging to the eigen value 2.}$$

The eigen value equation is

$$\begin{bmatrix} 0 & 1 & 1 \\ 1 & 0 & 1 \\ 1 & 1 & 0 \end{bmatrix}\begin{bmatrix} x_1 \\ y_1 \\ z_1 \end{bmatrix} = 2\begin{bmatrix} x_1 \\ y_1 \\ z_1 \end{bmatrix}$$

which gives

$$y_1 + z_1 = 2x_1$$
$$x_1 + z_1 = 2y_1$$
$$x + y_1 = 2z_1$$

Solving we get $x_1 = y_1 = z_1$.

If we take $x_1 = 1$, the normallified eigen vector belonging to the eigen value z is

$$x_1 = \frac{1}{\sqrt{3}}\begin{bmatrix} 1 \\ 1 \\ 1 \end{bmatrix}$$

Similarly we can proceed for $\lambda = -1$.

5. Show that

$$\begin{bmatrix} \cos\theta & -\sin\theta \\ \sin\theta & \cos\theta \end{bmatrix} = \begin{bmatrix} 1 & -\tan\frac{\theta}{2} \\ \tan\frac{\theta}{2} & 1 \end{bmatrix} \begin{bmatrix} 1 & -\tan\frac{\theta}{2} \\ -\tan\frac{\theta}{2} & 1 \end{bmatrix}^{-1}$$

6. For a given matrix [A], compute A′, A^{θ} and $A^{\dagger}$, where

$$[A] = \begin{bmatrix} 2-3i & 3-5i \\ 3 & 6+2i \\ 4+i & 3-2i \end{bmatrix}$$

In this case $[A'] = \begin{bmatrix} 2-3i & 3 & 4+i \\ 3-5i & 6+2i & 3-2i \end{bmatrix}$

$$[A^{\theta}] = \begin{bmatrix} 2+3i & 3+5i \\ 3 & 6-2i \\ 4-i & 3+2i \end{bmatrix}$$

$$[A^{\dagger}] = \begin{bmatrix} 2+3i & 3 & 4-i \\ 3+5i & 6-2i & 3-2i \end{bmatrix}$$

7. You are given a matrix. How will you find its

(i) Transpose matrix ;

(ii) Adjoint matrix ;

(iii) Complex conjugate ?

8. What is a Hermitian matrix ? Show that the following matrix is Hermitian.

$$\begin{bmatrix} 1 & 2-3i & 3-4i \\ 2+3i & 2 & 4-5i \\ 3+4i & 4+5i & 1 \end{bmatrix}$$

9. Show that the following matrix is skew Hermitian.

$$\begin{bmatrix} 3i & 1+2i & 2-2i \\ 1+2i & 4i & 4+3i \\ -2-2i & -4+3i & 0 \end{bmatrix}$$

10. Find the transpose [A′], conjugate $[A^{0}]$ and the conjugate transpose $[A^{\dagger}]$ of the given matrix

$$[A] = \begin{bmatrix} 5-3i & 3-5i \\ 3 & 6+2i \\ 4+i & 3-2i \end{bmatrix}$$

$$[A'] = \begin{bmatrix} 5-3i & 3 & 4+i \\ 3-5i & 6+2i & 3-2i \end{bmatrix}$$

$$[A^{\theta}] = \begin{bmatrix} 5+3i & 3+5i \\ 3 & 6-2i \\ 4-i & 3+2i \end{bmatrix}$$

$$[A^{\dagger}] = \begin{bmatrix} 5+3i & 3 & 4-i \\ 3+5i & 6-2i & 3+2i \end{bmatrix}.$$

11. Is the matrix

$$[A] = \begin{bmatrix} \frac{1}{\sqrt{2}} & \frac{i}{\sqrt{2}} \\ \frac{-i}{\sqrt{2}} & \frac{-1}{\sqrt{2}} \end{bmatrix} \text{ unitary ?}$$

Let us consider the product

$$[A]\,[A^{\dagger}] = \begin{bmatrix} \frac{1}{\sqrt{2}} & \frac{i}{\sqrt{2}} \\ \frac{-i}{\sqrt{2}} & \frac{-1}{\sqrt{2}} \end{bmatrix} \begin{bmatrix} \frac{1}{\sqrt{2}} & \frac{-i}{\sqrt{2}} \\ \frac{+i}{\sqrt{2}} & \frac{-1}{\sqrt{2}} \end{bmatrix}$$

$$= \begin{bmatrix} \frac{1}{2}+\frac{1}{2} & 0 \\ 0 & \frac{1}{2}+\frac{1}{2} \end{bmatrix} = \begin{bmatrix} 1 & 0 \\ 0 & 1 \end{bmatrix} = [I].$$

Hence the matrix [A] is unitary.

CHAPTER VI

MECHANICS I

6.1. Newton's Laws of Motion

(a) Mechanics of a single particle :

Velocity and acceleration of a particle in (i) plane polar co-ordinates—tangential and normal components ; radial and cross radial components. (ii) Cylindrical polar and (iii) spherical polar co-ordinate systems. Time and path integral of forces, work, energy, impulse ; conservative force and concept of potential. Motion under different types of forces ; Forces linerly dependent on velocity, conservation of linear and angular momenta. Motion of a system with variable mass (Rocket motion).

Expression for velocity and acceleration of a particle :

6.2. In Cartesian co-ordinates

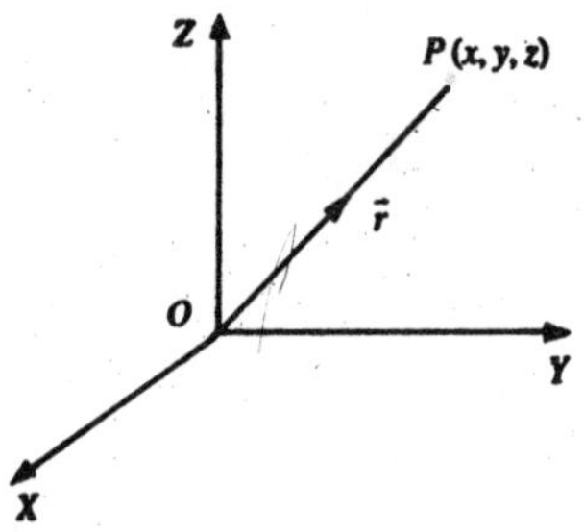

Fig. 6.1

Let the position vector of a particle P at any instant be specified by $\vec{r}$ with reference to the origin O (Fig. 6.1).

$\therefore$ $\vec{r} = \hat{i}x + \hat{j}y + \hat{k}z$ where $\hat{i}$, $\hat{j}$ and $\hat{k}$ are unit vectors along rectangular co-ordinate axes OX, OY and OZ respectively. The velocity vector $\vec{v}$ of the particle is given by

$$\vec{v} = \frac{d\vec{r}}{dt} = \hat{i}\frac{dx}{dt} + \hat{j}\frac{dy}{dt} + \hat{k}\frac{dz}{dt}.$$

Acceleration

$$\vec{f} = \frac{d\vec{v}}{dt} = \hat{i}\frac{d^2x}{dt^2} + \hat{j}\frac{d^2y}{dy^2} + k^2\frac{d^2z}{dt^2} \qquad (6.1)$$

$$|\vec{v}| = \sqrt{\left(\frac{dx}{dt}\right)^2 + \left(\frac{dy}{dt}\right)^2 + \left(\frac{dz}{dt}\right)^2}$$

$$|\vec{f}| = \sqrt{\left(\frac{d^2x}{dt^2}\right)^2 + \left(\frac{d^2y}{dt^2}\right)^2 + \left(\frac{d^2z}{dt^2}\right)^2}$$

6.3. Tangential and Normal Component of velocity and acceleration

Let us consider a particle P moving in a plane in which OX and OY axes are fixed. The position vector of the particle is $\vec{r} = \vec{OP}$ (Fig. 6.2).

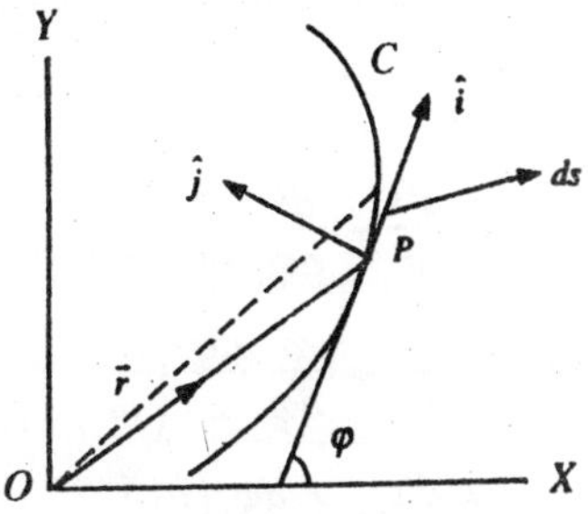

Fig. 6.2

The velocity $\vec{v} = \frac{d\vec{r}}{dt}$. If the path followed by the particle is the curve C, then dr is an infinitesimal displacement along C, so that $d\vec{r} = \hat{i}\,ds$, where ds is an elementary length along C and $\hat{i}$ is unit vector tangential to C.

Thus $\vec{v} = \hat{i}\frac{ds}{dt}$ or $|\vec{v}| = \frac{ds}{dt}$.

In other words, the velocity of a moving particle has a direction tangent to the path and has a magnitude $\frac{ds}{dt}$.

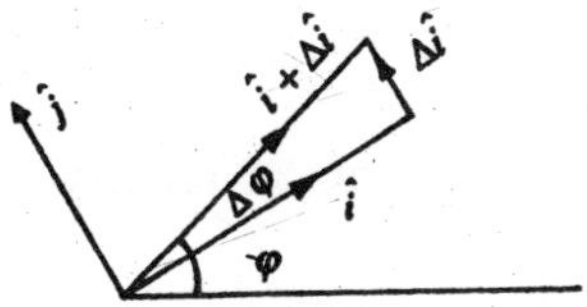

Fig. 6.3

Let $\hat{j}$ be unit vector normal to $\vec{r}$ and let φ be the inclination of $\vec{r}$ to X axis. It is clear that $\hat{i}$ and $\hat{j}$ are functions of φ. Figure 6.3 shows the vectors $\hat{i}$ and $\hat{i} + \Delta\hat{i}$ transferred to the common origin.

Since both of them are unit vectors, the triangle formed by the three vectors $\hat{i}$, $\hat{i} + \Delta\hat{i}$ and $\Delta\hat{i}$ is isosceles. The magnitude of $\Delta\hat{i}$ can be obtained as

$$\frac{|\Delta\hat{i}|}{\sin\Delta\varphi} = \frac{|\hat{i}|}{\sin\left(90° - \frac{\Delta\varphi}{2}\right)} = \frac{1}{\cos\frac{\Delta\varphi}{2}}$$

$$\therefore\ |\Delta\hat{i}| = 2\sin\frac{\Delta\varphi}{2} = \Delta\varphi \text{ in the limit.}$$

Hence in the limit, the magnitude $\frac{\Delta i}{\Delta\varphi} = 1$. Thus $\frac{d\hat{i}}{d\varphi}$ is a unit vector, pointing in the limiting case in the direction defined by $\Delta\hat{i}$, where $\Delta\varphi \to 0$. This direction is clearly that of $\hat{j}$ and so

$$\frac{d\hat{i}}{d\varphi} = \hat{j}.$$

With similar arguments $\dfrac{d\hat{j}}{d\varphi} = \hat{i}$

The acceleration $\vec{a}$ is defined as

$$\vec{a} = \frac{d\vec{v}}{dt} = \frac{d}{dt}\left(\hat{i}\,\frac{ds}{dt}\right) = \hat{i}\,\frac{d^2s}{dt^2} + \frac{d\hat{i}}{dt}\,\frac{ds}{dt}$$

$$= \hat{i}\,\frac{d^2s}{dt^2} + \frac{d\hat{i}}{d\varphi}\,\frac{d\varphi}{dt}\,\frac{ds}{dt}$$

$$= \hat{i}\,\frac{d^2s}{dt^2} + \frac{d\hat{i}}{d\varphi}\,\frac{d\varphi}{ds}\left(\frac{ds}{dt}\right)^2$$

$$= \hat{i}\,\frac{d^2s}{dt^2} + \hat{j}\,\frac{d\varphi}{ds}\left(\frac{ds}{dt}\right)^2$$

The radius of curvature of the curve C at the point concerned is given by $\rho = \dfrac{ds}{d\varphi}$.

Hence $\vec{a} = \hat{i}\,\dfrac{dv}{dt} + \hat{j}\,\dfrac{v^2}{\rho}$ (6.2)

Thus the acceleration of a moving point has components $\dfrac{dv}{dt}$ along the tangent and $\dfrac{v^2}{\rho}$ along the normal to the path.

6.4. Radial and transverse (cross radial) components of velocity and acceleration

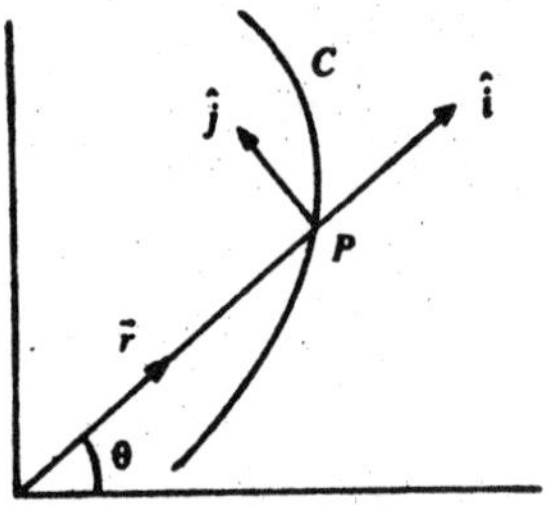

Fig. 6.4

Let us consider a particle P moving in a plane ; its position being described by the polar co-ordinates (r, θ). Let $\hat{i}$ and $\hat{j}$ be unit vectors along and perpendicular to OP (Fig. 6.4).

Proceeding as in the previous article we have

$$\frac{d\hat{i}}{d\theta} = \hat{j} \text{ and } \frac{d\hat{j}}{d\theta} = -\,\hat{i}$$

Now $\vec{r} = \hat{i}\,r \;\therefore\; \vec{v} = \dfrac{d\vec{r}}{dt} = \hat{i}\,\dfrac{dr}{dt} + r\,\dfrac{d\hat{i}}{dt}$

$$= \hat{i}\,\dot{r} + r\,\frac{d\hat{i}}{d\theta}\,\frac{d\theta}{dt}$$

$$= \hat{i}\,\dot{r} + r\,\theta\,\hat{j} \qquad (6.3)$$

Thus the components of $\vec{v}$ along and perpendicular to the radius vector are and $r\dot{\theta}$ respectively.

The acceleration of the moving particle is given by

$$\vec{a} = \frac{d\vec{v}}{dt} = \frac{d}{dt}(\hat{i}\dot{r}) + \frac{d}{dt}(r\dot{\theta}\hat{j})$$

$$= \hat{i}\frac{d^2\vec{r}}{dt^2} + \dot{r}\frac{d\hat{i}}{dt} + \dot{r}\dot{\theta}\hat{j} + r\ddot{\theta}\hat{j} + r\dot{\theta}\frac{d\hat{j}}{dt}$$

$$= \hat{i}\ddot{r} + \dot{r}\frac{d\hat{i}}{d\theta}\frac{d\theta}{dt} + \dot{r}\dot{\theta}\hat{j} + r\ddot{\theta}\hat{j} + \dot{r}\dot{\theta}\frac{d\hat{j}}{d\theta}\frac{d\theta}{dt}$$

$$= \hat{i}\ddot{r} + \dot{r}\hat{j}\dot{\theta} + \dot{r}\dot{\theta}\hat{j} + r\ddot{\theta}\hat{j} - r\dot{\theta}^2\hat{i}$$

$$= \hat{i}(\ddot{r} - r\dot{\theta}^2) + \hat{j}(2\dot{r}\dot{\theta} + r\ddot{\theta})$$

$$= \hat{i}(\ddot{r} - r\dot{\theta}^2) + \hat{j}\frac{1}{r}(r^2\dot{\theta}) \tag{6.4}$$

Hence the components of acceleration along and perpendicular to the radius vector are $\ddot{r} - r\dot{\theta}^2$ and $\frac{1}{r}\frac{d}{dt}(r^2\dot{\theta})$ respectively.

6.5. Components of velocity and acceleration in cylindrical co-ordinates

The instanteneous position vector of the moving point P is $\vec{OP} = \vec{r}$, its cylindrical co-ordinates are (R, φ, Z) (Fig. 6.5).

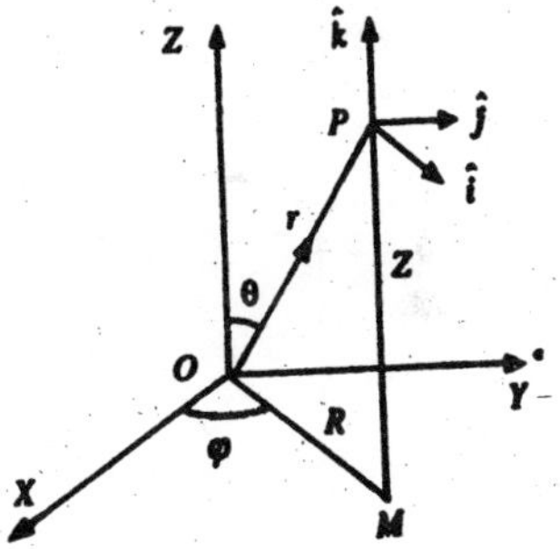

Fig. 6.5

$\hat{i}, \hat{j}, \hat{k}$ are unit vectors at P in the directions in which just one of the co-ordinates (R, φ, Z) increases, the other two remaining constant. $\hat{k}$ is constant in magnitude and direction. The direction of $\hat{i}$ and $\hat{j}$ do not depend on R and Z, but they depend on φ. Hence as in the previous article

$$\frac{d\hat{i}}{d\varphi} = \hat{j} \text{ and } \frac{d\hat{j}}{d\varphi} = -\hat{i}$$

Now $\vec{r} = \hat{i}R + \hat{k}z$

$$\therefore \vec{v} = \frac{d\vec{r}}{dt} = \frac{d\hat{i}}{dt}R + \hat{i}\frac{dR}{dt} + \hat{k}\frac{dz}{dt}$$

$$= \frac{d\hat{i}}{d\varphi}\frac{d\varphi}{dt} R + \hat{i}\dot{R} + \hat{k}\frac{dz}{dt}$$

$$= \hat{j} R\dot{\varphi} + \hat{i}\dot{R} + \hat{k}\dot{z}$$

$$= \hat{i}\dot{R} + \hat{j} R\dot{\varphi} + \hat{k}\dot{z} \tag{6.5}$$

Hence the velocity components are $\dot{R}$, $R\dot{\varphi}$ and $\dot{z}$ respectively

Again $\vec{a} = \frac{d\vec{v}}{dt} = \frac{d}{dt}\{\hat{i}\dot{R} + \hat{j}R\dot{\varphi} + \hat{k}\dot{z}\}$

$$= \frac{d\hat{i}}{dt}\dot{R} + \hat{i}\ddot{R} + \frac{d\hat{j}}{dt}R\dot{\varphi} + \hat{j}\dot{R}\dot{\varphi} + \hat{j}R\ddot{\varphi} + \hat{k}\ddot{z}$$

$$= \frac{d\hat{i}}{d\varphi}\frac{d\varphi}{dt}\dot{R} + \hat{i}\ddot{R} + \frac{d\hat{j}}{d\varphi}\frac{d\varphi}{dt}R\dot{\varphi} + \hat{j}\dot{R}\dot{\varphi} + \hat{j}R\ddot{\varphi} + \hat{k}\ddot{z}$$

$$= \hat{j}\dot{R}\dot{\varphi} + \hat{i}\ddot{R} - \hat{i}R\dot{\varphi}^2 + \hat{j}\dot{R}\dot{\varphi} + \hat{j}R\ddot{\varphi} + \hat{k}\ddot{z}$$

$$= \hat{i}(\ddot{R} - R\dot{\varphi}^2) + \hat{j}(2\dot{R}\dot{\varphi} + R\ddot{\varphi}) + \hat{}\ddot{z}$$

$$= \hat{i}(\ddot{R} - R\dot{\varphi}^2) + \hat{j}\frac{1}{R}\frac{d}{dt}(R^2\dot{\varphi}) + \hat{k}\ddot{z} \tag{6.6}$$

Hence the acceleration components are

$\ddot{R} - R\dot{\varphi}^2$, $\frac{1}{R}\frac{d}{dt}(R^2\dot{\varphi})$ and $\ddot{z}$ respectively.

6.6. Components of velocity and acceleration in spherical polar co-ordinates

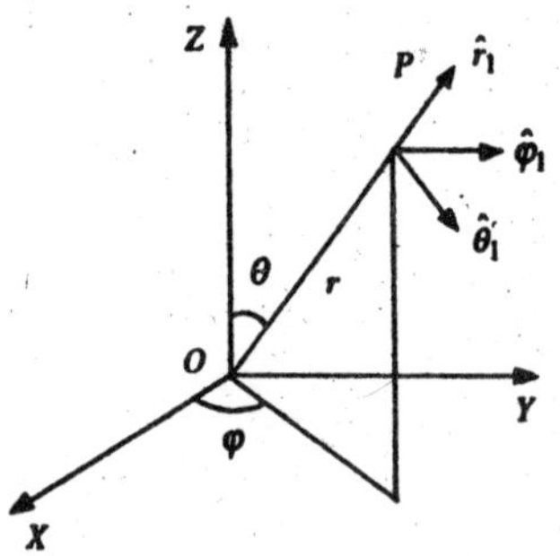

Fig. 6.6

The instanteneous position vector of the moving point is $\vec{OP} = \vec{r}$; its polar co-ordinates are (r, θ, φ) (Fig. 6.6). $\hat{r}_1, \hat{\theta}_1, \hat{\varphi}_1$ are unit vectors at P in the derections in which each of just one of the co-ordinates (r, θ, φ) increases, the other remaining constant. If $\hat{i}, \hat{j}, \hat{k}$ are unit vectors along OX, OY and OZ respectively,

$$\vec{r} = \hat{i}x + \hat{j}y + \hat{k}z = \hat{i}r\sin\theta\cos\varphi + \hat{j}r\sin\theta\sin\varphi + \hat{k}r\cos\theta$$

$$\therefore \frac{\partial\vec{r}}{\partial r} = \hat{i}\sin\theta\cos\varphi + \hat{j}\sin\theta\sin\varphi + \hat{k}\cos\theta$$

$$\left|\frac{\partial \vec{r}}{\partial r}\right| = 1 \quad \therefore\ \hat{r}_1 = \frac{\dfrac{\partial \vec{r}}{\partial r}}{\left|\dfrac{\partial \vec{r}}{\partial r}\right|} = i \sin\theta \cos\varphi + \hat{j} \sin\theta \sin\varphi + \hat{k} \cos\theta$$

$$\frac{\partial \vec{r}}{\partial \theta} = \hat{i}\, r \cos\theta \cos\varphi + \hat{j}\, r \cos\theta \sin\varphi - \hat{k}\, r \sin\theta$$

$$\left|\frac{\partial \vec{r}}{\partial \theta}\right| = r \quad \therefore\ \hat{\theta}_1 = \frac{\dfrac{\partial \vec{r}}{\partial \theta}}{\left|\dfrac{\partial \vec{r}}{\partial \theta}\right|} = \hat{i} \cos\theta \cos\varphi + \hat{j} \cos\theta \sin\varphi - \hat{k} \sin\theta$$

$$\frac{\partial \vec{r}}{\partial \varphi} = -\hat{i}\, r \sin\theta \sin\varphi + \hat{j}\, r \sin\theta \cos\varphi$$

$$\left|\frac{\partial \vec{r}}{\partial \phi}\right| = r \sin\theta \quad \therefore\ \hat{\varphi}_1 = \frac{\dfrac{\partial \vec{r}}{\partial \varphi}}{\left|\dfrac{\partial \vec{r}}{\partial \varphi}\right|} = -\hat{i} \sin\varphi + \hat{j} \cos\varphi$$

Now $\vec{r} = r\hat{r}_1$

$$\therefore\ \vec{v} = \frac{d\vec{r}}{dt} = \frac{dr}{dt}\hat{r}_1 + r\frac{d\hat{r}_1}{dt}$$

$$= \dot{r}\hat{r}_1 + r\left\{\hat{i}\cos\theta\,\dot{\theta}\cos\varphi - \hat{i}\sin\theta\sin\varphi\,\dot{\varphi} + \hat{j}\cos\theta\,\dot{\theta}\sin\varphi + \hat{j}\sin\theta\cos\varphi\,\dot{\varphi} - \hat{k}\sin\theta\,\dot{\theta}\right\}$$

$$= \dot{r}\hat{r}_1 + r\dot{\theta}\,(\hat{i}\cos\theta\cos\varphi + \hat{j}\cos\theta\sin\varphi - \hat{k}\sin\theta) + r\sin\theta\,\dot{\varphi}\,(-\hat{i}\sin\varphi + \hat{j}\cos\varphi)$$

$$= \dot{r}\hat{r}_1 + r\dot{\theta}\,\hat{\theta}_1 + r\sin\theta\,\dot{\varphi}\,\hat{\varphi}_1 \qquad (6.7)$$

Similarly,

$$\vec{a} = \frac{d\vec{r}}{dt} = \hat{r}_1\,(\ddot{r} - r\dot{\theta}^2 - r\sin^2\theta\,\dot{\varphi}^2) + \hat{\theta}_1\,(2\dot{r}\dot{\theta} + r\ddot{\theta} - r\sin\theta\cos\theta\,\dot{\varphi}^2$$

$$+ \hat{\varphi}_1\,(2\dot{r}\sin\theta\,\dot{\varphi} + 2r\sin\theta\,\dot{\theta}\,\dot{\varphi} + r\sin\theta\,\ddot{\varphi} \qquad (6.8)$$

EXERCISE

If A_ρ, A_φ and A_z are the components of a vector $\vec{A}$ in cylindrical co-ordinates, prove that

$$\frac{d\vec{A}}{dt} = \left(\frac{dA_\rho}{dt} - A_\varphi\frac{d\varphi}{dt}\right)\hat{\rho} + \left(\frac{dA_\varphi}{dt} + A_\rho\frac{d\varphi}{dt}\right)\hat{\varphi} + \frac{dA_z}{dt}\hat{z}.$$

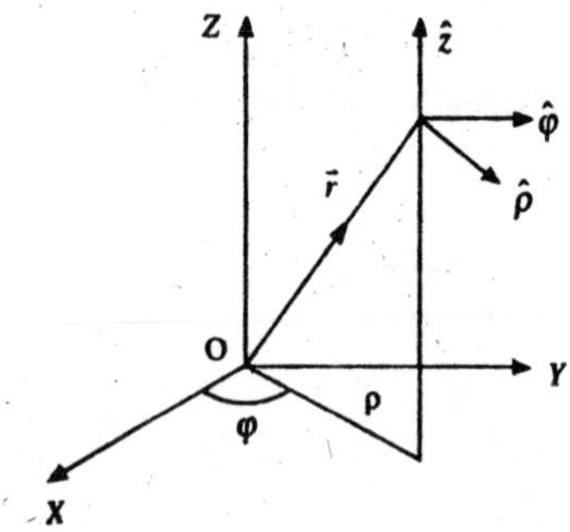

Fig. 6.7

$$\vec{r} = \hat{i}x + \hat{j}y + \hat{k}z$$

$$= \hat{i}\rho\cos\varphi + \hat{j}\rho\sin\varphi + \hat{k}z \qquad (6.8(a))$$

$$\text{Now } \hat{\rho} = \frac{\dfrac{\partial\vec{r}}{\partial\rho}}{\left|\dfrac{\partial\vec{r}}{\partial\rho}\right|} = \frac{\hat{i}\cos\varphi + \hat{j}\sin\varphi}{\sqrt{\cos^2\varphi + \sin^2\varphi}}$$

$$= \hat{i}\cos\varphi + \hat{j}\sin\varphi$$

$$\therefore \dot{\hat{\rho}} = -\hat{i}\sin\varphi\,\dot{\varphi} + \hat{j}\cos\varphi\,\dot{\varphi}$$

$$= \dot{\varphi}\,(-\hat{i}\sin\varphi + \hat{j}\cos\varphi)$$

$$= \dot{\varphi}\,\hat{\varphi}$$

Since

$$\hat{\varphi} = \frac{\dfrac{\partial\vec{r}}{\partial\varphi}}{\left|\dfrac{\partial\vec{r}}{\partial\varphi}\right|} = \frac{-\hat{i}\rho\sin\varphi + \hat{j}\rho\cos\varphi}{\sqrt{\rho^2\sin^2\varphi + \rho^2\cos^2\varphi}} = -\hat{i}\sin\varphi + \hat{j}\cos\varphi$$

$$\hat{z} = \frac{\dfrac{\partial\vec{r}}{\partial z}}{\left|\dfrac{\partial\vec{r}}{\partial z}\right|} = \hat{k}\ ; \quad \dot{\hat{\varphi}} = -\hat{i}\cos\varphi\,\dot{\varphi} - \hat{j}\sin\varphi\,\dot{\varphi}$$

$$= \dot{\varphi}\,(-\hat{i}\cos\varphi - \hat{j}\sin\varphi) = -\dot{\varphi}\,\hat{\rho}$$

Again $\vec{A} = \hat{\rho}A_\rho + \hat{\varphi}A_\varphi + \hat{k}A_z$

$$\therefore \frac{d\vec{A}}{dt} = \dot{\hat{\rho}}A_\rho + \rho\frac{dA_\rho}{dt} + \dot{\hat{\varphi}}A_\varphi + \hat{\varphi}\frac{dA_\varphi}{dt} + \hat{z}\frac{dA_z}{dt}$$

$$= \dot{\varphi}\,\hat{\varphi}\,A_\rho + \hat{\rho}\frac{dA_\rho}{dt} - \dot{\varphi}\hat{\rho}\,A_\rho + \hat{\varphi}\frac{dA_\rho}{dt} + \hat{z}\frac{dA_z}{dz}$$

$$= \hat{\rho}\left(\frac{dA_\rho}{dt} - A_\varphi\frac{d\varphi}{dt}\right) + \hat{\varphi}\left(\frac{dA_\rho}{dt} + \dot{\varphi}A_\varphi\right) + \hat{z}\frac{dA_z}{dt}. \qquad (6.9)$$

6.7. Newton's Laws of Motion

Newton's second law of motion embodies the essential physics involved in the mechanics of a particle. For a single particle the law has the form

$$\vec{F} = \frac{d\vec{p}}{dt} \qquad (6.10)$$

where $\vec{F}$ is the force acting on the particle and $\vec{p}$ is the linear momentum of the particle defined as follows. Let s be the curve traced by the particle in its motion and $\vec{r}$ the radius vector from the origin to the particle. The velocity vector is defined formally by the equation

$$\vec{v} = \frac{d\vec{r}}{dt} \tag{6.11}$$

where

$$\frac{d\vec{r}}{dt} = \underset{\delta t \to 0}{\text{Lt}} \frac{\vec{r}_2 - \vec{r}_1}{\delta t} = \underset{\delta t \to 0}{Lt} \frac{\delta \vec{s}}{\delta t} = \frac{d\vec{s}}{dt}. \tag{6.12}$$

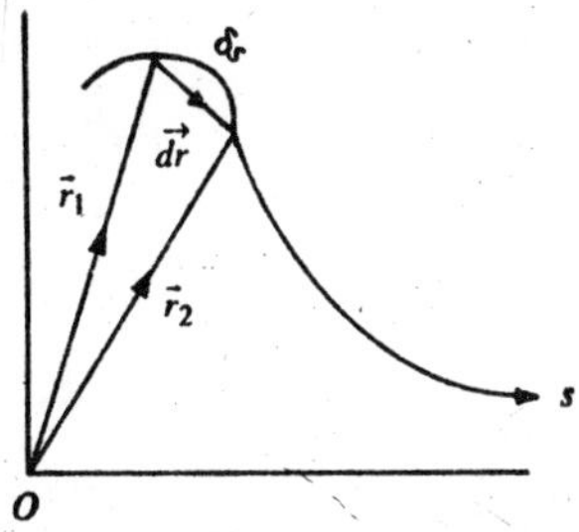

Fig. 6.8

Equation (6.12) suggests that $\vec{v}$ is tangential to the curve. The linear momentum $\vec{p}$ is defined from the velocity as

$$\vec{p} = m\vec{v} \tag{6.13}$$

so that we have from (6.10) $\vec{F} = \frac{d}{dt}(m\vec{v})$. (6.14)

In most cases the mass of the particle is constant and equation (6.14) gives

$$\vec{F} = m\frac{d\vec{v}}{dt} = m\vec{f} \tag{6.15}$$

where $\vec{f}$ is called the acceleration and is defined as

$$\vec{f} = \frac{d\vec{v}}{dt} = \frac{d^2\vec{r}}{dt^2}. \tag{6.16}$$

It is clear that if the total force $\vec{F}$ acting on the particle is zero, then $\dot{\vec{p}} = 0$ i.e., the linear momentum is conserved. This is the conservation theorem for linear momentum.

The angular momentum of the moving particle about O is defined as

$$\vec{L} = \vec{r} \times \vec{p} \tag{6.17}$$

where $\vec{r}$ is the radius vector from O to the particle.

We now define the moment of the force or the torque about O as

$$\vec{\Gamma} = \vec{r} \times \vec{F} \tag{6.18}$$

$$= \vec{r} \times \frac{d}{dt}(m\vec{v}) \tag{6.19}$$

Now $\frac{d}{dt}(\vec{r} \times m\vec{v}) = \vec{v} \times m\vec{v} + \vec{r} \times \frac{d}{dt}(m\vec{v})$

$$= \vec{r} \times \frac{d}{dt}(m\vec{v}), \text{ since } \vec{v} \times \vec{v} = 0$$

From (6.19) we $\vec{\Gamma} = \frac{d}{dt}(\vec{r} \times m\vec{v}) = \frac{d\vec{L}}{dt}$ (6.20)

If the total torque $\vec{\Gamma} = O$, $\frac{d\vec{L}}{dt} = O$ and $\vec{L} =$ constant i.e., the angular momentum is conserved. This is **the conservation theorem for angular momentum.**

The work done by the external force upon the particle in going from point 1 to point 2 $= W_{12} = \int_1^2 \vec{F}.\vec{ds}$

$$= m\int_1^2 \frac{d\vec{v}}{dt}.\vec{ds} = m\int_1^2 \frac{d\vec{v}}{dt}.\vec{v}\,dt = \frac{m}{2}\int_1^2 \frac{d}{dt}(v)^2\,dt$$

$$= \frac{1}{2}m(v_2^2 - v_1^2) = T_2 - T_1 \text{ (say)} \tag{6.21}$$

The scalar quantity $\frac{1}{2}mv^2$ is called the kinetic energy of the particle, so that the work done is the change in Kinetic energy.

If the force field is such that the work done around a closed path is zero, i.e., if $\oint \vec{F}.\vec{ds} = 0$, (6.22)

the force is said to be conservative.

Physically, it is clear that the system will not be conservative if friction and other dissipative forces are present, for $\vec{F}.\vec{ds}$ is always positive due to friction and the integral will not be zero.

By stokes' theorem $\oint \vec{F}.\vec{ds} = \iint \vec{\nabla} \times \vec{F}.\vec{dS}$. Hence by equation (6.22), the condition for the force to be conservative reduces to $\vec{\nabla} \times \vec{F} = 0$. (6.23).

Since the curl of gradient of a scalor always vanishes (i.e., $\vec{\nabla} \times \vec{\nabla}\varphi = 0$), $\vec{F}$ must therefore be expressible as the gradient of a scalar

$$\text{i.e., } \vec{F} = -\vec{\nabla}V. \tag{6.24}$$

The scalar function V is called the potential or potential energy.

In equation (6.24) we can add to V any quantity constant in space without affecting the result. Hence the zero level of V is arbitrary. For a conservative system the work done by the force W_{12}

$$= \int_1^2 \vec{F}.\vec{ds} = -\int_1^2 \vec{\nabla}V.\vec{ds} = V_1 - V_2 \tag{6.25}$$

Combining equations (6.25) and (6.21), we have

$$T_2 - T_1 = V_1 - V_2$$

$$\text{or,} \quad T_1 + V_1 = T_2 + V_2 \qquad (6.26)$$

which states in symbols the energy conservation theorem.

If the forces acting on the particle are conservative, then the total energy of the particle is constant.

Again the time integral of the force =

$$\text{Impulse} = \int_{t_1}^{t_2} \vec{F}\, dt = \int_{t_1}^{t_2} m \frac{d\vec{v}}{dt}\, dt = m(\vec{v}_2 - \vec{v}_1) =$$

change in the linear momentum.

EXERCISE

To prove that all central forces are conservative :

The force acting on a particle P moving under or centre of force is given by $\vec{F} = -f(r)\,\vec{r}$, where $\vec{r} = \overrightarrow{OP}$

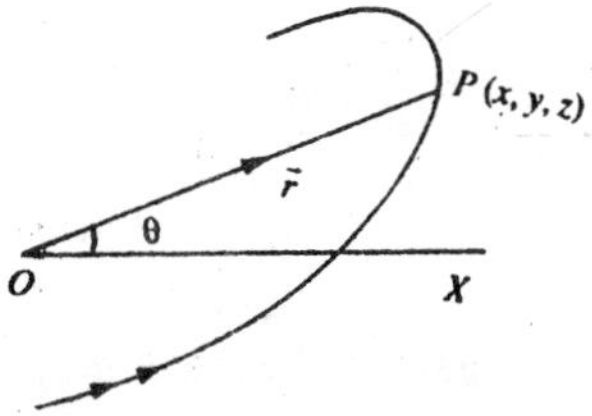

Fig. 6.9

Now curl $\vec{F} = \vec{\nabla} \times \vec{F}$

$$= -\left(\hat{i}\frac{\partial}{\partial x} + \hat{j}\frac{\partial}{\partial y} + \hat{k}\frac{\partial}{\partial r}\right) \times \{f(r)\,(\hat{i}x + \hat{j}y + \hat{k}z)\}$$

$$= -\left(\hat{i}\frac{\partial}{\partial x} + \hat{j}\frac{\partial}{\partial y} + \hat{k}\frac{\partial}{\partial z}\right) \times \{\varphi(r^2)\,(\hat{i}x + \hat{j}y + \hat{k}z)\}$$

$$= -\left\{\hat{k}\frac{\partial}{\partial x}\left[\varphi(r)^2 y\right] + \hat{j}\frac{\partial}{\partial x}\left[\varphi(r^2) z\right] - \hat{k}\frac{\partial}{\partial y}\left[\varphi(r^2) x\right] + \hat{i}\frac{\partial}{\partial y}\left[\varphi(r^2) z\right] + \hat{j}\frac{\partial}{\partial z}\left[\varphi(r^2) x\right] - i\frac{\partial}{\partial z}\left[\varphi(r^2) y\right]\right\}$$

$$= -\{\hat{k}\,2xy - \hat{j}\,2zx - \hat{k}\,2xy + \hat{i}\,2yz + \hat{j}\,2zx - \hat{i}\,2yz\}\,\varphi'(r^2) = 0$$

Since curl $\vec{F}$ is zero, the force $\vec{F}$ must be conservative.

6.8. Motion of a system with variable mass (Rocket motion)

The principle underlying the motion of rockets is the principle of conservation of momentum and Newton's Second Law of Motion.

In the rocket the fuel and the oxidising agent are carried in the aircraft necessary for the purpose of combustion and no air is drawn in from the surroundings.

The oxidising agent oxygen is carried in the liquid form or in the form of oxidisers like H_2O_2, HNO_3 etc.

The fuel may be in the form of liquid hydrogen, paraffin or solid like gun powder.

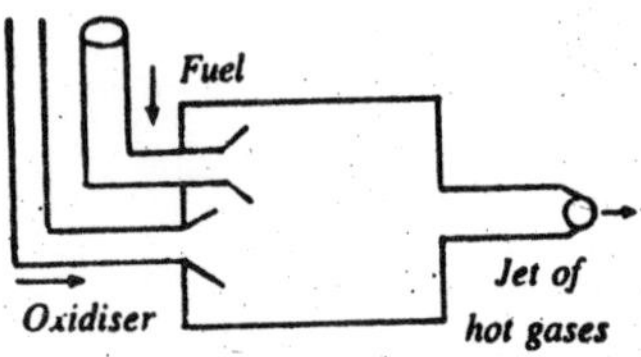

Fig. 6.10

Let m be the mass of (rocket + fuel) at any instant t and let its velocity be v relative to any fixed frame of reference. In time dt, the mass of hot gases ejected from the tail (Fig. 6.10) of the rocket be dm say. Let the exhaust velocity of the gases relative to the rocket ejected from the back of the rocket be u. Then the velocity of exhaust relative to the fixed frame of reference $= v - u$ $(\vec{V}_{BA} = \vec{V}_B - \vec{V}_A)$.

The momentum of the rocket at any instant $t = mv$.

The rate of change of momentum

$$= \frac{d}{dt}(mv) = m\frac{dv}{dt} + v\frac{dm}{dt}. \tag{6.27}$$

The rate of change of mass of the rocket exhaust $= -\dfrac{dm}{dt}$.

The rate of change of momentum of the rocket exhaust $= -\dfrac{dm}{dt}(v - u)$.

(6.28)

The net rate of change of momentum of the system

$$= m\frac{dv}{dt} + v\frac{dm}{dt} - \frac{dm}{dt}(v - u) = m\frac{dv}{dt} + u\frac{dm}{dt}.$$

This must be equal to the resultant external force acting on the system.

$$\therefore\ F = m\frac{dv}{dt} + u\frac{dm}{dt} \tag{6.29}$$

Case I : When the rocket is out of any external field $F = 0$

$$\therefore\ m\frac{dv}{dt} + u\frac{dm}{dt} = 0 \tag{6.30}$$

or, $dv = -\dfrac{u}{m}dm.$

Integrating

$$v = -u \log_e m + c_1. \tag{6.31}$$

Initially $v = 0$, where $m = m_0$

$$\therefore \quad c_1 = u \log_e m_0$$

$$\therefore \quad v = -u \log_e m + u \log_e m_0$$

$$= u \log_e \left(\frac{m_0}{m}\right). \tag{6.32}$$

If $v = v_0$ when $m = m_0$

$$c_1 = v_0 + u \log m_0.$$

Equation (6.31) then becomes

$$v = v_0 + u \log \left(\frac{m_0}{m}\right) \tag{6.33}$$

Case II : When the gravitational forces are taken into account, $F = -mg$.

Equation 6.29 reduces to

$$m \frac{dv}{dt} + u \frac{dm}{dt} = -mg \tag{6.34}$$

or, $$\frac{dv}{dt} + \frac{u}{m} \frac{dm}{dt} = -g.$$

Integrating w. r. t. 't'

$$v + u \log_e m = -gt + c_2.$$

Initially $v = 0$, when $m = m_0$, $t = 0$

$$\therefore \quad c_2 = u \log_e m_0.$$

Substituting $v + u \log_e m = -gt + u \log_e m_0$

or, $$v = u \log_e \left(\frac{m_0}{m}\right) - gt. \tag{6.35}$$

If at $t = 0$, $m = m_0$, $v = v_0$ (multi stage rocket)

$$c_2 = v_0 + u \log_e m_0$$

$$\therefore \quad v = v_0 + u \log_e \left(\frac{m_0}{m}\right) - gt \tag{6.36}$$

EXERCISE

A 8000 kg rocket is set for a vertical firing. If the exhaust speed is 800 m/sec., how much gas must be ejécted per second to supply the thrust needed (1) to overcome the weight of the rocket (2) to give the rocket an initial upward acceleration of 9.8 m/sec^2

Let u be the velocity of exhaust and $\frac{dm}{dt}$ the rate of the gas ejected.

The thrust is given by $u\,\frac{dm}{dt}$.

(i) Hence $u\,\frac{dm}{dt} = mg$

$$\text{or,}\quad \frac{dm}{dt} = \frac{g}{u}\,m = \frac{9.8}{800} \times 8000 = 98 \text{ kg/sec.}$$

(ii) If f be the upward acceleration of the rocket, we have its equation of motion

$$u\,\frac{dm}{dt} - mg = mf$$

$$\text{or,}\quad u\,\frac{dm}{dt} = m\,(g + f)$$

$$\text{or,}\quad \frac{dm}{dt} = \frac{m}{u}\,(g + f) = \frac{8000}{800}\,(9.8 + 9.8)$$

$$= 19.6 \times 10 = 196 \text{ kg/sec.}$$

The stages of a two stage rocket seperately weigh 100 kg and 10 kg and contain 800 kg and 90 kg of fuel respectively. What is the final velocity that can be achieved with an exhaust velocity of 1.5 km/sec. ($\log_e 10 = 2.3$ and $\log_{10} 2 = 0.3$).

From equation (6.33), the final velocity of a rocket is given by $v = v_0 + u \log_e \left(\frac{m_0}{m}\right)$.

For the first stage $m_0 = 800 + 90 + 100 + 10 = 1000$ kg

$$m = 90 + 10 + 100 = 200 \text{ kg},$$

$$v_0 = 0, \qquad u = 1.5 \text{ km/sec.}$$

The velocity of the rocket after all the fuel of the 1st. stage is burnt is given by $v_1 = v_0 + u \log_e \left(\frac{m_0}{m}\right)$

$$= 0 + 1.5 \log_e \frac{1000}{200} = 1.5 \times 2.3\ (\log_{10}10 - \log_{10}2)$$

$$= 1.5 \times 2.3 \times (1 - .3) = 1.5 \times 2.3 \times .7$$

$$= 2.415\ \text{km/sec.}$$

= initial velocity for the 2nd stage

For the 2nd stage

$$m_o = 10 + 90 = 100\ \text{kg}$$

$$m = 10\ \text{kg.}$$

The final velocity at the end of the 2nd stage $= v = 2.415 + 1.5 \log_e \frac{100}{10}$

$$= 2.415 + 1.5 \times 2.3 \times 1 = 2.865\ \text{km/sec.}$$

(b) **Motion of a system of particles :**

Centre of mass and its motion ; simple collision problems (one and two dimensional) ; linear momentum, angular momentum; torque on a system of particles. Energy of a system of particle ; Conservation principles in a system of particles.

6.9. Centre of mass and its motion

Let us consider a system of n particles of masses $m_1, m_2, m_3 \ldots\ldots m_n$ situated at points $A_1, A_2, A_3 \ldots\ldots A_n$. The position vectors of the points with reference to some arbitrary origin O are $\vec{r_1}, \vec{r_2}, \vec{r_3} \ldots\ldots \vec{r_n}$.

The vector sum of the 1st moment of the masses about O is $\sum_i m_i \vec{r_i}$, where m_i is the mass of the i.th particle and $\vec{r_i}$ is the position vector of the ith particle about O. The centre of mass of the system of particles is defined as the point about which the vector sum of the 1st moment of the masses of the system is zero.

To locate the point let us consider a point C, the position vector of which relative to O is given by (Fig. 6.11).

$$\vec{OC} = \frac{\sum_i m_i \vec{r_i}}{\sum_i m_i} \qquad (6.37)$$

Hence the vector sum of the first moment of the masses about C is $\sum_i m_i (\vec{r_i} - \vec{OC}) = \sum_i m_i \vec{r_i} - \sum m_i \vec{OC} = 0$ by virtue of equation (6.37). Thus C denotes the

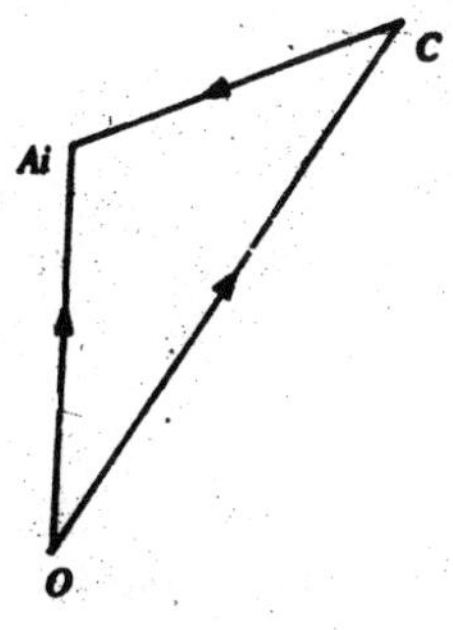

Fig. 6.11

centre of mass of the system and equation (6.37) gives the position vector of C relative to the arbitrary origin O. To **establish the uniqueness of the position of the centre of mass**, let us assume that there are two centres of mass C and C' Fig. (6.12) and let the position vectors of the particles relative to these centres of mass be $\vec{r}_1, \vec{r}_2, \vec{r}_3 ..... \vec{r}_n$ and $\vec{r'}_1, \vec{r'}_2, ... \vec{r'}_n$ respectively.

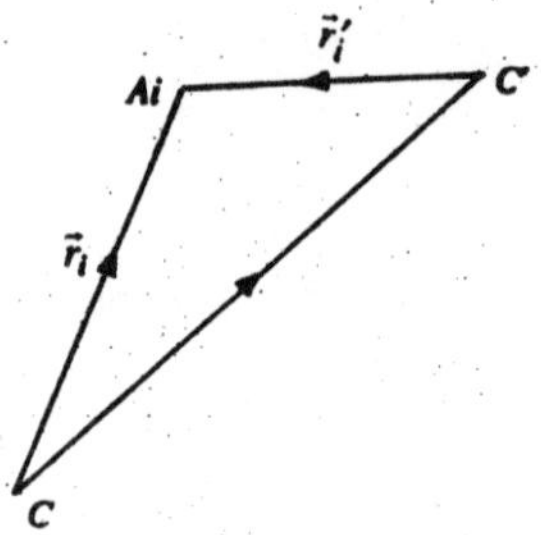

Fig. 6.12

By definition

$$\Sigma\, m_i \vec{r}_i = 0 \text{ and } \Sigma\, m_i \vec{r'}_i = 0 \tag{6.38}$$

But $\vec{r}_i = \vec{r'}_i + \overrightarrow{CC'}$. When the value of $\vec{r}_i$ is substituted in (6.38), we get $\overrightarrow{CC'} = 0$.

This shows that C and C' coincide. Thus the uniqueness of the centre of mass is established.

6.10. Motion of the centre of mass

In considering the motion of a system of particles, we must distinguish between the external forces acting on the system due to sources outside the system, and the internal forces on say, some particle i due to all other particles in the system. The equation of motion of the ith particle can be written as

$$\sum_j \vec{F}_{ji} + \vec{F}_i(e) = \dot{\vec{p}}_i \tag{6.39}$$

where $\vec{F}_i(e)$ stands for the external force, $\vec{F}_{ji}$ is the internal force on the ith particle due to the jth particle. Summed over all the particles, the above equation takes the form

$$\frac{d^2}{dt^2} \sum_i m_i \vec{r}_i = \sum_i \vec{F}_i(e) + \sum_{\substack{i,j \\ i \neq j}} \vec{F}_{ji} \,. \tag{6.40}$$

The first term on the right is simply the total external force $\vec{F}(e)$, while the second term vanishes, since from Newton's third law for each pair $\vec{F}_{ji} + \vec{F}_{ij}$ is zero.

To proceed further let us define a vector $\vec{R}$

$$\vec{R} = \frac{\sum_i m_i \vec{r}_i}{\sum_i m_i} = \frac{\sum_i m_i \vec{r}_i}{M} \tag{6.41}$$

$\vec{R}$ obviously represent the position vector of the centre of mass.

Equation (6.40) reduces to

$$M\frac{d^2\vec{R}}{dt^2}=\sum \vec{F}_i(e)=\vec{F}(e) \tag{6.42}$$

which can be interpreted as "the centre of mass moves as if the total external force were acting concentrated on the centre of mass".

An illustration of the above equation is the motion of an exploding shell, the centre of mass of the fragments travel as if shell were still in a single piece. The same principle is used in jet and rocket propulsion. In order that the motion of the centre of mass remains unaffected the ejection of the exhaust gases at high velocity is counterbalanced by the forward motion of the vehicle.

From equation (6.41) the total linear momentum of the system,

$$\vec{p}=\sum_i m_i\frac{d\vec{r}_i}{dt}=M\frac{d\vec{R}}{dt} \tag{6.43}$$

is the total mass of the system times the velocity of the centre of mas.

$$\text{If } \vec{F}(e)=0\,;\; M\frac{d\vec{R}}{dt}=\vec{p}=\text{constant.}$$

Consequently, the equation of motion of (6.43) can be restated as the conservation theorem of linear momentum of the system of particles : If the total external force is zero, the total linear momentum is conserved.

We obtain the total angular momentum of the system of particles by forming the cross product $\vec{r}\times\vec{p}$ and summing over i.

Thus $\vec{L}=\sum_i \vec{r}_i\times\vec{p}_i$

$$\therefore\; \dot{\vec{L}}=\sum_i\frac{d}{dt}(\vec{r}_i\times\vec{p}_i).$$

If the operation is performed in equation (6.39) there results

$$\dot{\vec{L}}=\sum_i\left\{\vec{v}_i\times\vec{p}_i+\vec{r}_i\times\dot{\vec{p}}_i\right\}=\sum_i\vec{r}_i\times\dot{\vec{p}}_i$$

$$\because\; \vec{v}_i\times\vec{p}_i=0$$

$$=\sum_i\vec{r}_i\times\vec{F}_i(e)+\sum_{ij}\vec{r}_i\times\vec{F}_{ji} \tag{6.44}$$

The last term on the right in (6.44) can be considered as a sum of pairs of the form $\vec{r}_i\times\vec{F}_{ji}+\vec{r}_j\times\vec{F}_{ij}=(\vec{r}_i-\vec{r}_j)\times\vec{F}_{ji}$. But $\vec{r}_i-\vec{r}_j$ is identical with the vector $\vec{r}_{ij}$ from j to i (Fig. 6.13). Hence from the law of vector product $\vec{r}_{ij}\times\vec{F}_{ji}=0$. Equation 6.44 may be written as $\frac{d\vec{L}}{dt}\sum_i\vec{r}_i\times\vec{F}_i(e)=\vec{\Gamma}_e.$ **(6.45)**

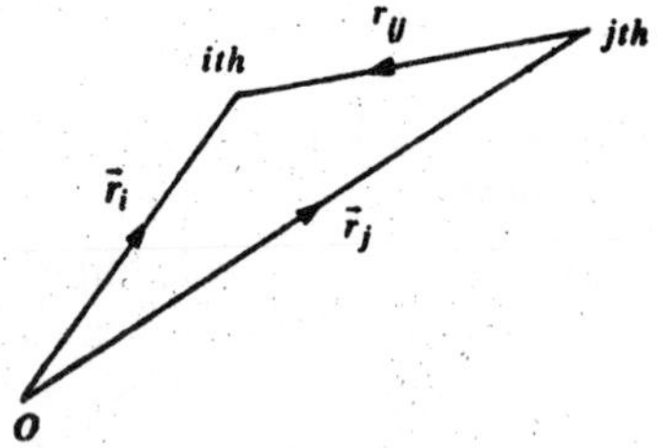

Fig. 6.13

Thus the time derivative of the total angular momentum is equal to the moment of the external force about the given point or the torque. Corresponding to equation (6.45) we have the conservation theorem of the total angular momentum : If the applied external torque is zero, the total angular momentum will be conserved.

6.11. Total energy of a system of particles

The Kinetic energy of a system is defined as the sum of the kinetic energies of its constituent particles. The formal expression is

$$T = \frac{1}{2}\sum_i m_i\left(\dot{x}_i^2 + \dot{y}_i^2 + \dot{z}_i^2\right) \tag{6.46}$$

Differentiating w.r.t. 't'

$$\dot{T} = \sum_i m_i\,(\dot{x}_i\,\ddot{x}_i + \dot{y}_i\,\ddot{y}_i + \dot{z}_i\,\ddot{z}_i)$$

$$= \sum_i (X_i\,\dot{x}_i + Y_i\,\dot{y}_i + Z_i\,\dot{z}_i) \tag{6.47}$$

where X_i, Y_i, Z_i are the compcnents of the total force, external and internal, acting on the ith particle. If W be the total work done by the forces from time t_o time t, we have from (6.47) $\dot{T} = \dot{W}$ (6.48)

Thus for a system of particles the rate of change of kinetic energy is equal to the rate of working of all the forces external and internal.

When the system is conservative, with potential evergy V, we have $\dot{W} = -\dot{V}$.

Hence from (6.48) $\dot{T} + \dot{V} = 0$

or, $T + V = E =$ constant (6.49)

which is the principle of conservation of energy, E being the total energy.

Kinetic energy of a system of particles is also given by $T = \frac{1}{2}\sum_i m_i v_i^2$.

If $\vec{v}$ be the velocity of the centre of mass C and $\vec{v}_i$ that of the ith particle relative to the fixed point O (Fig. 6.14), then the velocity of the ith particle relative to the centre of mass is given by $\vec{v}'_i = \vec{v}_i - \vec{v}$.

Hence the total K. E. of the system is given by

$$T = \frac{1}{2} \sum_i m_i (\vec{v}'_i + \vec{v})^2$$

$$= \frac{1}{2} \sum_i m_i \{ v_i'^2 + v^2 + 2 \vec{v}'_i . \vec{v} \}$$

$$= \frac{1}{2} M v^2 + \frac{1}{2} \sum_i m_i v'^2_i \qquad (6.50)$$

$\because \sum_i m_i \vec{v}'_i . \vec{v} = \vec{v} . \frac{d}{dt} \left(\sum_i m_i \vec{r}'_i \right) = 0$ by equation (6.38).

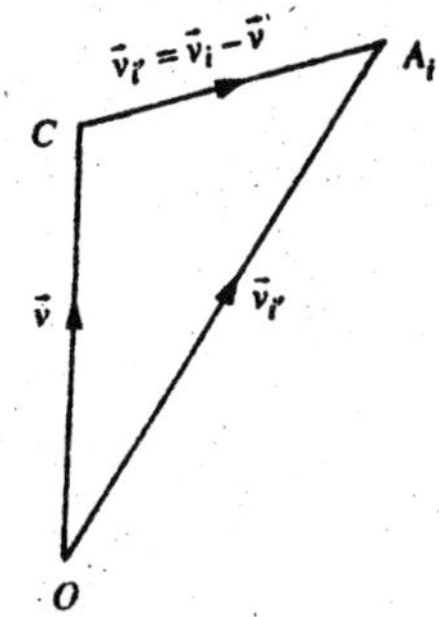

Fig. 6.14

The first term in equation (6.50) represents the K. E. of the centre of mass and the second term the sum of the kinetic energies of all the particles of the system relative to the centre of mass.

6.12. Simple Collision Problems

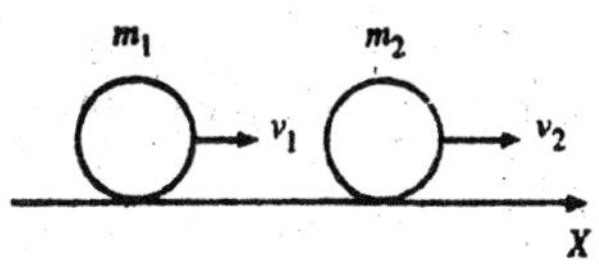

Fig. 6.15

Let two particles of massess m_1 and m_2 travelling along x-axis collide with each other. If v_1, v_2 and v'_1, v'_2 respectively be their velocities before and after collision (Fig 6.15), we have from Newton's Collision Rule

$$v'_1 - v_2' = e (v_2 - v_1) \qquad (6.51)$$

where $e =$ coefficient of restitution.

Applying the principle of conservation of momentum

$$m_1 v_1 + m_2 v_2 = m_1 v_1' + m_2 v_2' \qquad (6.52)$$

Solving the above two equations,

$$v'_1 = \frac{(m_1 - em_2) v_1 + m_2 (1 + e) v_2}{m_1 + m_2} \qquad (6.53)$$

$$v'_2 = \frac{m_1 (1 + e) v_1 + (m_2 - m_1) v_2}{m_1 + m_2}. \qquad (6.54)$$

(a) For perfectly inelastic collision $e = 0$

$$v'_1 = \frac{m_1 v_1 + m_2 v_2}{m_1 + m_2}, \qquad v'_2 = \frac{m_1 v_1 + m_2 v_2}{m_1 + m_2}.$$

After collision the particles move with the same velocity i.e., they move as a single particle.

(b) For perfectly elastic collision, $e = 1$

and $$v'_1 = \frac{(m_1 - m_2)\, v_1 + 2\, m_2 v_2}{m_1 + m_2}, \qquad v'_2 = \frac{2\, m_1 v_1 + (m_1 - m_2)\, v_2}{m_1 + m_2}.$$

To show that in a perfectly elastic collision, the principle of conservation of energy is obeyed.

$$\text{Total K. E. after collision} = \frac{1}{2} m_1 v_1'^2 + \frac{1}{2} m_2 v_2'^2$$

$$= \frac{1}{2} m_1 \left\{ \frac{(m_1 - m_2)\, v_1 + m_2 v_2}{m_1 + m_2} \right\}^2 + \frac{1}{2} m_2 \left\{ \frac{2\, m_1 v_1 + (m_2 - m_1)\, v_2}{m_1 + m_2} \right\}^2$$

$$= \frac{1}{2} m_1 v_1^2 + \frac{1}{2} m_2 v_2^2 = \text{total K. E. before collision.}$$

Oblique Collision :

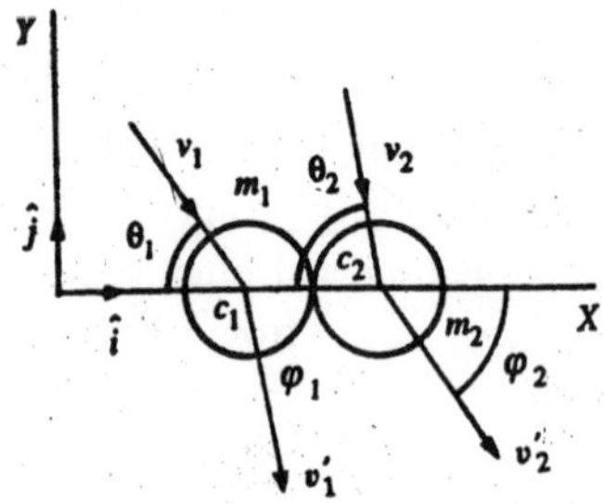

Fig. 6.16

Let v_1, v_2 and v'_1, v'_2 be the velocities of the spheres before and after collision (Fig. 6.16). Let us suppose that XY plane is the plane of v_1 and v_2 and at the instant of impact x-axis passes through the centres c_1 and c_2 of the spheres.

Applying the principle of conservation of momentum,

$$m_1 \vec{V_1} + m_2 \vec{V_2} = m_1 \vec{V_1'} + m_2 \vec{V_2'} \quad (6.55)$$

If v_1 be the magnitude of $\vec{V_1}$ etc. we have

$$\vec{V_1} = v_1 (\cos\theta_1\, \hat{i} - \sin\theta_1\, \hat{j}) \quad (6.56)$$

$$\vec{V_2} = v_2 (\cos\theta_2\, \hat{i} - \sin\theta_2\, \hat{j}) \quad (6.57)$$

$$\vec{V_1'} = v'_1 (\cos\varphi_1\, \hat{i} - \sin\varphi_1\, \hat{j}) \quad (6.58)$$

$$\vec{V_2'} = v'_2 (\cos\varphi_2\, \hat{i} - \sin\varphi_2\, \hat{j}). \quad (6.59)$$

Substituting in (6.55) and equating the coefficients of $\hat{i}$ and $\hat{j}$ we have

$$m_1 v_1 \cos\theta_1 + m_2 v_2 \cos\theta_2 = m_1 v'_1 \cos\varphi_1 + m_2 v'_2 \cos\varphi_2 \quad (6.60)$$

$$m_1 v_1 \sin\theta_1 + m_2 v_2 \sin\theta_2 = m_1 v'_1 \sin\varphi_1 + m_2 v'_2 \sin\varphi_2. \quad (6.61)$$

By Newton's collision rule we have relative velocity after impact along x-axis

$= - e$ (relative velocity before impact along x-axis)

i.e., $\vec{V_1'} \cdot \hat{i} - \vec{V_2'} \cdot \hat{i} = -e(\vec{V_1} \cdot \hat{i} - \vec{V_2} \cdot \hat{i})$ (6.62)

or, $v_1' \cos\varphi_1 - v_2' \cos\varphi_2 = -e(v_1 \cos\theta_1 - v_2 \cos\theta_2)$ (6.63)

Since the tangential velocities before and after impact are equal

$\vec{V_1} \cdot \hat{j} = \vec{V_1'} \cdot \hat{j}$ or, $v_1 \sin\theta_1 = v_1' \sin\varphi_1$ (6.64)

$\vec{V_2} \cdot \hat{j} = \vec{V_2'} \cdot \hat{j}$ or, $v_2' \sin\theta_2 = v_2' \sin\varphi_2$ (6.65)

From equations (6.60) and (6.61) we have

$$\vec{v'}_1 \cos\varphi_1 = \frac{(m_1 - m_2 e)\, v_1 \cos\theta_1 + m_2(1+e)\, v_2 \cos\theta_2}{m_1 + m_2}$$

$$\vec{v'}_2 \cos\varphi_2 = \frac{m_1(1+e)\, v_1 \cos\theta_1 + (m_2 - m_1 e)\, v_2 \cos\theta_2}{m_1 + m_2}.$$

From equations (6.64) and (6.65)

$$\vec{V_1'} = v_1'(\cos\varphi_1 \hat{i} - \sin\varphi_1 \hat{j})$$

$$= \frac{(m_1 - m_2 e)\, v_1 \cos\theta_1 \hat{i} + m_2(1+e)\, v_2 \cos\theta_2 \hat{i}}{m_1 + m_2} - v_1 \sin\theta_1 \hat{j}$$

$$\vec{V_2'} = v_2'(\cos\varphi_2 \hat{i} - \sin\varphi_2 \hat{j})$$

$$= \frac{m_1(1+e)\, v_1 \cos\theta_1 \hat{i} + (m_1 - m_2 e)\, v_2 \cos\theta_2 \hat{i}}{m_1 + m_2} - v_2 \sin\theta_2 \hat{j}.$$

EXERCISE

1. (a) By what fraction is the K. E. of a neutron of mass m_1 decreased in a head-on elastic collision with an atomic mass m_2 initially at rest ?

(b) Calculate the percentage loss of energy when the nucleous is that of

(i) hydrogen $\left(\frac{m_2}{m_1} = 1\right)$, (ii) Carbon $\left(\frac{m_2}{m_1} = 12\right)$, (iii) lead $\frac{m_2}{m_1} = 206$.

[C. U. 1996]

Ans.

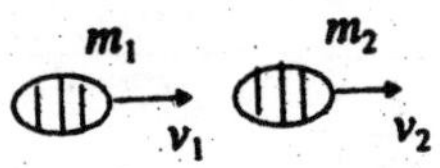

(a) Let the velocities before and after collision be denoted by undashed and dashed letters respectively.

The direction of the velocities are along the same straightline.

For perfectly elastic collision

$$v'_1 = \frac{(m_1 - m_2)\, v_1 + 2\, m_2 v_2}{m_1 + m_2} = \frac{(m_1 - m_2)\, v_1}{m_1 + m_2} \qquad \dots \qquad (1)$$

$$v'_2 = \frac{2\, m_1 v_1 + (m_1 - m_2)\, v_2}{m_1 + m_2} = \frac{2\, m_1\, v_1}{m_2 + v_2}. \qquad \dots \qquad (2)$$

Fraction of the K. E. last

$$= \frac{\frac{1}{2} m_1 v_1^2 - \frac{1}{2} m_1 v_1'^2}{\frac{1}{2} m_1 v_1^2} = 1 - \left(\frac{v'_1}{v_1}\right)^2 = 1 - \left(\frac{m_1 - m_2}{m_1 + m_2}\right)^2$$

$$= \frac{4\, m_1 m_2}{(m_1 + m_2)^2} = \frac{4\dfrac{m_1}{m_2}}{\left(1 + \dfrac{m_1}{m_2}\right)^2}.$$

(b) (i) when $\frac{m_2}{m_1} = 1$

Percentage loss of K. E.

$$= \frac{4}{(1+1)^2} \times 100\% = 100\%.$$

(ii) When $\frac{m_2}{m_1} = 12$

$$\text{Percentage loss of K. E.} = \frac{\frac{4}{12} \times 100}{\left(1 + \frac{1}{12}\right)^2} = \frac{4}{12} \times \frac{12^2}{13^2} \times 100$$

$$= \frac{4 \times 12 \times 100}{13^2} = \frac{48 \times 100}{169} = 28.4\%.$$

(iii) When $m_2/m_1 = 206$

$$\text{Percentage loss of K. E.} = \frac{\frac{4}{206} \times 100}{\left(1 + \frac{1}{206}\right)^2} = \frac{4 \times 100 \times 206}{(207)^2} = 1.92\%.$$

2. A particle moves along a semicircle from *A* to *B*. The force acting on it at any instant is directly proportional to the distance from *A* and is directed

towards A. Show that the work done on it is given by $W = F_0R$, where $F_0 =$ force acting on it at B and $R =$ radius of the circle. [C. U. (Hons.)]

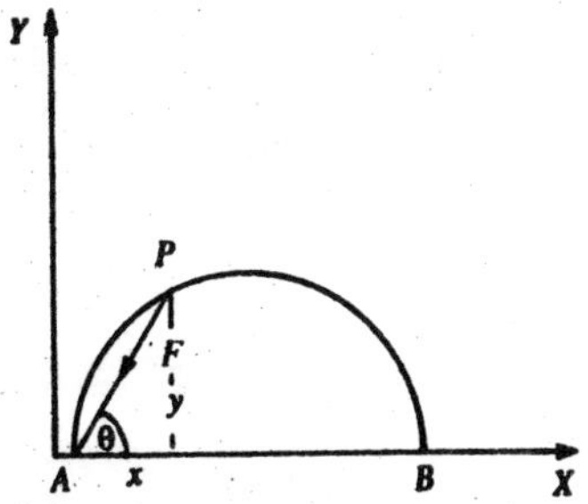

Fig. 6.17

Ans. With reference to the adjascent figure $(x - R)^2 + y^2 = R^2$

or, $x^2 + y^2 = 2xR$

where (x, y) is the co-ordinates of any position of this point with reference to A as this origin.

Now $F = K\sqrt{x^2 + y^2} = K\sqrt{2xR}$

Hence $F_0 = K\sqrt{2R.2R} = 2KR$ or, $K = \dfrac{F_0}{2R}$

$$\therefore\ F = \frac{F_0}{2R}\sqrt{2Rx} = \frac{F_0}{\sqrt{2R}}\sqrt{x}$$

Now work is done for y displacement.

$$\text{Hence } W = \int F\cos\theta\, dx = \int \frac{F_0}{\sqrt{2R}}\sqrt{x}\,\frac{x}{\sqrt{x^2 + y^2}}\,dx$$

$$-\int_0^{2R} \frac{F_0}{\sqrt{2R}}\sqrt{x}\,\frac{x}{\sqrt{2xR}}\,dx = \frac{F_0}{2R}\,\frac{(2R)^2}{2} = F_0 R.$$

3. If T be the instanteneous K. E. of a moving body of mass m, show that $\dfrac{d}{dt}(mT) = \vec{F}.\vec{p}$ where $\vec{F}$ is the force acting on the body and $\vec{p}$ is linear momentum. [C. U. (Hons.)]

Ans. $T = \frac{1}{2}mv^2 = \frac{1}{2}m\vec{v}.\vec{v}$

$$\therefore\ \frac{dT}{dt} = m\frac{d\vec{v}}{dt}.\vec{v} = \vec{F}.\vec{v}$$

$$\frac{d}{dt}(mT) = \frac{d}{dt}\left(\frac{1}{2}m^2\vec{v}.\vec{v}\right)$$

$$= \frac{d}{dt}\left(\frac{1}{2}m\vec{v}.m\vec{v}\right) = \frac{d}{dt}(m\vec{v}).m\vec{v} = \vec{F}.\vec{p}.$$

4. A cannon has its maximum range given by R_{max}. Prove that when the maximum range is actually realised (1) the height reached in such a case is $\frac{1}{4}R_{max}$ and the total time of flight is $2\sqrt{\dfrac{R_{max}}{2g}}$. [C. U. (Hons.) 1995]

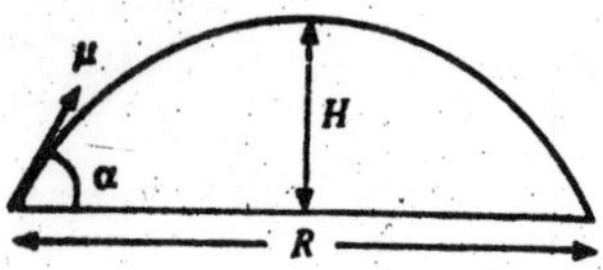

Fig. 6.18

Ans. Let u be the velocity of projection, α the angle of projection and T the time of flight.

Let H be maximum height attained and R the range.

$$\therefore u \cos\alpha \,.\, T = R$$

and $u \sin\alpha \; T - \frac{1}{2} g T^2 = 0$

$$\text{or,} \quad T = \frac{2u \sin\alpha}{g} \qquad \ldots \quad (1)$$

$$\text{Also } R = u \cos\alpha \,.\, \frac{2\,u \sin\alpha}{g} = u^2 \frac{\sin 2\alpha}{g}. \qquad \ldots \quad (2)$$

$$\text{Hence } R_{max} = \frac{u^2}{g} \text{ when } \alpha = 45° \qquad \ldots \quad (3)$$

At the highest point vertical component of velocity is zero.

$$\therefore 0 = u^2 \sin^2\alpha - 2\,gH$$

$$\text{or, } H = \frac{u^2 \sin^2\alpha}{2g} = \frac{u^2}{4g} \quad \text{when } \alpha = 45°$$

$$= \frac{R_{max}}{4}.$$

$$\text{Again, } T = \frac{2\,u \sin 45°}{g} = \frac{\sqrt{2}u}{g} = \frac{\sqrt{2}}{g} \sqrt{R_{max}\, g}$$

$$= 2\sqrt{\frac{R_{max}}{2g}}.$$

5. (a) A particle of mass m moves along the x-axis under the influence of a conservative force field having a potential $V(x)$. If the particle is located at positions x_1 and x_2 at respective times t_1 and t_2, prove that if E is the total energy, $t_2 - t_1 = \sqrt{\frac{m}{2}} \int_{x_1}^{x_2} \frac{dx}{\sqrt{E - V(x)}}$. **[C. U. (Hons.) 1995]**

Ans. Instanteneous K. E. of the particle is given by

$$\frac{1}{2} mv^2 = E - V(x)$$

$$\text{or,} \quad v = \sqrt{\frac{2}{m}} \sqrt{E - V(x)}$$

$$\text{or, } \frac{dx}{dt} = \sqrt{\frac{2}{m}}\sqrt{E - V(x)} \qquad \dots (1)$$

$$\text{or, } \frac{dx}{\sqrt{E - V(x)}} = \sqrt{\frac{2}{m}}\, dt$$

Integrating

$$\int_{t_1}^{t_2} dt = \sqrt{\frac{m}{2}} \int_{x_1}^{x_2} \frac{dx}{\sqrt{E - V(x)}}$$

(b) If in the above problem $V = \frac{1}{2}kx^2$ and the particle starts from rest at $x = a$, find the equation of motion in forms of x and t.

From equation (1)

$$\int \frac{dx}{\sqrt{E - \frac{1}{2}kx^2}} = \int \sqrt{\frac{2}{m}}\, dt \quad \text{or, } \sqrt{\frac{2}{k}} \int \frac{dx}{\sqrt{\frac{2E}{k} - x^2}} = \sqrt{\frac{2}{m}} \int dt$$

$$\text{or, } \sin^{-1}\left(\frac{x}{\sqrt{\frac{2E}{k}}}\right) = \sqrt{\frac{k}{m}}\, t + s.$$

$$\therefore x = \sqrt{\frac{2E}{k}} \sin\left(\sqrt{\frac{k}{m}}\, t + \delta\right) \qquad \dots (2)$$

$$\frac{dx}{dt} = \sqrt{\frac{2E}{x}}\sqrt{\frac{k}{m}} \cos\left(\sqrt{\frac{k}{m}}\, t + \delta\right).$$

At $t = 0$, $x = a$, $\frac{dx}{dt} = 0$

$$\therefore \cos\delta = 0 = \cos \pi/2$$

$$\text{or, } \delta = \pi/2.$$

$$\text{Hence } x = \sqrt{\frac{2E}{K}} \sin\left(\sqrt{\frac{k}{m}}\, t + \pi/2\right) = \sqrt{\frac{2E}{K}} \cos \sqrt{\frac{k}{m}}\, t$$

At $t = 0$, $x = a$

$$\therefore a = \sqrt{\frac{2E}{K}} \quad \text{or, } E = \frac{ka^2}{2}$$

$$\therefore x = a \cos \sqrt{\frac{ka^2}{2m}}\, t.$$

6. A frictionless circular disc of mass M is rotating with a constant angular velocity ω about an axis passing through its centre and perpendicular to its

plane. With what velocity relative to the disc will a fly of mass m at a distance r from the centre of the disc will move so that it will not fall off.

[C. U. (Hons.)]

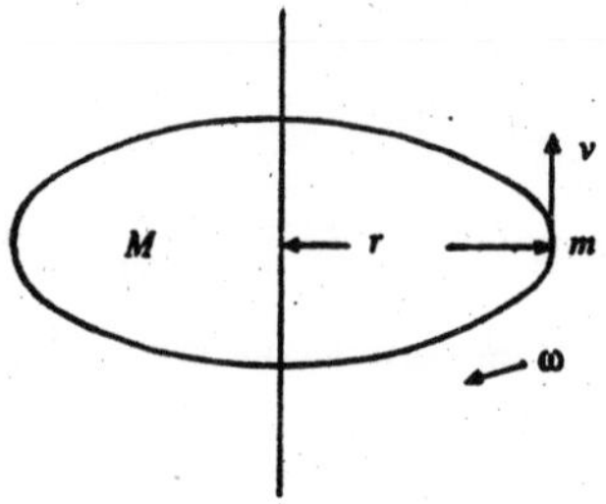

Fig. 6.19

Ans. Let v be the velocity of the fly. Applying the principle of conservation of angular momentum

$$mvr + \left(\frac{1}{2}Mr^2 + mr^2\right)\omega = 0$$

or, $$v = \frac{(M + 2m)\,r}{2m}\,\omega$$

$$\therefore\ V = \omega r$$

Hence the required relative velocity $= v + V$

$$= \omega r - \frac{(M + 2m)}{2m}\,\omega r = \frac{M\omega r}{2m}$$

CHAPTER VII

ROTATIONAL MOTION

Angular velocity, angular acceleration and angular momentum ; Principle of conservation of angular momentum ; Moment of inertia, Radius of gyration ; Energy of rotating bodies ; Parallel and perpendicular axes theorems in Moment of inertia. Calculation of Moment of inertia in simple symmetric systems : Ellipsoid of inertia ; Principal axes and principal moments of inertia : Setting up of Principal Axes in simple symmetric cases like sphere, cylinder right circular cone and rectangular plates, kinetic energy and angular momentum of rotating rigid bodies.

7.1. Angular velocity of a particle

Let a particle be rotating about an axis in a circular path of radius r. If the particle moves through an angle $d\theta$ in going from P to Q in time δt (Fig. 7.1), then the angular velocity of the particle is defined by $\omega = \underset{\delta t = 0}{\text{Lt}} \dfrac{\delta\theta}{\delta t} = \dfrac{d\theta}{dt} = \dot{\theta}$.

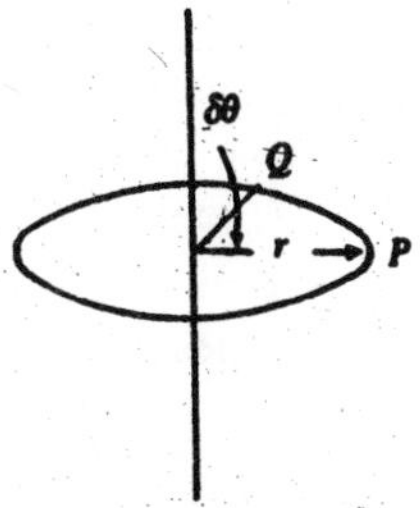

Fig. 7.1

If v be linear velocity of the particle tangential to the circular path it traverses

$$v = \underset{\delta t \to 0}{\text{Lt}} \frac{PQ}{\delta t} = \underset{\delta t \to 0}{\text{Lt}} \frac{\delta s}{\delta t} = \frac{ds}{dt} = \underset{\delta t \to 0}{\text{Lt}} \frac{r\,\delta\theta}{\delta t} = r\omega \qquad (7.1)$$

Angular acceleration (α) is defined as the rate of change of angular velocity i.e.

$$\alpha = \frac{d\omega}{dt} = \frac{d^2\theta}{dt^2}.$$

7.2. Moment of inertia (I)

Let a rigid body rotating about an axis be supposed to be made up of particles of masses m_1, m_2, ... at distances r_1, r_2, ... from the axis of

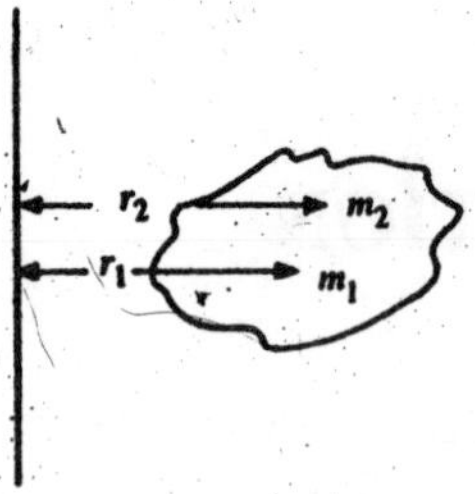

Fig. 7.2

rotation, then $m_1r_1^2 + m_2r_2^2 + ... = \sum_i m_i r_i^2$ i.e., the summation of the second moment of the masses about the axis is known as the moment of inertia of the body about the axis.

Radius of gyration : If the whole mass of the body is supposed to be concentrated at one point so as to have the same moment of inertia about the axis as the body, then the distance of the point from the axis is called the radius of gyration of the body i.e., if $\sum_i m_i r_i^2 = Mk^2$, where M is the total mass of the body, then K is called the radius of gyration of the body about the axis or $k^2 = \dfrac{\sum_i m_i r_i^2}{\sum_i m_i}$.

7.3. Theorem of perpendicular axes (two dimensions)

Let us consider a lamina body and two rectangular axes OX and OY lying in the plane of the lamina. OZ is a third axis perpendicular to the plane of the lamina and passing through O. If I_x, I_y and I_z be the moments of inertia of the lamina about OX, OY and OZ axes then this theorem states that

$$I_x + I_y = I_z.$$

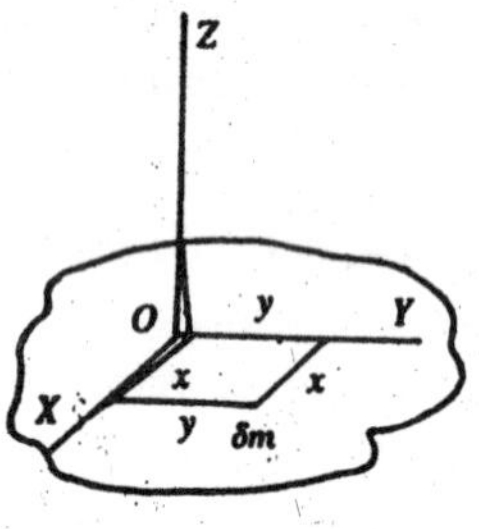

Fig. 7.3

Proof : Let us consider an element of the lamina of mass δm having co-ordinates (x, y) (Fig. 7.3).

Evidently $I_x = \sum \delta m . y^2$, $I_y = \sum \delta m . x^2$.

$$\therefore \quad I_x + I_y = \sum \delta m (x^2 + y^2) = I_z.$$

Theorem of perpendicular axes (three dimensions) :

Statement : If I_x, I_y, I_z are the moments of inertia of a rigid body about three mutually perpendicular concurrent axes OX, OY, OZ and I_o its moment of inertia about the origin O, then $I_x + I_y + I_z = 2I_o$.

Proof : Let us consider an element P of the body of mass δm having co-ordinates (x, y, z) (Fig. 7.4).

Evidently $I_x = \Sigma\ \delta m\ PN^2 =$

$\Sigma\ \delta m\ (PM^2 + MN^2) = \Sigma\ \delta m\ (y^2 + z^2)$

$I_y = \Sigma\ \delta m\ (PR^2) = \Sigma\ \delta m\ (x^2 + z^2)$

$I_z = \Sigma\ \delta m\ (PQ^2) = \Sigma\ \delta m\ (x^2 + y^2)$

Adding

$I_x + I_y + I_y = 2\ \Sigma\ \delta m\ (x^2 + y^2 + z^2)$

$= 2\ \Sigma\ \delta m\ P_o^2 = 2\ I_o$

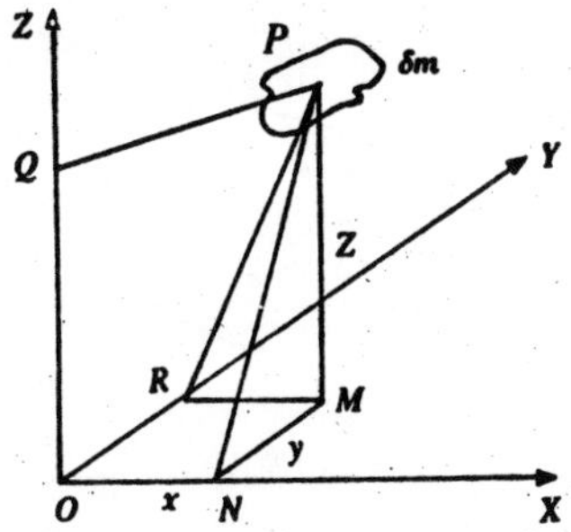

Fig. 7.4

Theorem of parallel Axes (General case) :

Statement : If I be the moment of inertia of a body about an axis and I_c that about a parallel axis through c.g. then $I = I_c + Ma^2$, where $M =$ mass of the body, $a =$ distance between the two axes.

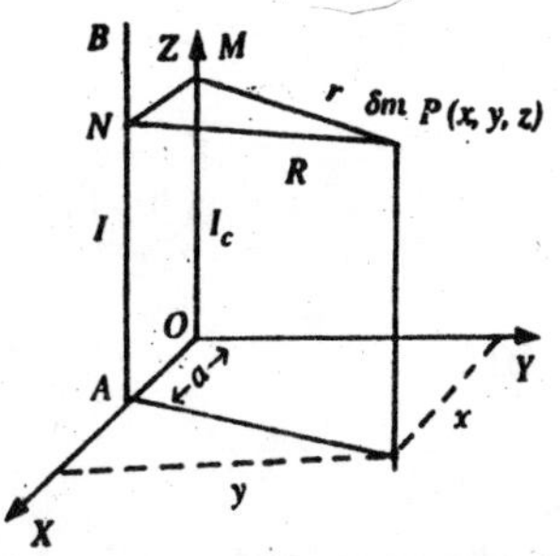

Fig. 7.5

Proof : P is an element of mass δm having co-ordinates (x, y, z). O is the c.g. of the body OZ is the axis about which the moment of inertia is I_c and AB is the parallel axis about which the moment of inertia is I. $AO = a$ (Fig. 7.5).

Now

$I = \Sigma\ \delta m\ (PN^2) = \Sigma\ \delta m\ \{(x - a)^2 + y^2\}$

$= \Sigma\ \delta m\ (x^2 + y^2) - \Sigma\ \delta m\ 2\,ax + \Sigma\ \delta m\ a^2$

$= \Sigma\ \delta m.\ PM^2 + a^2\ \Sigma\ \delta m - 2a\ \Sigma\ \delta m\ x$

$= I_c + Ma^2.$

Since $\dfrac{\Sigma\ \delta m x}{\Sigma \delta m} = \bar{x} = 0$ as O is the c.g. of the body.

7.4. Kinetic energy of a rotating body

Let a rigid body rotating about an axis with an angular velocity ω be supposed to be constituted of particles of masses m_1, m_2, ... at distances r_1, r_2 ..... from the axis (Fig. 7.6).

The K. E. of the whole body

$= \frac{1}{2}\ m_1 v_1^2 + \frac{1}{2} m_2 v_2^2 + \ .....$

$= \frac{1}{2} \omega^2\ (m_1 r_1^2 + m_2 r_2^2 + \ .....$

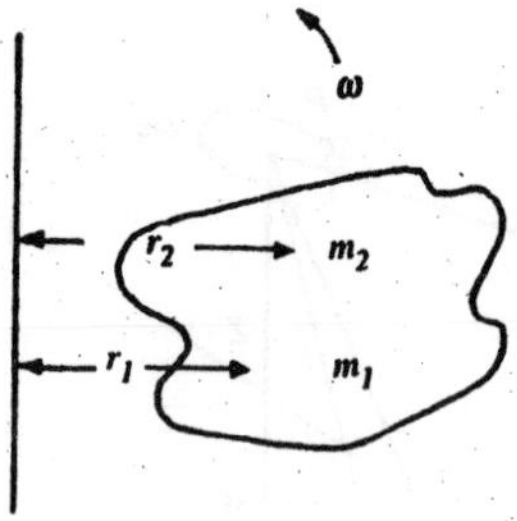

Fig. 7.6

$= \frac{1}{2}\omega^2 I$, where I is the moment of inertia of the body about the axis.

7.5. Calculation of moment of inertia in case of simple symmetric bodies

(a) **Thin rod (uniform) of mass M and length l :**

(i) About an axis passing through one of its ends and perpendicular to its length.

Considering an element of length dx at a distance x from the axis, its moment of inertia

$= \delta m\, x^2 = \frac{M}{l}\, dx\, x^2.$

Hence the moment of inertia of the whole rod

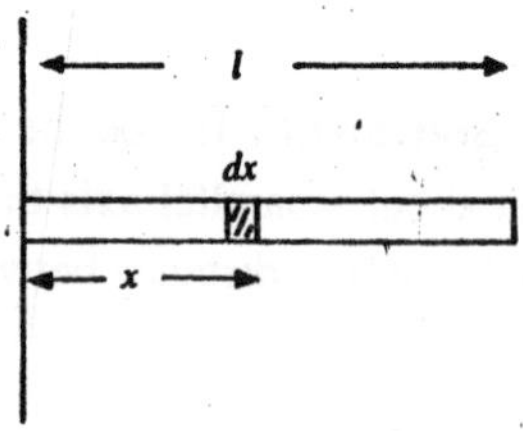

Fig. 7.7

$$I = \Sigma\, \frac{M}{l} x^2\, dx = \frac{M}{l} \int_0^l x^2 dx = \frac{M}{l}\, \frac{l^3}{3} = \frac{1}{3} Ml^2 \tag{7.2}$$

(ii) When the axis passes through its middle point we have by the theorem of parallel axes

$$I' = I - \frac{Ml^2}{4} = \frac{1}{3} Ml^2 - \frac{1}{4}\, Ml^2 = \frac{1}{12}\, Ml^2 \tag{7.3}$$

(b) **Rectangular Plate :**

(i) About an axis passing through the middle point and parallel to its length. (Mass = M, length = l, breadth = b)

let us divide the plate into thin strips parallel to its breadth. Moment of inertia of each any of such strips = $\frac{1}{12}\, mb^2$.

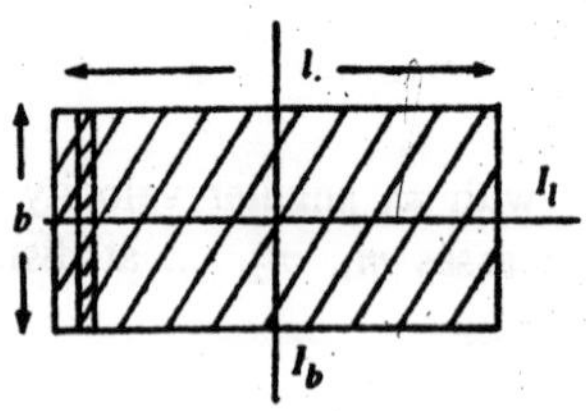

Fig. 7.8

Summing up for all such strips.

$$I_l = \Sigma\, \frac{1}{12}\, mb^2 = \frac{b^2}{12}\, \Sigma\, m = \frac{Mb^2}{12} \tag{7.4}$$

(ii) In a similar manner when the axes is parallel to the breadth

$$I_b = \frac{Ml^2}{12} \tag{7.5}$$

By the theorem of perpendicular axes the moment of inertia of the plate about an axis passing through its middle point and perpendicular to its plane is given by $I = I_l + I_b = \frac{M}{12}(l^2 + b^2)$. (7.6)

(c) **Rectangular Parallelopiped** (Mass = M, length = l, breadth = b, height = h) :

(i) About an axis passing through its middle point and parallel to its length.

A section of the parallelopiped by a plane perpendicular to its length will be a rectangular slab of length h and breadth b. Hence by the previous article

$$I_l = \Sigma \frac{m}{12}(h^2 + b^2) = \frac{M}{12}(h^2 + b^2) \tag{7.7}$$

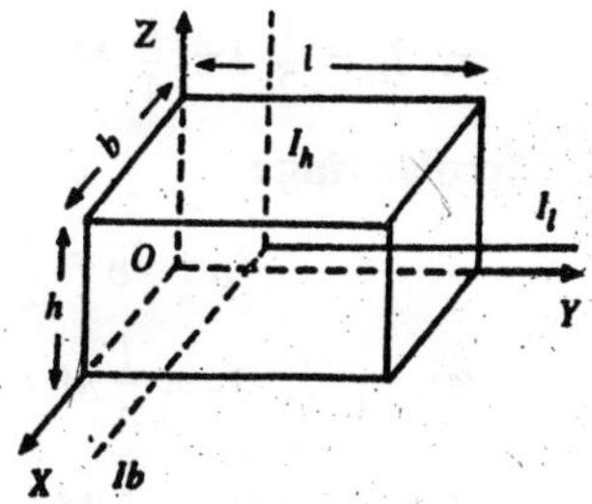

Fig. 7.9

Similarly (ii) $I_b = \frac{M}{12}(l^2 + h^2)$ (7.8)

and $I_h = \frac{M}{12}(l^2 + b^2)$ (7.9).

(d) **Circular disc and ring** (Mass = M, radius = r) :

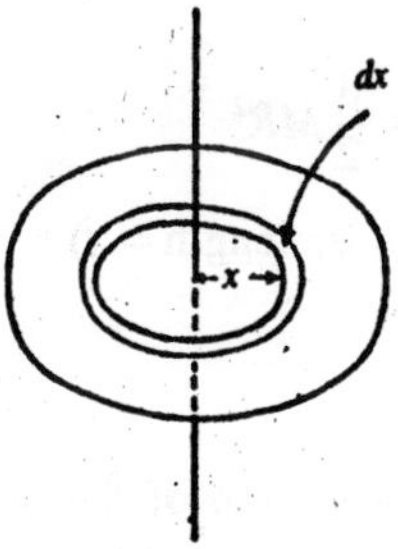

Fig. 7.10

(i) **About an axis passing through its centre and perpendicular to its plane :** Let us consider a thin concentric annular ring of radius x and thickness dx (Fig. 7.10). If m be the mass of the ring, its moment of inertia about the given axes is $mx^2 = 2\pi x dx \sigma x^2$, σ = mass per unit area.

Hence the moment of inertia of the whole disc

$$I = \int_0^r 2\pi x^3 dx \sigma = 2\pi\sigma\frac{r^4}{4} = \pi r^2 \sigma \frac{r^2}{2} = \frac{Mr^2}{2} \tag{7.10}$$

For a circular ring of mass M, internal radius r_1 and external radius r_2

$$I = 2\pi\sigma\int_{r_1}^{r_2} x^3 dx = \frac{2\pi\sigma}{4}(r_2^4 - r_1^4) = \frac{\pi(r_2^2 - r_1^2)}{2}\sigma(r_2^2 + r_1^2)$$

$$= \frac{M(r_2^2 + r_1^2)}{2} \tag{7.11}$$

For a thin ring $r_1 = r_2 = R$ say

$$\therefore \; I = MR^2. \tag{7.12}$$

(ii) About a diameter :

Circular disc : If I_D be the moment of inertia about a diameter I_D will also be the moment of inertia about a perpendicular diameter, then by the theorem of perpendicular axes $I_D + I_D = I$ (two dimensions)

$$\text{or,} \quad I_D = \frac{1}{2} I = \frac{1}{4} Mr^2 \tag{7.13}$$

Circular ring :

$$I_D + I_D = \frac{1}{2} M (r_1^2 + r_2^2)$$

$$\therefore \; I_D = \frac{1}{4} M (r_1^2 + r_2^2)$$

For a thin ring $I_D = \frac{1}{2} Mr^2$.

(iii) About a tangent :

Circular disc : If I_T be the moment of inertia about a tangent, then by the theorem of parallel axes $I_T = I_D + Mr^2 = \frac{1}{4} Mr^2 + Mr^2 = \frac{5}{4} Mr^2$ (7.14)

Circular ring : $I_T = I_D + Mr_2^2 = \frac{M}{4} (r_1^2 + 5r_2^2)$

For a thin ring $I_T = I_D + MR^2 = \frac{1}{2} MR^2 + MR^2 = \frac{3}{2} MR^2$

(e) **Right circular cylinder** (Mass = M, radius = r, length = l) :

Solid cylinder :

(i) About the axis of the cylinder :

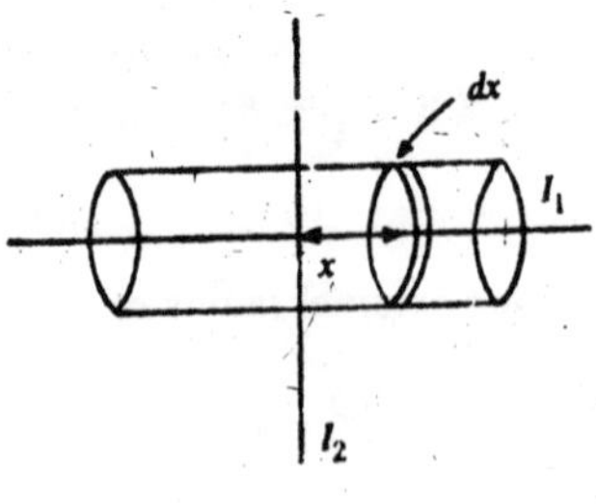

Fig. 7.11

Let us make a section of the cylinder by a plane perpendicular to its axis. The section will be a circular disc of radius r. Let it be at a distance x from the centre and let its thickness be dx (Fig. 7.11).

Its moment of inertia about the axis of the cylinder $= \frac{1}{2} mr^2$ (m = its mass). Summing up for all such discs, the total moment of inertia of the whole cylinder

$$I_1 = \Sigma \frac{1}{2} mr^2 = \frac{r^2}{2} \Sigma m = \frac{1}{2} Mr^2. \qquad (7.15)$$

(ii) About an axis passing through the middle point and perpendicular to the axis of the cylinder :

Let us consider a diameter of the circular disc considered above which is parallel to the given axis. By the theorem of parallel axes the moment of inertia of the circular disc about the given axis is $\frac{1}{4} mr^2 + mx^2$.

Summing up for all the discs the moment of inertia of the whole cylinder

$$I_2 = \Sigma \frac{1}{4} mr^2 + \Sigma mx^2$$

$$= \frac{1}{4} r^2 \Sigma m + \Sigma \pi r^2 dx \rho x^2 \qquad \text{where } \rho = \text{mass/vol.}$$

$$= \frac{Mr^2}{4} + \pi r^2 \rho 2 \int_0^{l/2} x^2 dx = \frac{Mr^2}{4} + 2\pi r^2 \rho \left[\frac{x^3}{3}\right]_0^{l/2}$$

$$= \frac{Mr^2}{4} + 2\pi r^2 \rho \frac{l^3}{24} = \frac{Mr^2}{4} + \pi r^2 \rho l \frac{l^2}{12}$$

$$= M\left(\frac{l^2}{12} + \frac{r^2}{4}\right).$$

(iii) For a cylindrical shell of internal radius r_1 and external radius r_2

$$I_2 = \Sigma \frac{1}{4} m (r_1^2 + r_2^2) + \Sigma \pi (r_2^2 - r_1^2) dx \rho x^2$$

$$= \frac{1}{4} (r_1^2 + r_2^2) \Sigma m + \pi (r_2^2 - r_1^2) \rho . 2 \int_0^{l/2} x^2 dx$$

$$= \frac{M}{4} (r_1^2 + r_2^2) + \frac{Ml^2}{12}.$$

(f) Solid sphere and spherical shell (Mass = M, Radius = R) :

(i) Moment of inertia about a diameter :

Let us consider a section of the solid sphere by a plane perpendicular to the diameter. The section will be a circular disc. Let it be at a distance x from the centre and have a thickness dx. If m be the mass of this disc, its moment of inertia about the diameter

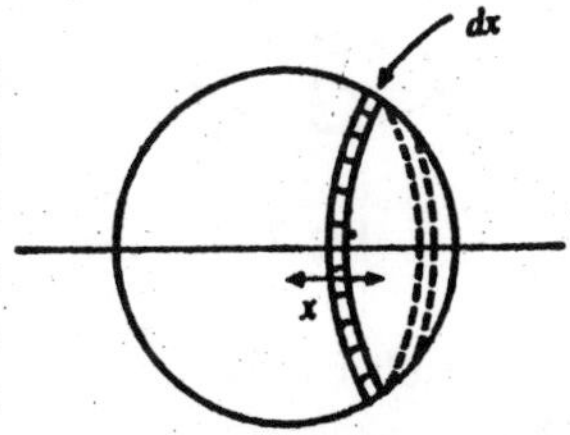

Fig. 7.12

$$= \frac{1}{2} m (R^2 - x^2) = \frac{1}{2} \pi (R^2 - x^2)\, dx\, \rho\, (R^2 - x^2), \qquad \rho = \text{mass/vol.}$$

$$= \frac{1}{2} \pi\rho\, (R^2 - x^2)^2\, dx.$$

Hence the moment of inertia of the whole sphere about the diameter

$$I_D = \frac{1}{2} \pi\rho \times 2 \int_0^R (R^2 - x^2)^2\, dx = \pi\rho \left[R^4 x - 2R^2 \frac{x^3}{3} + \frac{x^5}{5} \right]_0^R$$

$$= \pi\rho\, \frac{15 + 3 - 10}{15} R^5 = \frac{8}{15} \pi\rho\, R^5 = \frac{4}{3} \pi R^3 \rho\, \frac{2}{5} R^2$$

$$= M \frac{2}{5} R^2 = \frac{2}{5} MR^2.$$

For a spherical shell of internal radius R_1 and external radius R_2

$$I_D = \frac{2}{5} M_2 R_2^2 - \frac{2}{5} M_1 R_1^2 = \frac{2}{5} \left\{ \frac{4}{3} \pi R_2^3 \rho\, R_2^2 - \frac{4}{3} \pi R_1^3 \rho\, R_1^2 \right\}$$

$$= \frac{2}{5} \pi\rho\, (R_2^5 - R_1^5) \frac{4}{3} = \frac{2}{5} \pi \frac{M (R_2^5 - R_1^5)^{4/3}}{\frac{4}{3} \pi (R_2^3 - R_1^3)}$$

$$= \frac{2}{5} M \frac{R_2^4 + R_2^3 R_1 + R_2^2 R_1^2 + R_2 R_1^3 + R_1^4}{R_2^2 + R_2 R_1 + R_1^2}$$

For a thin spherical shell $R_1 \to R_2 \to r$

$$\therefore\ I_D = \frac{2}{5} M \frac{5r^4}{3r^2} = \frac{2}{3} Mr^2$$

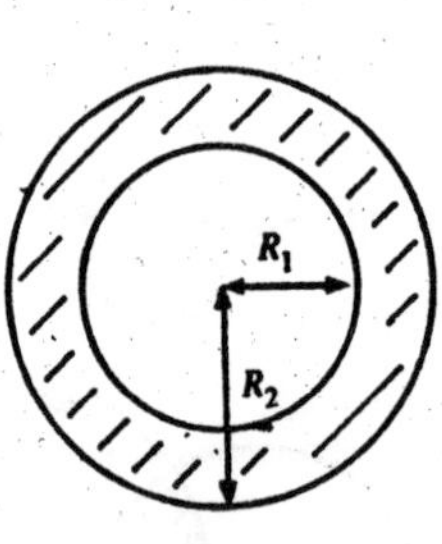

Fig. 7.13

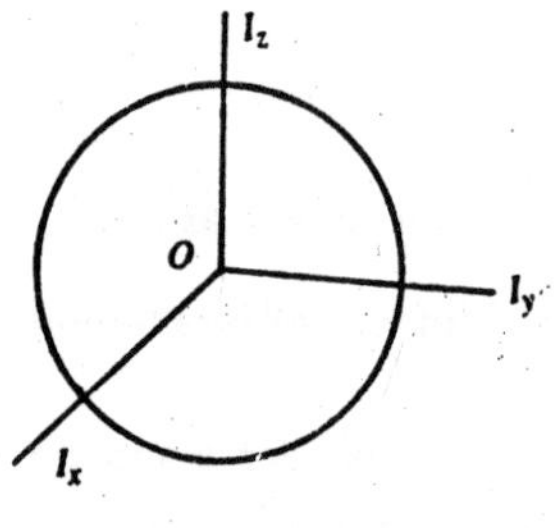

Fig. 7.14

Same result can be obtained by applying the theorem of perpendicular axes (three dimensions)

$$I_x + I_y + I_z = 2I_o$$

or, $3 I_D = 2 I_o = 2 Mr^2$

$\therefore \; I_D = \frac{2}{3} Mr^2$

(ii) Moment of inertia about a tangent :

For a solid sphere $I_T = I_D + MR^2$ (theorem of parallel axes) $= \frac{2}{5} MR^2 + MR^2 = \frac{7}{5} MR^2$

For thin spherical shell $I_T = \frac{2}{3} Mr^2 + Mr^2 = \frac{5}{3} Mr^2$

For thick spherical shell $I_T = I_D + Mr_2{}^2$.

(g) Elliptic disc (Mass = M, Semi major axis = a, Semi minor axis = b) :

(i) About major axis :

Let us consider a thin strip of length y above major axis and perpendicular to it at a distance x from the origin and of thickness dx (Fig. 7.15).

Moment of inertia of the strip about major axis $\frac{1}{3} my^2$. Considering its counter part below the major axis the total moment of inertia $= \frac{2}{3} my^2$

$= \frac{2}{3} ydx\, \sigma y^2$, σ = mass per unit area

Summing up for all such strips $I_a = \Sigma \frac{2}{3} y^3\, dx\, \sigma$

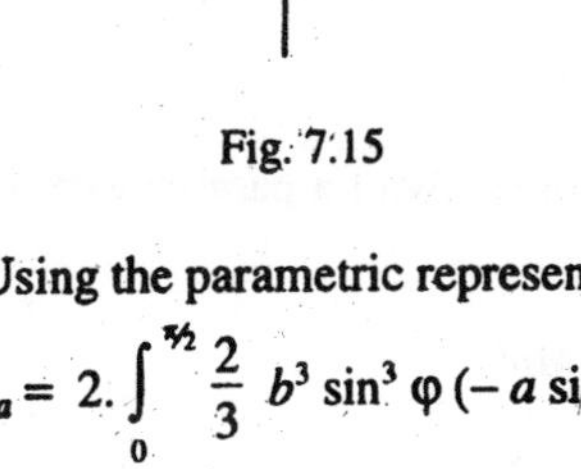

Fig. 7.15

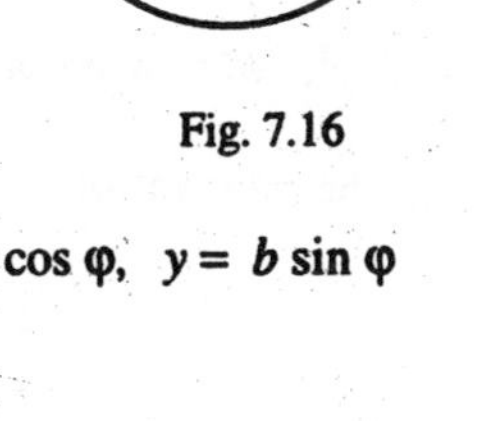

Fig. 7.16

Using the parametric representation $x = a \cos \varphi$, $y = b \sin \varphi$

$$I_a = 2.\int_0^{\pi/2} \frac{2}{3} b^3 \sin^3 \varphi \,(-a \sin \varphi \, d\varphi)\, \sigma$$

$$= \frac{4}{3} b^3 a \int_0^{\pi/2} \sin^4 \varphi \, d\varphi \, \sigma = \frac{4}{3} ab^3 \, \frac{3.1}{4.2} \, \pi/2 \, \sigma$$

$$= \frac{\pi\, ab^3}{4}\, \sigma = \pi\, ab\, \sigma\, \frac{b^2}{4} = \frac{Mb^2}{4}$$

In a similar manner

(ii) $I_b = \frac{Ma^2}{4}$.

(iii) By the theorem of perpendicular axes (two dimensions) moment of inertia of an elliptic disc about an axis passing through its centre and perpendicular to its plane is given by

$$I = I_a + I_b = \frac{M}{4}\,(a^2 + b^2).$$

(h) Triangular lamina :

(i) About an axis passing through the vertex and parallel to the base :

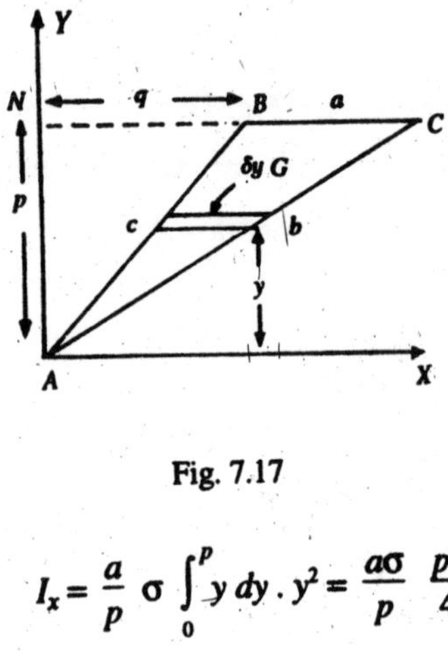

Fig. 7.17

Let ABC be a uniform triangular lamina of surface density σ. AX and AY are two rectangular axes passing through the vertex A. The base BC is parallel to AX (Fig. 7.17).

Let $AN = p$ be the length of the perpendicular from A on BC. Let us consider a strip of co-ordinate y and width dy its length $= \frac{a}{p}\, y$.

Moment of inertia of the strip about AX is

$$I_x = \frac{a}{p}\, \sigma \int_0^p y\, dy \,.\, y^2 = \frac{a\sigma}{p}\, \frac{p^4}{4}$$

$$= \frac{1}{2}\, a\, p\, \sigma\, \frac{p^2}{2}$$

$$= \frac{1}{2}\, Mp^2, \quad M = \text{mass of the lamina.}$$

If I_G be the moment of inertia of the lamina about a parallel axis through the c.g.,

$$I_G + M\, \frac{4}{9}\, p^2 = I_x = \frac{1}{2}\, Mp^2$$

$$\therefore\; I_G = \frac{1}{2}\, Mp^2 - \frac{4}{9}\, Mp^2 = \frac{1}{18}\, Mp^2$$

Also $\quad I_{BC} = \frac{1}{18} Mp^2 + \frac{1}{9} Mp^2 = \frac{1}{6} Mp^2$

If the co-ordinates of the centre of the strip is (x, y), we have $\frac{y}{x} = \frac{p}{q + \frac{a}{2}}$

where $q = BN$.

The length of the strip $= \frac{a}{p} y$

By the theorem of parallel axis

$$I_y = \int_0^p \frac{ay}{p} \sigma \left\{ \frac{1}{3} \left(\frac{ay}{2p} \right)^2 + \left(\frac{q + \frac{a}{2}}{p} \right)^2 y^2 \right\} dy$$

$$= \frac{a\sigma}{p} \left\{ \frac{1}{48} a^2 p^2 + \left(q + \frac{a}{2} \right)^2 \frac{p^2}{4} \right\}$$

$$= 2m \left\{ \frac{a^2}{12} + \frac{q^2 + aq}{4} \right\} = m \left\{ \frac{a^2}{6} + \frac{aq + q^2}{2} \right\}$$

By the theorem of perpendicular axes (two dimensions), moment of inertia of the triangular lamina about an axis passing through A and perpendicular to the plane of the lamina is given by $I = I_x + I_y$.

$$= m \left\{ \frac{p^2}{2} + \frac{a^2}{2} + \frac{aq + q^2}{2} \right\} = m \left\{ \frac{b^2 + a^2}{4} - \frac{a^2}{12} \right\}$$

Since $b^2 = p^2 + (a + q)^2$ and $c^2 = p^2 + q^2$

(i) Moment of inertia of a right circular cone :

(i) About the axis of the cone (solid) (Mass $= M$, radius of base $= a$, height $= h$) :

Let us consider a section of the cone by a plane perpendicular to its axis. The section will be a circular disc (Fig. 7.18). Moment of inertia of this disc about the axis of the cone $= \frac{1}{2} mr^2$.

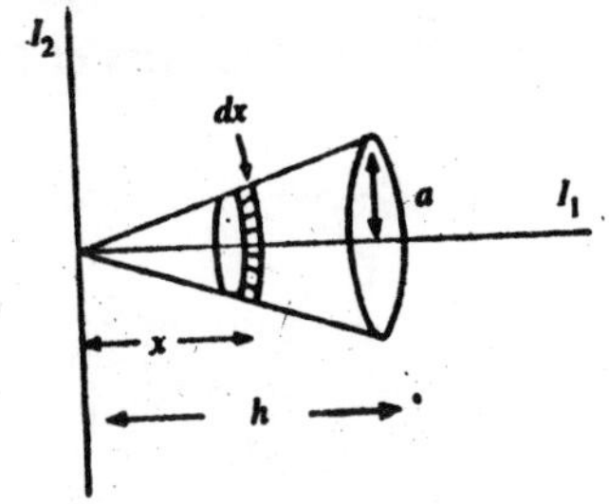

Fig. 7.18

Hence the moment of inertia of the whole cone

$$I_1 = \Sigma \frac{1}{2} mr^2 = \Sigma \frac{1}{2} \pi r^2 \, dx \, r^2 \rho$$

$$= \frac{\pi \rho}{2} \Sigma r^4 dx$$

Now $\frac{r}{x} = \frac{a}{h}$ $\quad \therefore r = \frac{a}{h} x$ where ρ = mass/vol.

x = distance the disc from the vertex ;

dx = its thickness.

Substituting, $I_1 = \frac{\pi\rho}{2} \frac{a^4}{h^4} \int_0^h x^4 dx = \frac{\pi\rho\, a^4}{2h^4} \frac{h^5}{5}$

$$= \frac{\pi a^4 h\rho}{10} = \frac{1}{3} \pi a^2 h\rho \frac{3}{10} a^2 = \frac{3}{10} Ma^2$$

(ii) About an axis passing through the vertex and perpendicular to the axis of the cone :

Let us consider a diameter of the circular disc mentioned above which is parallel to the given axis. Moment of inertia of the disc about any diameter is $\frac{1}{4}mr^2$. By the theorem of parallel axis its moment of inertia about the given axis is $\frac{1}{4} mr^2 + mx^2$. Summing up for all such discs the moment of inertia of the whole cone.

$$I_2 = \Sigma \frac{1}{4} mr^2 + \Sigma mx^2$$

$$= \frac{3}{20} Ma^2 + \Sigma \pi r^2 dx \rho x^2 = 3 \frac{Ma^2}{20} + \frac{\pi \rho a^2}{h^2} \int_0^h x^4 dx$$

$$= 3 \frac{Ma^2}{20} + \frac{\pi \rho a^2}{h^2} \frac{h^5}{5} = \frac{3 Ma^2}{20} + \frac{1}{3} \pi a^2 h\rho \frac{3}{5} h^2$$

$$= \frac{3}{20} Ma^2 + \frac{3}{5} Mh^2$$

7.6. Ellipsoid of Inertia ; Principal moments of inertia, Principal Axes

Definition of Products of Inertia : If (x, y, z) be the co-ordinates of an element of mass δm of a rigid body, then $\Sigma\, \delta m\, yz = I_{yz}$, $\Sigma\, \delta m\, zx = I_{zx}$ and $\Sigma\, \delta m\, xy = I_{xy}$ are called the products of inertia about yz, zx and xy axes respectively.

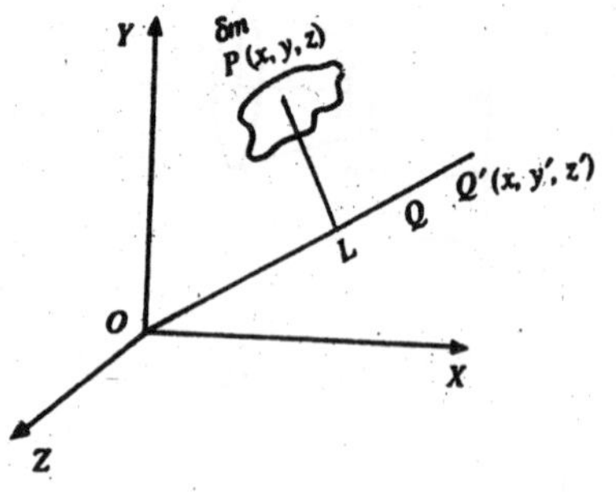

Fig. 7.19

The moments and products of inertia of a rigid body about three mutually perpendicular and concurrent axes OX, OY & OZ are given to be I_{xx}, I_{yy}, I_{zz} and I_{yz}, I_{zx} and I_{xy} respectively. Let us determine the moment of inertia of the body about any axis OQ having direction cosines (l, m, n). P is an element of mass δm of the body having co-ordinates (x, y, z) PL is perpendicular from P on OQ. The moment of inertia of the body about OQ is given by

$$I = \Sigma\ \delta m\ PL^2 = \Sigma\ \delta m\ (OP^2 - OL^2)$$

$$= \Sigma\ \delta m\ \left\{(x^2 + y^2 + z^2) - (lx + my + nz)^2\right\}$$

$$= \Sigma\ \delta m \left\{(1 - l^2)\, x^2 + (1 - m^2)\, y^2 + (1 - n^2)\, z^2 - 2\, mn\, yz - 2\, nl\, zx - 2\, lm\, xy\right\}$$

$$= \Sigma\ \delta m\ \left\{(m^2 + n^2)\, x^2 + (n^2 + l^2)\, y^2 + (l^2 + m^2)\, z^2 - 2\, mn\, yz - 2\, nl\, zx - 2\, lm\, xy\right\}$$

$$\because\ l^2 + m^2 + n^2 = 1$$

$$= l^2\ \Sigma\ \delta m\, (y^2 + z^2) + m^2\ \Sigma\ \delta m\ (z^2 + x^2) + n^2\ \Sigma\ \delta m\ (x^2 + y^2) - 2\, mn\ \Sigma\ \delta m\ yz - 2\, nl\ \Sigma\ \delta m\ zx - 2\, lm\ \Sigma\ \delta m\, xy$$

$$= I_{xx} l^2 + I_{yy}\, m^2 + I_{zz}\, n^2 - 2\, I_{yz}\, mn - 2\, I_{zx}\, nl - 2\, I_{xy}\, lm \qquad (7.16)$$

If we search out a point $Q'\ (x', y', z')$ on OQ such that the moment of inertia of the body about OQ is inversely proportional to $OQ'^{\,2}$

i.e. $$I = \frac{k}{OQ'^{\,2}} = I_{xx}\, l^2 + I_{yy}\, m^2 + I_{zz}\, n^2 - 2\, I_{yz}\, mn - 2\, I_{zx}\, nl - 2\, I_{xy}\, lm$$

or, $$I_{xx}\, x'^{\,2} + I_{yy}\, y'^{\,2} + I_{zz}\, z'^{\,2} - 2\, I_{yz}\, y'z' - 2\, I_{zx}\, z'x' - 2\, I_{xy}\, x'y' = K \qquad (7.17)$$

The above is the general equation of the ellipsoid known as the ellipsoid of inertia or momental ellipsoid of the body about O.

There is thus for anybody an imaginary ellipsoid about a point ; such that the moment of inertia of the body about an axis joining this point to any point on the surface of the ellipsoid is inversely proportional to the square of the radius vector.

For every ellipsoid there exists a set of three mutually perpendicular axes which when taken as the axes of co-ordinates the resulting equation will have no terms involving yz, zx and xy. These axes are called the principal axes of the ellipsoid. The equation of the ellipsoid of inertia referred to its principal axes as the axes of co-ordinates is $I'_{xx}\, x^2 + I'_{yy}\, y^2 + I'_{zz}\, z^2 = K'$ (7.18)

(x' etc. in equation (7.17) has been replaced by x.)

I'_{xx} etc. are the moments of inertia of the body about the principal axes and are called the principal moments of inertia of the body.

The principal axes of the body will be symmetrical with respect to the body and hence the products of inertia of the body about them will be zero.

7.7. Principal Axes and Principal moments of inertia in case of some bodies

(i) Solid Sphere :

For a solid sphere any three mutually perpendicular diameters will be the principal axes about the centre (Fig. 7.20).

In this case $I_{xx} = I_{yy} = I_{zz} = \frac{2}{5} Mr^2$.

The ellipsoid of inertia will be a sphere.

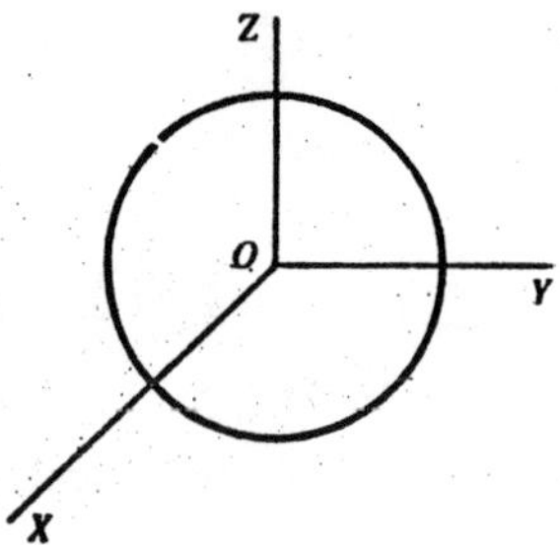

Fig. 7.20

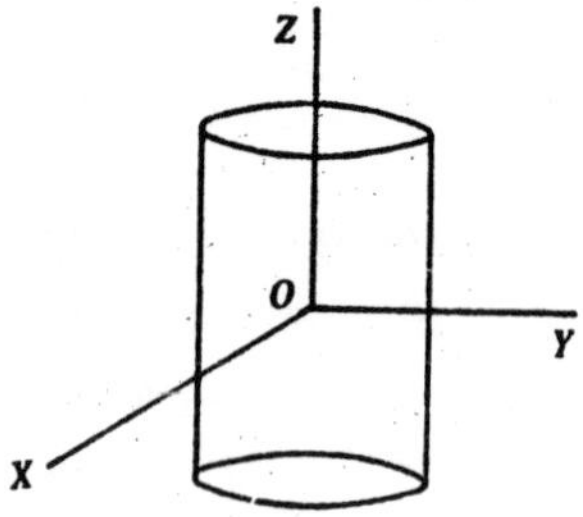

Fig. 7.21

(ii) Solid Cylinder :

In the case of the solid cylinder, the axis of the cylinder and two mutually perpendicular axes through its middle point will be the principal axes about the middle point (Fig. 7.21).

In this case $I_{xx} = I_{yy} = M\left(\frac{l^2}{12} + \frac{r^2}{4}\right)$

and $I_{zz} = \frac{1}{2} Mr^2$

The ellipsoid of inertia will be a spheroid.

(iii) Rectangular Parallelopiped :

In the case of a rectangular parallelopiped three axes parallel to the length, breadth and height through its middle point will be the principal axes (Fig. 7.22).

In this case $I_{xx} = \frac{M}{12}(l^2 + h^2)$, $I_{yy} = \frac{M}{12}(b^2 + h^2)$, $I_{zz} = \frac{M}{12}(l^2 + b^2)$.

The ellipsoid of inertia will be an ellipsoid in general.

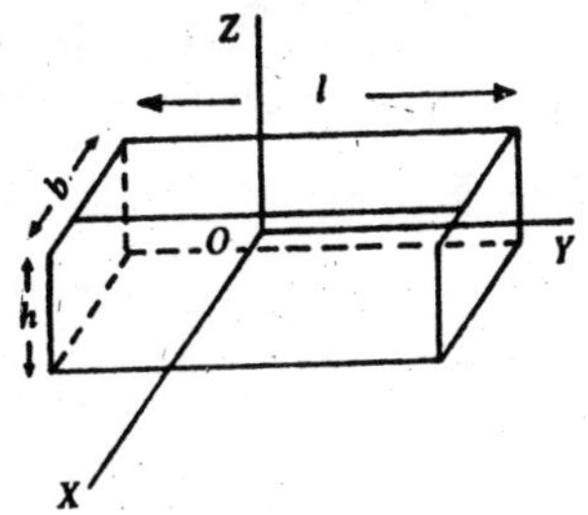

Fig. 7.22

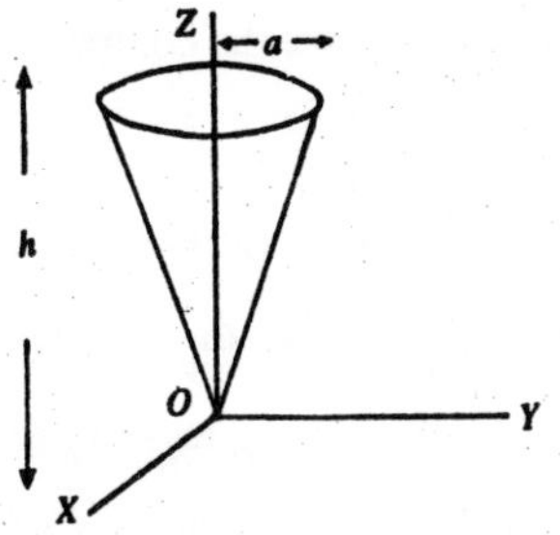

Fig. 7.23

(iv) **Solid Cone :**

In the case of a solid cone, the axis of the cone and any two mutually perpendicular axes perpendicular to it will be principal axes about the vertex (Fig. 7.23).

In this case $I_{xx} = I_{yy} = \frac{3}{20} Ma^2 + \frac{3}{5} Mh^2$ and $I_{zz} = \frac{3}{10} Ma^2$

In this case the ellipsoid of inertia will be a spheroid.

7.8. Angular momentum of a rigid body rotating about an axis

If $\vec{r}$ be the position vector of a particle of the body w.r.t. a fixed point O on the axis of rotation, m the mass of the particle and $\vec{v}$ its velocity, the angular momentum of the body is given by $\vec{L} = \Sigma\, \vec{r} \times m\vec{v}$ (7.19)

If $\vec{\omega}$ be the angular velocity of the body,

$$\vec{v} = \vec{\omega} \times \vec{r}$$

$$\therefore\ \vec{L} = \Sigma\, \vec{r} \times m\,(\vec{\omega} \times \vec{r})$$

$$= \Sigma\, m \left\{ (\vec{r}.\vec{r})\, \vec{\omega} - (\vec{r}.\vec{\omega})\, \vec{r} \right\} \qquad (7.20)$$

$$= \Sigma\, m \left\{ r^2\, \vec{\omega} - (x\omega_x + y\omega_y + z\omega_z)\, \vec{r} \right\}$$

$$= \Sigma\, m \left\{ (x^2 + y^2 + z^2)\, \vec{\omega} - (x\omega_x + y\omega_y + z\omega_z)\, \vec{r} \right\}$$

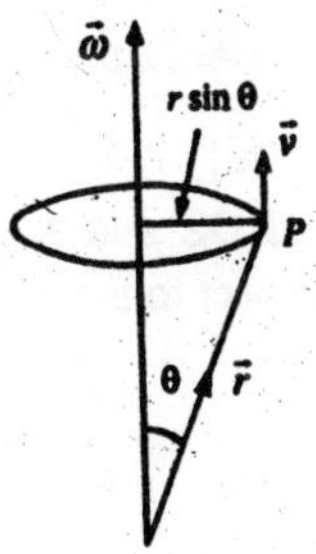

Fig. 7.24

where ω_x, ω_y, ω_z and x, y, z are the components of $\vec{\omega}$ and $\vec{r}$ respectively.

$$= \Sigma\, m \left\{ (x^2 + y^2 + z^2)\, (\hat{i}\,\omega_x + \hat{j}\,\omega_y + \hat{k}\,\omega_z) - (x\,\omega_x + y\,\omega_y + z\,\omega_z) \left\{ (\hat{i}\,x + \hat{j}\,y + \hat{k}\,z) \right\} \right.$$

$\hat{i}, \hat{j}, \hat{k}$ being unit vectors along OX, OY, OZ.

If L_x, L_y, L_z are the components of $\vec{L}$ along OX, OY and OZ

$$\left.\begin{aligned} L_x &= \Sigma m \{(x^2+y^2+z^2)\,\omega_x - (x\,\omega_x + y\,\omega_y + z\,\omega_z)\,x\} \\ &= \omega_x \,\Sigma\, m\,(y^2+z^2) - \omega y\,\Sigma\, m\,xy - \omega_z\,\Sigma\, m\,zx \\ &= I_{xx}\,\omega_x - I_{xy}\,\omega_y - I_{zx}\,\omega_z \\ \text{Similarly } L_y &= I_{yy}\,\omega_y - I_{xy}\,\omega_x - I_{yz}\,\omega_z \\ \text{and } L_z &= I_{zz}\,\omega_z - I_{zx}\,\omega_x - I_{yz}\,\omega_y \end{aligned}\right\} \quad (7.21)$$

where I_{xx} etc. are the moments of inertia and I_{yz} are the products of inertia about OX, OY and OZ axes.

If OX, OY, OZ are the principal axes of the body, I_{xy} etc. = 0. The angular momentum about the principal axes are

$$L_1 = I_1\,\omega_1, \quad L_2 = I_2\,\omega_2, \quad L_3 = I_3\,\omega_3$$

$$\therefore \vec{L} = I_1\,\omega_1\,\hat{a}_1 + I_2\,\omega_2\,\hat{a}_2 + I_3\,\omega_3\,\hat{a}_3 \quad (7.22)$$

where $\hat{a}_1$ etc. are unit vectors along the principal axes, ω_1 etc. are the angular velocity components about them and I_1, I_2, I_3 are the principal moments of inertia of the body.

7.9. Kinetic energy of a rigid body rotating about an axis

Kinetic energy of the point P (Fig. 7.24) is

$$\frac{1}{2}\,mv^2 = \frac{1}{2}\,m\,(\vec{\omega} \times \vec{r})^2$$

Hence the total K. E. of the whole body is $\Sigma\,\frac{1}{2}\,m\,(\vec{\omega} \times \vec{r})^2$.

If $\hat{i}, \hat{j}, \hat{k}$ are unit vectors along three mutually perpendicular axes OX, OY and OZ and λ, μ, ν are the direction cosines of the axis of rotation with respect to them, then

$$\vec{r} = \hat{i}x + \hat{j}y + \hat{k}z$$

and $$\begin{aligned} \vec{\omega} &= \hat{i}\,\omega_x + \hat{j}\,\omega_y + \hat{k}\,\omega_z \\ &= \hat{i}\,\omega\,\lambda + \hat{j}\,\omega\,\mu + \hat{k}\,\omega\,\nu \end{aligned}$$

$$\begin{aligned} \therefore \vec{\omega} \times \vec{r} &= (\hat{i}\,\omega_x + \hat{j}\,\omega_y + \hat{k}\,\omega_z) \times (\hat{i}x + \hat{j}y + \hat{k}z) \\ &= \hat{k}\,\omega_x\,y - \hat{j}\,\omega_x\,z - \hat{k}\,\omega_y\,x + \hat{i}\,\omega_y\,z + \hat{j}\,\omega_z\,x - \hat{i}\,\omega_z\,y \\ &= \hat{i}\,(\omega_y\,z - \omega_z\,y) + \hat{j}\,(\omega_z\,x - \omega_x\,z) + \hat{k}\,(\omega_x\,y - \omega_y\,x) \end{aligned}$$

Hence $(\vec{\omega} \times \vec{r})^2 = (\vec{\omega} \times \vec{r}) . (\vec{\omega} \times \vec{r})$

$$= (\omega_y z - \omega_z y)^2 + (\omega_z x - \omega_x z)^2 + (\omega_x y - \omega_y x)^2$$

So that the total K. E. of the whole body

$$= \frac{1}{2} \Sigma\, m\, (\vec{\omega} \times \vec{r})^2 = T$$

$$= \frac{1}{2} \Sigma\, m \left\{ \omega_x^{\,2} (y^2 + z^2) + \omega_y^{\,2} (z^2 + x^2) + \omega_z^{\,2} (x^2 + y^2) \right\}$$

$$- 2yz\, \omega_y \omega_z - 2\, zx\, \omega_z \omega_x - 2\, xy\, \omega_x \omega_y \}$$

$$= \frac{1}{2} I_{xx} \omega_x^{\,2} + \frac{1}{2} I_{yy} \omega_y^{\,2} + \frac{1}{2} I_{zz} \omega_z^{\,2} - I_{yz} w_y \omega_z - I_{zx} \omega_z w_x$$

$$- I_{xy} \omega_x \omega_y \qquad (7.23)$$

$\because\ \omega_x = \omega\lambda, \quad \omega_y = \omega\mu, \quad \omega_z = \omega\nu$

The above may be written as

$$T = \frac{1}{2} \omega^2 \left\{ I_{xx} \lambda^2 + I_{yy} \mu^2 + I_{zz} \nu^2 - 2 I_{yz} \mu \nu - 2 I_{zx} \nu \lambda - 2 I_{xy} \lambda \mu \right\}$$

$$= \frac{1}{2} I\omega^2 \qquad (7.24) \text{ by equation } (7.16)$$

Alternatively, defining the products of inertia as

$I_{xy} = I_{yx} = - \int \delta m\ xy$ etc. equation (7.23) may be written as

$$T = \frac{1}{2} I_{xx} \omega_x^{\,2} + \frac{1}{2} I_{yy} \omega_y^{\,2} + \frac{1}{2} I_{zz} \omega_z^{\,2} + I_{yz} \omega_y \omega_z + I_{zx} \omega_z \omega_x + I_{xy} \omega_x \omega_y$$

$$= \frac{1}{2} \sum_x \sum_y I_{xy} \omega_x \omega_y \qquad (7.24)$$

Using matrix notation

$$T = \frac{1}{2} \omega^T I \omega \quad \text{or,}$$

writing in details

$$T = \frac{1}{2} \begin{bmatrix} \omega_x \\ \omega_y \\ \omega_z \end{bmatrix}^T \begin{bmatrix} I_{xx} & I_{xy} & I_{xz} \\ I_{yx} & I_{yy} & I_{yz} \\ I_{zx} & I_{zy} & I_{zx} \end{bmatrix} \begin{bmatrix} \omega_x \\ \omega_0 \\ \omega_z \end{bmatrix} . \qquad (7.25)$$

when $[\quad]^T$ represents transposed matrix.

EXERCISE

1. (a) Find the moments and products of inertia of a uniform square plate of side a about x, y and z axes, x and y axes being taken as the adjacent sides of the plate and z perpendicular to the plane.

(b) Find the principal moments of inertia in this case. [C.U. (Hons.) '96, '99]

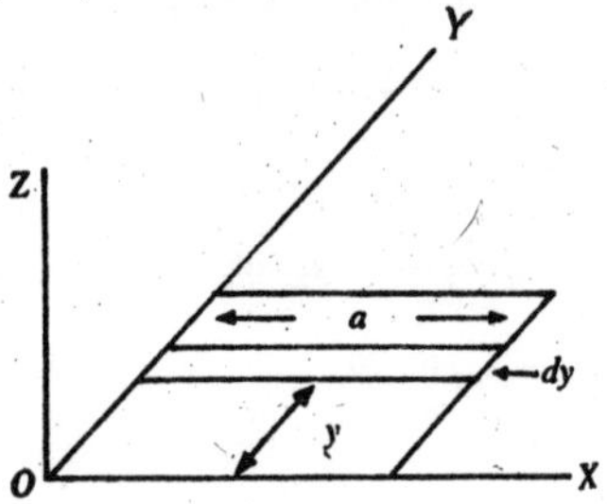

Fig. 7.25

Ans. (a) Let us consider a thin strip parallel to x axis and distance y from it and having a thickness dy.

$$I_x = \int_0^a a\,dy\,\sigma\,y^2 = a\sigma\left[\frac{y^3}{3}\right]_0^a = \sigma a\,\frac{a^3}{3}$$

$$= \frac{1}{3}\,a^2\sigma a^2 = \frac{Ma^2}{3}.$$

Similarly $\quad I_y = \int_0^a a\,dx\,\sigma\,x^2 = \dfrac{Ma^2}{3}.$

By this theorem of perpendicular axes

$$I_z = I_x + I_y = \frac{2Ma^2}{3}.$$

Taking an element of area $dxdy$ at (x, y).

$$I_{xy} = \Sigma\,\sigma\,dxdy\;xy$$

$$= \sigma\int_0^a x\,dx\int_0^a y\,dy = \sigma\,\frac{a^2}{2}\,\frac{a^2}{2} = \frac{Ma^2}{4}$$

Evidently $I_{yz} = I_{zx} = 0$.

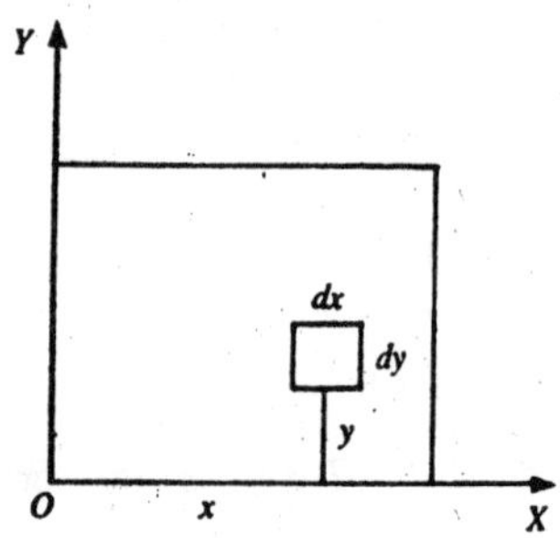

Fig. 7.26

(b) Two of the principal axes will be axes passing through the centre and parallel to the sides. The third one will be perpendicular to its plane and passing through the centre. Thus $I_{xx} = I_{yy} = \dfrac{Ma^2}{12}$

and $I_{zz} = I_{xx} + I_{yy} = \dfrac{Ma^2}{6}$ (Fig. 7.27).

(c) What are principal moments of inertia ? Show how you will find these quantities from a knowledge of the inertia tensor in an arbitrary frame of reference.

Ans. The nine quantities $I_{xx}, I_{xy} ..... I_{xz}$ can be written in an array often called a matrix or tensor given by

$$\begin{bmatrix} I_{xx} & I_{xy} & I_{xz} \\ I_{yx} & I_{yy} & I_{yz} \\ I_{zx} & I_{zy} & I_{zz} \end{bmatrix} \quad \ldots \quad (1)$$

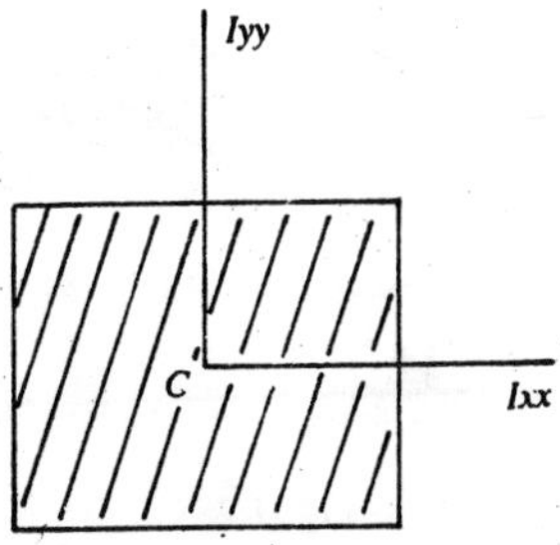

Fig. 7.27

Each quantity is called an element of the matrix or tensor. The diagonal consisting of the elements I_{xx}, I_{yy}, I_{zz} is called the principal or the main diagonal. Since $I_{xy} = I_{yx}$; $I_{yz} = I_{zy}$ and $I_{zx} = I_{xz}$. The matrix is often referred to as a symmetric matrix or tensor. A set of three mutually perpendicular axes through the point O about which the products of inertia are zero are called the principal axes of inertia or briefly the principal axes of the body. If the rigid body rotates about any principal axis, the direction of the angular momentum is the same as that of the angular velocity. Thus

$$\vec{L} = I\vec{\omega} \quad \ldots \quad (2)$$ where I is a scalor.

Thus we have

$$\left.\begin{aligned} (I_{xx} - I)\,\omega_x + I_{xy}\,\omega_y + I_{xz}\,\omega_2 &= 0 \\ I_{yx}\,\omega_x + (I_{yy} - I)\,\omega_y + I_{yz}\,\omega_2 &= 0 \\ I_{zx}\,\omega_x + I_{zy}\,\omega_y + (I_{zz} - I)\,\omega_2 &= 0 \end{aligned}\right\} \quad \ldots \quad (3)$$

In order that equation (3) have solutions other than the trivial one $\omega_x = 0$, $\omega_y = 0$, $\omega_z = 0$, we require that

$$\begin{vmatrix} I_{xx} - I & I_{xy} & I_{xz} \\ I_{yx} & I_{yy} - I & I_{yz} \\ I_{zx} & I_{zy} & I_{zz} - I \end{vmatrix} = 0 \quad \ldots \quad (4)$$

This leads to a cubic equation in I having three real roots I_1, I_2, I_3 called the principal moments of inertia.

2. Acceleration of a body rolling down an inclined place

A body of mass m and radius r is rolling down a smooth inclined plane inclined at an angle θ with the horizontal. The body starts from rest at A and goes to B in time t describing a distance s down the plane. If v be the linear velocity ω the angular velocity of the body at B, we have from the principle of coservation of energy.

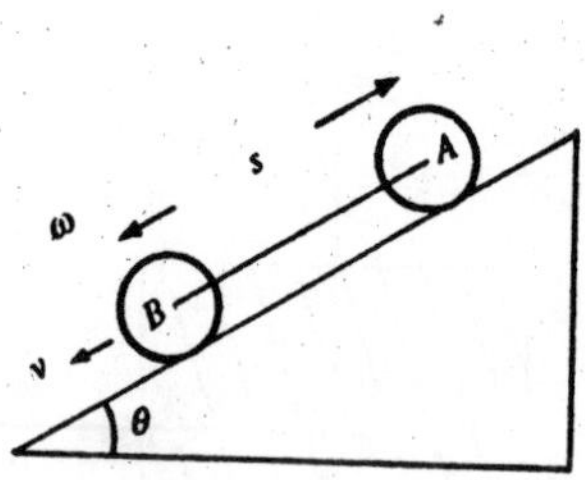

Fig. 7.28

Ans. $\frac{1}{2} I\omega^2 + \frac{1}{2} mv^2 = mg\, s \sin\theta$

or, $\frac{1}{2} mk^2 \cdot \frac{v^2}{r^2} + \frac{1}{2} mv^2 = mg\, s \sin\theta$

Differentiating w. r. t. 't'

$$\frac{1}{2}\frac{mk^2}{r^2} 2v \frac{dv}{dt} + \frac{1}{2} m\, 2v \frac{dv}{dt}$$

$$= mg \frac{ds}{dt} \sin\theta$$

k = radius of gyration about the axis of rotation

or, $\frac{mk^2}{r^2} v f + mvf = mg\, v \sin\theta$

or, $f = \dfrac{g \sin\theta}{1 + \dfrac{k^2}{r^2}}$

3. A circular disc of mass M and radius a can turn without friction about its axis which is horizontal. One end of a light string is attached to a point on the rim. The string is coiled round the rim and carries at its free and a mass m. Prove that when the motion is allowed to take place the angular acceleration of the wheel is

$\dfrac{mga}{Mk^2 + ma^2}$, where k is the radius of gyration of the wheel about its axis.

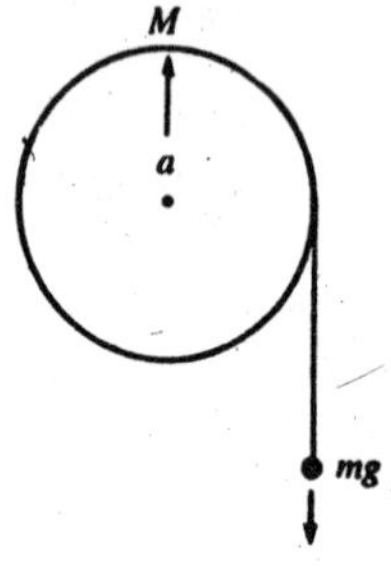

Fig. 7.29

Ans. Let α be the angular acceleration

$\therefore\ I\alpha =$ total torque $= mga$

or, $(Mk^2 + ma^2)\,\alpha = mga$

$\therefore\ \alpha = \dfrac{mga}{Mk^2 + ma^2}$.

4. A sphere of radius $2r$ and density d has an internal spherical cavity of radius r the diameter of which is the radius of the sphere. Find the moment of inertia of the body about an axis perpendicular to the common diameter of the sphere and the cavity and at a distance x from the centre of the sphere.

Ans. Let I = moment of inertia

I = Moment of inertia of the sphere – Moment of inertia of the cavity

$$= \frac{2}{5} M (2r)^2 + Mx^2 - \left\{ \frac{2}{5} mr^2 + m (r - x)^2 \right\}$$

$$= \frac{2}{5}\frac{4}{3} \pi (2r)^3 d (2r)^2 + \frac{4}{3} \pi (2r)^3 dx^2$$

$$- \left\{ \frac{2}{5}\frac{4}{3} \pi r^3 dr^2 + \frac{4}{3} \pi r^3 d (r - x)^2 \right\}$$

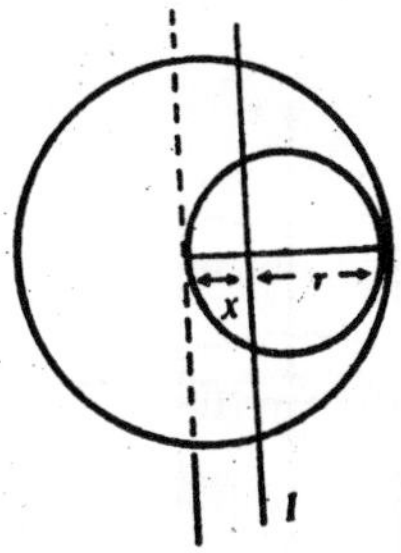

Fig. 7.30

or, $15 I = 256 \pi r^5 d + 160 \pi r^3 x^2 d$
$- 8 \pi r^5 d - 20 \pi r^3 d (r - x)^2$

$= 256 \pi r^5 d + 160 \pi r^3 x^2 d - 8 \pi r^5 d$
$- 20 \pi r^5 d + 40 \pi r^4 x d - 20 \pi r^3 x^2 d$

$= 228 \pi r^5 d + 140 \pi r^3 x^2 d + 40 \pi r^4 x d$

$= 4 \pi r^3 d \{ 57 r^2 + 35 x^2 + 10 rx \}$

5. **A cylinder has a mass M, length L and radius r. Find the ratio of L to r if the moment of inertia about an axis through the centre and perpendicular to the length is minimum.**

$$M = \pi r^2 L \rho \qquad \therefore \; dM = \rho\pi (2r dr L + r^2 dL) = 0$$

Ans. Again $I = M \left(\frac{L^2}{12} + \frac{r^2}{4} \right)$ $\qquad \therefore \; \frac{dL}{dr} = - \frac{2L}{r}$

I will be minimum if $dI = 0$

Now $dI = M \left(\frac{2L\, dL}{12} + \frac{2r\, dr}{4} \right) = 0$

or, $\frac{dL}{dr} = - \frac{3r}{L} = - \frac{2L}{r}$

or, $\frac{L^2}{r^2} = \frac{3}{2}$

$\therefore \; \frac{L}{r} = \frac{\sqrt{3}}{\sqrt{2}}$.

6. **A solid cylinder is rotating about its axis of symmetry aligned to vertical direction. Measuring z from the bottom of the cylinder along this axis, the density of the body is given by**

Ans. $\rho (z) = \rho_0 \left(1 - \frac{1}{2} \frac{z}{l} \right)$, **where ρ_0 is a constant and l the length of the cylinder. Calculate the moment of inertia of the cylinder, its radius being R.**

[C. U. (Hons.) 1982]

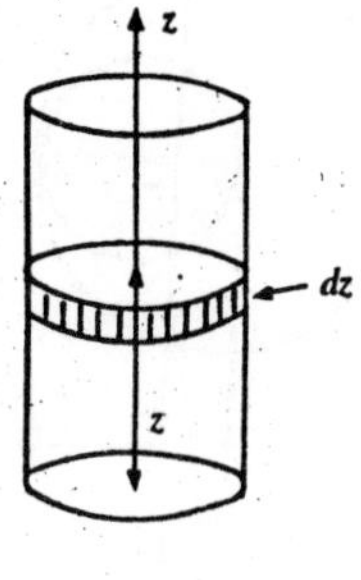

Fig. 7.31

Moment of inertia of a circular disc of thickness dz at a distance z from the bottom is given by

$$\frac{1}{2}\pi R^2\, dz\ \rho(z)\ R^2 = \frac{1}{2}\pi R^4 \rho_0 \left(1 - \frac{1}{2}\frac{z}{l}\right) dz$$

Summing up for all such discs moment of inertia of the cylinder

$$= \frac{1}{2}\pi R^4\ \rho_0 \int_0^l \left(1 - \frac{1}{2}\frac{z}{l}\right) dz$$

$$= \frac{1}{2}\pi R^4\ \rho_0 \left\{l - \frac{l^2}{4l}\right\} = \frac{1}{2}\pi R^4 \rho_0 \frac{3}{4} l = \frac{3}{8}\pi R^4\ \rho_0\, l$$

7. Find the moment of inertia of a cube about any axis passsing through its middle point.

Ans. Let us consider three mutually perpendicular axes through the middle point and parallel to the edges of this cube. Evidently these are principal axes of the cube. Hence the moment of inertia of the cube about any axis passing through the middle point is given by

$$I = I_1 l^2 + I_2 m^2 + I_3 n^2,$$

l, m, n being the direction cosine of the axis

$$= \frac{Ma^2}{6}\left(l^2 + m^2 + n^2\right) = \frac{Ma^2}{6}.$$

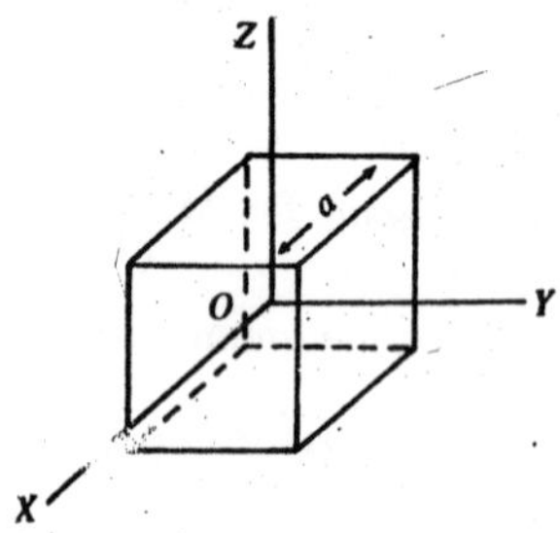

Fig. 7.32

8. Calculate the principal moments of inertia of a right circular cylinder and hence find the value of the moment of inertia of the cylinder about one of its generators. **[C. U. (Hons.) 86]**

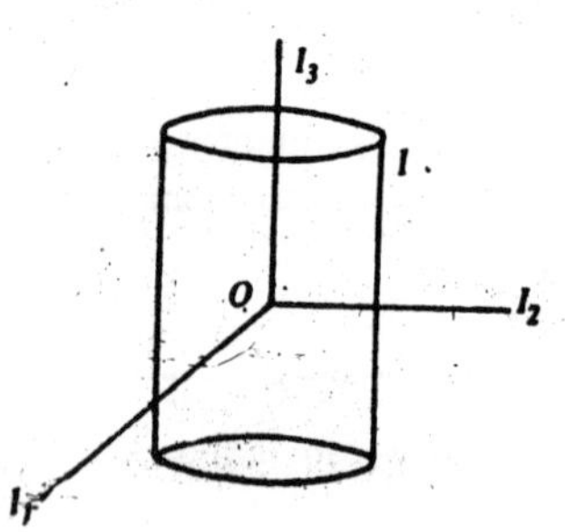

Fig. 7.33

Hints : $I_1 = I_2 = M\left(\frac{l^2}{12} + \frac{r^2}{4}\right)$,

$$I_3 = \frac{1}{2} Mr^2$$

$$I = I_3 + Mr^2 = \frac{3}{2} Mr^2$$

CHAPTER VIII
GRAVITATION

Newton's Law of Gravitation, Gravitational potential and intensity, application of Gauss's Theorem and Laplace Equation in simple symmetric problems. Theory of compound pendulum ; Motion under central forces ; Nature of the orbits in attractive inverse square fields ; Kepler's Laws of Planetary Motion.

8.1. Newton's Law of Gravitation

Statement : Every particle of matter in the universe attracts every other particle with a force which varies directly as the product of the masses of the two particles and inversely as the square of the distance between them, the line of action of the force being the line joining the two particles. If m_1 and m_2 be masses of the two particles separated by a distance r, the force of attraction between them is given by

$$F = G\frac{m_1 m_2}{r^2} \qquad (8.1)$$

where G is a constant, known as Universal Constant of Gravitation.

Gravitational intensity : Intensity at any point in a gravitational field may be defined in magnitude and direction by the force experienced by a unit mass placed at that point, provided its presence does not alter the original field.

Thus the intensity at a point P due to a particle of mass m at O is given by

$$F = \frac{Gm}{oP^2} = \frac{Gm}{r^2} \quad \text{in the direction } Po$$

(Fig. 8.1).

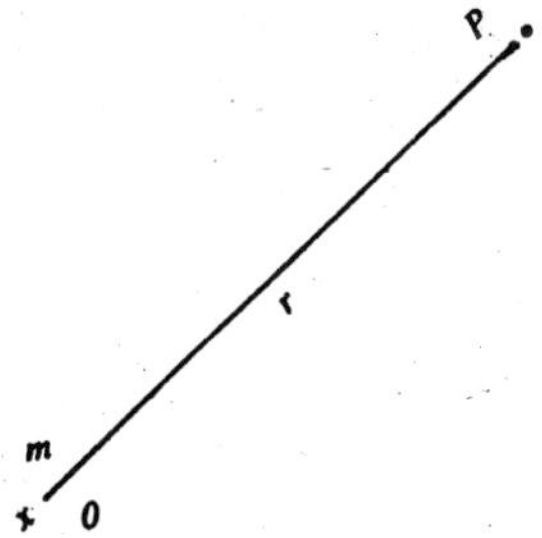

Fig. 8.1

In vector notation, $\vec{F} = -\frac{Gm}{r^2}\hat{r}$

$$= -\frac{Gm}{r^3}\vec{r} \qquad (8.2)$$

where $\vec{r} = \overrightarrow{OP}$, $\hat{r}$ is unit vector along OP.

Gravitational Potential : Potential at any point in a gravitational field is defined to be numerically equal to the work done in bringing a unit mass from infinity upto the point concerned.

Thus if V be the potential at P,

$$V = \int_{\infty}^{r} \frac{Gm}{x^2}\, dx = -\frac{Gm}{r} \tag{8.3}$$

Relation between intensity and potential.

Let V and $V + dV$ be the potentials at $P(r)$ and $Q(r + dr)$, Evidently

$$dV = -F\,dr \quad \text{or,} \quad F = -\frac{dV}{dr} \tag{8.4}$$

In vector notation $\vec{F} = -\hat{r}\dfrac{dV}{dr}$ (8.5)

Combining equation 8.2 and 8.3 we can write in a more general form

$$\vec{F} = -\vec{\nabla} V$$

Now $\vec{\nabla} \times \vec{F} = \vec{\nabla} \times \vec{\nabla}\left(\dfrac{Gm}{r}\right) = 0$. Hence gravitational field is a conservative field of force.

8.2. Gravitational potential and intensity due to a thin spherical shell (Mass $= M$, radius $= a$)

(i) **At internal point :**

P is a point inside a thin spherical shell at a distance r from its centre. Let us make a section of the shell by a plane perpendicular to OP.

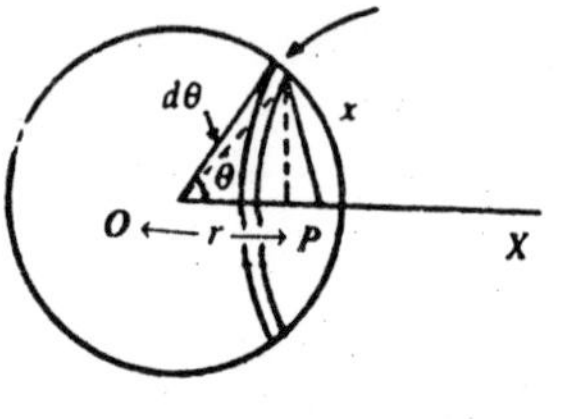

Fig. 8.2

The section will be a circular ring (Fig. 8.2). Every element on this ring is at the same distance x from P. Hence the potential at P due to this ring is

$$\delta V = -\frac{Gm}{x}$$ where x is the distance shown

$$= -\frac{G}{x}\, 2\pi a \sin\theta\, a d\theta\, \sigma \qquad \sigma = \text{mass/area}$$

(The radius of the ring subtends an angle θ and its width subtends an angle $d\theta$ at the centre.)

Now $x^2 = a^2 + r^2 - 2ar\cos\theta$

Differentiating,

$$2\,x dx = 2\,ar \sin\theta \, d\theta$$

Substituting $\delta V = -\dfrac{G}{x}\, 2\pi a \,\dfrac{x\,dx\,\sigma}{r} = -\dfrac{2\pi G\sigma}{r}\, dx\, a.$

For the whole shell, potential at P is given by

$$V = -\frac{2\pi G\sigma a}{r}\int_{a-r}^{a+r} dx = -\frac{2\pi G\sigma a}{r}(a + r - a + r)$$

$$= -4\pi Ga\sigma = -\frac{4\pi a^2 \sigma G}{a} = -\frac{GM}{a}. \qquad (8.6)$$

Thus at an internal point potential has the same value everywhere and is equal to the potential on the surface.

If F be the magnitude of the intensity at P,

$|\vec{F}| = \dfrac{dV}{dr} = 0$. Hence the inside of a thin spherical shell is a field free space.

(ii) **At external point :**

Proceeding as in the previous case

$$V = -\frac{2\pi Ga\sigma}{r}\int_{r-a}^{r+a} dx$$

$$= -\frac{2\pi Ga\sigma}{r}[x]_{r-a}^{r+a}$$

$$= -\frac{2\pi Ga\sigma}{r}\{r + a - r + a\}$$

$$= -\frac{4\pi a^2 \sigma G}{r} = -\frac{GM}{r} \qquad (8.7)$$

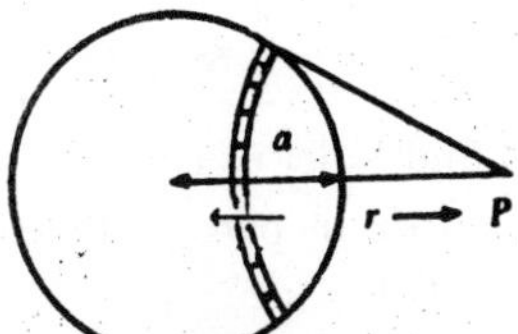

Fig. 8.3

$$|\vec{F}| = \frac{dV}{dr} = \frac{d}{dr}\left(-\frac{GM}{r}\right) = \frac{GM}{r^2} \qquad (8.8)$$

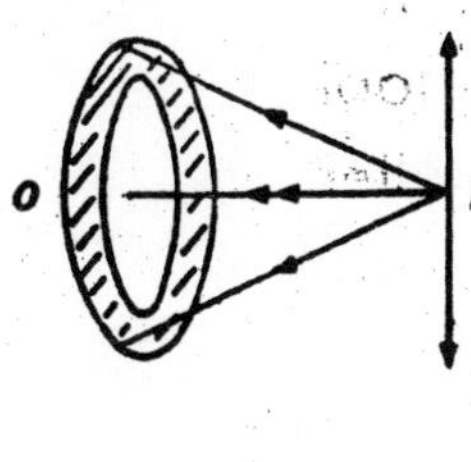

Fig. 8.4

Considering two diametrically opposite elements on the ring (Fig. 8.4), it is evident that the resultant intensity will be towards PO.

This is true for every pair of elements. Hence for the whole shell the resultant intensity will be towards PO. Equations (8.7) and (8.8) show that a thin spherical shell behaves at an external point as if its whole mass is concentrated at the centre.

8.3. Gravitational potential and intensity due to a solid sphere

(M = Mass, a = Radius)

(i) At internal point :

P is a point inside a solid sphere at a distance r from the centre. Let us consider a concentric spherical surface through P. It divides the solid sphere into two parts. Part I is a solid sphere of radius r for which P is a point on the surface. Part II is thick shell of internal radius r and external radius a (Fig. 8.5).

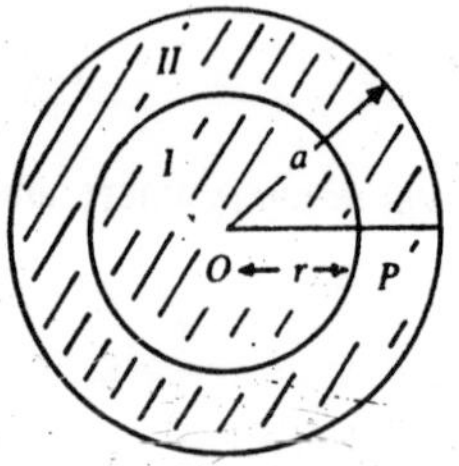

Fig. 8.5

The total potential at $P = V = V_1 + V_2$.

V_1 can be found by dividing part I into a number of thin spherical shells for each of which P is an external point. Hence the mass of part I may be supposed to be concentrated of its centre. Hence

$$V_1 = -\frac{Gm_1}{r} = -\frac{G}{r}\frac{4\pi}{3}r^3\rho$$

$$= -\frac{4}{3}\pi r^2 \rho\, G, \text{ where } \rho = \text{mass/vol.}$$

To find V_2 let us consider a thin concentric spherical shell of radius x and thickness dx in part II. Potential at P due to this shell (internal point) is

$$\delta V_2 = -\frac{Gm}{x}$$

$$= -G\frac{4\pi x^2 dx\rho}{x} = -4\pi\rho Gx dx$$

Hence $V_2 = -4\pi\rho G\int_r^a x\,dx = -2\pi\rho G(a^2 - r^2)$

Total potential at P, $V = V_1 + V_2$

$$= -\frac{4}{3}\pi r^2\rho G - 2\pi\rho G(a^2 - r^2)$$

$$= -\frac{4}{3}\pi\rho G\left\{r^2 + \frac{3}{2}(a^2 - r^2)\right\}$$

$$= -\frac{4}{3}\frac{\pi a^3\rho G}{a^3}\left\{r^2 + \frac{3}{2}(a^2 - r^2)\right\}$$

$$= -\frac{GM}{a^3}\left\{\frac{3}{2}a^2 - \frac{r^2}{2}\right\} = -\frac{GM}{2a^3}(3a^2 - r^2) \qquad (8.9)$$

Intensity at P can be found out either by differentiating equation (8.9) or by considering the fact that intensity due to part II is nil and that due to part I can be found out by supposing its mass to be concentrated at the centre.

Thus $$\left|\vec{F}\right| = \frac{Gm_1}{r^2} = \frac{G4\pi r^3\rho}{3r^2} = \frac{4}{3}\pi r G\rho$$

$$= \frac{4}{3}\pi r G\,\frac{M}{4/3\,\pi a^3} = \frac{GM}{a^3}\,r \quad \propto r \qquad (8.10)$$

From symmetry, the direction of intensity is towards PO.

(ii) **At external point :**

The solid sphere may be supposed to be made up of a number of thin concentric spherical shells. For each of these shells P is an external point.

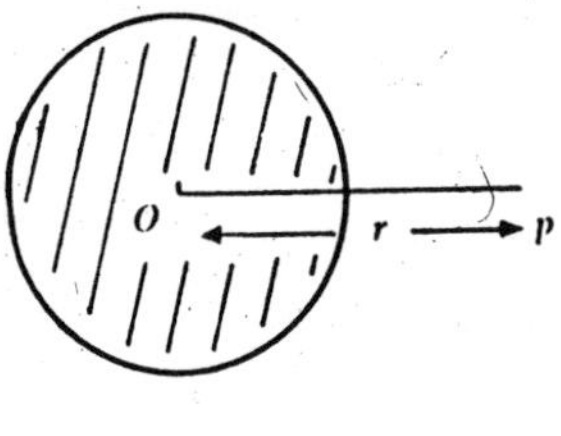

Fig. 8.6

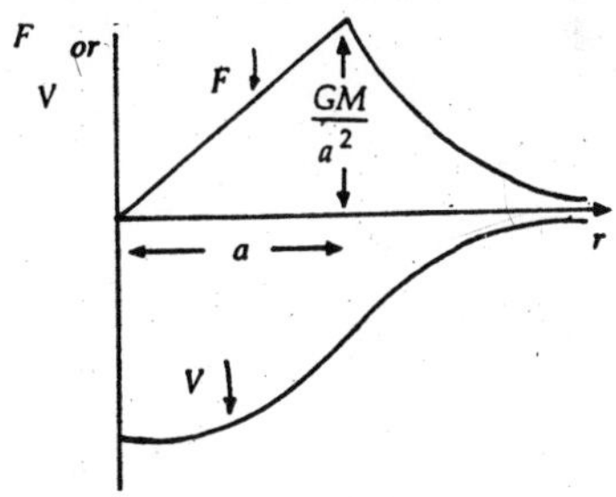

Fig. 8.7

Hence the potential or intensity at P can be found by supposing the mass of the solid sphere to be concentrated at the centre.

Thus $V = -\frac{GM}{r}$ and $|F| = \frac{GM}{r^2}$, the direction of the intensity is along PO. Fig. 8.7 shows the variation of intensity and potential due to a solid sphere with the distance from the centre.

8.4. Gravitational self-energy of a solid sphere

This means the work done to construct the solid sphere in view of the gravitational force of attraction among the parts.

Potential at a point at a distance r from the centre inside a solid sphere is $-\frac{GM}{2a^3}(3a^2 - r^2)$ = work done in bringing a unit mass from infinity to the point concerned.

Hence the work done in constructing a thin spherical shell of radius r and thickness dr is $\frac{1}{2}\left(-\frac{GM}{2a^3}\right)(3a^2 - r^2)\, 4\pi r^2 dr\, \rho$. Multiplication by $\frac{1}{2}$ is due to the fact that in the above consideration work done in bringing particle 1 in the field of particle 2 has been taken into account twice in the act of bringing the particle 2 in the field of particle 1.

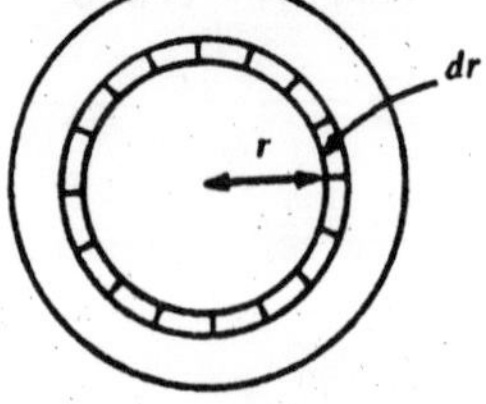

Fig. 8.8

Hence the total work done in constructing the sphere

$$= -\frac{4\pi GM\rho}{4a^3}\int_0^a (3a^2 - r^2)\, r^2 dr$$

$$= -\frac{\pi GM\rho}{a^3}\left\{3a^2\frac{a^3}{3} - \frac{a^5}{5}\right\} = -\frac{\pi GM}{a^3}\cdot\frac{4}{5}a^5\rho$$

$$= \frac{GM}{a^3}\frac{4}{5}\pi a^5\rho = -\frac{GM}{a^3}\cdot\frac{4}{3}\pi a^3\rho \times \frac{3}{5}a^2$$

$$= -\frac{3}{5}\frac{GM^2}{a}. \tag{8.11}$$

8.5. Gauss's theorem on Gravitation

Statement : If N be the normal intensity at any point of the element dS of any closed surface, measured positively along the outward drawn normal, due to any attracting mass, then $\int N dS = -4\pi GM$, where M is the attracting mass within, the integral being taken over the whole surface.

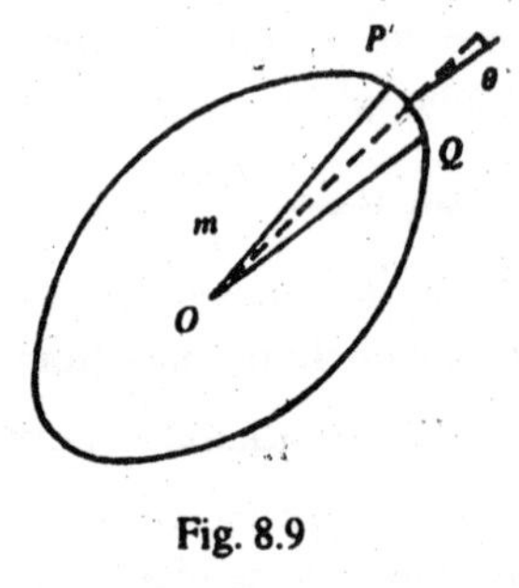

Fig. 8.9

Proof : The component of NdS due to a mass m at O on the element PQ is $= G\frac{m}{OP^2}\cos\theta\, dS$ along the inward drawn normal (Fig. 8.9).

Along the outward drawn normal the component of NdS is $-G\frac{m}{OP^2}\cos\theta\, dS = -Gm\,\delta\omega$

where $\delta\omega$ is the solid angle subtended by the element PQ at O.

Hence the total normal attraction along the outward drawn normal for the whole surface

$$= \int NdS = -Gm \sum \delta\omega = -Gm\, 4\pi$$

For all the masses inside

$$\int NdS = -G\, 4\pi \sum m$$
$$= -4\pi GM$$

8.6. Poisson's and Laplace's equation

Let us now take as the closed surface a small rectangular parallelopiped whose middle point P is the point (x, y, z) and whose edges DA, AB and AH are parallel to OX, OY and OZ and of length δx, δy, δz respectively.

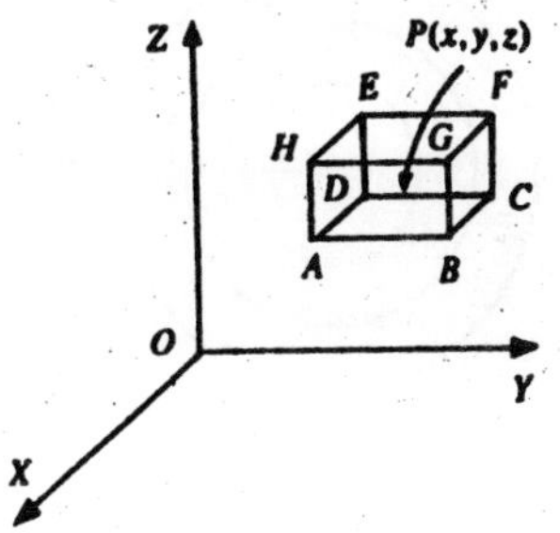

Fig. 8.10

If V be the potential at P, the intensity component along the + Ve direction of x axis $\frac{\delta V}{\delta x}$.

Hence the component of $\int NdS$ due to the face

$$ABGH = \left\{\frac{\partial V}{\partial x} + \frac{\partial}{\partial x}\left(\frac{\partial V}{\partial x}\right)\frac{\delta x}{2}\right\}\delta y\, \delta z$$

Considering the parallel back face the contribution is

$$-\left\{\frac{\partial V}{\partial x} - \frac{\delta x}{2}\frac{\partial}{\partial x}\left(\frac{\partial V}{\partial x}\right)\right\}\delta y\, \delta z$$

Adding $\int NdS$ over the two faces taken together $= \frac{\partial^2 V}{\partial x^2}\delta x\ \delta y\ \delta z$.

Similarly considering the faces perpendicular to OY and OZ, the contributions are $\frac{\partial^2 V}{\partial y^2}\delta x\ \delta y\, \delta z$ and $\frac{\partial^2 V}{\partial z^2}\delta x\ \delta y\ \delta z$ respectively.

If the parallelopiped be inside the attracting mass we have by Gauss's theorem

$$\left(\frac{\partial^2 V}{\partial x^2} + \frac{\partial^2 V}{\partial y^2} + \frac{\partial^2 V}{\partial z^2}\right)\delta x\ \delta y\ \delta z = -4\pi G\ \delta x\, \delta y\, \delta z\ \rho$$

or, $\nabla^2 V = -4\pi\rho G$ (8.12)

This is Poisson's equation. If there be no attracting mass within, $M = 0$

$$\therefore \quad \nabla^2 V = 0 \tag{8.13}$$

This Laplace's equation.

8.7. Application of Gauss's Theorem

(i) Intensity due a thin spherical shell :

(a) At external point *P* :

P is a point at a distance r from the centre and external to a thin spherical shell of radius a. If F be the intensity at P, then the same will be the intensity on the surface of a concentric spherical surface through P. It is also evident that its direction will be everywhere perpendicular to the surface of the sphere. By Gauss's theorem

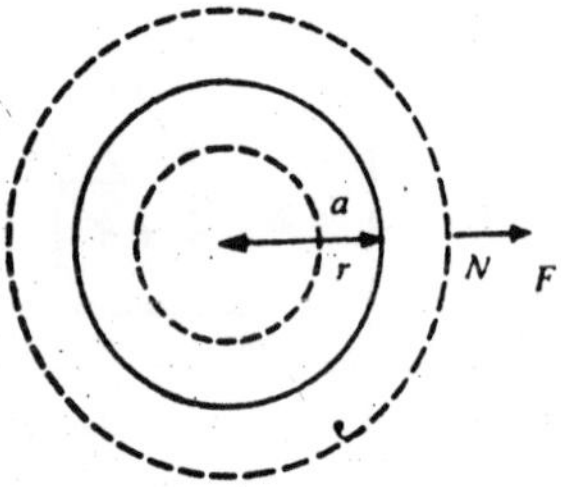

Fig. 8.11

$$\int N dS = -4\pi GM$$

or, $\quad -F\, 4\pi r^2 = -4\pi GM$

$\because \quad \vec{F} = -\vec{N}$

or, $\quad F = \dfrac{GM}{r^2}$.

(b) If the point P be inside the thin spherical shell $M = 0$

$$\therefore \quad F = 0$$

8.8. Application of Laplace's equation

(a) Potential at an external point due to a thin spherical shell :

We have at an external point $\nabla^2 V = 0$

i.e., $$\frac{1}{r^2}\frac{\partial}{\partial r}\left(r^2\frac{\partial V}{\partial r}\right) + \frac{1}{r^2 \sin\theta}\frac{\partial}{\partial \theta}\left(\sin\theta\frac{\partial V}{\partial \theta}\right) + \frac{1}{r^2\sin^2\theta}\frac{\partial^2 V}{\partial \varphi^2} = 0$$

in three dimensional polar co-ordinates.

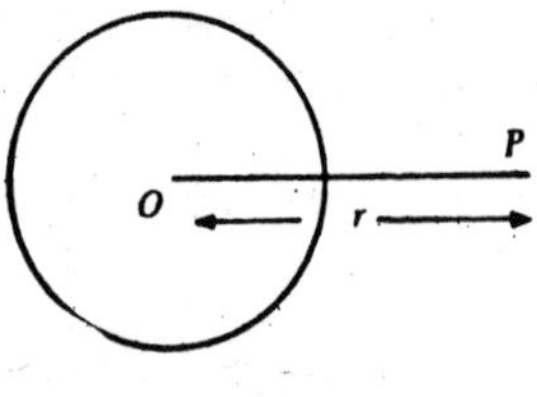

Fig. 8.12

In this particular case $\dfrac{\partial V}{\partial \theta} = \dfrac{\partial V}{\partial \varphi} = 0$

$$\therefore \quad \frac{1}{r^2}\frac{\partial}{\partial r}\left(\frac{r^2 \partial v}{\partial r}\right) = 0$$

or, $$\frac{\partial}{\partial r}\left(r^2\frac{\partial V}{\partial r}\right) = 0$$

Integrating $r^2 \frac{\partial V}{\partial r} = k$

or, $\frac{\partial V}{\partial r} = \frac{k}{r^2}$

Integrating again

$$V = -\frac{k}{r} + k'$$

At $r \to \infty$, $V \to 0$ $\therefore k' = 0$

$\therefore V = -\frac{k}{r} = -\frac{GM}{r}$ comparing with the potential due point mass M.

8.9. Theory of Compound Pendulum

A compound pendulum consists of a rigid body of arbitrary shape and size capable of vibration about a horizontal axis passing through one of its points. Fig. 8.13 shows a section of the body by a vertical plane passing through its c.g. (G). The horizontal axis passes through the point A of the body. If θ be the instantaneous angular displacement of the line AG and I the moment of inertia of the body about the axis of suspension then the differential equation of motion :

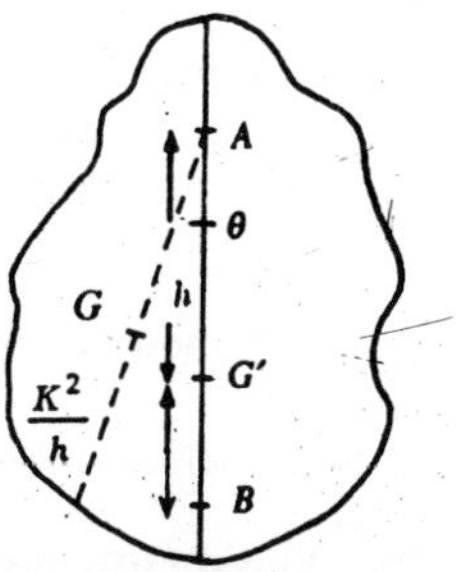

Fig. 8.13

Moment of the couple $= I \frac{d^2\theta}{dt^2} = Mgh \sin\theta$.

where $M =$ mass of the body, $h =$ distance of the point of suspension from the c.g.

or, $\frac{d^2\theta}{dt^2} = \frac{Mgh \sin\theta}{I}$.

Now $I = M(h^2 + k^2)$ where k is the radius of gyration of the body about a parallel axis through the c.g.

Substituting above

$$\frac{d^2\theta}{dt^2} = \frac{Mgh \sin\theta}{M(h^2 + k^2)} = \frac{gh\,\theta}{(h^2 + k^2)} \qquad (3.14)$$

if θ is small.

The above equation shows this motion to be simple harmonic having a time period given by $T = 2\pi\sqrt{\dfrac{h^2+k^2}{gh}} = 2\pi\sqrt{\left(h+\dfrac{k^2}{h}\right)\Big/g}$ (8.15)

$h + k^2/h$ is called the length of the equivalent simple pendulum. The point A at a distance h above c.g. is called the centre of suspension. On the otherhand, the point B at a distance k^2/h below the c.g. is called the centre of oscillation. These two points are asymmetrically situated with respect to the c.g.

To show that these two points are interchangeable :

Let T_1 and T_2 be the time periods about the points of suspension and oscillation respectively.

Evidently $T_1 = 2\pi\sqrt{\dfrac{h^2+k^2}{gh}}$

and $T_2 = 2\pi\sqrt{\left(\dfrac{k^2}{h}+\dfrac{k^2}{\frac{k^2}{h}}\right)\Big/g} = 2\pi\sqrt{\dfrac{h^2+k^2}{gh}}$

So $T_1 = T_2$. Hence the name reversible pendulum.

If L be the distance between the two points about which the time periods are the same one of the points being above and the other below the c.g. and placed asymmetricaly with respect to it then $T = 2\pi\sqrt{\dfrac{L}{g}}$ (8.16)

Equation 8.16 may be employed to determine the value of g.

8.10. Motion under central forces

(a) In two dimensions the differential equations of motion of a particle are

$$\ddot{r} - r\dot{\theta}^2 = \text{acceleration along the radius vector} \quad (8.17)$$

and $\dfrac{1}{r}\dfrac{d}{dt}(r^2\dot{\theta})$ = acceleration perpendicular to the radius vector (8.18)

Motion under a centre of force means the latter component of acceleration is absent.

(b) Nature of the orbits in inverse square attractive force and Kepler's Laws :

Kepler in 1618 published three laws, from purely astronomical observations.

These laws are as follows :

1. Each planet moves in a elliptic path with the sun at one of its focii.

2. The line joining the planet to the sun traces out equal areas in equal times i.e. the areal velocity of the planet is constant.

3. The square of the periodic time of describing the ellipse is directly proportional to the cube of the major axis of the orbit.

If we assume that the force of attraction between a planet and the sun we can arrive at these laws of Kepler.

Let us now discuss the motion under inverse square attractive force.

In case the law of force is central and inverse square attractive we have from equation (8.17)

$$\ddot{r} - r\dot{\theta}^2 = -P = -\frac{\mu}{r^2} \tag{8.19}$$

Putting $r = \frac{1}{u}$ and $h = r^2\dot{\theta}$, the above equation reduces to

$$\frac{d^2u}{d\theta^2} + u = \frac{P}{h^2u^2} \tag{8.20}$$

$$\left[\dot{r} = -\frac{1}{u^2}\frac{du}{d\theta}\dot{\theta} = -h\frac{du}{d\theta}\right.$$

$$\ddot{r} = -h\frac{d^2u}{d\theta^2}\dot{\theta} = -h^2u^2\frac{d^2u}{d\theta^2}$$

$$\therefore\ \ddot{r} - r\dot{\theta}^2 = -h^2u^2\frac{d^2u}{d\theta^2} - \frac{1}{u}h^2u^4 = -h^2u^2\left(\frac{d^2u}{d\theta^2} + u\right)\Bigg] \tag{8.21}$$

$$\dot{r}^2 + r^2\dot{\theta}^2 = h^2\left(\frac{du}{d\theta}\right)^2 + \frac{1}{u^2}h^2u^4 = h^2\left\{\left(\frac{du}{d\theta}\right)^2 + u^2\right\} \tag{8.22}$$

$$\text{K. E. per unit mass} = T = \frac{1}{2}(\dot{r}^2 + r^2\dot{\theta}^2)$$

$$= \frac{h^2}{2}\left\{\left(\frac{du}{d\theta}\right)^2 + u^2\right\}$$

or, $$E - V = \frac{h^2}{2}\left\{\left(\frac{du}{d\theta}\right)^2 + u^2\right\},$$

E = Total energy/mass
V = Potential energy/mass

or, $$\left(\frac{du}{d\theta}\right)^2 + u^2 = \frac{2(E - V)}{h^2} \tag{8.23}$$

From equation (8.20) $$\frac{d^2u}{d\theta^2} + u = \frac{P}{h^2u^2} = \frac{\mu}{h^2} \tag{8.24}$$

Solution of the above differential equation is

$$u = \underset{\text{(P.I.)}}{\frac{\mu}{h^2}} + \underset{\text{(C.F.)}}{c \cos(\theta - \theta_0)} \tag{8.25}$$

θ and θ_0 are constants of integration.

$$\text{Again P.E. per unit mass} = \int_{\infty}^{r} \frac{\mu}{r^2}\, dr$$

$$= -\frac{\mu}{r} = -\mu u \tag{8.26}$$

From (8.25), (8.26) and (8.23)

$$c^2 \sin^2(\theta - \theta_0) + \frac{\mu^2}{h^4} + c^2 \cos^2(\theta - \theta_0) + \frac{2\mu c}{h^2} \cos(\theta - \theta_0)$$

$$= \frac{2}{h^2}\left\{E + \frac{\mu^2}{h^2} + \mu c \cos(\theta - \theta_0)\right\}$$

$$\text{or,} \quad c^2 = \frac{\mu^2}{h^4} + \frac{2E}{h^2} \qquad \text{or,} \quad c = \frac{\mu}{h^2}\sqrt{\left(1 + \frac{2Eh^2}{\mu^2}\right)}$$

If we choose the initial line such that $\theta_0 = 0$, equation (8.25) yields

$$u = \frac{\mu}{h^2}\left\{1 + \sqrt{1 + \frac{2Eh^2}{\mu^2}} \cos\theta\right\} \tag{8.27}$$

Comparing with the polar equation of a conic $\frac{1}{r} = u = \frac{1}{l}(1 + e\cos\theta)$,

$l =$ semi latus rectum, $e =$ eccentricity.

we have for the path to be ellipse

$$\sqrt{1 + \frac{2Eh^2}{\mu^2}} < 1 \quad \text{i.e. } E \text{ is negative.}$$

for ellipse $e < 1$, for parabola $e = 1$, for hyperbola $e > 1$

$$\text{and semi-latus rectum} = \frac{h^2}{\mu} = \frac{b^2}{a} = a(1 - e^2)$$

$$= a\left(1 - 1 - \frac{2Eh^2}{\mu^2}\right) = -\frac{2aEh^2}{\mu^2} \qquad \left\{ \begin{array}{l} a = \text{semi-major axis} \\ b = \text{semi-minor axis} \end{array} \right. \tag{8.28}$$

$$\therefore\ a = \frac{\mu}{2E} \qquad \text{or,} \quad E = -\frac{\mu}{2a}$$

Again $E = \frac{1}{2}v^2 - \frac{\mu}{r}$ or, $v^2 = 2E + \frac{2\mu}{r}$

$$= -\frac{\mu}{a} + \frac{2\mu}{r}$$

$$= \mu\left(\frac{2}{r} - \frac{1}{a}\right) \tag{8.29}$$

If the particle is projected at a distance r_0 with a velocity v_0, then

$$v_0^2 = \mu\left(\frac{2}{r_0} - \frac{1}{a}\right) < \frac{2\mu}{r_0} < \text{escape velocity.}$$

Hence the path will be an ellipse which proves the first law of Kepler.

Since the force of interaction is along the line joining the planet and the sun, there is no acceleration perpendicular to the radius vector. Hence from equation (8.18)

$$\frac{1}{r}\frac{d}{dt}(r^2\dot{\theta}) = 0.$$

i.e. $r^2\dot{\theta} = h =$ constant.

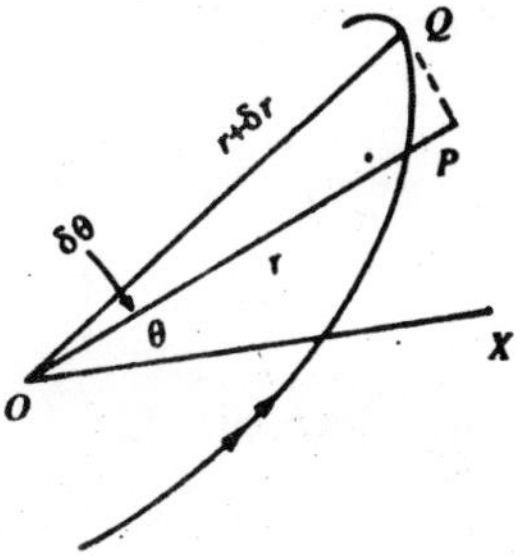

Fig. 8.14

If the particle goes from P to Q in time δt, rate of description of area by the radius

$$\text{vector} = \underset{\delta t \to 0}{\text{Lt}} \frac{\frac{1}{2} r.(r+\delta r)\sin\delta\theta}{\delta t} \quad \text{(Fig. 8.14).}$$

$$= \underset{\delta t \to 0}{\text{Lt}} \frac{\frac{1}{2} r(r+\delta r)\delta\theta}{\delta t} = \frac{1}{2}r^2\frac{d\theta}{dt} = \frac{1}{2}r^2\dot{\theta} = \frac{h}{2} = \text{constant.}$$

Thus the arial velocity is constant which proves the second law of Kepler.

If τ be the periodic time of describing the ellipse

$$\tau = \frac{2\pi ab}{h} = \frac{2\pi a}{h} h\sqrt{\frac{a}{\mu}} = \frac{2\pi}{\sqrt{\mu}} a^{3/2} \tag{8.30}$$

$\therefore$ $\tau^2 \propto a^3$ which proves the third law of Kepler.

(c) Newton's Law from Kepler's Laws :

From equation (8.29)

$$v^2 = \frac{2\mu}{r} - \frac{\mu}{a}$$

or, $$\dot{r}^2 + r^2\dot{\theta}^2 = \frac{2\mu}{r} - \frac{\mu}{a}$$

Differentiating w. r. t. 't'

$$2\dot{r}\ddot{r} + 2r\dot{r}\dot{\theta}^2 + 2r^2\dot{\theta}\ddot{\theta} = -\frac{2\mu}{r^2}\dot{r}$$

But $\frac{d}{dt}(r^2\dot{\theta}) = 0$

or, $2r\dot{r}\dot{\theta} + r^2\ddot{\theta} = 0$ or, $r^2\ddot{\theta} = -2r\dot{r}\dot{\theta}$

Substituting above

$$2\dot{r}\ddot{r} + 2r\dot{r}\dot{\theta}^2 - 4r\dot{r}\dot{\theta}^2 = -\frac{2\mu}{r^2}\dot{r}$$

or, $$\ddot{r} - r\dot{\theta}^2 = -\frac{\mu}{r^2} \tag{8.31}$$

which is Newtonian Law.

EXERCISE

1. (a) A mass m lies on the perpendicular through the centre of a uniform thin circular plate of radius a, at a distance b from the centre. Find the force of attraction between the plate and the mass m.

(b) Show that the angular momentum is conserved for motion under a central force. Hence or otherwise show that the motion under a central force is uniplanner.

(c) An artificial satellite revolves in a circular orbit about the earth at height H above the surface. Determine (i) the orbital speed and (ii) the orbital period so that a man in the satellite is in a state of weightlessness.

[C. U. (Hons.) 1995]

Ans. (a) Attraction at m due to a thin annular ring of radius x and thickness dx is

$$\frac{2\pi x\,dx\,\sigma\,Gm}{x^2 + b^2},$$ where σ = mass per unit area.

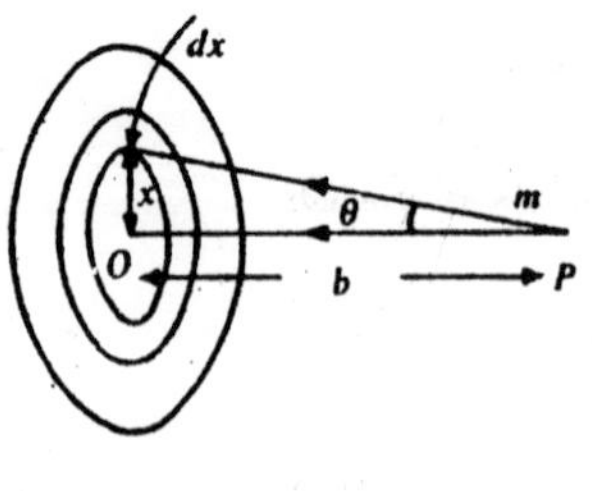

Fig. 8.15

Its component along

$$PO = \frac{2\pi\sigma Gm\,xdx}{(x^2 + b^2)}\cos\theta$$

$$= \frac{2\pi\sigma\,Gm\,xdx}{x^2 + b^2}\,\frac{b}{\sqrt{x^2 + b^2}}$$

component perpendicular to PO will be cancelled due to all the elements of the ring. Summing up for all such rings the total attraction at m due to the whole disc

$$= \int_0^a \frac{2\pi\sigma Gmb\, xdx}{(x^2+b^2)^{3/2}}$$

$$= \pi\sigma Gmb \left[\frac{-2}{\sqrt{x^2+b^2}}\right]_0^a$$

$$= \pi\sigma Gmb \left\{\frac{2}{b} - \frac{2}{\sqrt{a^2+b^2}}\right\}$$

$$= \frac{\pi a^2\sigma Gmb}{a^2}\left\{\frac{2}{b} - \frac{2}{\sqrt{a^2+b^2}}\right\} = \frac{2GMm}{a^2}\left\{1 - \frac{b}{\sqrt{a^2+b^2}}\right\}$$

where $M =$ mass of the disc.

(b) For motion under a centre of force, acceleration perpendicular to the radius vector is zero.

$$\therefore \frac{1}{r}\frac{d}{dt}(r^2\dot{\theta}) = 0 \qquad \text{or,} \quad r^2\theta = \text{constant}$$

or, Angular momentum $= mr^2\dot{\theta} = $ constant.

For motion under a centre of force $\ddot{r} - r\dot{\theta}^2 = f(r)$

which gives the equation of the paths in (r, θ) i.e. uniplanner.

(c) Condition for weightlessness is

$$\frac{mv^2}{R+H} = \frac{GMm}{(R+H)^2} \qquad \text{or,} \quad v^2 = \frac{GM}{(R+H)}$$

$$\text{or,} \quad v = \sqrt{\frac{GM}{(R+H)}}$$

$M =$ Mass of the earth
$R =$ Radius of the earth

Also $v = 2\pi(R+H)/T$

$$\text{or,} \quad T = \frac{2\pi(R+H)}{v} = 2\pi(R+H)\sqrt{\frac{R+H}{GM}} = \frac{2\pi}{\sqrt{GM}}(R+H)^{3/2}$$

2. (a) Prove that a central field is conservative and find the corresponding potential energy of a particle in that field.

(b) Writing $u = \frac{1}{r}$, show that the conservation of energy equation becomes

$$\left(\frac{du}{d\theta}\right)^2 + u^2 = \frac{-2(E-V)}{mh^2}$$, where h is twice the areal velocity and V is the potential energy.

(c) Differentiating the above relation obtain

$$\frac{d^2u}{d\theta^2} + u = \frac{-f\left(\frac{1}{u}\right)}{mh^2u^2}$$

(d) If a planet is to revolve around the sun in an elliptical orbit with sun at a focus show that the force of attraction must follow the inverse square law.

(e) Assuming the inverse square force to be $f(r) = -\frac{\mu}{r^2}$ $(\mu > 0)$, prove that the speed of the particle in the above force field is given by

$$v^2 = \frac{k}{m}\left(\frac{2}{r} \pm \frac{1}{a}\right) \quad (a \text{ is a constant.}).$$

the lower and the upper signs correspond to an elliptic and hyperbolic path respectively.

(f) Hence deduce the value of escape velocity. **[C. U. (Hons.) 1996]**

Ans. (a) The force $\vec{F}$ acting on a particle P moving under a centre of force O, is given by $\vec{F} = -f(r)\vec{r}$, where $\vec{r} = \vec{OP}$.

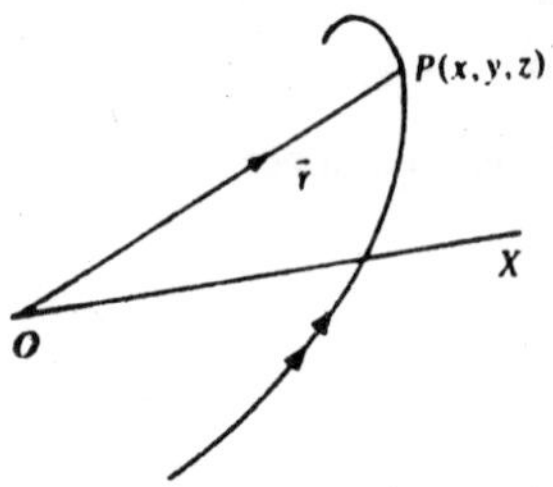

Fig. 8.16

Now curl $\vec{F} = \vec{\nabla} \times \vec{F}$

$$= -\left(\hat{i}\frac{\partial}{\partial x} + \hat{j}\frac{\partial}{\partial y} - \hat{k}\frac{\partial}{\partial r}\right) \times \{f(r)\,(\hat{i}x + \hat{j}y + \hat{k}z)\}$$

$$= -\left(\hat{i}\frac{\partial}{\partial x} + \hat{j}\frac{\partial}{\partial y} + \hat{k}\frac{\partial}{\partial z}\right) \times \{\varphi(r^2)\,(\hat{i}x + \hat{j}y + \hat{k}z)\}$$

$$= -\hat{k}\frac{\partial}{\partial x}\{\varphi(r^2)\,y\} + \hat{j}\frac{\partial}{\partial x}\{\varphi(r^2)\,z\}$$

$$+ \hat{k}\frac{\partial}{\partial y}\{\varphi(r^2)\,x\} - \hat{i}\frac{\partial}{\partial y}\{\varphi(r^2)\,z\} - \hat{j}\frac{\partial}{\partial z}\{\varphi(r^2)\,x\} + \hat{i}\frac{\partial}{\partial z}\{\varphi(r^2)\,y\}$$

$$= -\{\hat{k}\,2xy - \hat{j}\,2xz - \hat{k}\,2xy + \hat{i}\,2yz + \hat{j}\,2zx - \hat{i}\,2yz\}\,\varphi'(r) = 0$$

Since curl $\vec{F} = 0$, the force F must be conservative.

Writing $\vec{F} = -\vec{\nabla}V$

$$-\hat{i}\frac{\partial V}{\partial x} - \hat{j}\frac{\partial V}{\partial y} - \hat{k}\frac{\partial V}{\partial r} = -(\hat{i}x + \hat{j}y + \hat{k}z)\,\varphi(r^2)$$

$$\therefore \frac{\partial V}{\partial x} = x\varphi(r^2)$$

or, $V = \int x\, \varphi (x^2 + y^2 + z^2)\, dx = \dfrac{\psi (x^2 + y^2 + z^2)}{2} + c$

where $d(\psi) = \varphi$

(b) See equation (8.23) taking $m =$ mass of the particle.

(c) See equation (8.24).

(d) See equation (8.31).

(e) See equation (8.29).

(f) For escape $a \to \infty$

$$\therefore\ v^2 = \frac{2\mu}{r} \qquad \text{or,}\ v = \sqrt{\frac{2\mu}{r}} = \text{escape velocity.}$$

3. The density ρ at a distance r from the centre of sphere of radius a is given by

$\rho = \dfrac{M}{\pi a^3}\left(\dfrac{3}{2} - \dfrac{r}{a}\right)$. Show that M is the mass of the sphere. Also if g is the intensity of gravity at the surface of the sphere and powers of $\dfrac{h}{a}$ above the first are neglected, show that at a height h above the surface the intensity is $g\left(1 - \dfrac{2h}{a}\right)$, but that its value at a depth h below the surface is unaltered.

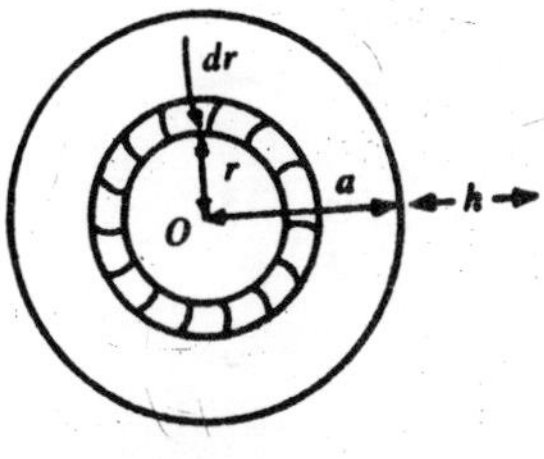

Fig. 8.17

Ans. Considering a thin spherical shell of radius r and thickness dr, intensity at a point at a distance h from the surface of the sphere is given by

$$F = \frac{4\pi r^2\, dr\, \rho G}{(a+h)^2}$$

$$= \frac{4\pi G M}{\pi a^3}\left(\frac{3}{2} - \frac{r}{a}\right)\frac{r^2\, dr}{(a+h)^2}$$

Integrating, the intensity due to the whole sphere

$$= \frac{4\pi GM}{(a+h)^2\, \pi a^3}\left\{\int_0^a \frac{3}{2} r^2\, dr - \int_0^a \frac{r^3}{a}\, dr\right\}$$

$$= \frac{4GM}{(a+h)^2\, a^3}\left\{\frac{3}{2}\frac{a^3}{3} - \frac{a^4}{4a}\right\}$$

$$= \frac{4GM}{(a+h)^2}\left\{\frac{1}{2}-\frac{1}{4}\right\} = \frac{GM}{(a+h)^2} = \frac{GM}{a^2}\left(1+\frac{h}{a}\right)^{-2}$$

$$= \frac{GM}{a^2}\left(1-\frac{2h}{a}\right) = g\left(1-\frac{2h}{a}\right). \qquad \because\ g = \frac{GM}{a^2}$$

Again mass of the sphere $= \int_0^a 4\pi r^2\, dr\, \rho = \int_0^a 4\pi r^2 \frac{M}{\pi a^3}\left(\frac{3}{2}-\frac{r}{a}\right) dr$

$$= \frac{4M}{a^3}\left\{\int_0^a \frac{3}{2} r^2\, dr - \int_0^a \frac{r^3}{a}\, dr\right\} = \frac{4M}{a^3}\left\{\frac{3}{2}\,\frac{a^3}{3} - \frac{a^4}{4a}\right\}$$

$$= \frac{4M\,a^3}{a^3}\left(\frac{1}{2}-\frac{1}{4}\right) = M.$$

Intensity at depth h below the surface is given by

$$F = \int_0^{a-h} \frac{4\pi r^2\, dr \rho G}{(a-h)^2}.$$

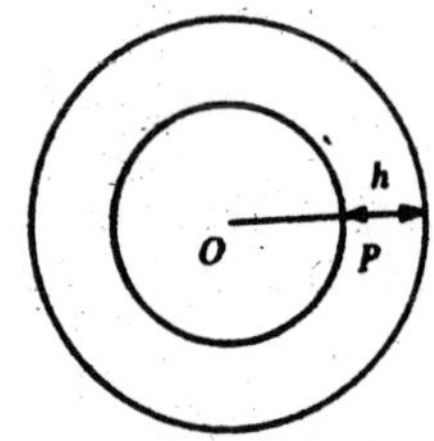

Fig. 8.18

[Intensity due to a thin spherical shell at an intenal point is zero.]

$$= \int_0^{a-h} \frac{4\pi G}{(a-h)^2}\,\frac{M}{\pi a^3}\left(\frac{3}{2}-\frac{r}{a}\right) r^2\, dr$$

$$= \frac{4\pi GM}{(a-h)^2 \pi a^3}\left\{\frac{3}{2}\,\frac{(a-h)^3}{3} - \frac{(a-h)^4}{4a}\right\}$$

$$= \frac{4GM}{(a-h)^2\, a^3}(a-h)^3\left\{\frac{1}{2}-\frac{a-h}{4a}\right\}$$

$$= \frac{4GM}{a^3}(a-h)\,\frac{2a-a+h}{4a}$$

$$= \frac{4GM}{4a^4}(a-h)(a+h)$$

$$= \frac{GM}{a^2}\left(1-\frac{h^2}{a^2}\right) = \frac{GM}{a^2} = g \text{ Proved.}$$

4. Under the influence of a central force at O, a particle moves in a circular orbit which passes through O. Find the law of force.

Ans. In polar co-ordinates the equation of a circle of radius a passing through 0 is $r = 2a\cos\theta$.

Thus $u = \frac{1}{r} = \frac{\sec\theta}{2a}$

$$\frac{du}{d\theta} = \frac{\sec\theta \tan\theta}{2a}$$

$$\frac{d^2u}{d\theta^2} = \frac{\sec\theta \sec^2\theta + \sec\theta \tan\theta \tan\theta}{2a}$$

$$= \frac{\sec^3\theta + \sec\theta \tan^2\theta}{2a}$$

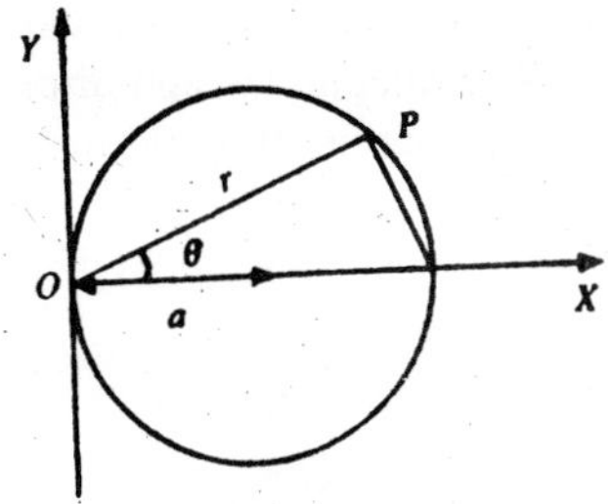

Fig. 8.19

From equation (8.20)

$$\frac{d^2u}{d\theta^2} + u = \frac{P}{h^2u^2} = \frac{\sec\theta + \sec^3\theta + \sec\theta \tan^2\theta}{2a}$$

$$= \frac{\sec\theta (1 + \tan^2\theta) + \sec^3\theta}{2a} = \frac{\sec^3\theta}{a}$$

$$\therefore\ P = h^2u^2 \frac{\sec^3\theta}{a} = h^2u^2 \frac{(2au)^3}{a} = 8a^2 h^2 u^5$$

$$\therefore\ P \propto \frac{1}{r^5}$$

5. A particle moves in a circular orbit obeying the inverse square law. Show that for orbits of difference radii the angular momentum of the particle about the centre of mass varies as the square root of the radius and the total energy varies inversely as the radius. **[C. U. (Hons.)]**

Ans. As given $\frac{mv^2}{r} = \frac{\mu}{r^2}$ $\quad m =$ mass, $\mu =$ constant

or, $m\omega^2 r^2 = \frac{\mu}{r}$

or, $\frac{(mr^2\omega)^2}{mr^2} = \frac{\mu}{r}$ $\quad$ or, $(I\omega)^2 = \mu\, mr.$

Thus angular momentum $L = I\omega = \sqrt{\mu\, mr} \propto \sqrt{r}$

Now the total energy $E = \frac{1}{2} mv^2 - \frac{\mu}{r}$

$$= \frac{\mu}{2r} - \frac{\mu}{r} = -\frac{\mu}{2r}$$

$$\therefore\ E \propto \frac{1}{r}$$

6. If a spherical mass of radius a and density d, has an eccentric spherical cavity of radius b, show that the attraction at any two points in the cavity is the same in magnitude and direction and find its value if the distance of the centre of the cavity from the centre of the sphere is c.

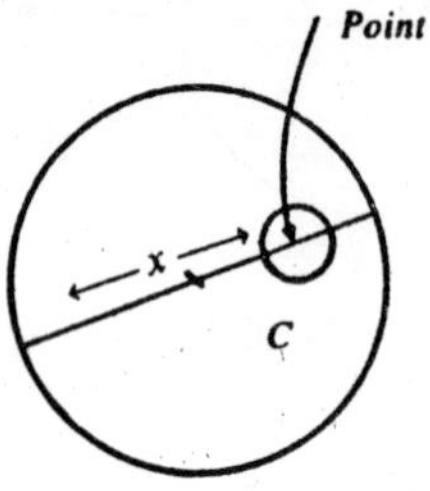

Fig. 8.20

Ans. Let us consider a point in the cavity at a distance x from the centre of the sphere. Two cases may arise.

Case I ($x < c$) :

The intensity at the point concerned is

$$\frac{GM}{a^3}x + \frac{Gm}{b^3}(c - x) = \frac{4\pi\, Gdx}{3} + \frac{4}{3}\pi Gd\,(c - x)$$

$$= \frac{4}{3}\pi\, Gdc.$$

Case II ($x > c$) :

The intensity at the point concerned is

$$\frac{4}{3}\pi\, Gdx - \frac{4}{3}\pi\, dG\,(x - c) = \frac{4}{3}\pi\, dGc.$$

Hence at both the points the field intensity has the same magnitude and direction (towards the centre of the sphere of radius a).

7. Gravitational attraction between two parts of the same body : If the two parts of the body be denoted by A and B as shown in the adjacent figure, the attraction of the whole body on A is the attraction of B on A, plus the attraction of A on itself. But this latter attraction is zero. Hence the attraction between two parts of the same body is the same as the attraction of the whole on one part.

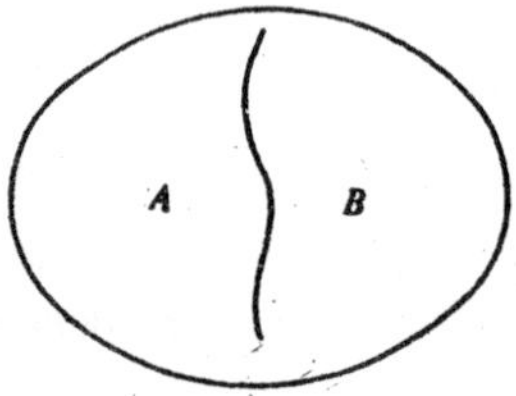

Fig. 8.21

8. A uniform sphere of radius a and density ρ is divided into two parts by a plane at a distance b from the centre. Prove that the mutual attraction between the two parts is $\frac{1}{3}\pi^2 G\rho^2\,(a^2 - b^2)^2$.

Ans. If P is a point at a distance y from the plane XOZ, the attraction per unit mass at P is directed along PO and has a magnitude $\frac{4}{3}\pi\rho Gr$ where $OP = r$. The component of this normal to the plane XOZ is $\frac{4}{3}\pi\rho Gy$ and this has the same value at all points in a plane through P normal to OY. The attraction on the circular disc between y and y + dy is

$\frac{4}{3}\pi\rho gy \times \pi\rho\,(a^2 - y^2)\,dy,$ since $\sqrt{a^2 - y^2}$ is the radius of the disc.

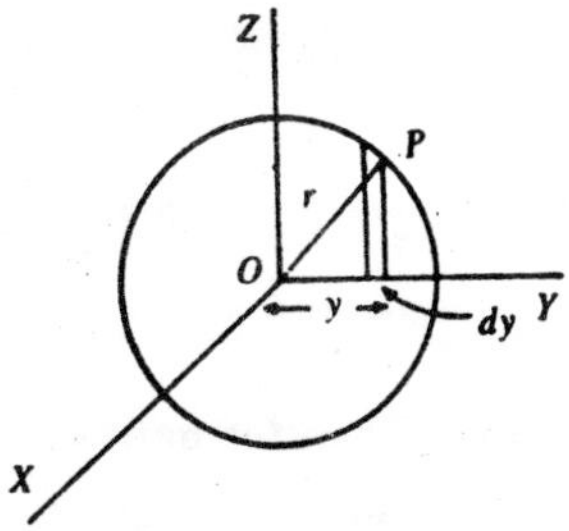

Fig. 8.22

Hence the required attraction is

$$\frac{4}{3}\pi^2 G\rho^2 \int_b^a y\,(a^2 - y^2)\,dy = \frac{4}{3}\pi^2\rho^2 G\left\{\int_b^a a^2 y\,dy - \int_b^a y^3\,dy.\right\}$$

$$= \frac{4}{3}\pi^2\rho^2 G\left\{a^2\,\frac{(a^2 - b^2)}{2} - \frac{(a^4 - b^4)}{4}\right\}$$

$$= \frac{4}{3}\pi^2\rho^2 G\left\{\frac{(a^2 - b^2)}{2}\left(a^2 - \frac{a^2 + b^2}{2}\right)\right\} = \frac{4}{3}\pi^2\rho^2 G\,(a^2 - b^2)\,\frac{(a^2 - b^2)}{4}$$

$$= \frac{1}{3}\pi^2\rho^2 G\,(a^2 - b^2)^2$$

CHAPTER IX

MECHANICS II

Galilian Transformation : Inertial and Non-Inertial Frames.

9.1. A system relative to which the motion of any object can be described is called **a frame of reference**. There are in general two frames of reference—(1) Intertial and (2) Non-Inertial.

The frames with respect to which the body, not acted upon by external force, is unaccelerated, i.e. either at rest or moving with uniform velocity are called inertial frames.

Let a moving body have co-ordinates (x, y, z) relative to any co-ordinate system. If no external force acts on it,

$$\text{then } m\frac{d^2x}{dt^2} = 0, \quad m\frac{d^2y}{dt^2} = 0, \quad m\frac{d^2z}{dt^2} = 0$$

$$\text{or,} \quad \frac{d^2x}{dt^2} = \frac{d^2y}{dt^2} = \frac{d^2z}{dt^2} = 0$$

$$\text{i.e.} \quad \frac{dx}{dt} = \text{constant}, \quad \frac{dy}{dt} = \text{constant}, \quad \frac{dz}{dt} = \text{constant}.$$

In other words, in the absence of any external force, a body in motion continues to move in a straight line with constant velocity, which is Newton's First Law. As a consequence we may say "An inertial frame is one in which the law of inertia or Newton's First Law is valid.

Conversely speaking, the frame of reference with respect to which the body, not acted upon by external force is accelerated is called a non-inertial frame.

Any frame of reference moving with constant velocity relative to an inertial frame is also an inertial frame. Since the acceleration of the body in both frames is zero, while the velocity of the body is different but uniform.

A frame of reference fixed in stars is an inertial frame. A co-ordinate system fixed in the earth is not an inertial frame, since the earth rotates round the sun and also about its own axis. A body on the earth is neither at rest nor moving in a straight line with constant velocity though no forces act on it.

9.2. Galilian Transformation

Let us consider two inertial frames S and S', the latter being in uniform motion with respect to the former and moves with a velocity v along the + ve direction of x-axis (Fig. 9.1). The space and time co-ordinates of any event referred to S and S' are (x, y, z, t) and (x', y', z', t') respectively. The relationship between the two systems can be established from the two fold hypothesis—(i) Time is absolute and (ii) Space is absolute.

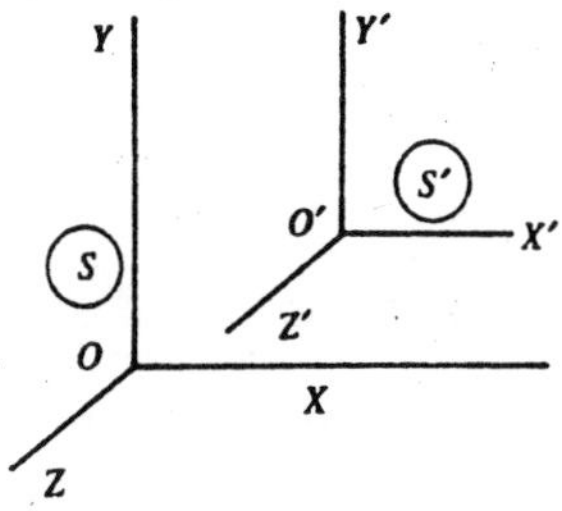

Fig. 9.1

The former gives

$$t_2 - t_1 = (t'_2 - t'_1) \tag{9.1}$$

The latter gives

$$(x_2 - x_1)^2 + (y_2 - y_1)^2 + (z_2 - z_1)^2 = (x'_2 - x'_1)^2 + (y'_2 - y'_1)^2 + (z'_2 - z'_1)^2 \tag{9.2}$$

Equation (9.1) signifies that the time interval between same two events will be same in the two systems. Equation (9.2) signifies that the distance between two points will be same in the two systems. The above equations represent what is known as Galilian transformation. In the most general case when v has any direction, the relation between S and S' systems will be represented by

$$\left.\begin{aligned} x' &= x - v_x t \\ y' &= y - v_y t \\ z' &= z - v_z t \\ t' &= t \end{aligned}\right\} \tag{9.3}$$

where v_x, v_y, v_z are the velocity components of the S' frame relative to the S frame. In the special case when the axes of the two systems are parallel and v is along x-axis of S, the above equation reduce to

$$\left.\begin{aligned} x' &= x - vt \\ y' &= y \\ z' &= z \\ t' &= t \end{aligned}\right\} \tag{9.4}$$

9.3. Moving Frame

(a) Frame of reference with translational motion :

Let S be an inertial frame and S' a frame of reference which has relative to S a translational motion.

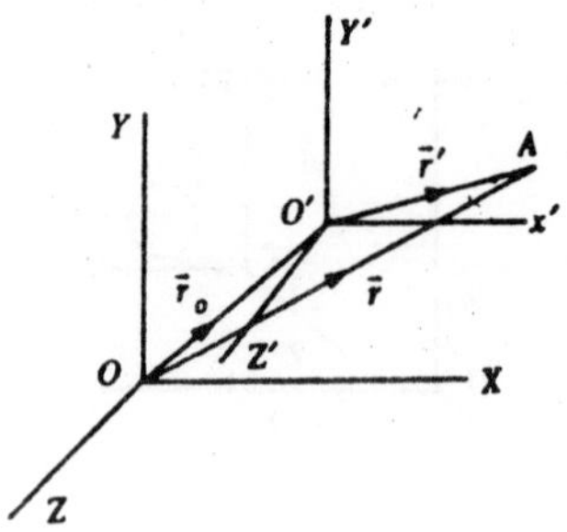

Fig. 9.2

A particle 'A' has position vectors $\vec{r} = \vec{OA}$ and $\vec{r}' = \vec{O'A}$ with respect to the origins O and O' of the two frames (Fig. 9.2), then

$\vec{r} = \vec{r}_0 + \vec{r}'$ (9.5) where $\vec{r}_0 = \vec{OO'}$

Differentiating with respect to 't'

$$\vec{v} = \vec{v}_0 + \vec{v}' \tag{9.6}$$

Differentiating again

$$\vec{f} = \vec{f}_0 + \vec{f}' \tag{9.7}$$

where $\vec{f}$ and $\vec{f}'$ are the acceleration of the particle in the fixed and moving systems respectively and f_0 is the acceleration of the moving system relative to the fixed system.

Multiplying both sides of equation (9.7) by m the mass of the particle

$$m\vec{f} = m\vec{f}_0 + m\vec{f}' \tag{9.8}$$

or, $\vec{F} = m\vec{f}_0 + \vec{F}'$, where F and F' are the forces acting on the particle in the fixed and moving systems respectively.

Rewriting, $\vec{F}' = \vec{F} - m\vec{f}_0$ (9.9)

Thus the force $\vec{F}$ acting on the particle in fixed frame has been transformed into $\vec{F} - m\vec{f}_o$ in the moving frame, the fictitious form $m\vec{f}_o$ appearing is due to the acceleration of the moving system. When a car suddenly starts or stops, the force experienced by a passenger is this fictitious force.

(b) Frame of reference with rotational motion :

Let S be an inertial frame and S' a frame of reference which is rotating with an angular velocity ω with respect to S.

Let $\vec{A}$ be a vector whose components are A_x, A_y, A_z in the fixed frame S (Fig. 9.3).

Hence $\vec{A} = \hat{i} A_x + \hat{j} A_y + \hat{k} A_z$ (9.10)

where $\hat{i}, \hat{j}, \hat{k}$ are unit vectors along OX, OY and OZ respectively. Similarly if Ax', Ay' and Az' are the components of $\vec{A}$ along $O'X'$, $O'Y'$ and $O'Z'$ respectively and $\hat{i}', \hat{j}', \hat{k}'$ are unit vectors along these directions

$$\vec{A} = \hat{i}' Ax' + \hat{j}' Ay' + \hat{k}' Az' \tag{9.11}$$

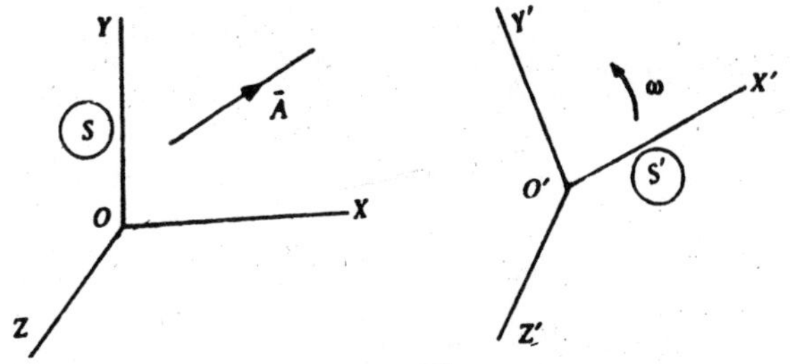

Fig. 9.3

Differentiating equation (9.10) with respect to t

$$\frac{d\vec{A}}{dt} = \hat{i}\,\frac{dAx}{dt} + \hat{j}\,\frac{dAy}{dt} + \hat{k}\,\frac{dAz}{dt} \qquad (9.12)$$

since the unit vectors $\hat{i}$, $\hat{j}$, $\hat{k}$ remain unchanged in the fixed system. But in the rotating frame these unit vectors change with time,

Hence from (9.11)

$$\frac{d\vec{A}}{dt} = \hat{i}'\,\frac{dAx'}{dt} + \hat{j}'\,\frac{dAy'}{dt} + \hat{k}'\,\frac{dAz'}{dt} + Ax'\,\frac{d\hat{i}'}{dt} + Ay'\,\frac{d\hat{j}'}{dt} + Az'\,\frac{d\hat{k}'}{dt} \qquad (9.12(a))$$

The term $\hat{i}'\,\frac{dAx'}{dt} + \hat{j}'\,\frac{dAy'}{dt} + \hat{k}'\,\frac{dAz'}{dt} = \frac{d'\vec{A}}{dt}$ (9.12(b))

represents the rate of change of $\vec{A}$ in the rotating frame.

Next we prove that the rate of change with time of vector $\vec{B}$ fixed in the rotating frame w. r. t. the fixed frame is given by $\frac{d\vec{B}}{dt} = \vec{\omega} \times \vec{B}$.

Let the rotating frame be revolving about the axis OP and the angle between $\vec{B}$ and $\vec{\omega}$ is θ. In the interval δt, the change in $\vec{B}$ is $\overrightarrow{CD}$ (Fig. 9.4).

Now $CD = B \sin\theta \ (\omega\, \delta t) = \delta B$

Hence the rate of change of $B = \frac{dB}{dt}$

$$= \underset{\delta t \to 0}{\text{Lt}} \frac{\delta B}{\delta t} = B \sin\theta\, \omega$$

In vector rotation $\frac{d\vec{B}}{dt} = B \sin\theta\, \omega\, \hat{n}$, $\hat{n}$ is unit vector along $\overrightarrow{CD}$. Hence $\frac{d\vec{B}}{dt} = \vec{\omega} \times \vec{B}$ (9.13)

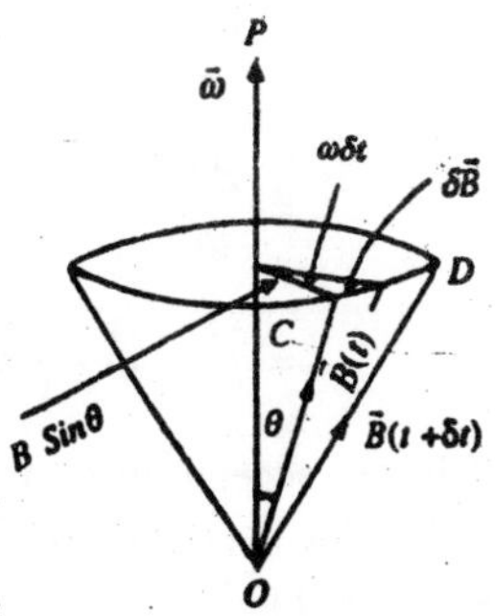

Fig. 9.4

Applying (9.13) to $\frac{d\hat{i}'}{dt}$, $\frac{d\hat{j}'}{dt}$ and $\frac{d\hat{k}'}{dt}$ of equation (9.12(a))

$$\frac{d\vec{A}}{dt} = \frac{d'\vec{A}}{dt} + Ax'\,(\vec{\omega}\times\hat{i}') + Ay'\,(\vec{\omega}\times\hat{j}') + Az'\,(\vec{\omega}\times\hat{k}')$$

$$= \frac{d'\vec{A}}{dt} + \vec{\omega}\times\vec{A}. \tag{9.14}$$

By the operators $\frac{d}{dt}$ and $\frac{d'}{dt}$ are denoted the time rate of range in the fixed and rotating frames respectively, while $\vec{\omega}X$ denotes the rate of change as observed from the fixed frame due to rotation of the rotating frame.

Hence expressed as a mathematical operator

$$\frac{d}{dt} = \frac{d'}{dt} + \vec{\omega}\times \tag{9.15}$$

Applying equation (9.15) to the position vector $\vec{r}$ of a moving particle p,

$$\frac{d\vec{r}}{dt} = \frac{d'\vec{r}}{dt} + \vec{\omega}\times\vec{r}$$

or, $$\vec{v} = \vec{v}' + \vec{\omega}\times\vec{r} \tag{9.16}$$

where $\vec{v}$ and $\vec{v}'$ are the velocities of the moving particle as observed from the fixed and rotating frames respectively.

Again the acceleration of the moving point in the fixed frame

$$\vec{f} = \frac{d}{dt}\left(\frac{d\vec{r}}{dt}\right) = \frac{d^2\vec{r}}{dt^2}$$

$$= \frac{d}{dt}\left\{\frac{d'\vec{r}}{dt} + \vec{\omega}\times\vec{r}\right\} = \left\{\frac{d'}{dt} + \vec{\omega}\,x\right\}\left\{\frac{d'\vec{r}}{dt} + \vec{\omega}\times\vec{r}\right\}$$

$$= \frac{d'^2\vec{r}}{dt^2} + \frac{d'}{dt}(\vec{\omega}\times\vec{r}) + \vec{\omega}\times\frac{d'\vec{r}}{dt} + \vec{\omega}\times(\vec{\omega}\times\vec{r})$$

$$= \frac{d'^2\vec{r}}{dt^2} + \vec{\omega}\times(\vec{\omega}\times\vec{r}) + 2\vec{\omega}\times\frac{d'\vec{r}}{dt} + \frac{d'\vec{\omega}}{dt}\times\vec{r}$$

$$= \vec{f}' + \vec{\omega}\times(\vec{\omega}\times\vec{r}) + 2\vec{\omega}\times\vec{v}' + \frac{d'\vec{\omega}}{dt}\times\vec{r} \tag{9.17}$$

f' is the acceleration of the particle in the rotating frame ; $\vec{\omega}\times(\vec{\omega}\times\vec{r})$ is called the centripetal acceleration and $2\vec{\omega}\times\vec{v}'$ is called the Coriolis acceleration.

Multiplying both sides of equation (9.17) by m (the mass of the particle)

$$\vec{F} = \vec{F}' + m\left\{\vec{\omega}\times(\vec{\omega}\times\vec{r}) + 2\vec{\omega}\times\frac{d\vec{r}}{dt} + \frac{d\vec{\omega}}{dt}\times\vec{r}\right\}$$

$$\text{or, } \vec{F}' = \vec{F} - m\left\{\vec{\omega}\times(\vec{\omega}\times\vec{r}\right\} + 2m\vec{v}'\times\vec{\omega} - m\frac{d'\vec{\omega}}{dt}\times\vec{r} \qquad \mathbf{(9.18)}$$

It appears from equation (9.18) that when the fictitious forces $-m\vec{\omega}\times(\vec{\omega}\times\vec{r})$, $2m\vec{v}'\times\vec{\omega}$ and $-m\dfrac{d'\vec{\omega}}{dt}\times\vec{r}$ are added to the applied force $\vec{F}$ we get the force $\vec{F}'$ acting on the body in the rotating frame. The first one is the centrifugal force and second one the coriolis force. The third term became absent if $\dfrac{d'\vec{\omega}}{dt}$ is zero ; i.e. the angular velocity of rotation of the frame is constant.

Coriolis force due to rotation of the earth :

Terrestrial measurements are usually made with respect to a co-ordinate system fixed in the earth, which is rotating uniformly with a constant angular velocity relative to the inertial frame. Hence in addition to the centrifugal force, a moving particle on the surface of the earth experiences the coriolis force $2m\vec{v}'\times\vec{\omega}$. In the northern hemisphere where $\vec{\omega}$ points out of the ground, the coriolis force, $2m\vec{v}'\times\vec{\omega}$ tends to deflect the particle shot along the earth's surface to the right of its direction of travel. The coriolis deflection reverses its direction in the southern hemisphere and is zero at the equator where $\vec{\omega}$ is horizontal.

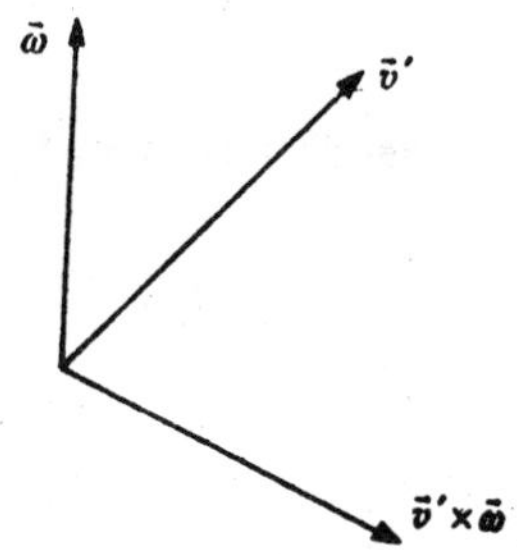

Fig. 9.5

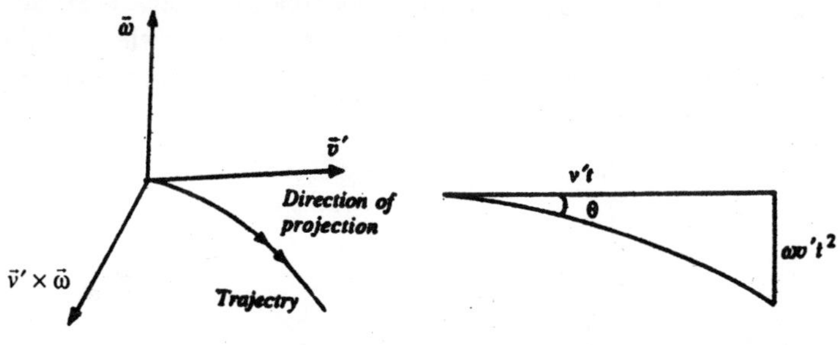

(a) (b)

Fig. 9.6

The maximum coriolis acceleration will be at the north pole where $\vec{\omega}$ is vertical. This maximum is $2\omega v'$.

Hence linear deflections after a time t is $\frac{1}{2}(2\omega v')\,t^2 = \omega v' t^2$ and the angular deflection $\theta = \dfrac{\omega v' t^2}{v' t} = \omega t$ (Fig. 9.6). (9.19)

But $\omega = \dfrac{2\pi}{T} = \dfrac{2\times 3.14}{24\times 3600} = 7.28\times 10^{-5}$ radius/sec.

For a time of flight of 100 secs.,

$$\theta = 7.28\times 10^{-5}\times 100 = 7.28\times 10^{-3} \text{ radius}$$
$$\approx 0.4^\circ.$$

Coriolis force has another important application in connection with the direction of windflow. Wind is air in motion and the direction of this motion will be from higher to lower pressure. However in the northern hemisphere, the coriolis force wants to deflect the wind to the right (Fig. 9.7).

Coriolis force on a freely falling body :

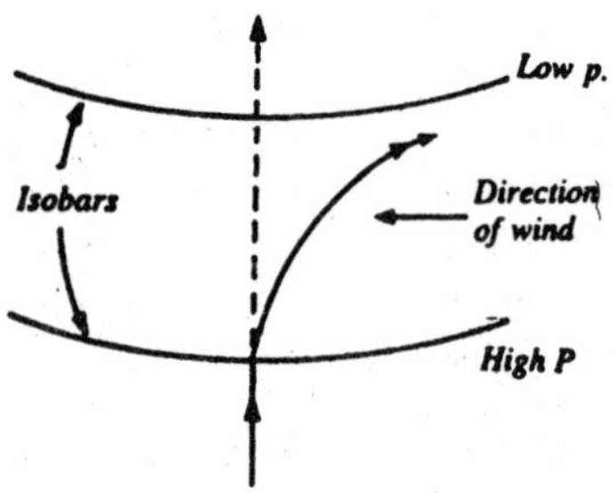

Fig. 9.7

If the only impressed force acting on a particle is the force F_g due to gravitational attraction of the earth, then from equation (9.18)

$$m\vec{g}' = \vec{F}_g - m\vec{\omega}\times(\vec{\omega}\times\vec{r}) - 2m\,\vec{\omega}\times\vec{v}' \qquad (9.20)$$

The acceleration g of gravity as measured by a pendulum which has no appreciable velocity relative to the earth is not, however, the acceleration due to F_g alone, but rather due to the resultant of $\vec{F}_g$ and the centrifugal force $-m\vec{\omega}\times(\vec{\omega}\times\vec{r})$. Hence $m\vec{g} = \vec{F}_g - m\vec{\omega}\times(\vec{\omega}\times\vec{r})$.

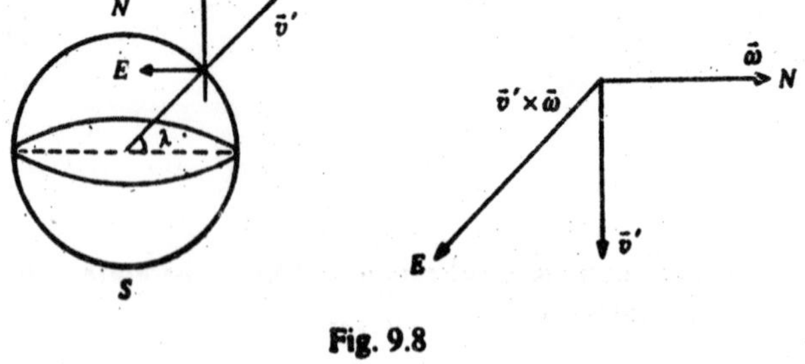

Fig. 9.8

Substituting in (9.20) the apparent acceleration relative to the earth of a freely falling body

$$\vec{g}' = \vec{g} - 2\,\vec{\omega} \times \vec{v}'$$

$$= \vec{g} + 2\vec{v}' \times \vec{\omega} \qquad (9.21)$$

Let us consider a body falling freely from rest (Fig. 9.8). The velocity $\vec{v}'$ is at all times nearly downwards, and as $\vec{\omega}$ is directed from south to north, the vector $2\vec{v}' \times \vec{\omega}$ is directed eastward. Taking the origin at the point from which the body is dropped from rest with x-axis pointing eastward, the falling body is subjected to an eastward accelerations in addition to the vertically downward acceleration g. The eastward acceleration is given by $\frac{d^2x}{dt^2} = 2\omega v' \cos\lambda$, where λ is the latitude of the place.

To a sufficient degree of approximation $\vec{v}'$ is equal to the product g by the time t during which the body has been falling, so

$$\frac{d^2x}{dt^2} = 2\omega gt \cos\lambda.$$

Integrating, $x = \frac{\omega g t^3 \cos\lambda}{3}$, since at $t = 0$ $\frac{dx}{dt} = 0$, $x = 0$.

But the height h through which the body has fallen is very closely

$$h = \frac{1}{2} g t^2.$$

Eliminating t, the eastward deflection is given by

$$x = \frac{1}{3}\omega g \cos\lambda \left(\frac{2h}{g}\right)^{3/2}$$

$$= \frac{1}{3}\omega \cos\lambda \sqrt{\frac{8h^3}{g}} \qquad (9.22)$$

which is proportional to the cosine of the latitude and the three halves power of the height.

Problems :

1. A projectile is fired at a lititude λ. The velocity v_0 is directed towards the west and at an angle θ with the horizontal. Prove that if the terms containing ω^2 are neglected, the time taken to reach the maximum height is

$$\frac{v_0 \sin\theta}{g} - \frac{2\omega\, v_0^2 \cos\lambda \sin\theta \cos\theta}{g^2},$$

Let $\hat{i}$, $\hat{j}$, $\hat{k}$ be unit vectors along the east, north and the vertical (Fig. 9.9).

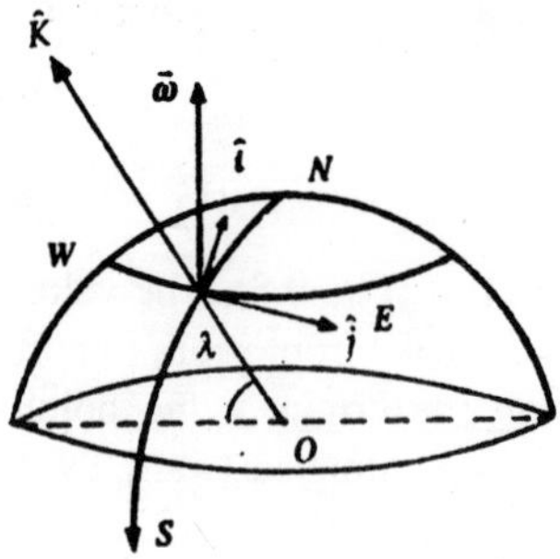

Fig. 9.9

$$\therefore \ \vec{v_0} = - v_0 \cos\theta \, \hat{i} + v_0 \sin\theta \, \hat{k}$$

$$\vec{\omega} = \omega \cos\lambda \, \hat{j} + \omega \sin\lambda \, \hat{k}$$

Hence

$$\vec{v_0} \times \vec{\omega} = (- v_0 \cos\theta \, \hat{i} + v_0 \sin\theta \, \hat{k}) \times (\omega \cos\lambda \, \hat{j} + \omega \sin\lambda \, \hat{k})$$

$$= - v_0 \, \omega \cos\theta \cos\lambda \, \hat{k} + v_0 \, \omega \cos\theta \sin\lambda \, \hat{j} - v_0 \, \omega \sin\theta \cos\lambda \, \hat{i}$$

(Initial) Horizontal comp. of coriolis acceleration.

$$= 2\omega v_0 \sqrt{\cos^2\theta \sin^2\lambda + \sin^2\theta \cos^2\lambda}$$

(Initial) vertical comp. of coriolis acceleration

$$= - 2v_0 \, \omega \cos\theta \cos\lambda$$

vertical comp. of the velocity of projection = $v_0 \sin\theta$.

If t be time to the greatest height

$$0 = v_0 \sin\theta - (g + 2v_0 \, \omega \cos\theta \cos\lambda) \, t$$

or, $$t = \frac{v_0 \sin\theta}{g + 2v_0 \, \omega \cos\theta \cos\lambda}$$

$$= \frac{v_0 \sin\theta}{g} \left(1 - \frac{2v_0 \omega \cos\theta \cos\lambda}{g} \right)$$

$$= \frac{v_0 \sin\theta}{g} - \frac{2\omega v_0^2 \sin\theta \cos\theta \cos\lambda}{g^2}.$$

2. A projectile located at a latitude λ is fired with a velocity v_0 in southward direction at an angle θ with the horizontal. Find its position after time t and prove that it is deflected towards the east of the original plane by an amount

$$\frac{1}{3} \omega g \cos\lambda \, t^3 - \omega v_0 \sin (\theta + \lambda) \, t^2$$

with the same notation as in problem 1.

$$\vec{v_0} = - v_0 \cos\theta \, \hat{j} + v_0 \sin\theta \, \hat{k}$$

$$\vec{\omega} = \omega \cos\lambda \, \hat{j} + \omega \sin\lambda \, \hat{k}$$

After a time t

$$\vec{v} \times \vec{\omega} = (- v_0 \cos\theta \, \hat{j} + v_0 \sin\theta \, \hat{k} - gt \, \hat{k}) \times (\omega \cos\lambda \, \hat{j} + \omega \sin\lambda \, \hat{k})$$

$$= -\,v_0\,\omega\cos\theta\,\sin\lambda\,\hat{i} - v_0\,\omega\sin\theta\,\cos\lambda\,\hat{i} + gt\,\omega\cos\lambda\,\hat{i}$$

$$= -\,\hat{i}\,\{(\cos\theta\,\sin\lambda + \sin\theta\,\cos\lambda)\,v_0\,\omega - gt\cos\lambda\,\omega\}$$

Hence coriolis acceleration $= 2\,\hat{i}\,\{gt\,\omega\cos\theta - v_0\,\omega\sin(\theta+\lambda)\}$

or, $$\frac{d^2x}{dt^2} = 2gt\omega\cos\lambda - 2v_0\,\omega\sin(\theta+\lambda)$$

or, $$\frac{dx}{dt} = \frac{2gt^2\omega}{2}\cos\lambda - 2v_0\,\omega\sin(\theta+\lambda)\,t + A$$

or, $$x = \frac{gt^3\,\omega\cos\lambda}{3} - v_0\,\omega\sin(\theta+\lambda)\,t^2 + At + B$$

At $t = 0,\ \dfrac{dx}{dt} = 0,\quad x = 0$

$\therefore\ A = B = 0$

$$\therefore\ x = \frac{g\omega\sin\lambda}{3}\,t^3 - v_0\omega\sin(\theta+\lambda)\,t^2.$$

3. A particle is projected vertically upwards with a velocity v_o. Calculate the distance of the point when the motion is terminated from the point of projection.

Coriolis acceleration $= 2\,(\vec{v}\times\vec{\omega})$ towards west

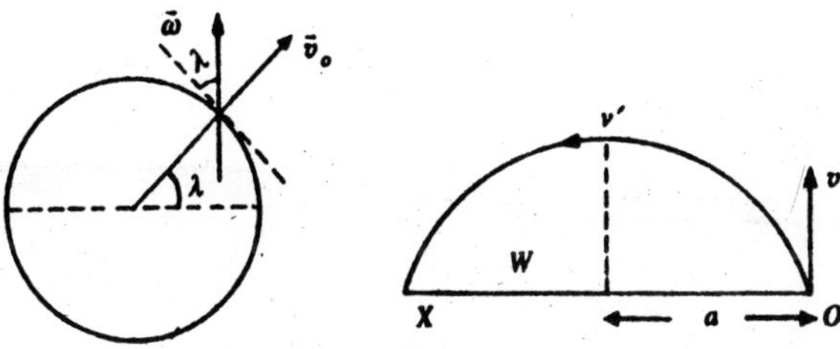

Fig. 9.10

$$\therefore\ \frac{d^2x}{dt^2} = 2\,(v_0 - gt)\,\omega\cos\lambda$$

Integrating $$\frac{dx}{dt} = 2\left(v_0\,t - \frac{gt^2}{2}\right)\omega\cos\lambda$$

and $$x = 2\left(\frac{v_0\,t^2}{2} - \frac{gt^3}{6}\right)\omega\cos\lambda$$ towards west

At the highest point $v_0 - gt' = 0$

$$\text{or, } t' = \frac{v_0}{g}$$

$$\therefore\ a = 2\left\{v_0 \frac{v_0^{\,2}}{2g^2} - g\frac{v_0^{\,3}}{6g^3}\right\}.\ \omega\cos\lambda = \frac{2v_0^{\,3}}{g_0^{\,2}}\left\{\frac{1}{2} - \frac{1}{6}\right\}\omega\cos\lambda$$

$$= \frac{2}{3}\frac{v_0^{\,3}}{g^2}\ \omega\cos\lambda = \text{ westward shift during rise.}$$

Same shift will be during the fall. Horizontal velocity at the top most point

$$= V' = 2\left(v_0 t' - \frac{gt'^2}{2}\right)\omega\cos\lambda$$

$$= 2\left(v_0\frac{v_0}{g} - g\frac{v_0^2}{2g^2}\right)\omega\cos\lambda = \frac{v_0^2}{g}\ \omega\cos\lambda \text{ towards west}$$

Coriolis acceleration towards east $= 2\,gt\,\omega\cos\lambda$

$$\therefore\ \frac{d^2x}{dt^2} = -\,2\,gt\,\omega\cos\lambda\ \text{(west)}$$

Integrating $\frac{dx}{dt} = gt^2\omega\cos\lambda + c = gt^2\,\omega\cos\lambda + \frac{v_0^2}{g}\,\omega\cos\lambda$

$$\text{and } x = -\frac{gt^3}{3}\ \omega\cos\lambda + \frac{v_0^2\,\omega\cos\lambda}{g}\,t + 0$$

$$= -\frac{g}{3}\,\omega\cos\lambda\ \frac{v_0^3}{g^3} + \frac{v_0^2\,\omega\cos\lambda}{g}\ \frac{v_0}{g} = \frac{v_0^3}{g^2}\,\omega\cos\lambda\left(1 - \frac{1}{3}\right)$$

$$= \frac{2}{3}\ \frac{v_0^3\,\omega\cos\lambda}{g^2}\ \text{(west)}$$

Hence the total shift $= \frac{4}{3}\frac{v_0^3}{g^2}\ \omega\cos\lambda$ towards west

Effect of centrifugal force due to earth's rotation :

From equation (9.18) it is evident that the effective weight of a body of mass m due to earth's rotation is given by

$$m\vec{g}_e = m\vec{g} - m\,\vec{\omega}\times(\vec{\omega}\times\vec{r}) \qquad (9.23)$$

where $\vec{g}$ is the acceleration due to gravity considering earth's attraction alone, and $\vec{g}_e$ is the effective acceleration due to gravity taking into account earth's rotation.

Now $\vec{\omega}\times\vec{r} = \hat{n}\ \omega r\sin(90° - \lambda)$ where λ is the latitude of the place and $\hat{n}$ is unit vector perpendicular to the plane containing $\vec{\omega}$ and $\vec{r}$ (Fig. 9.11).

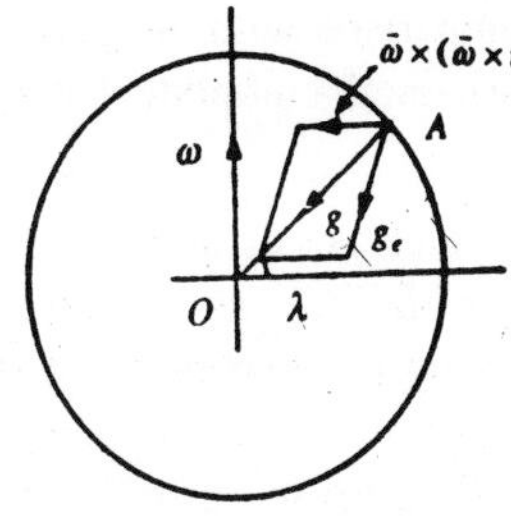

Fig. 9.11

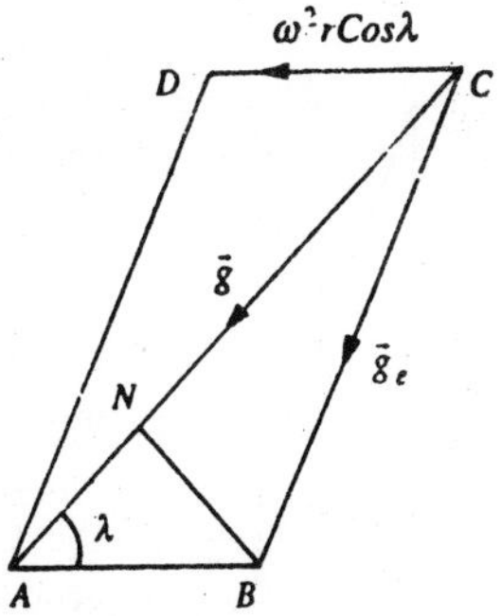

Fig. 9.12

Hence $\vec{\omega} \times (\vec{\omega} \times \vec{r})$ will point inwards at A and has a magnitude $\omega^2 r \cos\lambda$.

To a first approximation

$$\begin{aligned} g_e &= CB \approx CN \\ &= CA - AN \\ &= g - \omega^2 r \cos^2\lambda \end{aligned}$$

Hence the effect of the diurnal rotation of the earth is to **reduce g** by $\omega^2 r \cos^2\lambda$ (Fig. 9.12).

9.4. The Focault Pendulum

Coriolis force arising out of the rotation of the earth was demonstrated by an ingeneous device designed by Foucault. The arrangement consisted of a heavy bob (large inertia) of mass about 28 kg. suspended by a long string of length about 72 meters. Due to large inertia the pendulum can continue to swing freely for a long time. The successive oscillations of the pendulum showed that the plane of vibration of it rotated with respect to the floor. In the absence of the earth's rotation the plane of oscillation would have remained fixed. But if the earth rotates the plane of oscillation should precess about the axis. This will be evident from the following discussion.

Let us choose the origin of co-ordinates vertically below the point of support (Fig. 9.13) and $\vec{r}$, the position vector of the pendulum bob. For small amplitudes of oscillation this vector will be practically horizontal. In the northern hemisphere the angular velocity vector $\vec{\omega}$ will point outwards in a general direction making an angle θ with the vertical. If T be the tension of the string, the equation of motion of the bob is given by

$$m \frac{d^2 \vec{r}}{dt^2} = \vec{T} + m\vec{g}_e - 2m\vec{\omega} \times \frac{d\vec{r}}{dt} \qquad (9.24)$$

The vertical component of the Coriolis force is negligible in comparison with the gravitational force, but the horizontal component can affect the nature of motion.

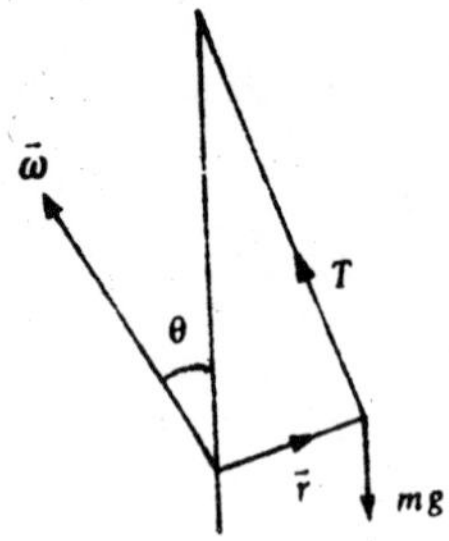

Fig. 9.13

To proceed further let us introduce a new co-ordinate system rotating about the vertical axis through the point of suspension at such an angular velocity that the horizontal component of the Coriolis force add up to zero. Let the new co-ordinate system rotate about the vertical with constant angular velocity $\overset{*}{\omega}\hat{k}$, where $\hat{k}$ is a unit vector along the vertical. Denoting the differentiation about the new axis by $\frac{d^*}{dt}$ we have

$$\frac{d'\vec r}{dt} = \frac{d^*\vec r}{dt} + \overset{*}{\vec\omega} \times \vec r = \frac{d^*\vec r}{dt} + \overset{*}{\omega}\hat{k} \times \vec r \tag{9.25}$$

and
$$\frac{d'^2\vec r}{dt^2} = \frac{d'}{dt}\left\{\frac{d^*\vec r}{dt} + \overset{*}{\omega}\hat{k} \times \vec r\right\}$$

$$= \left\{\frac{d^*}{dt} + \overset{*}{\omega}\hat{k}\right\} \times \left\{\frac{d^*}{dt}\vec r + \overset{*}{\omega}\hat{k} \times \vec r\right\}$$

$$= \frac{d^{*2}\vec r}{dt^2} + \overset{*}{\omega}^2\hat{k} \times \{\hat{k} \times \vec r\} + 2\overset{*}{\omega}\hat{k} \times \frac{d^*\vec r}{dt} \tag{9.26}$$

Substituting in equation (9.24) we have

$$m\frac{d^{*2}\vec r}{dt^2} = \vec T + m\vec g_e - 2m\vec\omega \times \left(\frac{d^*\vec r}{dt} + \overset{*}{\omega}\hat{k} \times \vec r\right)$$

$$- m\overset{*}{\omega}^2\hat{k} \times (\hat{k} \times \vec r) - 2m\overset{*}{\omega}\hat{k} \times \frac{d^*\vec r}{dt}$$

$$= \vec T + m\vec g_e - 2m\overset{*}{\omega}\vec\omega \times (\hat{k} \times \vec r) - 2m(\vec\omega + \overset{*}{\omega}\hat{k}) \times \frac{d^*\vec r}{d^*t} - m\overset{*}{\omega}^2\hat{k} \times (\hat{k} \times \vec r)$$

But $\vec\omega \times (\hat{k} \times \vec r) = \hat{k}(\vec\omega \cdot \vec r) - \vec r(\vec\omega \cdot \hat{k})$

and $\hat{k} \times (\hat{k} \times \vec r) = \hat{k}(\hat{k} \cdot \vec r) - \vec r(\hat{k} \cdot \hat{k})$

Hence
$$m\frac{d^{*2}\vec r}{dt^2} = \vec T + m\vec g_e - 2m\overset{*}{\omega}\left\{\hat{k}(\vec\omega \cdot \vec r) - \vec r(\vec\omega \cdot \hat{k})\right\}$$

$$-2m(\vec\omega + \overset{*}{\omega}\hat{k} \times \frac{d^*\vec r}{dt} - m\omega^{*2}\left\{\hat{k}(\hat{k} \cdot \vec r) - \vec r\right\} \tag{9.27}$$

Every term on the right hand side except the last but one term lies in the vertical plane containing the pendulum.

For small amplitude $\frac{d^*\vec{r}}{dt}$ is almost horizontal. Hence to make the last but one term lie in the vertical plane by choosing $\omega + \overset{*}{\omega}\hat{k}$ horizontal. The necessary condition is that $\hat{k} \cdot (\vec{\omega} + \overset{*}{\omega}\hat{k}) = 0$

$$\text{or, } \omega\cos\theta + \overset{*}{\omega} = 0$$

$$\text{or, } \overset{*}{\omega} = -\omega\cos\theta \tag{9.28}$$

Thus the angular velocity of the rotating system is numerically equal to cosθ times the angular velocity of the rotating earth, where θ is the angle between the vertical and the axis of rotation of the earth. The vertical is along the direction of $-g_e$ which is nearly the same as the direction $-g$, so that θ is nearly equal to the co-latitude, that is the angle between $\vec{r}$ and $\vec{\omega}$. Since all the terms in equation (9.27) now lie in a vertical plane containing the pendulum, the acceleration of the bob in the precessing system given by $\frac{d^{*2} r}{dt^2}$ is always towards the vertical axis.

Hence the pendulum will continue to swing in the same vertical plane in the precessing co-ordinate system. Thus as seen from the earth the plane of oscillation of the pendulum precesses with angular velocity ω^* whose magnitude and sense is given by equation (9.28). The direction of this precession will be clockwise in the northern hemisphere.

At the equator $\omega^* = 0$ and hence the precession of the plane of oscillation vanishes. At the north pole $\omega^* = \omega$ and hence the precession is maximum. It is interesting to note that at the south pole $\omega^* = -\omega$. For an observer in space, the oscillation plane at the north pole remains fixed while the earth will appear to rotate anticlockwise beneath him.

EXERCISE

1. Distinguish between inertial and non-inertial frames of reference. Show that our earth is not an inertial frame.

2. Prove that a frame of reference moving with uniform velocity relative to an inertial frame is also inertial.

3. What is Galilian transformation ? Write down the expression for it. Use Galilian transformations to prove that the distance between two points is invariant in two inertial frames.

4. Show that a moving frame of reference gives rise to a force which is purely due to the acceleration of the system.

5. Explain the origin of centrifugal and Coriolis forces in a rotating frame.

6. Calculate the magnitude and direction of coriolis force acting on a particle of mass 100 gm, moving with a velocity $3\hat{i} + 4\hat{j} - 5\hat{k}$ cm/sec. relative to a rotating frame rotating with an angular velocity $10\,\hat{i} - 6\,\hat{j}\ sec^{-1}$.

[magnitude = 6×10^4 dynes, direction cosines $-15/41,\ 25/41\ -20/41$]

7. What is the effect of coriolis force on a particle falling freely under earth's gravity ?

8. Which of the banks of rivers in the northern hemisphere are sharper than others ?

9. If a projectile is fired due east from a point on the surface of the earth at a northern latitude λ with velocity v_0 and an angle α with the horizontal, show that the lateral deflection of the particle when it strikes the earth is

$$d = \frac{4v_0^3\omega}{g^2}\sin\lambda\sin^2\alpha\cos\alpha.$$

10. An object at the equator is thrown vertically upward with a speed of 50 mph. How far from its original position will it land ?

11. Show that the angular deviation of a plumbline bob from the true vertical at a point on the earth's surface at a latitude λ is

$$\frac{r_0\,\omega^2\sin\lambda\cos\lambda}{g - r_0\,\omega^2\cos^2\lambda},$$ where r_0 is the radius of the earth.

12. An object is shown downwards with initial speed v_0. Prove that after a time t, it is deflected east of the vertical by an amount

$$\omega\, v_o \sin\lambda t + \frac{1}{2}\,\omega g t^3 \sin\lambda.$$

13. What is Foucault Pendulum ? What does it prove ?

14. How long would it take for the plane of oscillation of a Foucault Pendulum to make one complete revolution if the pendulum is located (*a*) at the north pole, and (*b*) at a plane of co-latitude 45° ?

15. A long pendulum is set in motion at a point on the surface of the earth (rotating) with its momentum originally in the vertical plane containing the pendulum. Find the rate at which its plane of oscillation rotates. Discuss the motion completely.

16. A particle is fired horizontally along the surface of the earth. Show that the angular deviation from the direction of initial travel resulting from the Corolis force varies linearly with time at a rate $\omega\cos\varphi$ to a first approximation, where ω is the angular frequency of earth's rotation and φ is the co-latitude. What is the direction of deviation in the northern hemisphere?

CHAPTER X

MECHANICS II

10.1. Generalised co-ordinates, degrees of freedom and constraints

In dynamics we use in most cases the system of Cartesian Co-ordinates. But in discussing problems which are very complex, the application of Cartesian Co-ordinates involving Newton's Laws of Motion becomes useless. Laws of Dynamics have forms other than the Newtonian form and in complex cases their application makes the solution easier.

To be familiar with these forms we must have some elementary ideas regarding degrees of freedom and generalised co-ordinates.

10.2. Degrees of Freedom

Elements of a dynamical system are the particles and rigid bodies. The least number of independent variables by which the configuration of a dynamical system can be fully expressed is called its degrees of freedom.

If may also be defined as the number of independent squared terms entering into expression for energy of the system.

In three dimensions the degrees of freedom of a free particle are three. Because its position can be completely specified by three co-ordinates (x, y, z) or (r, θ, φ).

If the motion of the particle is confined in a plane, its position can be specified completely by (x, y) or (r, θ). In this case the particle has two degrees of freedom.

If the motion is confined in a straight line, the degree of freedom is one.

For two particles rigidly fixed with a fixed distance between them the degree of freedom is five. If the particles were free the degrees of freedom would have been six. Because six co-ordinates (x_1, y_1, z_1) and (x_2, y_2, z_2) would be necessary to specify them completely. Since the distance $\sqrt{(x_2 - x_1)^2 + (y_2 - y_1)^2 + (z_2 - z_1)^2}$ between them is constant, we can eliminate one of the six variables by it and the position of the two particles can be completely specified by five independent co-ordinates. The above equation puts condition of constraints in the dynamical system.

In general, in a dynamical system consisting of N particles and subjected to m constraints, the number of degrees of freedom is $n = 3N - m$.

Each constraint is related in some way to the co-ordinates. If this relation is known, then from each of these relations one co-ordinate can be eliminated.

10.2A. Constraints

A constraint may be defined as one which puts restriction on the freedom of a system. It can be classified as (a) Holonomic and (b) Non-Holonomic or (c) Scleronomous and (d) Rheonomous.

A holonomic constraint can be expressed as function of the co-ordinates of the system and time. On the other hand, a non-holonomic constraint can not be expressed as above. The most general form of holonomic constraint is

$$f\,(x_1, y_1, z_1, x_2, y_2, z_2 \ldots\ldots x_n, y_n, z_n, t) = 0. \qquad (10.1)$$

If there are m equations of constraints, the number of degrees of freedom will be $(3N - m)$.

It is to be remembered that the constraints of a system are consequences of the forces acting on the systems. Since the mechanism of these forces are not always known, it is necessary to eliminate these forces of constraints. This elimination is only possible when the constraints are holonomic. Non-holonomic constraints can not be eliminated in a general way.

The time dependence of the constraints divides them into two types—(1) Scleronomous and (2) Rheonomous.

The former class of constraints is dependent on time, while the latter is independent of time.

For a rigid body the constraint is scleronomous while that on a bead on a rotating wire loop is rheonomous.

In a diatomic molecule with a fixed distance between the molecules the number of degrees of freedom is evidently five. For a triatomic molecule having linear arrangement of atoms the number of degrees of freedom is $(3 \times 3 - 2) = 7$. But for a triangular arrangement of atoms it will be $3 \times 3 - 3 = 6$.

It is not essential that in all cases three dimensional co-ordinates are to be used in determining the number of degrees of freedom. For example, in the case of two particles joined rigidly, the co-ordinates of their centre of mass $(\bar{x}, \bar{y}, \bar{z})$ and to represent the orientation of their axis the angles θ, and φ are sufficient. These five independent co-ordinates also specify the position of the particles completely

The number of degrees of freedom of a free rigid body is six. Because to specify the body, the position of its c.g. and the direction of a straight line fixed in the body should be known. For c.g. $(\bar{x}, \bar{y}, \bar{z})$ and to fix the straight line three independent angles are required.

If the rigid body begins to rotate about a fixed point, the number of degrees of freedom becomes three. Because the direction of an axis fixed in the body will specify it.

While rotating about a fixed axis, the number of degrees of freedom of a rigid body is one, because the angular displacement of a straight line fixed in the body will determine the position of the body.

10.3. Generalised Co-ordinates

If the number of degrees of freedom be n, the configuration of the system can be completely described with the help of conveniently chosen n independent variables q_i ($i = 1, 2, 3 \ldots\ldots n$). These n variables are called generalised co-ordinates.

They may be space co-ordinates, angular co-ordinates or any other kind of variables or the functions of these variables. There is no hard and fast rule in selecting the generalised co-ordinates. They should be conveniently chosen in accordance with the system concerned.

10.4. Generalised Velocities, Forces and Momenta

If q_i ($i = 1, 2, 3 \ldots . n$) represent generalised co-ordinates,

$$\frac{dq_i}{dt} = \dot{q}_i \text{ etc.} \tag{10.2}$$

are called generalised velocities.

It should be kept in mind that each of the generalised co-ordinates q_i is a function of the cartesian co-ordinates of all the particles of the system, i.e.

$$q_i = q_i\ (x_1, y_1, z_1, x_2, y_2, z_2, \ldots\ldots)$$

with the help of the above equation a relation between the generalised velocities $\dot{q}_i$ and the ordinary velocities $\dot{x}_1, \dot{y}_1, \dot{z}_1$ etc. can be found out. Conversely, each cartesian co-ordinate can be expressed as functions of generalised co-ordinates i.e.

$x_i = x_i\ (q_1, q_2, q_3 \ldots . q_n)$. In a similar manner the cartesian velocity $\dot{x}_i$ is a function of the generalised velocities.

If $\vec{F}_i(e)$ be the external force on the ith particle of the system and $\delta\vec{r}_i$ its virtual displacement, the virtual work done on the ith. particle is $\vec{F}_i(e) . \delta\vec{r}_i$. If this is extended to all the particles of the system, the total virtual work done

$$\delta W = \sum_i \vec{F}_i(e) . \delta\vec{r}_i = \sum_i (X_i\,\delta x_i + Y_i\,\delta y_i + Z_i\,\delta z_i) \tag{10.3}$$

X, Y, Z are the components of $\bar{F}(e)$.

If the system is described in terms of generalised co-ordinates q_i ($i = 1, 2, 3 \ldots n$), the total work done (virtual)

$$\delta W = \sum_i Q_i \, \delta q_i . \tag{10.4}$$

In equation (10.3) X_i etc. represent the components of the forces along x direction etc. From analogy the multiplier Q_i of δq_i in equation (10.4) is called the generalised force.

In generalised co-ordinates the K. E. T is a function of the square of generalised velocities $\dot{q}_i \, \dot{q}_j$ and is in general a function of q_i etc.

$$\frac{\partial T}{\partial \dot{q}_i} = p_i \quad (i = 1, 2, 3 \ldots n) \tag{10.5}$$

are called generalised momenta.

If q_i represents length, p_i becomes linear momentum. If q_i represents angle, p_i becomes angular momentum.

10.5. Principle of virtual work, D'Alembert's Principle, Lagrange's Equation for conservative system

Let any number of forces be applied to the ith particle of a system at any instant of time. The virtual displacement of the system means a change in the configuration of the system as a result of any arbitrary change of the co-ordinates of the system consistant with the forces.

If the system is in equilibrium, the total force on each particle of the system is zero, i.e. $\vec{F}_i = 0$.

So that $\vec{F}_i . \delta\vec{r}_i = 0$, where $\delta\vec{r}_i$ is the virtual displacement of the ith particle.

The sum of these vanishing products over all the particles of the system must also be zero, i.e. $\sum_i \vec{F}_i . \delta\vec{r}_i = 0$. (10.6)

The force $\vec{F}_i$ acting on the ith particle is the sum of the external forces $\vec{F}_i(e)$ and the forces of constraints $\vec{f}_i$ (forces of constraints are those internal forces which restricts the motion of the system); Thus $\vec{F}_i = \vec{F}_i(e) + \vec{f}_i$.

Equation (10.6) may be written as

$$\sum_i \vec{F}_i(e) . \delta\vec{r}_i + \sum_i \vec{f}_i . \delta\vec{r}_i = 0. \tag{10.7}$$

If we take the forces of constraints normal to the direction of motion

$$\vec{f_i} \cdot \delta\vec{r_i} = 0$$

Thus $\sum_i \vec{F_i}(e) \cdot \delta\vec{r_i} = 0.$ (10.8)

Thus for equilibrium of a system, the virtual work of the applied forces is zero. This is known as the principle of virtual work. This principle has been modified into a new one by James Bernoulli and further developed by D'Alembert by taking $\vec{F_i} = \dot{\vec{p_i}}$. Then $\vec{F_i} - \dot{\vec{p_i}} = 0$ which states that for a moving system, the particles will be in equilibrium under a force equal to the actual (applied) force plus a reversed force (force of Kinetic reaction) $- \dot{\vec{p_i}}$.

Thus equation (10.6) may be written like

$$\sum_i (\vec{F_i} - \dot{\vec{p_i}}) \cdot \delta\vec{r_i} = 0 \tag{10.9}$$

or, $\sum_i (\vec{F_i}(e) - \dot{\vec{p_i}}) \cdot \delta\vec{r_i} + \sum_i \vec{f_i} \cdot \delta\vec{r_i} = 0.$

If the virtual work of the forces of constraints vanishes, then

$$\sum_i (\vec{F_i}(e) - \dot{\vec{p_i}}) \cdot \delta\vec{r_i} = 0 \tag{10.10}$$

when the forces of constraints disappear we may use $\vec{F_i}(e)$ in the place of $\vec{F_i}$ and say that for moving body the external forces are in dynamic equilibrium with the forces of Kinetic reaction. This is known as D'Alembert's Principle.

We now transform the principle into an expression involving generalised co-ordinates. The transformation from $\vec{r_i}$ to q_i can be done with the help of the relation

$$\vec{r_i} = \vec{r_i}\,(q_1, q_2, \ldots q_n, t).$$

Thus $\vec{v_i} = \dot{\vec{r_i}} = \sum_j \frac{\partial \vec{r_i}}{\partial q_j} \dot{q}_j + \frac{\partial \vec{r_i}}{\partial t}$ (10.11)

In the same manner,

$$\delta\vec{r_i} = \sum_j \frac{\partial \vec{r_i}}{\partial q_j} \delta q_j . \tag{10.12}$$

It is to be noticed that δt is not involved here, since a virtual displacement considers only displacement of co-ordinates.

In terms of generalised co-ordinates the virtual work of $\vec{F_i}$ becomes

$$\sum_i \vec{F_i} \cdot \delta\vec{r_i} = \sum_{i,j} \vec{F_i} \cdot \frac{\partial \vec{r_i}}{\partial q_j} \delta q_j$$

$$= \sum_j Q_j \,\delta q_j \tag{10.13}$$

where the Q_j are called generalised force components, defined as

$$Q_j = \sum_i \vec{F}_i \cdot \frac{\partial \vec{r}_i}{\partial q_j}. \tag{10.14}$$

It is interesting to note that just as the q's need not have the dimensions of length, so the Q's do not necessarily have the dimensions of force, but $Q_j\,\delta q_j$ must have the dimensions of work.

Turning back to the other terms of equation (10.10), which may be written as

$\sum_i \vec{\dot{p}}_i \cdot \delta\vec{r}_i = \sum_i m_i \vec{\ddot{r}}_i \cdot \delta\vec{r}_i$ and expressing $\delta\vec{r}_i$ by equation (10.12), this becomes

$$\sum_{i,j} m_i \vec{\ddot{r}}_i \cdot \frac{\partial \vec{r}_i}{\partial q_j}\,\delta q_j.$$

Again $$\sum_i m_i \vec{\ddot{r}}_i \cdot \frac{\partial \vec{r}_i}{\partial q_j} = \sum_i \left\{\frac{d}{dt}\left(m_i \vec{\dot{r}}_i \cdot \frac{\partial \vec{r}_i}{\partial q_j}\right) - m_i \vec{\dot{r}}_i \cdot \frac{d}{dt}\left(\frac{\partial \vec{r}_i}{\partial q_j}\right)\right\} \tag{10.15}$$

Now $$\frac{d}{dt}\left(\frac{\partial \vec{r}_i}{\partial q_j}\right) = \frac{\partial}{\partial q_j}\left(\frac{\partial \vec{r}_i}{\partial t}\right) = \frac{\partial \vec{v}_i}{\partial q_j}.$$

Further from equation (10.11)

$$\frac{\partial \vec{v}_i}{\partial \dot{q}_j} = \frac{\partial \vec{r}_i}{\partial q_j}$$

Substituting these changes in (10.15)

$$\sum_i m_i \vec{\ddot{r}}_i \cdot \frac{\partial \vec{r}_i}{\partial q_j} = \sum_i \left\{\frac{d}{dt}\left(m_i \vec{v}_i \cdot \frac{\partial \vec{v}_i}{\partial \dot{q}_j}\right) - m_i \vec{v}_i \cdot \frac{\partial \vec{v}_i}{\partial q_j}\right\}$$

Hence $$\sum_i \vec{\dot{p}}_i \cdot \delta\vec{r}_i = \sum_i m_i \vec{\ddot{r}}_i \cdot \delta\vec{r}_i$$

$$= \sum_{i,j} m_i \vec{\ddot{r}} \cdot \frac{\partial \vec{r}_i}{\partial q_j}\,\delta q_j$$

$$= \sum_j \left\{\frac{d}{dt}\left(\frac{\partial}{\partial \dot{q}_j} \sum_i \frac{1}{2} m_i v_i^2\right) - \frac{\partial}{\partial q_j}\left(\sum_i \frac{1}{2} m_i v_i^2\right)\right\}\delta q_j$$

Representing $\sum_i \frac{1}{2} m_i v_i^2 =$ K.E. of the system D'Alemberts's principle expressed by equation (10.10) becomes

$$\sum_j \left[\left\{\frac{d}{dt}\left(\frac{\partial T}{\partial \dot{q}_j}\right) - \frac{\partial T}{\partial q_j}\right\} - Q_j\right]\delta q_j = 0 \tag{10.16}$$

Since any virtual displacement δq_j is independent of δq_k, the only consequence is that

$$\frac{d}{dt}\left(\frac{\partial T}{\partial \dot{q}_j}\right) - \frac{\partial T}{\partial q_j} - Q_j = 0$$

$$\text{or,} \quad \frac{d}{dt}\left(\frac{\partial T}{\partial \dot{q}_j}\right) - \frac{\partial T}{\partial q_j} = Q_j . \tag{10.17}$$

When the system is conservative i.e. when the forces are derivable from a scalar potential function, V.

$$\vec{F_i} = -\vec{\nabla}_i V$$

In this case the generalised forces can be written as

$$Q_j = \sum_i \vec{F_i} \cdot \frac{\partial \vec{r_i}}{\partial q_j} = -\sum_i \vec{\nabla}_i V \cdot \frac{\partial \vec{r_i}}{\partial q_j}$$

which is nothing but the partial derivative of a function $-V(r_1, r_2, \ldots r_n)$ with respect to q_j

$$\left\{\left\{\left[\hat{i}\frac{\partial V}{\partial x} + \hat{j}\frac{\partial V}{\partial y} + \ldots\right] \cdot (\hat{i}\,dx + \hat{j}\,dy + \ldots) \frac{1}{\partial q_j} = \frac{\partial V}{\partial x}dx + \frac{\partial V}{\partial y}dy + \frac{\partial V}{\partial z}dz = dV(x, y, z)\right\}\right\}$$

It is to be noted that V is not a function of t.

Hence for a conservative system

$$Q_j = -\frac{\partial V}{\partial q_j} . \tag{10.18}$$

Equation (10.16) can then be written as

$$\frac{d}{dt}\left(\frac{\partial T}{\partial \dot{q}_j}\right) - \frac{\partial}{\partial q_j}(T - V) = 0.$$

Since the potential is a function of position only, must be independent of the generalised velocities.

$$\text{Hence} \quad \frac{d}{dt}\left\{\frac{\partial (T - V)}{\partial \dot{q}_j}\right\} - \frac{\partial}{\partial q_j}(T - V) = 0$$

Defining a new function L, known as the Lagrangian as $L = T - V$, the above becomes

$$\frac{d}{dt}\left(\frac{\partial L}{\partial \dot{q}_j}\right) - \frac{\partial L}{\partial q_j} = 0 \tag{10.19}$$

There are n such equations for $j = (1, 2, \ldots n)$

Equations (10.17) and (10.19) may both be called Lagrangian Equations. But the latter form valid for a conservative system is generally called the Lagrangian Equation.

Lagrange's Equations for Non-holonomic Systems

If there are m constraints of the form

$$\sum_j A_j\, dq_j + A\, dt = 0,\ \sum B_j\, dq_j + B dt = 0 \qquad \ldots \qquad \text{(a)}$$

From equation (a) we have

$$Q_j = \frac{d}{dt}\left(\frac{\partial T}{\partial \dot{q}_j}\right) - \frac{\partial T}{\partial q_j} = \sum_i m_i \ddot{\vec{r}}_i \cdot \frac{\partial \vec{r}_i}{\partial q_j},$$

It can be shown that

$$Q_j = \varphi_j - \lambda_1 A_j - \lambda_2 B_j + \ldots\ldots = 0.$$

Hence $$\frac{d}{dt}\left(\frac{\partial T}{\partial \dot{q}_j}\right) - \frac{\partial T}{\partial q_j} = \varphi_j + \lambda_1 A_j + \lambda_2 B_j + \ldots\ldots, j = 1, 2 \ldots\ldots n \qquad \ldots \qquad \text{(b)}$$

If the forces are derivable from a potential $\varphi_j = -\dfrac{\partial V}{\partial q_j}$, where V does not depend on $\dot{q}_j$. Thus (b) can be written as

$$\frac{d}{dt}\left(\frac{\partial L}{\partial \dot{q}_j}\right) - \frac{\partial L}{\partial q_j} = \lambda_1 A_j + \lambda_2 B_j + \ldots\ldots \qquad \ldots \qquad \text{(c)}$$

10.6. Hamiltonian and Hamilton's equations

Since V is independent if $\dot{q}_j$, the Lagrange's equation may be expressed in generalised momentum co-ordinates p_j as

$$p_j = \frac{\partial L}{\partial \dot{q}_j} \qquad (10.20)$$

$$\left[p = mv,\ T = \frac{1}{2} mv^2 \ \therefore\ \frac{\partial T}{\partial v} = mv = p\right]$$

In these equations p_j is expressed as a function of q_j's $\dot{q}_j$'s and t's.

Let us define a function H as

$$H = \sum_{j=1}^{n} \dot{q}_j \frac{\partial L}{\partial \dot{q}_j} - L \qquad (10.21)$$

The function $H = H(q, p, t)$ defined above is the Hamiltonian function.

Since $L = T(q, \dot{q}, t) - V(q)$,

where T is homogeneous and quadratic in generalised velocities, Euler's theorem of homogeneous function gives

$$\sum_{j=1}^{n} \dot{q}_j \frac{\partial T}{\partial \dot{q}_j} = 2T.$$

Hence equation (10.21) gives

$H = 2T - L = 2T - (T - V) = T + V$ which is the total energy Kinetic and Potential.

Combining equations (10.20) and (10.21)

$$H = \sum_{j=1}^{n} \dot{q}_j p_j - L. \tag{10.22}$$

Since L is a function of $q, \dot{q}, t$ equation (10.22) gives the total derivative of H as

$$dH = \sum_{j=1}^{n} p_j \, \delta\dot{q}_j - \sum_{j=1}^{n} \dot{q}_j \, \delta p_j - \sum_{j=1}^{n} \frac{\partial L}{\partial \dot{q}_j} \delta\dot{q}_j - \sum_{j=1}^{n} \frac{\partial L}{\partial q_j} \delta q_j - \frac{\partial L}{\partial t} \delta t \tag{10.23}$$

The 1st and the 3rd terms on the right-hand side cancel on account of (10.20). The remaining differentials δq_j, δp_j, δt being $(2n + 1)$ in number may be regarded independent and arbitrary. Hence equation (10.23) gives

$$\frac{\partial H}{\partial p_j} = \dot{q}_j; \quad \frac{\partial H}{\partial q_j} = -\frac{\partial L}{\partial q_j} \quad \text{and} \quad \frac{\partial H}{\partial t} = -\frac{\partial L}{\partial t}.$$

One implication of these equations is that if t does not occur explicitly in L, then neither does it occur in H. The second implication is more important : from n Lagrangian equations of motion (10.19) we obtain $2n$ equations of motion

$$\dot{q}_j = \frac{\partial H}{\partial p_j} \tag{10.24}$$

and $$-\frac{\partial H}{\partial q_j} = \frac{\partial L}{\partial q_j} = \frac{d}{dt}\left(\frac{\partial L}{\partial \dot{q}_j}\right) \text{ from equation (10.19)}$$

$$= \dot{p}_j \tag{10.25}$$

These are Hamilton's equations of motion, commonly known as Cannonical equations.

The total order of Hamilton's equation is the same as the total order of Lagrange's equations, namely $2n$. But whereas Lagrange's equations present us with n equations, each of second order, Hamilton's equations are $2n$ equations, each of the 1st order.

10.7. Applications of Lagrange's equations

(a) Simple Pendulum :

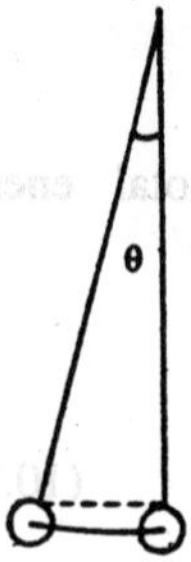

Fig. 10.1

The simple plane pendulum consists of a mass suspended from a certain point by a weightless rod of length l. It is assumed that the swings occur in one plane so that the pendulum has one degree of freedom. We take the angle of deflection θ of the pendulum from the vertical as the generalised co-ordinate. Obviously, the velocity of the bob of mass m is $l\dot{\theta}$, so that the K.E. $T = \frac{1}{2} m l^2\dot{\theta}^2$. The instanteneous P.E. determined by the height of the mass above its mean position is $V = mg\,l\,(1 - \cos\theta)$. Whence, the Lagrangian for the Pendulum is

$$L = T - V = \frac{1}{2} m l^2\dot{\theta}^2 - mgl\,(1 - \cos\theta).$$

Applying Lagrangian equation

$$\frac{d}{dt}\left(\frac{\partial L}{\partial \dot{\theta}}\right) - \left(\frac{\partial L}{\partial \theta}\right) = 0$$

or, $$\frac{d}{dt}(ml^2\dot{\theta}) + mgl\sin\theta = 0$$

or, $$\ddot{\theta} = -\,g/l\,\sin\theta \approx -\,g/l\,\theta, \text{ if } \theta \text{ is small.}$$

Solving, the time period is given by

$$t = 2\pi\sqrt{l/g}.$$

(b) Double Pendulum :

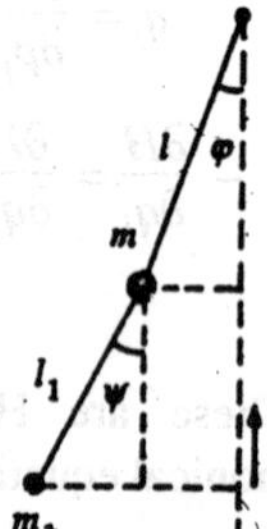

Fig. 10.2

A double pendulum consists of a mass m suspended by a weightless rod of length l. There is another hinge at the bottom of m from which another pendulum which is forced to oscillate in the same plane is suspended. Let the mass and the length of the second pendulum be m_1 and l_1 respectively, and its angle of deflection from the vertical is ψ. The angle of deflection of the first particle is φ. The displacement components of the second particle are

$x_1 = l \sin\varphi + l_1 \sin\psi$

$z_1 = l\,(1 - \cos\varphi) + l_1\,(1 - \cos\psi)$

Its velocity components are

$\dot{x}_1 = l \cos\varphi\ \dot{\varphi} + l_1 \cos\psi\,\dot{\psi}$

$\dot{z}_1 = l \sin\varphi\ \dot{\varphi} + l_1 \sin\psi\,\dot{\psi}$

K.E. of the second particle

$$T_1 = \frac{1}{2} m_1\,(\dot{x}_1^2 + \dot{z}_1^2)$$

$$= \frac{1}{2} m_1 \left\{ l^2 \dot{\varphi}^2 + l_1^2 \dot{\psi}^2 + 2 l l_1 \dot{\varphi}\dot{\psi} \cos(\varphi - \psi) \right\}$$

Its P.E. $V_1 = mgl\,(1 - \cos\varphi) + m_1 g l_1\,(1 - \cos\psi)$

Adding the corresponding quantities of the first particle we get the expression for the Lagrangian of the double pendulum in the form

$$L = \frac{m + m_1}{2} l^2 \dot{\varphi}^2 + \frac{m_1 l^2}{2} \dot{\psi}^2 + m_1 l l_1\,\dot{\varphi}\dot{\psi} \cos(\varphi - \psi)$$

$$- (m + m_1)\,gl\,(1 - \cos\varphi) - m_1 g l_1\,(1 - \cos\psi) \quad \ldots \quad \text{(A)}$$

in terms of generallised co-ordinates φ and ψ and generalised velocities $\dot{\varphi}$ and $\dot{\psi}$.

If we confine ourselves to small oscillations, φ and ψ should be considered as vanishingly small. In the above Lagrangian $\cos(\varphi - \psi)$ must be replaced by 1 and $\cos\varphi$ and $\cos\psi$ by $1 - \frac{\varphi^2}{\lfloor 2}$ and $1 - \frac{\psi^2}{\lfloor 2}$ respectively.

The Lagrangian will take the form

$$L = \frac{m + m_1}{2} l^2 \dot{\varphi}^2 + \frac{m_1 l_1{}^2}{2} \dot{\psi}^2 + m_1 l l_1\,\dot{\varphi}\dot{\psi}$$

$$- (m + m_1)\,gl\,\frac{\varphi^2}{2} - m_1 g l_1\,\frac{\psi^2}{2} \quad \ldots \quad \text{(B)}$$

The Lagrangian equations associated with φ and ψ are

$$\frac{d}{dt}\left(\frac{\partial L}{\partial \dot{\varphi}}\right) - \frac{\partial L}{\partial \varphi} = 0 \quad \text{and} \quad \frac{d}{dt}\left(\frac{\partial L}{\partial \dot{\psi}}\right) - \frac{\partial L}{\partial \psi} = 0.$$

Using the value of L given by (B), we have

$$\frac{d}{dt}\left\{ (m + m_1) l^2 \dot{\varphi} + m_1 l l_1 \dot{\psi} \right\} + (m + m_1)\,gl\,\varphi = 0$$

or, $(m+m_1)\, l\,\ddot{\varphi} + m_1 l_1 \ddot{\psi} + (m+m_1)\, g\,\varphi = 0$ (C)

and $\frac{d}{dt}\{m_1 l_1^2 \dot{\psi} + m_1 l l_1 \dot{\varphi}\} + m_1 g\, l_1 \psi = 0$

or, $l_1 \ddot{\psi} + l\ddot{\varphi} + g\,\psi = 0$ (D)

(c) Simple Pendulum with a moving support :

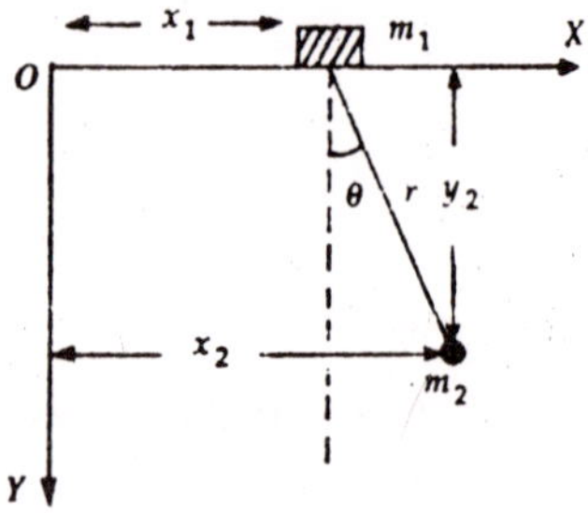

Fig. 10.3

A pendulum of mass m_2 is attached to a block of mass m_1 which is free to slide without friction along the horizontal x-axis. Assuming r to be constant and all the motion confined in the xy plane, the system has two degrees of freedom. Let us take x_1 and θ as the generalised co-ordinates. Starting with the co-ordinates x_1, x_2 and y_2, it is seen that

$$T = \frac{1}{2} m_1 \dot{x}_1^2 + \frac{1}{2} m_2 (\dot{x}_2^2 + \dot{y}_2^2)$$

Putting $x_2 = x_1 + r\sin\theta$ any $y_2 = r\cos\theta$

$$\dot{x}_2 = \dot{x}_1 + r\cos\theta\,\dot{\theta} \quad \text{and} \quad \dot{y}_2 = -\,r\sin\theta\,\dot{\theta}$$

Hence $T = \frac{1}{2} m_1 \dot{x}_1^2 + \frac{1}{2} m_2 (\dot{x}_1^2 + 2\dot{x}_1\, r\cos\theta\,\dot{\theta} + r^2\dot{\theta}^2)$

$$V = m_2 gr\,(1-\cos\theta)$$

$$\therefore\ L = \frac{1}{2} m_1 \dot{x}_1^2 + \frac{1}{2} m_2 (\dot{x}_1^2 + 2\dot{x}_1\, r\cos\theta\,\dot{\theta} + r^2\dot{\theta}^2) - m_2 gr\,(1-\cos\theta) \quad \ldots \quad \text{(A)}$$

The Lagrangian equations of motion associated with x_1 and θ are

$$\frac{d}{dt}\left(\frac{\partial L}{\partial \dot{x}_1}\right) - \frac{\partial L}{\partial x_1} = 0$$

and $\frac{d}{dt}\left(\frac{\partial L}{\partial \dot{\theta}}\right) - \frac{\partial L}{\partial \theta} = 0$

i.e. $\frac{d}{dt}\{(m_1+m_2)\,\dot{x}_1 + m_2 r\dot{\theta}\cos\theta\} = 0$ ... (B)

and $\frac{d}{dt}\{m_2 \dot{x}_1 r\cos\theta + m_2 r^2\dot{\theta}\} + m_2 \dot{x}_1 r\cos\theta\,\dot{\theta} + m_2 gr\sin\theta = 0$

or, $m_2 r \ddot{x}_1 \cos\theta - m_2 r \dot{x}_1 \sin\theta\, \dot{\theta} + m_2 r^2 \ddot{\theta} + m_2 \dot{x}_1 r \sin\theta\, \dot{\theta} + m_2 gr \sin\theta = 0$ ... (C)

If it is assumed that the motion is such that θ is always small, we can replace cosθ by unity and sinθ by θ. The equations of motion then becomes

$$(m_1 + m_2)\ \ddot{x}_1 + m_2\ r\ddot{\theta} = 0 \qquad \text{... (D)}$$

and $\ddot{x}_1 + r\ddot{\theta} = -\ g\theta$ ... (E)

(d) Motion of a symmetrical top :

Euler's Angles as Generalised Co-ordinates :

X, Y, Z are axes fixed in space. Let the system be given a rotation φ about *Z* axis (Fig. 10.4). *OX* and *OY* take the positions *O*ζ and *O*η respectively. *O*ζ is called the line of nodes.

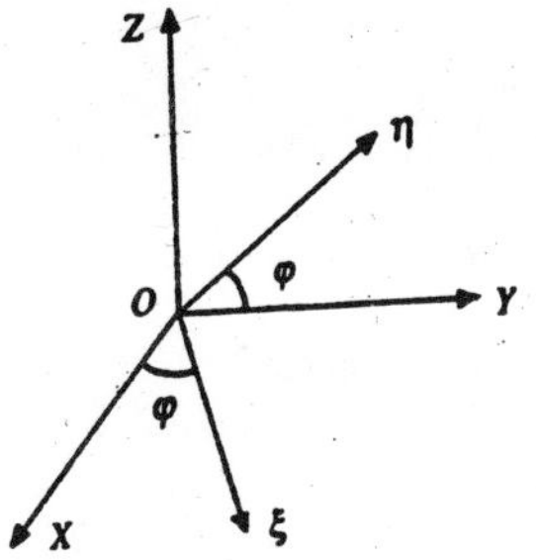

Fig. 10.4

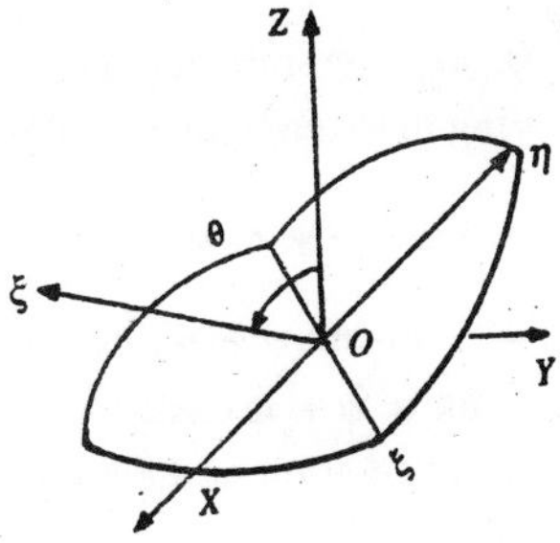

Fig. 10.5

In the next operation the *XY* plane is rotated about the line of nodes through an angle θ (Fig. 10.5). Due to this rotation *Y* axis and *Z* axis will occupy the new positions *O*η and *O*ξ respectively. Lastly the system is given a rotation through an angle ψ about the ξ axis (Fig. 10.6). ζ, η, ξ will now take the form α, β, γ. $\dot{\varphi}$, $\dot{\theta}$ and $\dot{\psi}$ represent angular velocities along *z*, ζ and ϒ axes respectively (Fig. 10.7).

If the body rotates about the axis γ, then $\dot{\psi}$ represents spin angular velocity. When $\dot{\theta} = 0$, the angular velocity $\dot{\varphi}$ represents the angular velocity with which the axis γ is rotating about *Z* axis. The angular velocity $\dot{\varphi}$ is called precessional angular velocity. The angular velocity $\dot{\theta}$ shows that γ will not remain on the surface of a cone. This motion is called nutation. The angles φ, θ, ψ are called Euler's angles.

Let a body be revolving with an angular velocity ω whose components about the principal axes are ω_α, ω_β and ω_Υ. Axes $(\alpha\ \beta\ \gamma)$ rotate with an

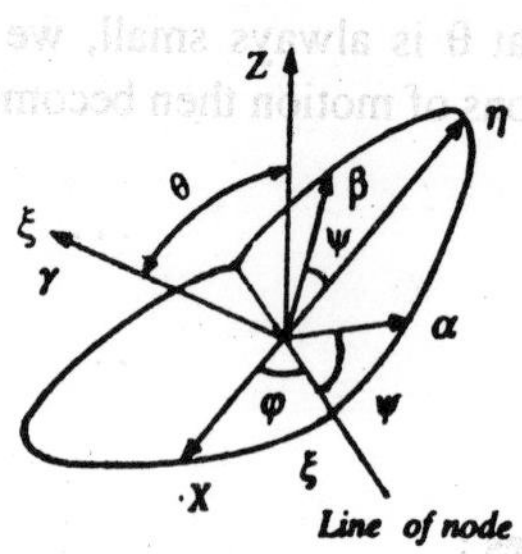

Fig. 10.6

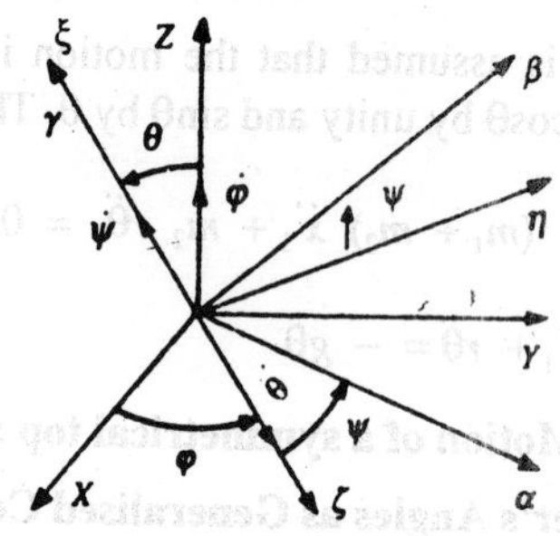

Fig. 10.7

angular velocity $\dot{\psi}$ relative to (ζ, η, ξ) axes. Hence from figure 10.7, we have

$$\omega_\zeta = \dot{\theta},\ \omega_\eta = \dot{\varphi}\sin\theta \text{ and } \omega_\xi = \dot{\psi} + \dot{\varphi}\cos\theta \qquad \ldots \text{(A)}$$

Resolving these components along $(\alpha\ \beta\ \Upsilon)$ axes we get

$$\left.\begin{aligned}\omega_\alpha &= \omega_\zeta \cos\psi + \omega_\eta \sin\psi = \dot{\theta}\cos\psi + \dot{\varphi}\sin\theta\sin\psi \\ \omega_\beta &= -\omega_\zeta \sin\psi + \omega_\eta\cos\psi = -\dot{\theta}\sin\psi + \dot{\varphi}\sin\theta\cos\psi\end{aligned}\right\} \ldots \text{(B)}$$

$$\omega_\gamma = \omega_\xi = \dot{\psi} + \dot{\varphi}\cos\theta$$

The K.E. of rotation of the body is given by

$$T = \frac{1}{2}\left\{ I_\alpha\,\omega_\alpha^2 + I_\beta\,\omega_\beta^2 + I_\Upsilon\,\omega_\Upsilon^2 \right\} \qquad \ldots \text{(C)}$$

To apply the above expression for K.E. to the motion of a symmetrical top, let us confine our attention to the diagram (10.8). The external torque acting on the rotating top is the gravitational one. For a symmetrical top $I_\alpha = I_\beta$.

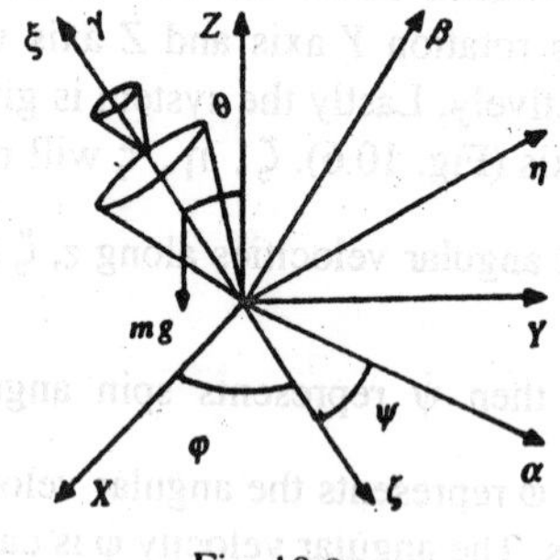

Fig. 10.8

Hence equation (C) reduces to

$$T = \frac{1}{2} I_\alpha (\dot{\theta}^2 + \dot{\varphi}^2\sin^2\theta) + \frac{1}{2} I_\Upsilon (\dot{\psi} + \dot{\varphi}\cos\theta)^2 \qquad \ldots \text{(D)}$$

The Lagrangian for the top is

$$L = \frac{1}{2} I_\alpha (\dot{\theta}^2 + \dot{\varphi}^2\sin^2\theta) + \frac{1}{2} I_\Upsilon (\dot{\psi} + \dot{\varphi}\cos\theta)^2 - mgl\cos\theta \ldots \text{(E)}$$

l is the distance of the c.g. from the tip.

In the absence of nutation $\dot{\theta} = 0$

$$\therefore \quad L = \frac{1}{2} I_\alpha \dot{\varphi}^2 \sin^2\theta + \frac{1}{2} I_\Upsilon (\dot{\psi} + \dot{\varphi}\cos\theta)^2 - mgl\cos\theta$$

Hence the Lagrange's equations associated with θ, φ & ψ are

$$\frac{d}{dt}\left(\frac{\partial L}{\partial \dot{\psi}}\right) - \frac{\partial L}{\partial \psi} = 0 \qquad \ldots \quad \text{(F)}$$

$$\frac{d}{dt}\left(\frac{\partial L}{\partial \dot{\varphi}}\right) - \frac{\partial L}{\partial \varphi} = 0 \qquad \ldots \quad \text{(G)}$$

$$\text{and} \quad \frac{\partial L}{\partial \theta} = 0 \qquad \ldots \quad \text{(H)}$$

or, $\frac{d}{dt}\left\{ I_\Upsilon (\dot{\psi} + \dot{\varphi}\cos\theta) \right\} = 0 \longrightarrow$ simplifying (F)

$$\text{i.e.} \quad \ddot{\psi} + \ddot{\varphi}\cos\theta = 0 \qquad \ldots \quad \text{(I)}$$

and $\frac{d}{dt}\left\{ I_\alpha \dot{\varphi}\sin\theta + I_\Upsilon (\dot{\psi} + \dot{\varphi}\cos\theta)\cos\theta \right\} = 0 \longrightarrow$ simplifying (G)

$$\text{i.e.} \quad I_\alpha \ddot{\varphi}\sin\theta + I_\Upsilon (\ddot{\psi} + \ddot{\varphi}\cos\theta)\cos\theta = 0$$

$$\left.\begin{aligned} \text{or,} \quad \ddot{\varphi}\sin\theta &= 0 \\ \therefore \quad \dot{\varphi} &= \text{constant} \\ \text{or,} \quad \theta &= 0 \end{aligned}\right\} \qquad \ldots \quad \text{(J)}$$

Lastly from (H)

$$I_\alpha \dot{\varphi}\, 2\sin\theta\cos\theta + I_\Upsilon (\dot{\psi} + \dot{\varphi}\cos\theta)(-\dot{\varphi}\sin\theta) + mgl\sin\theta = 0 \ldots \quad \text{(K)}$$

Alternatively, considering the general expression (E) for the Lagrangian and using the idea of cyclic co-ordinates we get, since φ and ψ are absent in the Lagrangian, the corresponding generalised momenta p_φ and p_ψ will be conserved.

$$\text{Now} \quad p_\varphi = \frac{\partial L}{\partial \dot{\varphi}} = I_\alpha \dot{\varphi}\sin^2\theta + I_\Upsilon (\dot{\psi} + \dot{\varphi}\cos\theta)\cos\theta \qquad \ldots \quad \text{(L)}$$

$$\text{and} \quad p_\psi = \frac{\partial L}{\partial \dot{\psi}} = I_\Upsilon (\dot{\psi} + \dot{\varphi}\cos\theta) = I_\Upsilon \omega_\Upsilon \qquad \ldots \quad \text{(M)}$$

The total energy E is given by

$$E = \frac{1}{2} I_\alpha \dot{\theta}^2 + \frac{1}{2} I_\alpha \dot{\varphi}^2 \sin^2\theta + \frac{1}{2} I_\Upsilon (\dot{\psi} + \dot{\varphi}\cos\theta)^2 + mgl\cos\theta \qquad \ldots \quad \text{(N)}$$

$$= \frac{1}{2} I_\alpha \dot{\theta}^2 + \frac{(p_\varphi - p_\psi \cos\theta)^2}{2 I_\alpha \sin^2\theta} + \frac{p_\psi^2}{2I_\Upsilon} + mgl \cos\theta$$

$$= \text{constant} \qquad \text{(O)}$$

Again $\frac{p_\psi^2}{2I_\Upsilon}$ being constant, we have

$$E' = E - \frac{p_\psi^2}{2I_\Upsilon} = \text{constant} = \frac{1}{2} I_\alpha \dot{\theta}^2 + V(\theta) \qquad \ldots \quad \text{')}$$

where $$V(\theta) = \frac{(p_\varphi - p_\psi \cos\theta)^2}{2I_\alpha \sin^2\theta} + mg \cos\theta \qquad \ldots \quad \text{(Q)}$$

is the effective P.E.

(e) **Planetary Motion :**

Let us consider a particle of mass m moving in a plane under an attractive force $\frac{m\mu}{r^2}$ directed to the origin of polar co-ordinates (r, θ).

The K.E. and P.E. of the particle are

$$T = \frac{1}{2} m (\dot{r}^2 + r^2\dot{\theta}^2)$$

$$V = \int_\infty^r \frac{m\mu}{r^2} dr = -\frac{m\mu}{r}$$

Hence the Lagrangian $L = T - V$

$$= \frac{1}{2} m (\dot{r}^2 + r^2 \dot{\theta}^2) + \frac{m\mu}{r} \qquad \ldots \quad \text{(A)}$$

Taking r and θ as generalised co-ordinates, the Lagrangian equations are

$$\frac{d}{dt}\left(\frac{\partial L}{\partial \dot{r}}\right) - \frac{\partial L}{\partial r} = 0 \qquad \ldots \quad \text{(B)}$$

$$\& \ \frac{d}{dt}\left(\frac{\partial L}{\partial \dot{\theta}}\right) - \frac{\partial L}{\partial \theta} = 0 \qquad \ldots \quad \text{(C)}$$

The former gives $\frac{d}{dt}(m\dot{r}) - \frac{\partial}{\partial r}\left(\frac{1}{2} mr^2\dot{\theta}^2 + \frac{m\mu}{r}\right) = 0$

$$\text{or, } \ddot{r} - r\dot{\theta}^2 = -\frac{\mu}{r^2} \qquad \ldots \quad \text{(D)}$$

The latter gives

$$\frac{d}{dt}(mr^2\dot{\theta}) = 0$$

$$\text{or, } \frac{d}{dt}(r^2\dot{\theta}) = 0 \qquad \dots \qquad (E)$$

which are famous differential equations of planetary motion.

(f) Two particles of equal masses m are connected by springs of equal spring constant K, so that the masses are free to slide on a frictionless table. The ends of the springs are attached to the fixed walls. To form the Lagrangian and equations of motion of the system.

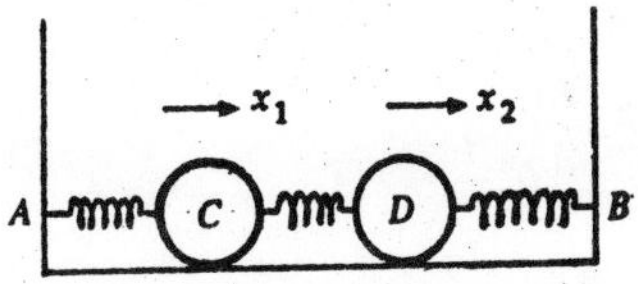

Fig. 10.9

Let x_1 and x_2 be the displacements of the two masses from their equilibrium positions at any instant of time t. The K.E. of system $= T = \frac{1}{2}m\dot{x}_1^2 + \frac{1}{2}m\dot{x}_2^2$. If k represents the spring constant of the springs, the P.E. of the system is given by

$V = \frac{1}{2}kx_1^2 + \frac{1}{2}k(x_2 - x_1)^2 + \frac{1}{2}kx_2^2$, since the extensions of AC, CD and DB are x_1, $(x_2 - x_1)$ and x_2 respectively. The Lagrangian of the system

$$L = T - V = \frac{1}{2}m(\dot{x}_1^2 + \dot{x}_2^2) - \frac{1}{2}kx_1^2 - \frac{1}{2}k(x_2 - x_1)^2 - \frac{1}{2}kx_2^2 \qquad \dots \qquad (A)$$

The Lagrangian equations of motion are

$$\frac{d}{dt}\left(\frac{\partial L}{\partial \dot{x}_1}\right) - \frac{\partial L}{\partial x_1} = 0 \qquad \dots \qquad (B)$$

$$\text{and } \frac{d}{dt}\left(\frac{\partial L}{\partial \dot{x}_2}\right) - \frac{\partial L}{\partial x_2} = 0 \qquad \dots \qquad (C)$$

$$\left.\begin{aligned} &\text{or, } m\ddot{x}_1 - k(x_2 - 2x_1) = 0 \\ &\text{and } m\ddot{x}_2 - k(x_1 - 2x_2) = 0 \end{aligned}\right\} \qquad \dots \qquad (D)$$

(g) Show that if the Lagrangian of a system does not contain time explicitly, the total energy of the system is conserved.

Let us consider a conservative system i.e. a system in which the forces are derivable from a potential function V dependent on position only. The Lagrangian is given by

$L = L(q_1, q_2, \ldots q_n; \dot{q}_1, \dot{q}_2, \ldots \dot{q}_n; t)$. If L does not contain t explicitly, then

$$L = L(q_1, q_2, \ldots q_n; \dot{q}_1, \dot{q}_2, \ldots \dot{q}_n)$$

The total time derivative of L is given by .

$$\frac{dL}{dt} = \frac{\partial L}{\partial q_1}\dot{q}_1 + \frac{\partial L}{\partial q_2}\dot{q}_2 + \ldots \frac{\partial L}{\partial \dot{q}_1}\ddot{q}_1 + \frac{\partial L}{\partial \dot{q}_2}\ddot{q}_2 + \ldots$$

$$= \sum_j \frac{\partial L}{\partial q_j}\dot{q}_j + \sum_j \frac{\partial L}{\partial \dot{q}_j}\ddot{q}_j \qquad \ldots \text{(A)}$$

Lagrangian equation is

$$\frac{d}{dt}\left(\frac{\partial L}{\partial \dot{q}_j}\right) - \frac{\partial L}{\partial q_j} = 0 \qquad \text{or,} \qquad \frac{d}{dt}\left(\frac{\partial L}{\partial \dot{q}_j}\right) = \frac{\partial L}{\partial q_j}$$

with this substitution equation (A) becomes

$$\frac{dL}{dt} = \sum_j \frac{d}{dt}\left(\frac{\partial L}{\partial \dot{q}_j}\right)\dot{q}_j + \sum_j + \frac{\partial}{\partial \dot{q}_j}\ddot{q}_j$$

$$= \sum_j \frac{d}{dt}\left(\frac{\partial L}{\partial \dot{q}_j}\dot{q}_j\right)$$

$$\therefore \frac{d}{dt}\left\{L - \sum \frac{\partial L}{\partial \dot{q}_j}\dot{q}_j\right\} = 0$$

which implies that $L - \sum \frac{\partial L}{\partial \dot{q}_j}\dot{q}_j = \text{constant}$ $\qquad \ldots$ (B)

Now $\frac{\partial L}{\partial \dot{q}_j} = \frac{\partial}{\partial \dot{q}_j}(T - V) = \frac{\partial T}{\partial \dot{q}_j}$

Again $T = \sum_{kj} A_{kj}\dot{q}_k \dot{q}_j$

$$\therefore \frac{\partial T}{\partial \dot{q}_j} = 2\sum_k A_{kj}\dot{q}_k \qquad \text{or,} \qquad \frac{\partial T}{\partial \dot{q}_j}\dot{q}_j = 2\sum_{kj} A_{kj}\dot{q}_k \dot{q}_j = 2T$$

Substituting in (B)

$$L - \sum \frac{\partial T}{\partial \dot{q}_j}\dot{q}_j = L - 2T = \text{constant}$$

or, $T - V - 2T =$ constant

or, $- T - V =$ constant

$\therefore \quad T + V =$ constant.

(h) To show that if the total applied force on a system of particles vanishes, the linear momentum is conserved.

Let us consider a conservative system so that the potential energy V is a function of position only and the K.E. is independent of position i.e.,

$$\frac{\partial V}{\partial \dot{q}_j} = 0 \text{ and } \frac{\partial T}{\partial q_j} = 0. \qquad \ldots \qquad (A)$$

The Lagrangian equation for such a system may be written as

$$\frac{d}{dt}\left(\frac{\partial L}{\partial \dot{q}_j}\right) - \frac{\partial L}{\partial q_j} = 0$$

or, $$\frac{d}{dt}\left(\frac{\partial L}{\partial \dot{q}_j}\right) + \left(\frac{\partial V}{\partial q_j}\right) = 0$$

So $$\dot{p}_j = \frac{d}{dt}\left(\frac{\partial L}{\partial \dot{q}_j}\right) = -\frac{\partial V}{\partial q_j} = Q_j \qquad \ldots \qquad (B)$$

If we can show that when generalised force Q_j represent the component of the total force, the generalised momentum p_j is the component of the total linear momentum along the direction of translation of the system, the equation (B) will represent the solution of the problem.

The generalised force Q_j is given by

$$Q_j = \sum_i \vec{F}_i \cdot \frac{\partial \vec{r}_i}{\partial q_j}.$$

If $\hat{n}$ is unit vector along the direction of translation of the system, then

$$\delta \vec{r}_i = \hat{n}\, \delta q_j$$

or, $$\frac{\delta \vec{r}_i}{\delta q_j} = \hat{n}.$$

Hence $$Q_j = \sum_i F_i \cdot \hat{n} = \hat{n} \cdot \vec{F} \qquad \ldots \qquad (C)$$

$=$ Total force along the direction of translation.

Now (K.E.) $$T = \frac{1}{2} \sum_i m_i v_i^2$$

$$= \frac{1}{2} \sum_i m_i \vec{v}_i \cdot \vec{v}_i$$

$$\therefore \quad p_j = \frac{\partial T}{\partial \dot{q}_j} = \sum_i m_i \vec{v}_i \cdot \frac{\partial \vec{v}_i}{\partial \dot{q}_j} = \sum_i m_i \vec{v}_i \cdot \frac{\partial \vec{r}_i}{\partial q_j} \text{ from equation (10.11)}$$

$$= \sum_i m_i \vec{v}_i \cdot \hat{n} = \hat{n} \cdot \sum_i m_i \vec{v}_i$$

= Total linear momentum along the direction of translation.

If $Q_j = 0$; $\dot{p}_j = 0$

i.e. $p_j =$ constant.

Hence if a given component of total applied force vanishes, the corresponding linear momentum is conserved.

(i) To show that if the component of the applied torque along the axis of rotation vanishes, the corresponding total angular momentum along the same direction is conserved.

Let us consider a conservative system so that the potential energy V is dependent on position only and the K.E. is independent of position, i.e.

$$\frac{\partial V}{\partial \dot{q}_j} = 0 \quad \text{and} \quad \frac{\partial T}{\partial q_j} = 0. \qquad \ldots \quad \text{(A)}$$

The Lagrangian equation for such a system is

$$\frac{d}{dt}\left(\frac{\partial L}{\partial \dot{q}_j}\right) - \frac{\partial L}{\partial q_j} = 0$$

$$\text{or,} \quad \frac{d}{dt}\left(\frac{\partial L}{\partial \dot{q}_j}\right) = \frac{\partial L}{\partial q_j} = -\frac{\partial V}{\partial q_j}$$

$$\text{i.e.} \quad \dot{p}_j = Q_j. \qquad \ldots \quad \text{(B)}$$

If we can show that with rotation co-ordinate q_j, the generalised force Q_j is the component of the total applied torque about the axis of rotation and the generalised momentum p_j is the component of the total angular momentum along the same axis, then equation (B) will be a solution of the problem.

The generalised force Q_j is given by

$$Q_j = \sum_i \vec{F}_i \cdot \frac{\partial \vec{r}_i}{\partial q_j}.$$

The magnitude of the change of position co-ordinate $r_i(q_j)$ due to a change in the rotation co-ordinate q_j is given by

$$|\delta \vec{r}_i| = AB\, \delta q_j = r \sin\theta\, \delta q_j$$

$$\therefore \quad \left|\frac{\delta \vec{r}_i}{\delta q_j}\right| = r_i \sin\theta. \qquad \ldots \quad \text{(C)}$$

If $\hat{n}$ is unit vector along the axis of rotation, equation (C) may be written as

$$\frac{\delta \vec{r}_i}{\delta q_j} = \hat{n} \times \vec{r}_i . \qquad \ldots \text{(D)}$$

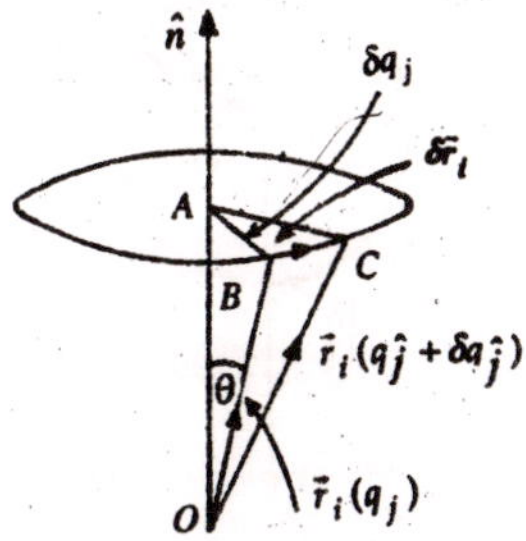

Fig. 10.10

Now the generalised force can be expressed as

$$Q_j = \sum_i \vec{F}_i . \hat{n} \times \vec{r}_i = \sum_i \hat{n} . (\vec{r}_i \times F_i)$$

$$= \sum_i \hat{n} . \vec{N}_i = \hat{n} . \sum_i N_i = \hat{n} . \vec{N} \qquad \ldots \text{(E)}$$

where $\vec{N} = \sum_i \vec{N}_i =$ total torque. Equation (E) shows that Q_j is the component of the total torque along the axis of rotation.

$$\text{Again } p_j = \frac{\partial L}{\partial \dot{q}_j} = \frac{\partial T}{\partial \dot{q}_j} = \sum_i m_i \vec{v}_i . \frac{\partial \vec{v}_i}{\partial \dot{q}_i}$$

$$= \sum_i m_i \vec{v}_i . \frac{\partial \vec{r}_i}{\partial q_j} \qquad \text{from equation (10.11)}$$

$$= \sum_i m_i \vec{v}_i . \hat{n} \times \vec{r}_i = \sum_i \hat{n} . \vec{r}_i \times m_i \vec{v}_i = \hat{n} . \sum \vec{L}_i$$

$$= \hat{n} . \vec{L} \qquad \ldots \text{(F)}$$

= component of the total angular momentum along the axis of rotation.

Hence the equation $\dot{p}_j = Q_j$ signifies that if $Q_j = 0$, $p_j =$ constant i.e. if the component of the applied torque along the axis of rotation vanishes, the component of the total angular momentum along the same direction will be constant.

(j) A simple pendulum of mass m whose point of support (i) moves uniformly in a vertical circle with constant frequency ω, (ii) oscillates horizontally in the plane of motion of the pendulum and according to the law $x = a \cos \omega t$, (iii) oscillates vertically according to the law $y = a \cos \omega t$. Form the Lagrangian and deduce the equation of motion.

(i) Let a be the radius of the circle and A the instanteneous position of the point of support.

The instanteneous co-ordinates of m are

$$x = \cos \omega t + l \sin\theta$$

$$y = a \sin \omega t + l \cos\theta.$$

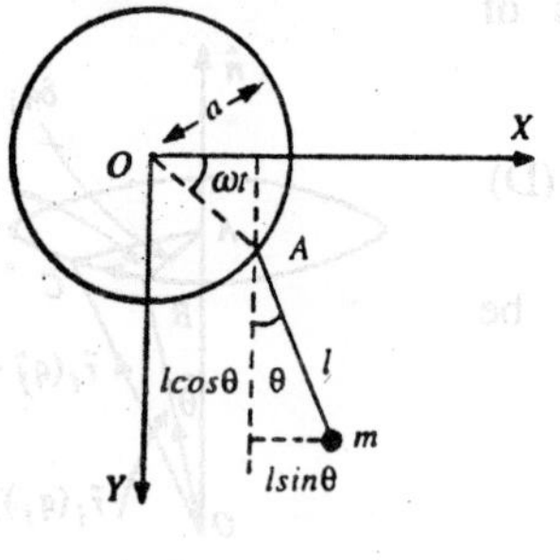

Fig. 10.11

The velocity components are

$$\dot{x} = -a\omega \sin \omega t + l \cos\theta\, \dot{\theta}$$

$$\dot{y} = a\omega \cos \omega t - l \sin\theta\, \dot{\theta}$$

$$\text{K.E. } (T) = \frac{1}{2}\, m\, (\dot{x}^2 + \dot{y}^2)$$

$$= \frac{1}{2} m \left\{ a^2\omega^2 + l^2\dot{\theta}^2 - 2al\omega\dot{\theta} \sin(\omega t + \theta) \right.$$

$$\text{P.E. } (V) = -mg\,(l\cos\theta + a \sin \omega t)$$

$$L = T - V = \frac{1}{2} m \left\{a^2\omega^2 + l^2\dot{\theta}^2 - 2al\omega\dot{\theta} \sin(\omega t + \theta)\right.$$
$$\left. + mg\,(l\cos\theta + a\sin\omega t)\right\}$$

Lagrangian equation motion is

$$\frac{d}{dt}\left(\frac{\partial L}{\partial \dot{\theta}}\right) - \frac{\partial L}{\partial \theta} = 0$$

or, $\frac{d}{dt}\left\{ml^2\,\dot{\theta} - mal\,\omega \sin(\omega t + \theta)\right\} + ma\,\omega l\,\dot{\theta}\,\cos(\omega t + \theta) + mgl \sin\theta = 0$

or, $l\ddot{\theta} - a\omega\,(\omega + \dot{\theta}) \cos(\omega t + \theta) + a\omega\,\dot{\theta}\,\cos(\omega t + \theta) + g\sin\theta = 0$

or, $l\ddot{\theta} - a\omega^2 \cos(\omega t + \theta) + g\sin\theta = 0$

(ii) The instanteneous co-ordinates of m are

$x = a\cos \omega t + l\sin\theta$

$y = l\cos\theta$

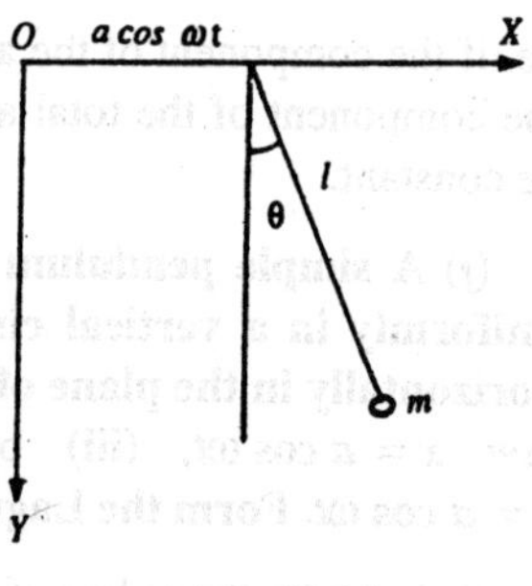

Fig. 10.12

$\therefore\ \dot{x} = -a\omega \sin \omega t + l\cos\theta\, \dot{\theta}$

$\dot{y} = -l\sin\theta\,\dot{\theta}$

$$\text{K.E. } (T) = \frac{1}{2} m\, (\dot{x}^2 + \dot{y}^2)$$

$$= \frac{1}{2} m\, \{a^2\omega^2 \sin^2 \omega t + l^2\dot{\theta}^2$$

$$- 2\, a\omega l \sin \omega t \cos\theta\, \dot{\theta}\}$$

$$\text{P. E. } (V) = -mgl\cos\theta$$

$$L = \frac{1}{2} ma^2\omega^2 \sin^2\omega t + \frac{1}{2} ml^2\dot{\theta}^2$$

$$- ma\omega l \sin\omega t \cos\theta\, \dot{\theta} + mgl \cos\theta$$

Lagrangian equation of motion is

$$\frac{d}{dt}\left(\frac{\partial L}{\partial \dot{\theta}}\right) - \frac{\partial L}{\partial \theta} = 0$$

or, $\frac{d}{dt}\{ ml^2\dot{\theta} - ma\omega l \sin\omega t \cos\theta\} - ma\omega l \sin\omega t \sin\theta\, \dot{\theta} + mgl \sin\theta = 0$

or, $l^2\ddot{\theta} - a\omega^2 l \cos\omega t \cos\theta + a\omega l \sin\omega t \sin\theta\, \dot{\theta} - a\omega l \sin\omega t \sin\theta\, \dot{\theta} + gl \sin\theta = 0$

or, $l\ddot{\theta} - a\omega^2 \cos\omega t \cos\theta + g \sin\theta = 0.$

(iii) The instanteneous co-ordinates of m are

$x = l \sin\theta$

$y = a \cos\omega t + l \cos\theta$

$\therefore \quad \dot{x} = l \cos\theta\, \dot{\theta}$

and $\dot{y} = -a\omega \sin\omega t - l \sin\theta\, \dot{\theta}$

$\therefore \quad T = \frac{1}{2} m (\dot{x}^2 + \dot{y}^2)$

$= \frac{1}{2} m \{l^2\dot{\theta}^2 + a^2\omega^2 \sin^2\omega t$

$+ 2 al\omega \sin\omega t \sin\theta\, \dot{\theta}\}$

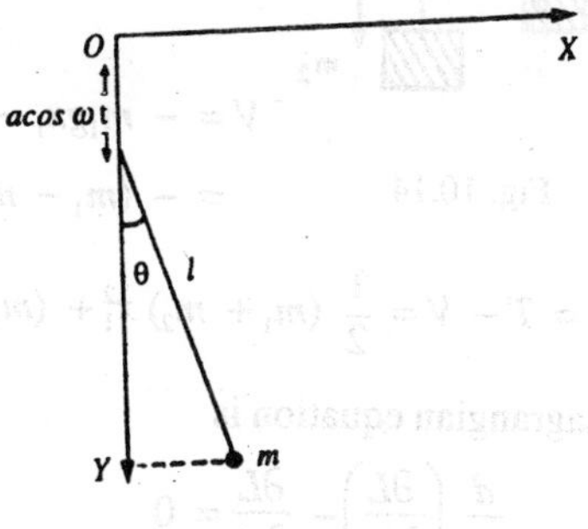

Fig. 10.13

$V = -mg(l\cos\theta + \alpha \cos\omega t)$

$L = T - V = \frac{1}{2}\{l^2\dot{\theta}^2 + a^2\omega^2 \sin^2\omega t + 2 al\omega \sin\omega t \sin\theta\, \dot{\theta}\}$

$+ mg(l \cos\theta + a \cos\omega t)$

Lagrangian equation of motion is

$$\frac{d}{dt}\left(\frac{\partial L}{\partial \dot{\theta}}\right) - \frac{\partial L}{\partial \theta} = 0$$

or, $\frac{d}{dt}\{ ml^2\dot{\theta} + mal\omega \sin\omega t \sin\theta\} - mal\dot{\omega} \sin\omega t \cos\theta\, \dot{\theta} + mgl \sin\theta = 0$

or, $l\ddot{\theta} + a\omega^2 \cos\omega t \sin\theta + a\omega \sin\omega t \cos\theta\, \dot{\theta} - a\omega \sin\omega t \cos\theta\, \dot{\theta} + g \sin\theta = 0$

or, $l\ddot{\theta} + a\omega^2 \cos\omega t \sin\theta + g \sin\theta = 0.$

(k) Two masses m_1 and m_2 are suspended over a frictionless pulley of radius a and connected by a flexible string of constant length l. Form the Lagrangian of the system and the Lagrangian equation.

(Atwood', machine)

Considering the centre of the pulley as the origin, y-axis horizontal and x-axis vertical, let x_1 and x_2 be the x-co-ordinates of the two masses. Equation of constraint for such a system is

$$x_1 + x_2 + \pi a = l$$

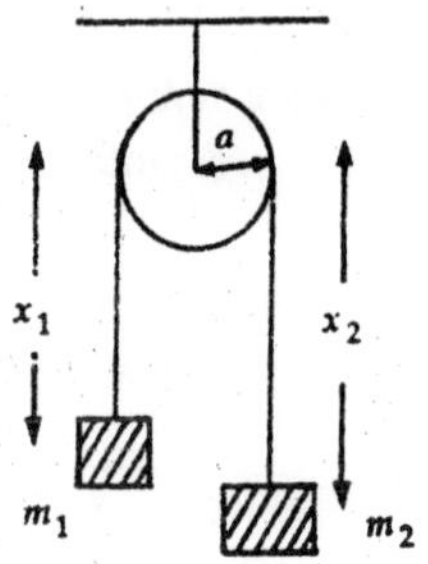

Fig. 10.14

$$\therefore \quad \dot{x}_2 = -\dot{x}_1$$

$$\text{and} \quad \ddot{x}_2 = -\ddot{x}_1$$

$$T = \frac{1}{2} m_1\dot{x}_1^2 + \frac{1}{2} m_2\dot{x}_2^2$$

$$= \frac{1}{2}(m_1 + m_2)\dot{x}_1^2$$

$$V = -m_1gx_1 - m_2gx_2 = -m_1gx_1 - m_2g(l - \pi a - x_1)$$

$$= -(m_1 - m_2)gx_1 + V_0$$

$$L = T - V = \frac{1}{2}(m_1 + m_2)\dot{x}_1^2 + (m_1 - m_2)gx_1 - V_0$$

Lagrangian equation is

$$\frac{d}{dt}\left(\frac{\partial L}{\partial \dot{x}_1}\right) - \frac{\partial L}{\partial x_1} = 0$$

$$\text{or,} \quad \frac{d}{dt}\{(m_1 + m_2)\dot{x}_1\} - (m_1 - m_2)g = 0$$

$$\text{or,} \quad \ddot{x}_1 = \frac{m_1 - m_2}{m_1 + m_2} g.$$

(l) A dipole of masses m_1 and m_2 and with opposite charges is placed in a uniform field. Form the Lagrangian of the system and the equation of motion for small oscillation.

$$\vec{R} = \frac{m_1\vec{r}_1 + m_2\vec{r}_2}{m_1 + m_2}$$

$$\vec{r} = \vec{r}_1 - \vec{r}_2$$

$$\vec{V} = \dot{\vec{R}} = \frac{m_1\vec{v}_1 + m_2\vec{v}_2}{m_1 + m_2}$$

$\vec{v} = \dot{\vec{r}} = \vec{v}_1 - \vec{v}_2$

Solving $\vec{v}_1 = \vec{V} + \frac{\mu}{m_1}\vec{v}$

and $\vec{v}_2 = \vec{V} - \frac{\mu}{m_2}\vec{v}$

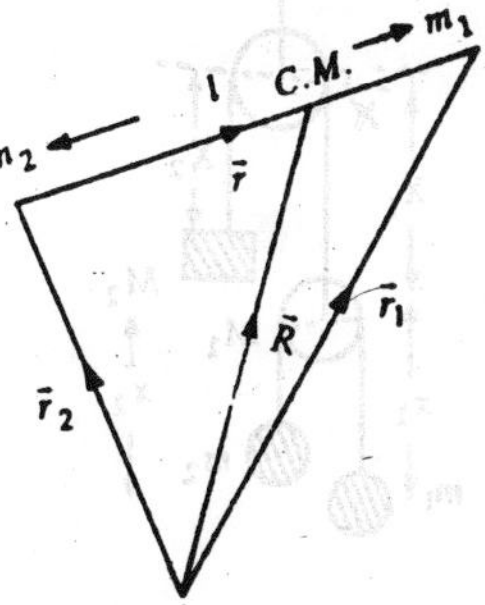

Fig. 10.15

$$T = \frac{1}{2}m_1 v_1^2 + \frac{1}{2}m_2 v_2^2$$

$$= \frac{1}{2}m_1\left\{V^2 + \frac{\mu^2}{m_1^2}v^2 + 2\frac{\mu}{m_1}\vec{V}.\vec{v}\right\}$$

$$+ \frac{1}{2}m_2\left\{V^2 + \frac{\mu^2}{m_2^2}v^2 - 2\frac{\mu}{m_2}\vec{V}.\vec{v}\right\}$$

$$= \frac{1}{2}(m_1 + m_2)V^2 + \frac{1}{2}\frac{m_1 m_2}{m_1 + m_2}v^2 = \frac{1}{2}MV^2 + \frac{1}{2}\mu v^2$$

$$= \frac{1}{2}M\dot{R}^2 + \frac{1}{2}\mu v^2$$

$$\mu = \frac{m_1 m_2}{m_1 + m_2} = \frac{m_1 m_2}{M} = \text{reduced mass}$$

For the dipole

$$v^2 = \dot{r}^2 + r^2\dot{\theta}^2 + r^2\sin^2\theta\,\dot{\varphi}^2 = l^2(\dot{\theta}^2 + \sin^2\theta\,\dot{\varphi}^2)$$

P.E. $= Eql\,(1 - \cos\theta)$

$$\therefore\ L = \frac{1}{2}M\dot{R}^2 + \frac{1}{2}\mu l^2(\dot{\theta}^2 + \sin^2\theta\,\dot{\varphi}^2) - Eql\,(1 - \cos\theta)$$

Taking R, θ, φ as the generalised co-ordinates, three Lagrangian equations will be obtained.

(m) A mass M_2 hangs at one end of a string which passes over a fixed frictionless non-rotating pulley. At the other end of that string there is a non-rotating of mass M_1 over which there is a string carrying masses m_1 and m_2.

(A) Set up the Lagrangian of the system (B). Find the acceleration of the mass M_2.

Let X_1 and X_2 to the distances of the masses M_1 and M_2 below the centre of the fixed pulley and x_1 and x_2 the distances of the masses m_1 and m_2 respectively below the centre of the movable pulley M_1.

Since the strings are of constant length, $X_1 + X_2 = \text{constant} = a$,

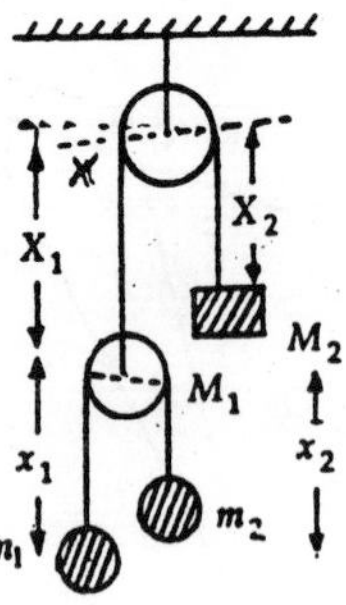

Fig. 10.16

and $x_1 + x_2 = \text{constant} = b$

Differentiating w.r.t. 't'

$$\dot{X}_1 = -\dot{X}_2 \quad \text{and} \quad \dot{x}_1 = -\dot{x}_2$$

Thus the velocity of $M_1 = \dot{X}_1$

" " " " $M_2 = -\dot{X}_1$

" " " " $m_1 = \dot{x}_1 + \dot{X}_1$

" " $m_2 = \dot{X}_1 + \dot{x}_2 = \dot{X}_1 - \dot{x}_1$

The K.E. of the system is

$$T = \frac{1}{2} M_1 \dot{X}_1^2 + \frac{1}{2} M_2 \dot{X}_1^2 + \frac{1}{2} m_1 (\dot{x}_1 + \dot{X}_1)^2 + \frac{1}{2} m_2 (\dot{X}_1 - \dot{x}_1)^2 \quad \ldots \text{(A)}$$

The total P.E. of the system measured from a horizontal plane through the centre of the fixed pulley as reference is

$$V = -M_1 g X_1 - M_2 g X_2 - m_1 g (X_1 + x_1) - m_2 g (X_1 + x_2)$$

$$= -M_1 g X_1 - M_2 g (a - X_1) - m_1 g (X_1 + x_1) - m_2 g (X_1 + b - x_1) \quad \ldots \text{(B)}$$

The Lagrangian of the system is

$$L = T - V = \frac{1}{2} M_1 \dot{X}_1^2 + \frac{1}{2} M_2 \dot{X}_1^2 + \frac{1}{2} m_1 (\dot{X}_1 + \dot{x}_1)^2 + \frac{1}{2} m_2 (\dot{X}_1 - \dot{x}_1)^2$$

$$+ M_1 g X_1 + M_2 g (a - X_1) + m_1 g (X_1 + x_1) + m_2 g (X_1 + b - x_1) \quad \ldots \text{(C)}$$

Lagrangian equations corresponding to X_1 and x_1 are

$$\frac{d}{dt}\left(\frac{\partial L}{\partial \dot{X}_1}\right) - \frac{\partial L}{\partial X_1} = 0 \quad \text{and} \quad \frac{d}{dt}\left(\frac{\partial L}{\partial \dot{x}_1}\right) - \frac{\partial L}{\partial x_1} = 0.$$

The former gives

$$\frac{d}{dt}\left\{ M_1 \dot{X}_1 + M_2 \dot{X}_1 + m_1 (\dot{X}_1 + \dot{x}_1) + m_2 (\dot{X}_1 - \dot{x}_1) \right\}$$

$$- \left\{ M_1 g - M_2 g + m_1 g + m_2 g \right\} = 0$$

or, $$(M_1 + M_2 + m_1 + m_2) \ddot{X}_1 + (m_1 - m_2) \ddot{x}_1 = (M_1 - M_2 + m_1 + m_2) g. \quad \ldots \text{(D)}$$

The latter gives

$$\frac{d}{dt}\{m_1(\dot{X}_1+\dot{x}_1)-m_2(\dot{X}_1-\dot{x}_1)-m_1g+m_2g\}=0$$

or, $(m_1-m_2)\ddot{X}_1+(m_1+m_2)\ddot{x}_1=(m_1-m_2)g$ ... (E)

Solving (C) and (D) simultaneously,

$$\ddot{X}_1=\frac{(M_1-M_2)(m_1+m_2)+4m_1m_2}{(M_1+M_2)(m_1+m_2)+4m_1m_2}$$

and $$\ddot{x}_1=\frac{2M_2(m_1-m_2)}{(M_1+M_2)(m_1+m_2)+4m_1m_2}.$$

(n) **Speed Governor :**

A particle of mass m_2 moves on a vertical axis and the whole system rotates about the axis with a constant angular velocity ω, under gravity. Let each segment be of length a and θ be instanteneous angle it makes with the vertical.

Let φ be the angle of rotation i.e. $\omega=\dot{\varphi}$

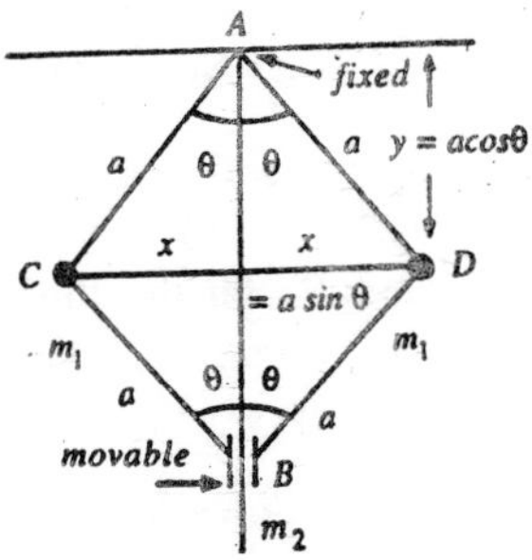

Fig. 10.17

$$T=2\times\frac{1}{2}m_1(\dot{x}^2+\dot{y}^2+x^2\dot{\varphi}^2)$$

$$+\frac{1}{2}m_2(2\dot{y})^2$$

$$=m_1a^2(\dot{\theta}^2+\sin^2\theta\,\dot{\varphi}^2)+2m_2a^2\sin^2\theta\,\dot{\theta}^2$$

$$V=-2m_1gy-m_2g(2y)$$

$$=-2(m_1+m_2)\,ag\cos\theta$$

$$L=T-V=m_1a^2(\dot{\theta}^2+\sin^2\theta\,\omega^2)$$

$$+2m_2a^2\sin^2\theta\,\dot{\theta}^2+2(m_1+m_2)\,ag\cos\theta$$

Lagrangian equation of motion is

$$\frac{d}{dt}\left(\frac{\partial L}{\partial\dot{\theta}}\right)-\frac{\partial L}{\partial\theta}=0$$

or, $$\frac{d}{dt}\left\{m_1a^2\,2\dot{\theta}+4m_2a^2\sin^2\theta\,\dot{\theta}\right\}-m_1a^2\,2\sin\theta\cos\theta\,\omega^2$$

$$-2m_2a^2\times 2\sin\theta\cos\theta\,\dot{\theta}^2+2(m_1+m_2)\,ga\sin\theta=0$$

$$\text{or,} \quad m_1 a\ddot{\theta} + 2m_2 a \{\sin^2\theta\, \ddot{\theta} + 2\sin\theta\cos\theta\, \dot{\theta}^2\} - m_2 a \sin\theta\cos\theta\, \omega^2$$

$$- 2m_2 a \sin\theta\cos\theta\, \dot{\theta}^2 + (m_1 + m_2)\, g \sin\theta = 0$$

10.8. Ignorable or Cyclic Ordinates

If the Lagrangian of a system does not contain a given co-ordinate q_j (although it may contain the corresponding velocity $\dot{q}_j$) then the co-ordinate is to be cyclic or ignorable. The Lagrange's equation of motion

$$\frac{d}{dt}\left(\frac{\partial L}{\partial \dot{q}_j}\right) - \frac{\partial L}{\partial q_j} = 0 \text{ reduces for a cyclic co-ordinate, to}$$

$$\frac{d}{dt}\left(\frac{\partial L}{\partial \dot{q}_j}\right) = 0 \quad \text{or} \quad \frac{d}{dt}(p_j) = 0$$

which means that p_j = constant.

Hence we can state as a general conservation theorem that the generalised momentum conjugate to a cyclic co-ordinate is conserved.

Again if $\dot{p}_j$ vanishes, it follows that $\frac{\partial H}{\partial q_j}$ is likewise zero. A Co-ordinate that is cyclic will thus also be absent from the Hamiltonian. Conversely if a generalised co-ordinate does not occur in H, the conjugate momentum is conserved. So far the situation is exactly as in the Lagrangian formulation. However if some co-ordinate, say q_n is cyclic, the Lagrangian as a function of q and $\dot{q}$ can be written as

$$L = L(q_1, q_2 \ldots\ldots q_{n-1}, \dot{q}_1, \dot{q}_2 \ldots\ldots \dot{q}_n, t).$$

All the generalised velocities still occur in the Lagrangian and in general will be a function of time. We still have to solve a problem of n degrees of freedom even though one degree of freedom correspond to a cyclic co-ordinate.

A cyclic co-ordinate in the Hamilt nian formulaion, on the other hand truly deserves its alternative descripti n as 'ignorable', for in the same situation p_n is some constant α, and H has the form

$$H = H(q_1, q_2, \ldots\ldots q_{n-1}, p_1, p_2 \ldots\ldots p_{n-1}, \alpha, t).$$

In effect the Hamiltonian now describes a problem involving only $(n - 1)$ co-ordinates, which may be solved completely ignoring the cyclic co-ordinate except as it is manifested in the constant of integration α, to be determined from initial conditions.

Examples :

1. Spherical Pendulum :

Let us consider a spherical pendulum consisting of a particle of mass m which moves under gravity on a smooth sphere of radius a. In terms of spherical polar co-ordinates a, θ, φ, with θ measured from the downward vertical, the kinetic and potential energies are

$$T = \frac{1}{2} ma^2 (\dot{\theta}^2 + \sin^2\theta\, \dot{\varphi}^2),$$

$$V = -\, mga \cos\theta$$

$$L = T - V = \frac{1}{2} ma^2 (\dot{\theta}^2 + \sin^2\theta\, \dot{\varphi}^2) + mga \cos\theta$$

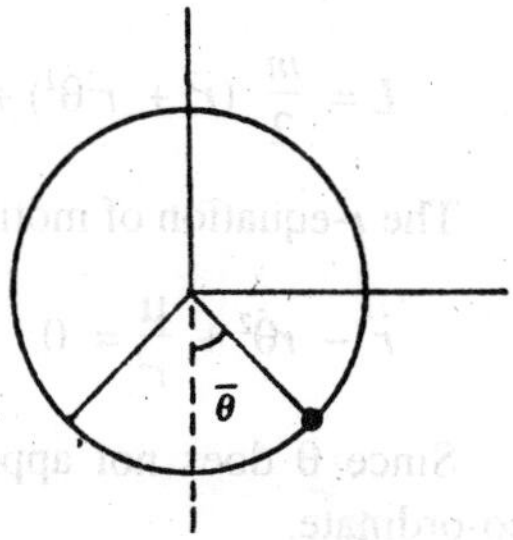

Fig. 10.18

Writing p_θ and p_φ for the generalised moment-a we have

$$p_\theta = \frac{\partial L}{\partial \dot{\theta}} = ma^2 \dot{\theta}$$

$$p_\varphi = \frac{\partial L}{\partial \dot{\varphi}} = ma^2 \sin^2\theta\, \varphi$$

$$\therefore\ \dot{\theta} = \frac{p_\theta}{ma^2}$$

and $\dot{\varphi} = \dfrac{p_\varphi}{ma^2 \sin^2\theta}$.

Accordingly the Hamiltonian is

$$H = T + V = \frac{1}{2ma^2} \left\{ p_\theta^2 + cosec^2\theta\, p_\varphi^2 \right\} - mag \cos\theta.$$

Since the Lagrangian does not contain the co-ordinate φ, the co-ordinate φ is ignorable and therefore p_φ is constant determined by initial conditions. We have then only two cannonical equations to solve ; they are

$$\dot{\theta} = \frac{\partial H}{\partial p_\theta} = \frac{p_\theta}{ma^2}$$

and $\dot{p}_\theta = -\dfrac{\partial H}{\partial \theta} = \dfrac{p_\varphi^2}{ma^2} \dfrac{\cos\theta}{\sin^3\theta} - mga \sin\theta$

where p_φ is a mere constant.

Eliminating p_θ

$$\ddot{\theta}\, ma^2 = \frac{p_\varphi^{\,2}}{ma^2}\, cosec^2\,\theta \cot\theta - mga \sin\theta$$

which is a second order differential equation for θ.

2. **Planetary Motion :**

As proved before

$$L = \frac{m}{2}\,(\dot{r}^2 + r^2\dot{\theta}^2) + \frac{m\mu}{r}$$

The r-equation of motion is

$$\ddot{r} - r\dot{\theta}^2 + \frac{\mu}{r^2} = 0.$$

Since θ does not appear in the Lagrangian function, it is an ignorable co-ordinate.

Hence $p_\theta = \dfrac{\partial L}{\partial \dot{\theta}} = mr^2\dot{\theta}$ = constant

or, $r^2\dot{\theta}$ = constant which is Kepler's 2nd law.

10.9. Applications of Hamilton's equations

(a) **Write down the Hamiltonian and equation of motion for a simple pendulum :**

$$\text{(K.E.) } T = \frac{1}{2}\, ml^2\,\dot{\theta}^2,$$

$$\text{(P.E.) } V = mgl\,(1 - \cos\theta)$$

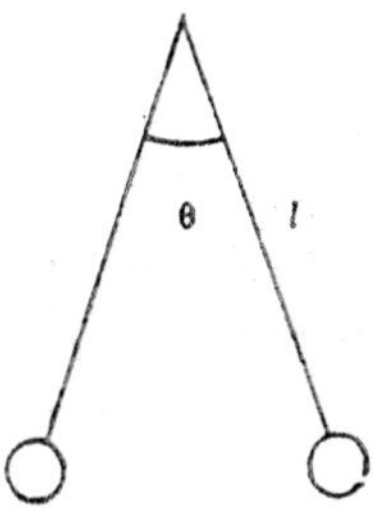

$$L = T - V = \frac{1}{2}\, ml^2\dot{\theta}^2 - mgl\,(1 - \cos\theta)$$

$$H = T + V = \frac{1}{2}\, ml^2\dot{\theta}^2 + mgl\,(1 - \cos\theta)$$

$$p_\theta = \frac{\partial L}{\partial \dot{\theta}} = ml^2\dot{\theta}$$

$$\text{or, } \dot{\theta} = \frac{p_\theta}{ml^2}$$

Fig. 10.19

$$\therefore\ H = \frac{1}{2}\, ml^2 \left(\frac{p_\theta}{ml^2}\right)^2 + mgl\,(1 - \cos\theta)$$

$$= \frac{p_\theta^2}{2ml^2} + mgl\ (1 - \cos\theta)$$

$$\therefore \quad \frac{\partial H}{\partial p_\theta} = \frac{p_\theta}{ml^2}\ ; \qquad \frac{\partial H}{\partial \theta} = mgl \sin\theta.$$

Hamilton's equations of motion are

$$\dot{\theta} = \frac{\partial H}{\partial p_\theta} = \qquad \dot{p}_\theta = -\frac{\partial H}{\partial \theta}$$

i.e. $\quad \dot{\theta} = \dfrac{p_\theta}{ml^2} \quad$ and $\quad \dot{p}_\theta = -\ mgl \sin\theta$

Eliminating p_θ, we have

$$ml^2\ddot{\theta} = -\ mgl \sin\theta$$

$$\ddot{\theta} = -\frac{g}{l}\sin\theta.$$

(b) Write down the Hamiltonian function and equation of motion of a Compound Pendulum :

In this case $L = \dfrac{1}{2} I\dot{\theta}^2 + mgh\cos\theta$

$$\therefore \quad p_\theta = \frac{\partial L}{\partial \dot{\theta}} = I\dot{\theta}$$

$$H = \frac{1}{2} I\dot{\theta}^2 - mgh\cos\theta$$

$$\therefore \quad H = \frac{1}{2} I\left(\frac{p_\theta}{I}\right)^2 - mgh\cos\theta = \frac{p_\theta^2}{2I} - mgh\cos\theta$$

$$\therefore \quad \frac{\partial H}{\partial p_\theta} = \frac{p_\theta}{I}\ ; \qquad \frac{\partial H}{\partial \theta} = mgh\sin\theta.$$

Hamilton's equations are

$$\dot{\theta} = \frac{\partial H}{\partial p_\theta} \quad \text{and} \quad \dot{p}_\theta = -\frac{\partial H}{\partial \theta}$$

i.e. $\quad \dot{\theta} = \dfrac{p_\theta}{I} \quad$ and $\quad \dot{p}_\theta = -\ mgh\sin\theta$

But $p_\theta = I\dot{\theta} \qquad \therefore\ \dot{p}_\theta = I\ddot{\theta}$

Hence $I\ddot{\theta} = -\,mgh\sin\theta \quad$ or, $\quad \ddot{\theta} + \dfrac{mgh}{I}\sin\theta = 0.$

CHAPTER XI

SMALL OSCILLATIONS

11.1. Small Oscillations

In many cases we come across system which perform oscillations of such small amplitude that only the fundamentals are excited. In the position of static equilibrium all the particles of the system will be at rest. We now consider small oscillations about the position of static equilibrium.

We confine our attention to conservative system only. The equilibrium position will be characterised by the position of minimum potential energy which is a function of only the generalised co-ordinates q. Since in the equilibrium position the generalised forces Q_j acting on the system will be zero, we have $Q_j = \left(\dfrac{\partial V}{\partial q_j}\right)_0 = 0$, the suffix zero indicating the equilibrium position. (11.1)

In the equilibrium position the co-ordinates of the particles of the system are $q_{01}, q_{02}, \ldots \ldots q_{0n}$.

Denoting by η_j the deviation of the generalised co-ordinate of the jth particle from the equilibrium position we have

$$q_j - q_{j0} = \eta_j . \tag{11.2}$$

Expressing the potential function by Mac Laurin's theorem,

$$V(q_1, q_2, \ldots . q_n) = V(q_{01}, q_{02}, \ldots . q_{0n}) + \sum_j \left(\frac{\partial V}{\partial q_j}\right)_0 \eta_j + \frac{1}{2}\sum_{j,k}\left(\frac{\partial^2 V}{\partial q_j \partial q_k}\right)_0 \eta_j \eta_k + \ldots\ldots$$

$$\text{or,}\quad V = V_0 + \sum \left(\frac{\partial V}{\partial q_j}\right)_0 \eta_j + \frac{1}{2}\sum_{jk}\left(\frac{\partial^2 V}{\partial q_j \partial q_k}\right)_0 \eta_j \eta_k + \tag{11.3}$$

The first term on the right hand side being a constant may be neglected with the proper choice of level, while the second term vanishes on account of equation (11.1). Thus we have $V = \dfrac{1}{2}\sum_{j,k}\left(\dfrac{\partial^2 V}{\partial q_j \partial q_k}\right)_0 \eta_j \eta_k$ to a first approximation.

$$= \frac{1}{2} \sum_{jk} V_{jk}\, \eta_j\, \eta_k \tag{11.4}$$

where $V_{jk} = \dfrac{\partial^2 V}{\partial q_j\, \partial q_k} = V_{kj}$.

Since the K.E. can be expressed as a positive quadratic function of generalised velocities,

$$T = \frac{1}{2} \sum_{jk} a_{jk}\, \dot{q}_j\, \dot{q}_k = \frac{1}{2} \sum_{jk} a_{jk}\, \dot{\eta}_j\, \dot{\eta}_k \tag{10.5}$$

Expanding by Mac Laurin's theorem we have

$$a_{jk}(q_1, q_2, \ldots \ldots q_n)$$

$$= a_{jk}(q_{01}, q_{02}, \ldots \ldots q_{0n}) + \sum_i \left(\frac{\partial a_{jk}}{\partial_i}\right) \eta_i + \ldots \ldots$$

$$= (a_{jk})_0 + \sum_i \left(\frac{\partial a_{jk}}{\partial q_i}\right) \eta_i + \ldots \ldots$$

Writing $(a_{jk})_0 = T_{jk}$, we have neglecting small quantities of higher order a result similar to equation (11.4) as

$$T = \frac{1}{2} \sum_{j,k} T_{jk}\, \dot{\eta}_j\, \dot{\eta}_k = \frac{1}{2} \sum_{k,j} T_{k,j}\, \dot{\eta}_k\, \dot{\eta}_j \tag{11.6}$$

The Lagrangian is given by $L = T - V$

$$= \frac{1}{2} \sum_{kj} \left\{ T_{kj}\, \dot{\eta}_k\, \dot{\eta}_j - V_{kj}\, \eta_k\, \eta_j \right\} \tag{11.7}$$

Lagrangian equations of motion derived from above are

$$\sum_j T_{kj}\, \ddot{\eta}_j + \sum_j V_{kj}\, \eta_j = 0 \tag{11.8}$$

The above can be represented in matraix notation as

$$T\ddot{\eta} + V\eta = 0 \tag{11.9}$$

where $T = \sum T_{kj}$ and $V = \sum V_{kj}$ and the matrices T and V are $n \times n$ matrices.

When written explicitly

$$\begin{bmatrix} T_{11} & T_{12} & \ldots & T_{1n} \\ T_{21} & T_{22} & \ldots & T_{2n} \\ \ldots & \ldots & \ldots & \ldots \\ T_{n1} & T_{n2} & \ldots & T_{nn} \end{bmatrix} \begin{bmatrix} \ddot{\eta}_1 \\ \ddot{\eta}_2 \\ \cdot \\ \cdot \\ \ddot{\eta}_n \end{bmatrix} + \begin{bmatrix} V_{11} & V_{12} & \ldots & V_{1n} \\ V_{21} & V_{22} & \ldots & V_{2n} \\ \ldots & \ldots & \ldots & \ldots \\ V_{n1} & V_{n2} & \ldots & V_{nn} \end{bmatrix} \begin{bmatrix} \eta_1 \\ \eta_2 \\ \cdot \\ \cdot \\ \cdot \\ \eta_n \end{bmatrix} = 0 \tag{11.10}$$

Let T^{-1} be the inverse of T matrix.

Hence from equation (11.9)

$$TT^{-1}\ddot{\eta} + VT^{-1}\eta = 0$$

or, $$\ddot{\eta} + \omega^2\eta = 0 \qquad (11.11)$$

where $\omega^2 = VT^{-1}$

$$\ddot{\eta} + \omega^2\eta = 0. \qquad (11.12)$$

In equation (11.12) η displays the part of a simple harmonic motion of the vector η having angular frequency ω. In this case η is called the eigen vector belonging to the eigen value ω^2.

Solution of equation (11.12) gives

$$\eta_j = ca_j e^{-i\omega t} \qquad (11.13)$$

Hence ca_j represents the complex amplitude of oscillation of each co-ordinate η_j. Substituting of the solution into the equations of motion leads to the following equation for the eigen frequencies

$$a_j V_{kj} - \omega^2 a_j T_{kj} = 0. \qquad (11.14)$$

Equations (11.14) constitute linear homogeneous equations for the aj's and consequently can have a solution only if the determinant

$$| V_{kj} - \omega^2 T_{kj} | = 0 \qquad (11.15)$$

is true. The above determental equation gives the normal frequencies.

Application :

(a) An elastic string of length $4l$ and fixed at the ends has three equal masses m attached to it at an equal spacing l. Neglecting the action of gravity discuss the normal modes of vibration.

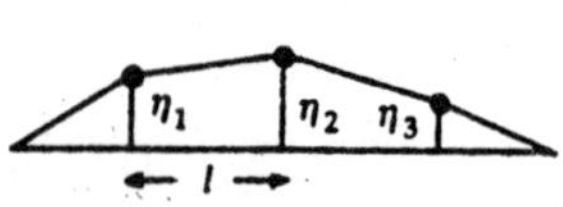

Fig. 11.1

Ans. In the case of three particles,

length $= 4l$

$k =$ Tension.

$$T = \frac{1}{2} m \{\dot{\eta}_1^2 + \dot{\eta}_2^2 + \dot{\eta}_3^2\} \qquad \ldots (1)$$

$$V = \frac{k}{2l} \{(\eta_1 - \eta_0)^2 + (\eta_2 - \eta_1)^2 + (\eta_3 - \eta_2)^2 + (\eta_4 - \eta_3)^2\} \qquad \ldots (2)$$

with $\eta_0 = \eta_4 = 0$

$$= \frac{k}{2l} \{2\eta_1^2 + 2\eta_2^2 + 2\eta_3^2 - 2\eta_1\eta_2 - 2\eta_2\eta_3\}$$

$$\therefore\ T_{kj} = \begin{bmatrix} 1 & 0 & 0 \\ 0 & 1 & 0 \\ 0 & 0 & 1 \end{bmatrix} \qquad V_{kj} = \begin{bmatrix} 2 & -1 & 0 \\ -1 & 2 & -1 \\ 0 & -1 & -2 \end{bmatrix} \quad \ldots \quad (3)$$

Choosing $m = 1$, $\frac{k}{l} = 1$.

The eigen frequencies are given by

$$| V_{kj} - \omega^2 T_{kj} | = 0$$

i.e. $$\begin{vmatrix} 2-\omega^2 & -1 & 0 \\ -1 & 2-\omega^2 & -1 \\ 0 & -1 & 2-\omega^2 \end{vmatrix} = 0$$

or, $(2-\omega^2)\{(2-\omega^2)(2-\omega^2) - 1\} + 1\{-1(2-\omega^2)\} = 0$

or, $(2-\omega^2)\{(2-\omega^2)^2 - 2\} = 0$

$\therefore\ \omega^2 = 2,\ 2 \pm \sqrt{2}$

In conventional units $\omega^2 = \frac{2k}{l},\ \frac{k}{l}(2 \pm \sqrt{2})$

$$\therefore\ \omega = \sqrt{\frac{2k}{l}},\quad \sqrt{\frac{k}{l}(2 \pm \sqrt{2})}.$$

Examples :

1. Case of two equal particles fixed in string of length = 3*l*, *k* = Tension :

$$\text{K.E.}(T) = \frac{1}{2} m \{\dot{\eta}_1^2 + \dot{\eta}_2^2\} \quad \ldots \quad (1)$$

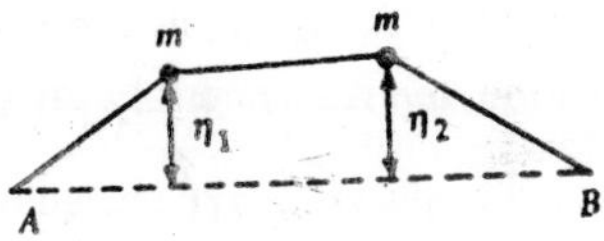

Fig. 11.2

$$V = \frac{k}{2l}\{(\eta_1 - \eta_0)^2 + (\eta_2 - \eta_1)^2 + (\eta_3 - \eta_2)^2\} \quad \ldots \quad (3)$$

with $\eta_0 = \eta_3 = 0$

$$= \frac{k}{2l}\{2\eta_1^2 + 2\eta_2^2 - 2\eta_1\eta_2\}$$

In matrix form $T = \frac{1}{2} m \begin{bmatrix} \dot{\eta}_1 & 0 \\ 0 & \dot{\eta}_2 \end{bmatrix} \begin{bmatrix} \dot{\eta}_1 \\ \dot{\eta}_2 \end{bmatrix}$

$$V = \frac{1}{2}\frac{k}{l} \begin{bmatrix} 2\eta_1 & -\eta_1 \\ -\eta_2 & 2\eta_2 \end{bmatrix} \begin{bmatrix} \eta_1 \\ \eta_2 \end{bmatrix}$$

Hence $T_{kj} = \begin{bmatrix} 1 & 0 \\ 0 & 1 \end{bmatrix}$; $V_{kj} = \begin{bmatrix} 2 & -1 \\ -1 & 2 \end{bmatrix}$

choosing $m = 1,\ k/l = 1$

The eigen frequencies are given by

$$| V_{kj} - \omega^2 T_{kj} | = 0$$

i.e. $$\begin{vmatrix} 2-\omega^2 & -1 \\ -1 & 2-\omega^2 \end{vmatrix} = 0$$

Its solution gives $(2-\omega^2)^2 - 1 = 0$

or, $2 - \omega^2 = \pm 1$ or, $\omega^2 = 3,\ 1$

In conventional units $\omega^2 = \dfrac{3k}{lm},\ \dfrac{k}{lm}$

$$\therefore\ \omega = \sqrt{\frac{3k}{lm}},\ \sqrt{\frac{k}{lm}}$$

2. Free vibrations of a linear triatomic molecule :

Let us consider the equilibrium configuration of the molecule such that two of its atoms each of mass m are symmetrically placed on each side of the third atom of mass M. All the three atoms are collinear. Assuming the motion to be along the line of the molecules and there being no interaction between the end atoms, let k be the force constant approximated by two springs of force joining the three atoms

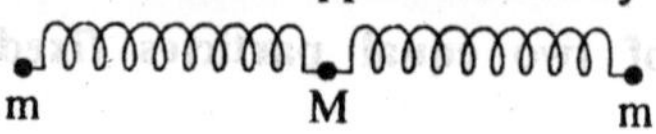

Taking $x_1,\ x_2,\ x_3$ as the displacement of each atom from the equilibrium position, the K.E. and P.E. are given by

$$T = \frac{1}{2} m\,(\dot{x}_1^2 + \dot{x}_3^2) + \frac{1}{2} M\dot{x}_2^2 \quad \ldots \quad (1)$$

$$V = \frac{1}{2} k\,(x_2 - x_1)^2 + \frac{1}{2} k\,(x_3 - x_2)^2$$

$$= \frac{1}{2} k \left\{ x_1^2 + 2x_2^2 + x_3^2 - 2x_1 x_2 - 2x_2 x_3 \right\} \quad \ldots \quad (2)$$

In this case the T_{kj} and V_{kj} matrices are

$$T_{kj} = \begin{bmatrix} m & 0 & 0 \\ 0 & M & 0 \\ 0 & 0 & m \end{bmatrix} \qquad V_{kj} = \begin{bmatrix} k & -k & 0 \\ -k & 2k & -k \\ 0 & -k & k \end{bmatrix} \quad \ldots \quad (3)$$

Eigen frequencies are given by

$$| V_{kj} - \omega^2 T_{kj} | = 0$$

i.e. $$\begin{vmatrix} k-\omega^2 m & -k & 0 \\ -k & 2k-M\omega^2 & -k \\ 0 & -k & k-\omega^2 m \end{vmatrix} = 0 \quad \ldots \quad (4)$$

which on expansion gives

$$\omega^2 (k-\omega^2 m)\left\{k(M+2m)-\omega^2 m M\right\}=0$$

i.e. $\omega^2 = 0, \quad \dfrac{k}{m}, \quad \dfrac{k(M+2m)}{mM}.$

The three normal frequencies are given by

$$\omega_1 = 0, \quad \omega_2 = \sqrt{k/m}, \quad \omega_3 = \sqrt{\frac{k}{m}\left(1+\frac{2m}{M}\right)} \quad \ldots \quad (5)$$

The first eigen value $\omega_1 = 0$ does not correspond to an oscillatory motion. Such a vanishing frequency is caused by the molecules translated rigidly along the axis. Let x_1, x_2, x_3 be the eigen vectors correspoding to ω_1, ω_2 and ω_3 and let the components of x_1 be x_{11}, x_{12} and x_{13}.

Then $\omega_1 = 0$ gives

$$\begin{bmatrix} k & -k & 0 \\ -k & 2k & -k \\ 0 & -k & k \end{bmatrix}\begin{bmatrix} x_{11} \\ x_{12} \\ x_{13} \end{bmatrix} = 0$$

{ { Eigen vectors can be obtained from

$(V-\omega^2 T)x = 0$ i.e.

$\left[V_{kj}-\omega^2 T_{kj}\right][x] = 0$ } }

Expanding $x_{11}-x_{12}=0, \quad -x_{11}+2x_{12}-x_{13}=0$

and $-x_{12}+x_{13}=0$

solving $x_{11}=x_{12}=x_{13}=\alpha$ say.

Which shows that each atom is equally displaced = $\underset{\rightarrow}{\text{m}}$ $\underset{\rightarrow}{\text{M}}$ $\underset{\rightarrow}{\text{m}}$.

$\omega_2 = \sqrt{\dfrac{k}{m}}$ gives

$$\begin{bmatrix} 0 & -k & 0 \\ -k & 2k-k(M/m) & -k \\ 0 & -k & 0 \end{bmatrix}\begin{bmatrix} x_{21} \\ x_{22} \\ x_{23} \end{bmatrix} = 0$$

Solving $x_{22}=0, \quad x_{21}+x_{23}=0$

or, $x_{21} = -x_{23} = \beta$ (say)

$\underset{\leftarrow}{\text{m}}$ M $\underset{\rightarrow}{\text{m}}$

The central atom is at rest and the end ones are displaced equally in opposite directions.

Lastly $\omega_3 = \sqrt{\frac{k}{m}\left(1 + \frac{2m}{M}\right)}$ gives

$$\begin{bmatrix} -\frac{2m}{M}k & -k & 0 \\ -k & -k\frac{M}{m} & -k \\ 0 & -k & -\frac{2mk}{M} \end{bmatrix} \begin{bmatrix} x_{31} \\ x_{32} \\ x_{33} \end{bmatrix} = 0$$

solving $K\frac{2m}{M} x_{31} + K x_{32} = 0$

$$K x_{31} + \frac{KM}{m} x_{32} + K x_{33} = 0$$

$$K x_{32} + \frac{2m}{M} K x_{33} = 0$$

which yield $x_{31} = x_{33} = \Upsilon$ ray.

$$x_{32} = -\frac{2m}{M} x_{31} = -\frac{2m}{M}\Upsilon$$

which shows that the displacements of

m $\longrightarrow$ $\longleftarrow$ M m $\longrightarrow$

end atoms are equal while that of the central atom is equal to the neither.

11.2. Transformation of generalised co-ordinates into normal co-ordinates

In generalised co-ordinates the K.E. and P.E. can be represented in the form

$$T = \frac{1}{2} \Sigma a_{kj} \dot{q}_j \dot{q}_k$$

$$V = \frac{1}{2} \Sigma b_{kj} q_j q_k$$

Putting $q'_j = c_{kj} q_k$, T and V transforms into

$$T = \frac{1}{2}\left\{\dot{q}_1'^2 + \dot{q}_2'^2 + \ldots\ldots\right\} = \frac{1}{2} \Sigma \dot{q}_j'^2$$

$$V = \frac{1}{2}\left\{\lambda_1 q_1'^2 + \lambda_2 q_2'^2 + \ldots\ldots\right\}.$$

$= \frac{1}{2} \Sigma \lambda_j q_j'^2$ where λ'_js are constant and are roots of $| b_{kj} - \lambda_j a_{kj} | = 0.$

The transformations of T and V in the second form are called normal forms and the co-ordinates q'_j's are normal co-ordinates.

The Lagranges equations are

$$\frac{d}{dt}(\dot{q}'_1) + \lambda_1 q'_1 = 0$$

or, $\ddot{q}'_1 + \lambda_1 q'_1 = 0$

Similarly $\ddot{q}'_2 + \lambda_2 q'_2 = 0 \ldots\ldots$

$$\ldots\ldots\ldots$$

$$\ddot{q}'_n + \lambda_n q'_n = 0$$

EXERCISE (C.U. 1987)

1. Consider two identical simple harmonic oscillators which are coupled together. Their K.E. and P.E. are given by

$$T = \frac{1}{2} m (\dot{q}_1^2 + \dot{q}_2^2) \quad \text{and} \quad V = \frac{1}{2} k (q_1^2 + q_2^2) - h q_1 q_2 .$$

(a) Find the Hamiltonian of the system.

(b) Find the frequencies of the normal modes.

(c) Introduce normal co-ordinates and set up the corresponding Lagrange's equations.

Ans.

(a) $H = \frac{1}{2} m (\dot{q}_1^2 + \dot{q}_2^2) + \frac{1}{2} k (q_1^2 + q_2^2) - h q_1 q_2$

(b) $T = \frac{1}{2} m (\dot{q}_1^2 + \dot{q}_2^2) = \frac{1}{2} m \begin{bmatrix} \dot{q}_1 & 0 \\ 0 & \dot{q}_2 \end{bmatrix} \begin{bmatrix} \dot{q}_1 \\ \dot{q}_2 \end{bmatrix}$

$$V = \frac{1}{2} k \begin{bmatrix} q_1 & -\frac{h}{k} q_1 \\ -\frac{h}{k} q_2 & q_2 \end{bmatrix} \begin{bmatrix} q_1 \\ q_2 \end{bmatrix}$$

$$\therefore \; T_{kj} = m \begin{bmatrix} 1 & 0 \\ 0 & 1 \end{bmatrix} ; \quad V_{kj} = k \begin{bmatrix} 1 & -\frac{h}{k} \\ -\frac{h}{k} & 1 \end{bmatrix}$$

Putting $m = 1$, $k = 1$, the eigen frequencies are given by $| V_{kj} - \omega^2 T_{kj} | = 0$

i.e. $\begin{vmatrix} 1 - \omega^2 & -\frac{h}{k} \\ -\frac{h}{k} & 1 - \omega^2 \end{vmatrix} = 0$

or, $(1-\omega^2)^2 - \dfrac{h^2}{k^2} = 0$ or, $1-\omega^2 = \pm\, h/k$

or, $\omega^2 = 1 - \dfrac{h}{k}$; $1 + \dfrac{h}{k}$ i.e. $\dfrac{k}{m}\left(1-\dfrac{h}{k}\right), \dfrac{k}{m}\left(1+\dfrac{h}{k}\right)$ in conventional units

$$\therefore\ \omega = \sqrt{\frac{k}{m}\left(1-\frac{h}{k}\right)},\quad \sqrt{\frac{k}{m}\left(1+\frac{h}{k}\right)}$$

(c) Putting $q_1 = x_1 + x_2$ so that $x_1 = \dfrac{q_1+q_2}{2}$

and $q_2 = x_1 - x_2$ and $x_2 = \dfrac{q_1-q_2}{2}$

We have $\dot{q}_1 = \dot{x}_1 + \dot{x}_2$

and $\dot{q}_2 = \dot{x}_1 - \dot{x}_2$

Hence $T = \dfrac{1}{2} m \left\{ (\dot{x}_1+\dot{x}_2)^2 + (\dot{x}_1-\dot{x}_2)^2 \right\} = m(\dot{x}_1^2+\dot{x}_2^2)$

$$V = \frac{1}{2} k \left\{ (x_1+x_2)^2 + (x_1-x_2)^2 \right\} - h(x_1+x_2)(x_1-x_2)$$

$$= k(x_1^2+x_2^2) - h(x_1^2-x_2^2) = (k-h)\,x_1^2 + (k+h)\,x_2^2$$

Since the expressions for K.E. and P.E. does not contain any cross product term, the co-ordinates x_1 and x_2 are normal.

$$L = T - V = m(\dot{x}_1^2+\dot{x}_2^2) - (k-h)\,x_1^2 - (k+h)\,x_2^2$$

Corresponding Lagrange's equations are—

$$\frac{d}{dt}\left(\frac{\partial L}{\partial \dot{x}_1}\right) - \frac{\partial L}{\partial x_1} = 0$$

i.e. $\dfrac{d}{dt}(2m\,\dot{x}_1) - 2(k-h)\,x_1 = 0$

or, $m\ddot{x}_1 + (k-h)\,x_1 = 0$

and $\dfrac{d}{dt}\left(\dfrac{\partial L}{\partial \dot{x}_2}\right) - \dfrac{\partial L}{\partial x_2} = 0$

i.e. $\dfrac{d}{dt}(2m\,\dot{x}_2) + 2(k+h)\,x_2 = 0$

or, $m\ddot{x}_2 + (k+h)\,x_2 = 0$

EXERCISE

1. A particle of mass m moves under the influence of gravity on the inner surface of a paraboloid of revolution $x^2 + y^2 = az$ which is assumed to be frictionless. Obtain the equations of motion.

Using cylindrical co-ordinates (ρ, φ, z)

$$L = \frac{1}{2} m (\dot{\rho}^2 + \rho^2 \dot{\varphi}^2 + \dot{z}^2) - mgz \quad \ldots (1)$$

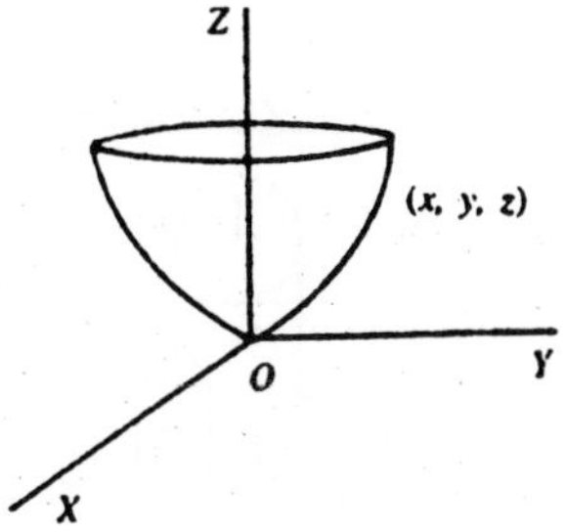

Fig. 11.3

Since $x^2 + y^2 = \rho^2$, the condition of constraint is $\rho^2 - az = 0$

Differentiating $2\rho\, d\rho - a\, dz = 0 \quad \ldots (2)$

Lagrange's equation can be written as

$$\frac{d}{dt}\left(\frac{\partial L}{\partial \dot{\rho}}\right) - \frac{\partial L}{\partial \rho} = \lambda_1 2\rho$$

$$\text{or,} \quad \frac{d}{dt}(m\dot{\rho}) - m\rho\dot{\varphi}^2 = 2\lambda_1 \rho \quad \ldots (a)$$

$$\frac{d}{dt}\left(\frac{\partial L}{\partial \dot{\varphi}}\right) - \frac{\partial L}{\partial \varphi} = 0 \quad \ldots (b)$$

$$\text{and} \quad \frac{d}{dt}\left(\frac{\partial L}{\partial \dot{z}}\right) - \frac{\partial L}{\partial r} = -\lambda_1 a \quad \ldots (c)$$

Simplifying $m(\ddot{\rho} - \rho\dot{\varphi}^2) = 2\lambda_1 \rho \quad \ldots (d)$

$$m \frac{d}{dt}(\rho^2 \dot{\varphi}) = 0 \quad \ldots (e)$$

$$m\ddot{z} = -mg - \lambda_1 a \quad \ldots (f)$$

We have also the constraint condition $2\rho\dot{\rho} - a\dot{z} = 0. \quad \ldots (g)$

2. Show from Lagrange's equations, that the orbit of a particle in a central force field is coplanner.

3. A body is sliding down an inclined plane which is moving horizontally with constant velocity. Find the position of the body as a function of time.

4. A particle of mass m moves without friction on a straight wire inclined at angle θ with the vertical axis. If the wire rotates about the vertical axis with constant angular velocity ω, form the Lagrange's equation of motion and its solution.

5. Find the acceleration of the 5 gm mass in the pulley system given in the figure 11.4.

[**Ans.** 71g/622]

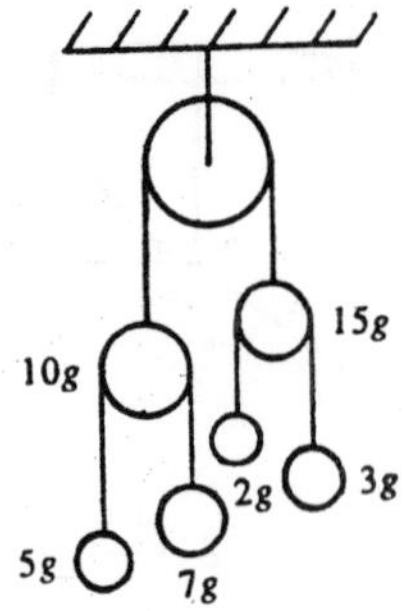

Fig. 11.4

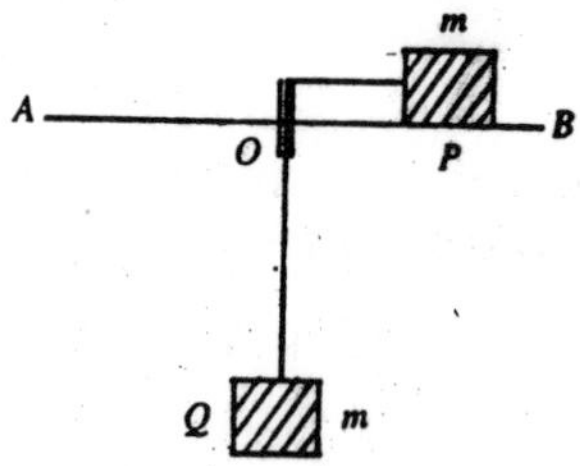

Fig. 11.5

6. In the figure 11.5 AB represents a frictionless plane having a small opening at O. A string of length l which passes through O has at its ends a particle P of mass m and a particle Q of equal mass which hanges freely. The particle P is given an initial velocity of magnitude v_0 at right angles to the string oP ($oP = a$).

Let r be the instanteneous distance of OP while θ is the angle between OP and some fixed line through O.

(a) Set up the Lagrangian of the system.

(b) Write a differential equation for P in terms of r.

(c) Find the speed of P to any position.

[**Ans.** (a) $L = \frac{1}{2} m\,(2\dot{r}^2 + r^2\dot{\theta}^2) + mg\,(l - r)$

(b) $\ddot{r} = a^2 v_0{}^2/r^2 - g$

(c) $\dot{r} = \sqrt{2av_0{}^2 + 2g\,(a - r) - 2a^2 v_0{}^2/r}$]

7. If a particle of mass m and charge e moves with velocity v in the electric field $\vec{E}$ and magnetie field $\vec{B}$, the force acting on it is given by $\vec{F} = e\,(\vec{E} + \vec{v} \times \vec{B})$.

In terms of a scalar potential φ and vector potential $\vec{A}$ the field can be expressed by the relations $\vec{E} = -\vec{\nabla}\varphi - \frac{\partial A}{\partial t}$, $\vec{B} = \vec{\nabla} \times \vec{A}$.

Prove that the Lagrangian defining the motion of the particle is $L = \frac{1}{2} mv^2 + e(\vec{A} + \vec{v}) - e\varphi$.

8. Discuss the sliding of a bead on a uniformly rotating wire in a force free field.

Hint : As shown in the figure,

Let $AP = r$

$\omega =$ angular vel. of rotation of the wire.

Then $x = r \sin\alpha \cos\omega t$

$y = r \sin\alpha \sin\omega t$

$z = h - r\cos\alpha$ when $h = oA$

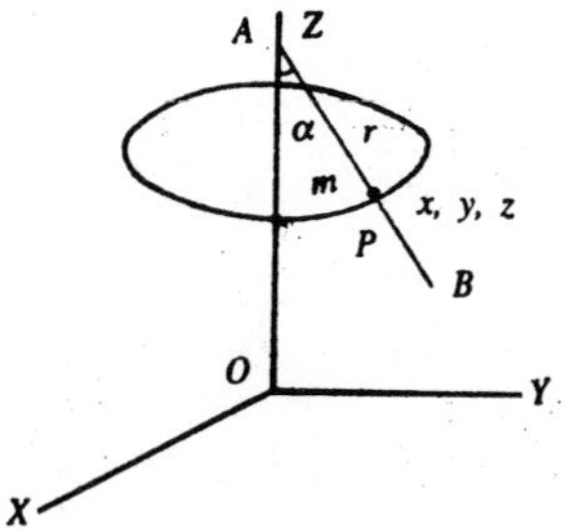

Fig. 11.6

$$T = \frac{1}{2} m (\dot{x}^2 + \dot{y}^2 + \dot{z}^2),$$

$$= \frac{1}{2} m (\dot{r}^2 + \omega^2 r^2 \sin^2\alpha)$$

$$V = mg (h - r\cos\alpha)$$

Then $L = T - V =$ etc. $\quad \frac{\partial}{\partial t}\left(\frac{\partial L}{\partial \dot{r}}\right)$ etc.

9. A sphere of radius a and mass m rests on the top of a fixed sphere of radius b. The first sphere is slightly displaced so that it rolls without slipping down the sphere. Obtain the Lagrangian and the equation of motion of the rolling sphere.

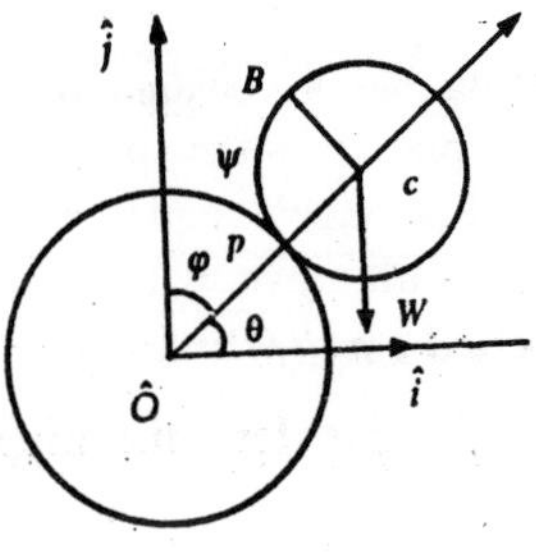

Fig. 11.7

Hint : Let us choose φ and ψ as generalised co-ordinates.

Since the sphere of radius $CP = a$ rolls without slipping on the sphere of radius $OP = b$,

we have $\quad b\frac{d\varphi}{dt} = a\frac{d\psi}{dt}$ i.e.

$b\varphi = a\psi$.

The K.E. of the rolling sphere is

$$T = \frac{1}{2} m\,(a+b)^2\,\dot{\varphi}^2 + \frac{1}{2} I\omega^2$$

$$= \frac{1}{2} m\,(a+b)^2\,\dot{\varphi}^2 + \frac{1}{2}\left(\frac{2}{5} ma^2\right)(\dot{\varphi}+\dot{\psi})^2.$$

Taking the horizontal plane through P as datum level the P.E. is

$$V = mg\,(a+b)\cos\varphi$$

Hence $L = T - V =$ etc.

Since $b\,\delta\varphi = a\,\delta\psi$

we have

$$\frac{d}{dt}\left(\frac{\partial L}{\partial \dot{\varphi}}\right) - \frac{\partial L}{\partial \varphi} = \lambda_1 b$$

and $$\frac{d}{dt}\left(\frac{\partial L}{\partial \dot{\psi}}\right) - \frac{\partial L}{\partial \psi} = -\lambda_1 a$$

i.e. $m\,(a+b)^2\,\ddot{\varphi} + \frac{2}{5} ma^2\,(\ddot{\varphi}+\ddot{\psi}) - mg\,(a+b)\sin\varphi = \lambda_1 b$

and $\frac{2}{5} ma^2\,(\ddot{\varphi}+\ddot{\psi}) = -\lambda_1 a$

10. The point of support of a simple pendulum of l and mass m is moved along a vertical line according to the equation

$$Y = y\,(t).$$

The motion of the pendulum is confined in a vertical plane.

(a) Show that the K.E. of the pendulum is given by

$$T = \frac{1}{2} m\,(l\,\dot{\theta})^2 + \frac{1}{2} m\,\dot{y}^2 + ml\,\dot{y}\,\dot{\theta}\,\sin\theta.$$

Derive the Lagrangian of the pendulum.

11. A vertical spring has a constant K and mass M. If a mass m is placed on the spring and set into motion, use Lagrange's equations to prove that the system will move with simple harmonic motion of period $2\pi\sqrt{(M+3m)/3k}$.

12. Set up the Lagrangian for the tripple pendulum shown.

Hint : $x_1 = l\sin\theta_1,\ y_1 = l\cos\theta_1$

$x_2 = l\sin\theta_1 + l\sin\theta_2,\ y_2 = l\cos\theta_1 + l\cos\theta_2$

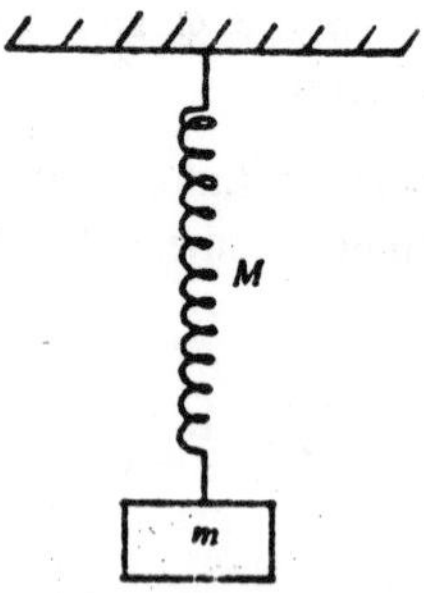

Fig. 11.8

$$x_3 = l\sin\theta_1 + l\sin\theta_2 + l\sin\theta_3$$
$$y_3 = l\cos\theta_1 + l\cos\theta_2 + l\cos\theta_3$$
$$T = \frac{1}{2}m(\dot{x}_1^2 + \dot{y}_1^2) + \frac{1}{2}m(\dot{x}_2^2 + \dot{y}_2^2) + \frac{1}{2}m(\dot{x}_3^2 + \dot{y}_3^2)$$
$$V = -mgl\cos\theta_1 - mgl(\cos\theta_1 + \cos\theta_2) - mgl(\cos\theta_1 + \cos\theta_2 + \cos\theta_3)$$

Form three Lagrange's equations with θ_1, θ_2 and θ_3.

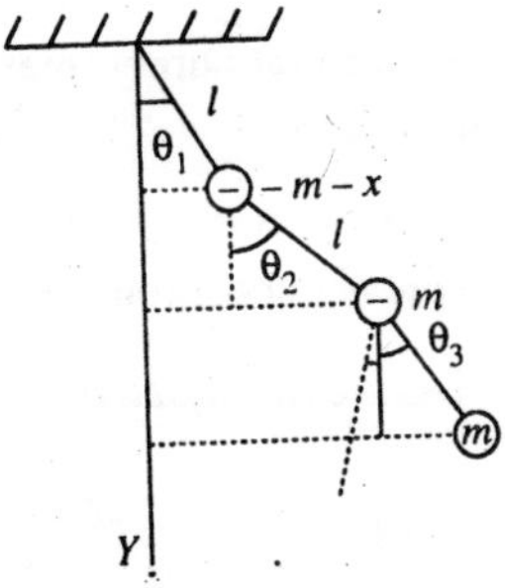

Fig. 11.9

13. In the figure 11.10, AB and BC represent two uniform rods having same length l and mass M smoothly hinged at B and at rest in a frictionless horizontal frame. An impulse is applied at C normal to BC in the direction indicated by the arrow, so that the normal velocity of the point C is v_0. Find the initial velocities of A and B and the magnitudes of the initial angular velocities of AB and BC about the centres of mass.

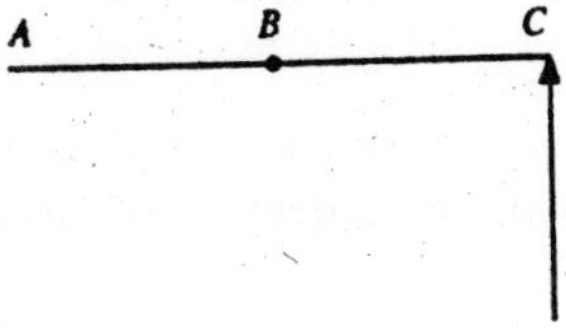

Fig. 11.10

Ans. (a) $v_0/7, -2v_0/7$, (b) $3v_0/7l, -9v_0/7l$.

14. Set up the Lagrangian for a particle of mass m falling freely in a uniform gravitational field and write Lagranges equations.

Hint : K.E. $= \frac{1}{2}m\dot{x}^2$

P.E. $= mg(h - x)$

$$L = \frac{1}{2}m\dot{x}^2 - mg(h - x)$$
$$\frac{d}{dt}\left(\frac{\partial L}{\partial \dot{x}}\right) - \frac{\partial L}{\partial x} = 0$$
or, $$\frac{d}{dt}(m\dot{x}) - mg = 0$$
or, $$\ddot{x} = g$$

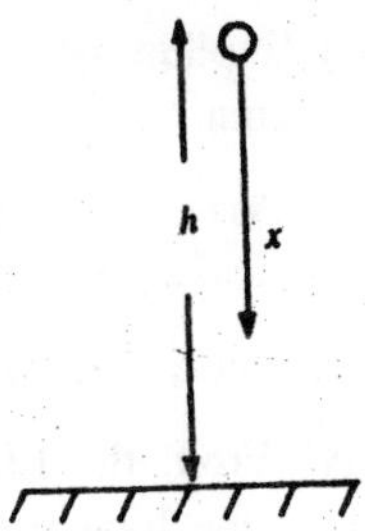

Fig. 11.11

15. (a) Explain the significance of the following equation $\vec{R} = \frac{\Sigma \vec{m}_i \vec{r}_i}{\Sigma m_i}$.

Prove that purely internal force have no effect on the motion of the centre of mass.

Ans. $\vec{R}$ is the position vector of the centre of mass ; m_i = mass of the ith particle $\vec{r_i}$ = position vector of the ith particle of the system

Derive the equation

$$M\frac{d^2\vec{R}}{dt^2} = \vec{F}(e),$$

This equation shows that the motion of the centre of mass is only affected by the external force $\vec{F}(e)$.

(b) The centre of mass of a dumbell (two masses connected by a weightless rod of length l) is constrained to move on a circle of radius R. Find the number of degrees of freedom.

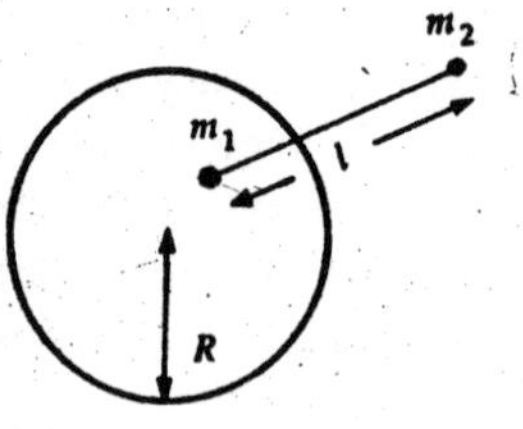

Fig. 11.12

Ans. Four. For two particles in space joined by a rod, the number of degrees of freedom $= 3 \times 2 - 1 = 5$.

Since the centre of mass is fixed in a circle, the number will be $5 - 1 = 4$.

(c) A particle of mass m moves in the gravitational field of a body of mass M. The gravitational force on the particle is $-\frac{GMm}{r^2}$ and is always towards M.

(i) Explain why you can use plane polar co-ordinates for writing the Lagrangian.

(ii) Write the Lagrangian. Is there any cyclic Co-ordinate ? If so what is the physical significance of the corresponding conserved quantity ?

(iii) Show that the radius vector traverses equal areas in equal times.

(iv) From the Lagrangian obtain this Hamiltonian. Is the Hamiltonian equal to the total energy ? Is the equality of the Hamiltonian and the total energy valid in general ?

Ans. (i) Planetary motion is uniplanner.

(ii) & (iii) See the problem of planetary motion.

(iv) Hamiltonian $= T + V = \frac{1}{2}m(\dot{r}^2 + r^2\dot{\theta}^2) + \frac{GM}{r}$

No. Only in the case of a conservative systems $H = T + V$.

[C. U. (Hons.) 1995]

16. (a) A rectangular thin plate of dimensions a and $2a$ spins about an axis along the diagonal at a constant angular velocity ω

given $I_x = \frac{1}{12} ma^2$; $I_y = \frac{1}{3} ma^2$.

(i) What is the direction and magnitude of the angular momentum ?

(ii) Calculate the torque on the axis of rotation.

(b) Perturbations in the molecular spins due to Coriolis forces appear as interactions between the rotational and vibrations levels of the molecule. —Explain.

(c) Discuss briefly Focault's pendulum experiment. Why is this experiment is famous in the history of Physics ?

Ans. (a) (i) $I_x = \frac{1}{12} ma^2$

$$I_y = \frac{1}{12} m(2a)^2 = \frac{1}{3} ma^2$$

$$I_z = I_x + I_y = \frac{1}{12} ma^2 + \frac{1}{3} ma^2$$

$$\omega_x = \omega \cos\theta = \omega \frac{a/2}{\sqrt{\frac{a^2}{4} + a^2}}$$

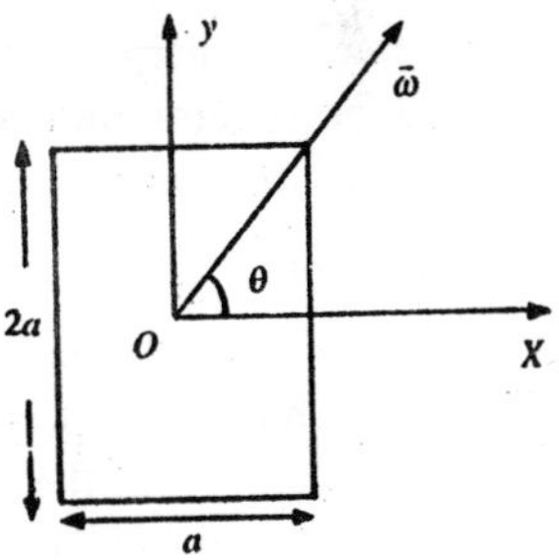

Fig. 11.13

$$= \omega \frac{a/2}{\sqrt{\frac{5a^2}{4}}} = \frac{\omega}{\sqrt{5}}$$

$$\omega_j = \omega \sin\theta = \frac{\omega}{\sqrt{5}} \frac{a}{\frac{a}{2}} = \frac{2\omega}{\sqrt{5}}, \quad \omega_2 = 0.$$

Angular momentum components are

$$L_x = I_x \omega_x = \frac{1}{12} ma^2 \frac{\omega}{\sqrt{5}} = \frac{ma^2\omega}{12\sqrt{5}}$$

$$L_y = I_y \omega_y = \frac{1}{3} ma^2 \frac{2\omega}{\sqrt{5}} = \frac{2ma^2\omega}{3\sqrt{5}}$$

$$L_z = 0$$

$$\therefore\ L = \sqrt{Lx^2 + Ly^2} = \sqrt{\frac{m^2 a^4 \omega^2}{144 \times 5} + \frac{4m^2 a^4 \omega^2}{9 \times 5}} = ma^2 \omega \sqrt{\frac{1 + 64}{144 \times 5}}$$

$$= ma^2 \omega \sqrt{\frac{\not{65}\ 13}{144 \times \not{5}}} = \frac{\sqrt{13}\ ma^2 \omega}{12}.$$

If the direction of the angular momentum makes an angle ψ with x-axis

$$\tan \psi = \frac{L_y}{L_x} = \frac{2}{3\sqrt{5}} \times 12\sqrt{5} = 8$$

$$\therefore\ \psi = \tan^{-1} 8$$

(ii) Torque $(T) = \dfrac{dL}{dt} = \dfrac{d}{dt}\left\{\dfrac{\sqrt{13}\ ma^2 \omega}{12}\right\} = 0$ **[C. U. (Hons.) 1995]**

(b) In polyatomic molecules two types of motions may occur symultaneously—the molecule rotates as a rigid body as a whole ; and the atoms vibrate about their mean positions. As a result of these vibrations the atoms are in motion relative to the molecule which form a rotating co-ordinate system. The consequent coriolis force will cause the atoms to move in directions perpendicular to the original vibrations. Perturbations in the molecular spectra due to coriolis forces thus appear as interactions between the rotation and vibrational levels of the molecules.

(c) See book.

17. (a) Two similar springs of constant 'k' hang vertically downward from a rigid support with two equal masses 'm' attached to them as shown in the diagram. Find the normal mode frequencies of small oscillation along the vertical direction only.

[C. U. (Hons.) 1995]

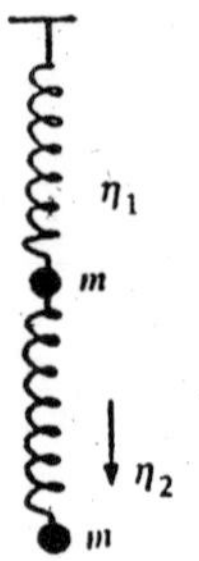

Fig. 11.14

Ans. Let η_1 and η_2 be the instanteneous extensions of the springs as shown.

The instanteneous kinetic and potential energies of the system are given by

$$T = \frac{1}{2} m \dot{\eta}_1^2 + \frac{1}{2} m\ (\dot{\eta}_1 + \dot{\eta}_2)^2 = \frac{1}{2} m\ \{2\dot{\eta}_1^2 + \dot{\eta}_2^2 + 2\dot{\eta}_1 \dot{\eta}_2\}$$

$$= \frac{1}{2}\ m \begin{bmatrix} 2\dot{\eta}_1 & \dot{\eta}_1 \\ \dot{\eta}_2 & \dot{\eta}_2 \end{bmatrix} \begin{bmatrix} \dot{\eta}_1 \\ \dot{\eta}_2 \end{bmatrix}.$$

$$V = \frac{1}{2} k\, \eta_1^2 + \frac{1}{2} k\, \eta_2^2 = \frac{k}{2} \begin{bmatrix} \eta_1 & 0 \\ 0 & \eta_2 \end{bmatrix} \begin{bmatrix} \eta_1 \\ \eta_2 \end{bmatrix}$$

The matrices T_{kj} and V_{kj} are

$$T_{kj} = \begin{bmatrix} 2m & m \\ m & m \end{bmatrix} ; \quad V_{kj} = \begin{bmatrix} K & 0 \\ 0 & K \end{bmatrix}$$

The normal frequencies are given by

$$| V_{kj} - \omega^2 T_{kj} | = 0$$

i.e. $\begin{vmatrix} k - 2m\omega^2 & -\omega^2 m \\ -\omega^2 m & k - \omega^2 m \end{vmatrix} = 0$

or, $(k - 2m\omega^2)(k - m\omega^2) - m^2\omega^4 = 0$

or, $k^2 - 2mk\omega^2 - mk\omega^2 + 2m^2\omega^4 - m^2\omega^4 = 0$

or, $k^2 - 3mk\omega^2 + m^2\omega^4 = 0$

or, $\omega^2 = \dfrac{3mk \pm \sqrt{9m^2k^2 - 4m^2k^2}}{2m^2}$

$$= \frac{3mk \pm mk\sqrt{5}}{2m^2} = \frac{k(3 \pm \sqrt{5})}{2m}$$

$$\therefore \omega = \sqrt{\frac{k}{2m}(3 \pm \sqrt{5})}$$

(b) Show that the Lagrangian equations of motion remain unchanged in form of you add to the Lagrangian a total time derivative of a function of generalised co-ordinates and time. **[C. U. (Hons.) 1995]**

Ans. Let us consider the composite Lagrangian—

$$L'(q, \dot{q}, t) = L(q, \dot{q}, t) + \frac{d}{dt}\varphi(q, t)$$

Now $\displaystyle\int_{t_1}^{t_2} L'(q, \dot{q}, t)\, dt$

$$= \int_{t_1}^{t_2} L(q, \dot{q}, t)\, dt + \int_{t_1}^{t_2} \frac{d}{dt}\varphi(q, t)\, dt$$

$$= \int_{t_1}^{t_2} L(q, \dot{q}, t)\, dt + \Big[\varphi(q, t)\Big]_{t_1}^{t_2}$$

$$= \int_{t_1}^{t_2} L(q, \dot{q}, t)\, dt + \varphi(q_2, t_2) - \varphi(q_1, t_1)$$

Hence $\delta\left\{\displaystyle\int_{t_1}^{t_2} L'(q, \dot{q}, t)\right\} - \delta\left\{\displaystyle\int_{t_1}^{t_2} L(q, \dot{q}, t) = 0\right\}$

indentically, since $\delta q^{(1)} = \delta q^{(2)} = 0$

Therefore $\delta L'(q, \dot{q}, t) = 0$ and $\delta L(q, \dot{q}, t) = 0$ lead to identical Lagrange's equations.

PART II

GENERAL PROPERTIES OF MATTER

CHAPTER I

ELASTICITY

Small deformations; Hooke's law; Interrelations of elastic constants for an isotropic solid; Torsional rigidity; Bending moment and shearing forces, Cantilever, Beams supported at the ends; Strain energy relations.

1.1. Fundamental definition

The parts of a physical body are not absolutely rigidly connected, but suffer small displacements relative to each other under the action of balanced external forces or couples. In the absence of any external force the constituent particles are at such distances apart that the resultant force on each particle is nil. This distance is called neutral distance. Any change of this distance caused by external forces will call into play either an attractive or a repulsive force between the particles. The net result of these forces balances the external force.

The forces of reaction, called into play are called stresses. Stress is measured as applied force per unit area.

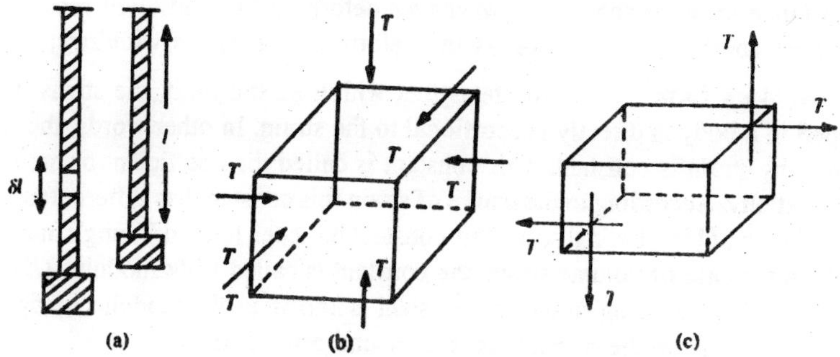

Fig. 1.1. (a) Longitudinal strain, (b) Volume strain and (c) Shearing strain.

The deformation produced whithin the body is known as strain. Strain is measured as deformation per unit dimension and has different names according to the nature of deformation. Thus when the deformation consists of a change of length, the strain is referred to as the longitudinal strain and is measured as change of length per unit length. Hence longitudinal or tensile strain $= \dfrac{\delta l}{l}$.

When the deformation consists of a change of volume only, the strain is referred to as the volume strain and is measured as change in volume per unit volume. Hence volume strain $= \dfrac{\delta \vartheta}{\vartheta}$.

When the deformation consists of a change of shape only, the strain is referred to as the shearing strain. Shearing strain is measured by angular deformation. The above three types of strain are shown diagrametically in Fig. 1.1.

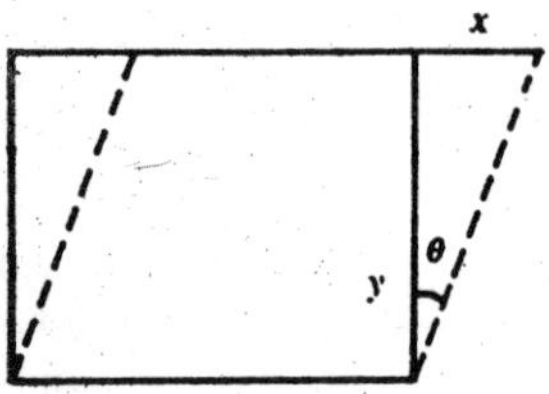

Fig. 1.2

As an example of shearing strain let us consider a square deformed into a rhombus as shown in Fig. 1.2. The angle θ measures shearing strain. For small deformation $\theta = \tan\theta = \dfrac{x}{y}$.

Elastic Limit : It is defined as the maximum stress that can be applied on a body so that on withdrawal of it the body recovers its original configuration completely.

Elasticity : It is defined as the property of matter by virtue of which it offers resistance to deformation and whenever deformed the body will recover its original configuration as soon as the deforming forces are withdrawn.

Hooke's Law : This law states that within elastic limit, the stress developed in a body is directly proportional to the strain. In other words, the ratio stress by strain is constant. This constant is called the coefficient or modulus of elasticity. According to the nature of strain this constant has different names. In the case of longitudinal strain this constant has been termed Young's modulus (Y). In the case of volume strain, the constant is called Bulk modulus (K) and in the case of shearing strain, the constant is known as the modulus of rigidity (n). Thus if F be the normal force over and area A, then

$$Y = \frac{\text{Longitudinal Stress}}{\text{Longitudinal Strain}} = \frac{F/A}{\delta l/l} \text{ and } K = \frac{\text{Volume Stress}}{\text{Volume Strain}} = \frac{F/A}{\frac{\delta v}{v}}.$$

Similarly if F be the tangential force over an area A, then $n = \dfrac{F/A}{\theta} = \dfrac{F/A}{\frac{x}{y}}$.

In general, elongational strain in any direction is accompanied by a compressional strain in a rightangular direction and vice versa. For small deformations the ratio of lateral strain to longitudinal strain is constant. This constant is called the Poission's ratio (σ). Thus for a wire length l and diameter d, if the increase in length be δl and consequent decrease in diameter be δd,

Then $\sigma = \dfrac{\delta d / d}{\delta l / l}$.

Axial Modulus (χ) : Axial modulus is defined as the ratio of longitudinal stress to longitudinal strain when there are other stresses present to prevent any change of dimension in the lateral direction.

1.2. Components of Stress and Strain

Let us consider a parallelopiped *ABCDEFGH* (Fig. 1.3) of dimensions δx, δy, δz with its six sides parallel to the axes of co-ordinates *OX, OY* and *OZ*.

If no transitional motion or rotation is to occur, the most general distribution of forces reduces to three normal stresses X_x, Y_y and Z_z and three different pairs of tangential stresses $X_y = Y_x$, $Z_x = X_z$ and $Y_z = Z_y$. The notation is such that the capital letters indicate the direction of forces and the subscripts denote the axes perpendicular to the faces across which the normal or tangential forces are acting.

[Total torque about

$$EF = X_y\ \delta y\ \delta z\ \delta x - Y_x \delta x\ \delta z\ \delta y = 0$$

$\therefore\ X_y = Y_x$ etc.]

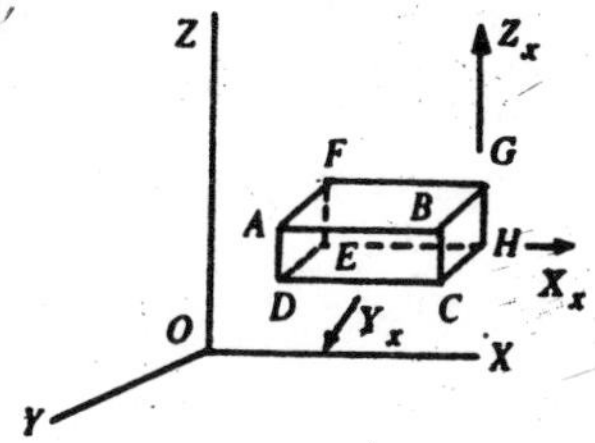

Fig. 1.3

N.B. : Force along Z direction will not produce any torque in that direction.

The strain may likewise be resolved into six components e_{xx}, e_{yy}, e_{zz}, and $e_{xy} = e_{yx}$, $e_{zx} = e_{xz}$, $e_{yz} = e_{zy}$, where the former constitute the strains produced by normal stresses and the latter by shearing stresses. Thus e_{yz} is the relative displacement of the planes perpendicular to *OY* and *OZ* respectively and initially at unit distance apart.

1.3. Interrelations between the elastic constants

(a) Relation between $y-k-\sigma$:

The Constants *Y, K, n,* σ and χ are known as elastic constants.

Since for isotropic bodies the directions of the axes of strain (the directions which remain perpendicular before and after straining) will be those of normal stresses, the most general stress at a point will be P_1, P_2, P_3 along OX, OY and OZ respectively. Hence $X_x = P_1$, $Y_y = P_2$ and $Z_z = P_3$, while $X_y = Y_z = Z_x = 0$.

The corresponding strains will be

$$e_{xx} = \frac{1}{Y}\{P_1 - \sigma(P_2 + P_3)\} \tag{1.1}$$

[Since P_2 will produce a compressional strain $\frac{P_2}{Y}\sigma$ along x and z directions etc.]

$$e_{yy} = \frac{1}{Y}\{P_2 - \sigma(P_1 + P_3)\} \tag{1.2}$$

$$e_{zz} = \frac{1}{Y}\{P_3 - \sigma(P_1 + P_2)\} \tag{1.3}$$

while $e_{xy} = e_{yz} = e_{zx} = 0$

Adding (1.1), (1.2) and (1.3)

$$Y(e_{xx} + e_{yy} + e_{zz}) = (P_1 + P_2 + P_3)(1 - 2\sigma) \tag{1.4}$$

Adding (1.2) and (1.3)

$$Y(e_{yy} + e_{zz}) = (P_2 + P_3)(1 - \sigma) - 2P_1\sigma \tag{1.5}$$

Eliminating $(P_2 + P_3)$ from (1.4) and (1.5)

$$Y(e_{xx} + e_{yy} + e_{zz}) = \left\{\frac{P_1 + Y(e_{yy} + e_{zz}) + 2P_1\sigma}{1 - \sigma}\right\}(1 - 2\sigma)$$

or, $Y(1-\sigma)(e_{xx} + e_{yy} + e_{zz}) = \{P_1(1+\sigma) + Y(e_{yy} + e_{zz})\}(1 - 2\sigma)$

or, $P_1(1+\sigma) = \dfrac{Y(1-\sigma)}{(1-2\sigma)}\left\{e_{xx} + (e_{yy} + e_{zz})\left(1 - \dfrac{1-2\sigma}{1-\sigma}\right)\right\}$

$$\therefore\ P_1 = \frac{Y(1-\sigma)}{(1-2\sigma)(1+\sigma)}\left\{e_{xx} + \frac{\sigma}{1-\sigma}(e_{yy} + e_{zz})\right\}$$

$$= \frac{1}{1-2\sigma}\left\{\frac{Ye_{xx}(1-\sigma)}{1+\sigma} + \frac{Y\sigma}{1+\sigma}(e_{yy} + e_{zz})\right\}$$

$$= \frac{1}{1-2\sigma}\left\{\frac{Y(1-2\sigma)}{1+\sigma}e_{xx} + \frac{Y\sigma}{1+\sigma}\left(e_{xx}+e_{yy}+e_{zz}\right)\right\}$$

$$= \lambda\delta + 2n'e_{xx} \tag{1.6}$$

Where $\delta = e_{xx} + e_{yy} + e_{zz}$

$$\lambda = \frac{Y\sigma}{(1+\sigma)(1-2\sigma)} \text{ and } 2n' = \frac{Y}{1+\sigma}$$

Similarly,

$$P_2 = \lambda\delta + 2n'e_{yy} \tag{1.7}$$

and $$P_3 = \lambda\delta + 2n'e_{zz} \tag{1.8}$$

Here $\delta = e_{xx} + e_{yy} + e_{zz}$ measures the dilation and to a first approximation s equal to the volume strain.

$\left[\frac{\delta v}{v} = \frac{a(1+e_{xx})\times b(1+e_{yy})\times c(1+e_{zz}) - abc}{abc}\right.$; a, b, c being the lengths of sides

of the parallelopiped $= \frac{abc + abc(e_{xx}+e_{yy}+e_{zz}) - abc}{abc} = e_{xx} + e_{yy} + e_{zz}$

neglecting small quantities of the second and higher orders.]

For a uniform compression or dilation,

$P_1 = P_2 = P_3 = P.$

Hence adding equations (1.6), (1.7) and (1.8)

$$3P = 3\lambda\delta + 2n'\delta \quad \text{or,} \quad \delta = \frac{3P}{3\lambda + 2n'} \tag{1.9}$$

But the bulk modulus $K = \frac{P}{\delta}$

Hence $$K = \frac{3\lambda + 2n'}{3}$$

$$= \frac{\frac{3Y\sigma}{(1+\sigma)(1-2\sigma)} + \frac{Y}{1+\sigma}}{3}$$

$$= \frac{3\sigma + 1 - 2\sigma}{3(1-2\sigma)} \cdot \frac{Y}{1+\sigma} = \frac{Y}{3(1-2\sigma)}$$

$$\text{or, } Y = 3k(1-2\sigma) \tag{1.10}$$

(b) Relation between $y-n-\sigma$:

If now a simple stress Q acts on the four sides of a cube $ABCDEFGH$ (Fig. 1.4) as shown, the stresses across the diagonal planes $ACHF$ and $DEGB$ will be compressive and extensive respectively and each of magnitude $\frac{\sqrt{2Q}}{\sqrt{2}.1} = Q$.

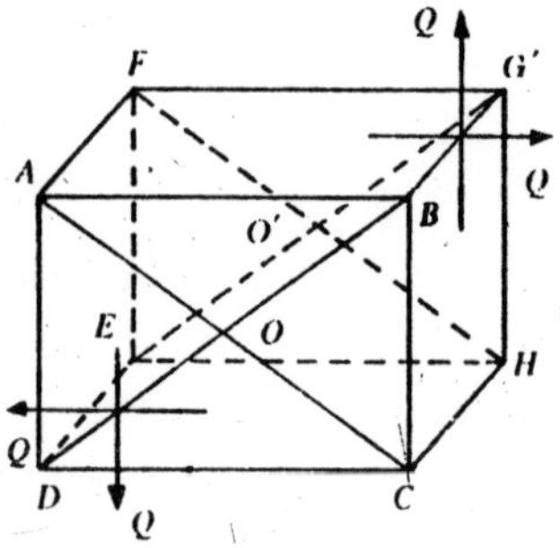

Fig. 1.4

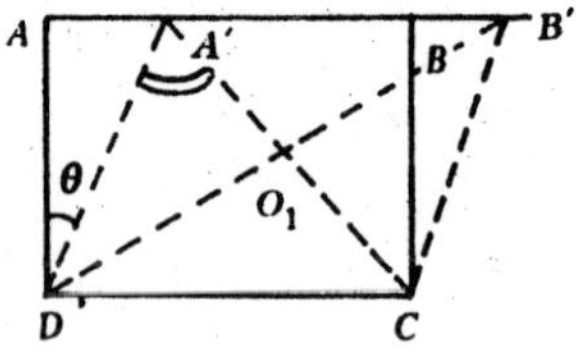

Fig. 1.5

Taking the axes OX, OY and OZ parallel to OB, OA and OO', we have $X_x = Q$, $Y_y = -Q$ and $Z_z = 0$ and the corresponding strains are

$$e_{xx} = -e_{yy} = (1+\sigma)\frac{Q}{Y} \text{ and } e_{zz} = 0.$$

Now $\tan DA'O_1 = \frac{DO_1}{A'O_1}$ (Fig. 1.5)

$$= \frac{(1+e_{xx})DO}{(1+e_{yy})AO} = \frac{1+e_{xx}}{1+e_{yy}}$$

where the dashed figures correspond to the strained cube.

The shearing strain $\theta = \angle DA'B' - \angle DAB$ (Fig. 1.5)

$$= 2\angle DA'O_1 - \frac{\pi}{2}$$

$$\text{or, } \tan\frac{\theta}{2} = \frac{\tan DA'O_1 - 1}{1 + \tan DA'O_1} = \frac{\frac{1+e_{xx}}{1+e_{yy}} - 1}{\frac{1+e_{xx}}{1+e_{yy}} + 1}.$$

$$= \frac{e_{xx} - e_{yy}}{e_{xx} + e_{yy} + 2} = \frac{e_{xx} - e_{yy}}{2} \quad \because e_{xx} \text{ etc. are small fractions.}$$

Hence for small deformation

$$\theta = e_{xx} - e_{yy} = 2(1+\sigma)\frac{Q}{Y}.$$

Now the rigidity modulus is defined by $n = \frac{Q}{\theta} = \frac{Y}{2(1+\sigma)}$

or, $Y = 2n(1+\sigma)$ (1.11)

Comparison show that $n = n'$

Eliminating (σ) between (1.10) and (1.11)

$$Y = \frac{9nk}{3k+n} \tag{1.12}$$

or, eliminating Y

$$\sigma = \frac{3k-2n}{6k+2n} \tag{1.13}$$

(c) Relation between χ, k and n :

A unit cube *ABCDEFGH* is subjected to a linear tensile strain along x-axis. To prevent the change of dimension along y and z directions stresses

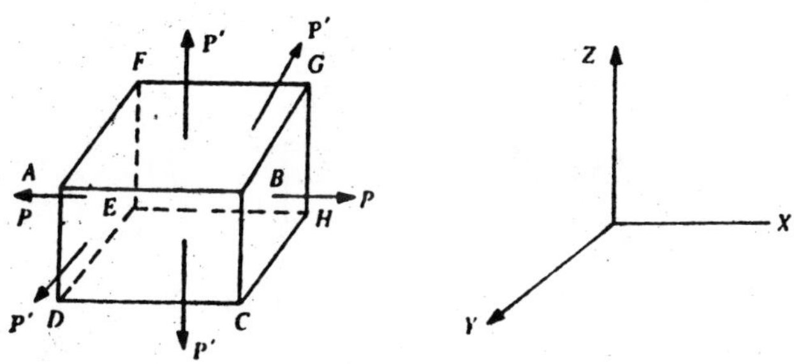

Fig. 1.6

P' are also applied along these two directions. The change of dimension along y-axis is $\frac{P'}{Y} - \sigma\frac{P'}{Y} - \frac{\sigma P}{Y}$, due to stresses P', P' and P along y, z and x-axes respectively. Since there is no change of dimension along y-axis,

$$\frac{P'}{Y}(1-\sigma) - \frac{P}{Y}\sigma = 0.$$

$$\therefore P' = \frac{P\sigma}{1-\sigma} \tag{1.14}$$

Extension along x direction is given by

$$\frac{P}{Y} - 2\frac{P'\sigma}{Y} = \frac{P}{Y} - \frac{2\sigma}{Y}\ \frac{P\sigma}{1-\sigma}$$

$$\therefore\ \chi = \frac{P}{\frac{P}{Y}\left(1 - \frac{2\sigma^2}{1-\sigma}\right)} = \frac{Y(1-\sigma)}{(1+\sigma)(1-2\sigma)}$$

$$= 3k\frac{1-\sigma}{1+\sigma} \qquad \because\ Y = 3k(1-2\sigma)$$

$$= 3k.\frac{3k+4n}{6K+2n}.\frac{6k+2n}{9k} = \frac{3k+4n}{3}$$

$$= k + \frac{4}{3}n \tag{1.15}$$

[$\because\ 3k(1-2\sigma) = 2n(1+\sigma)$

or, $3k - 2n = \sigma(2n+6k)$

$$\therefore\ \sigma = \frac{3k-2n}{2n+6k}$$

or, $1-\sigma = \dfrac{6k+2n-3k+2n}{6k+2n} = \dfrac{3k+4n}{6k+2n}$

and $1+\sigma = \dfrac{6k+2n+3k-2n}{6k+2n} = \dfrac{9k}{6k+2n}$]

1.4. Torsion of a cylinder and Torsional rigidity

A cylinder of length l, radius r and modulus of rigidity n is firmly clamped at one extremity and subjected to an axial couple at the other extremity. The relative twist between the two extremities is θ (radians). To find an expression for the moment of the internal couple developed, let us consider a thin concentric annular ring of radius x and thickness dx at the free extremity (Fig. 1.7). As a result of twist a point P on this ring is displaced to P', so that $\angle POP' = \theta$. The point Q corresponding to P in the other extremity is not at all displaced. Hence the shearing strain developed $= \dfrac{PP'}{PQ} = \dfrac{x\theta}{l}$. The corresponding shearing stress $= nx\,\theta/l$. Moment of this stress about the axis of the cylinder $= \frac{nx\,\theta}{l}x$. Magnitude of the moment calculated over the whole annular ring $= \frac{nx^2\theta}{l} 2\pi x dx$. Integrating over the whole section the total moment developed

$$= \frac{2\pi n\theta}{l}\int_0^r x^3 dx = \frac{\pi n\theta r^4}{2l}. \qquad (1.16)$$

N.B. : By measuring the applied couple, *l*, *r*, and θ, *n* can be determined.

Hence the moment developed per unit twist or the torsional rigidity $= \frac{\pi n r^4}{2l}$.

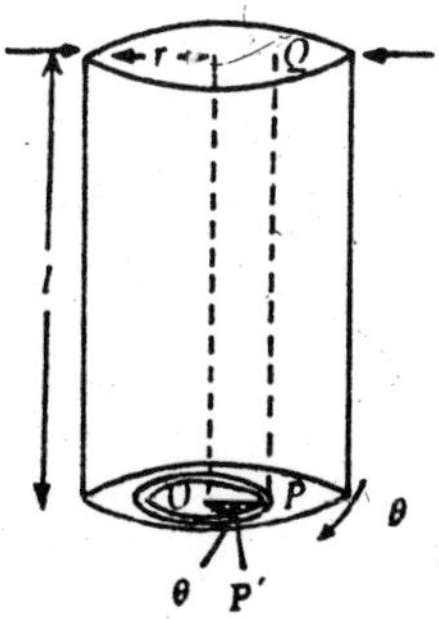

Fig. 1.7

1.5. Bending moment and shearing forces

Cantilever : A simple Cantilever consists of a beam *AB* clamped horizontally at one end *A* and subjected to a downward force (load) *W* at the other end *B*.

Let us consider the equilibrium of the portion *BC* of the beam (Fig. 1.8). The reaction at *C* due to the rest of the beam will be an upward force *W*. Thus the element *BC* is under the action of a shearing stress $W/_{bd}$ (*b* = breadth, *d* = depth of the beam) and a bending couple $W \times BC$. The beam will be bent from its natural shape and the end *B* will be depressed. But as the depression due to the bending couple is much greater than that due to the shearing stress we need consider the effect of the bending couple only. Due to bending some filaments along the length of the beam are elongated, while some other are compressed. But filaments occupying the midway region are neither elongated nor compressed and retain their unaltered length. These latter filaments constitute a surface called the neutral surface.

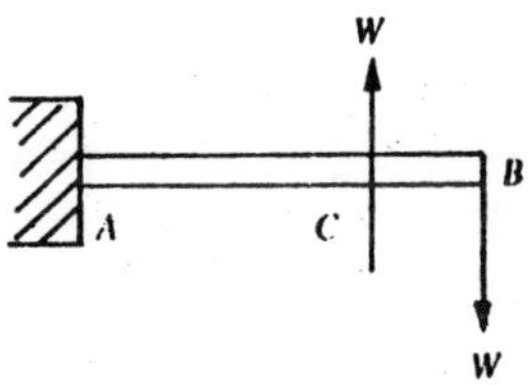

Fig. 1.8

Thus in Fig. 1.9, *PQ* is the section of the neutral surface by the plane of the paper. Filaments above *PQ* have been elongated, while those below *PQ* have been compressed. To find an expression for the moment of the internal couple developed (Internal Bending Moment), let us consider an element *EF* at a distance *Z* above the neutral surface. *QW* is drawn parallel to *EA*. Let *EW* = *s* = original length and *WF* = δs = extension of the element

Longitudinal strain of the filament $= \dfrac{\delta s}{s}$, and the longitudional stress $= Y\dfrac{\delta s}{s}$, Y being Young's modulus. If α be the area of cross-section of the filament, the stretching force on it $f = \dfrac{Y\delta s}{s}$. Moment of this force about the neutral surface $Y.\dfrac{\delta s}{s}\alpha Z$. Let O be centre of curvature and R the radius of curvature of the neutral surface. From similar triangles WFQ and PQO, $\dfrac{WF}{PQ} = \dfrac{QF}{QO}$ or, $\dfrac{\delta s}{s} = \dfrac{Z}{R}$.

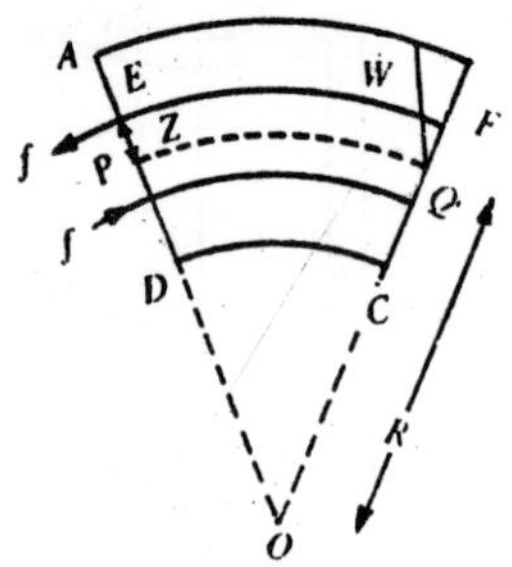

Fig. 1.9

Hence the above moment $= \dfrac{Y\alpha Z^2}{R}$.

Summing up for all the filaments (above and below the neutral surface), the total internal bending moment $= \dfrac{Y}{R}\sum \alpha z^2 = \dfrac{YAK^2}{R}$, where A = area of cross-section of the beam and K = radius of gyration of the cross-section of the beam about the neutral axis.

Calculation of the depression of the free end (For small bending of a weightless beam) :

A weightless beam of length l, breadth b and thickness d is clamped horizontally at one end A and loaded with a load $W(= Mg)$ at the other end B. Choosing the clamped end A as the origin, x-axis parallel to the undeflected position of the beam and y-axis vertically downwards, let (x, y) be co-ordinates of any point C of the beam in the deflected position. If δ be the depression of the free end B, its co-ordinates will be (l, δ) for small bending.

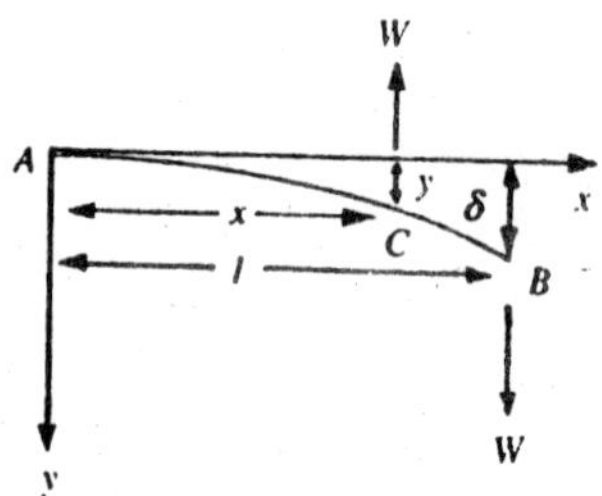

Fig. 1.10

Considering the equilibrium of the portion BC we have $\dfrac{YAK^2}{R} = W(l - x)$

Now $R = \dfrac{\left\{1+\left(\dfrac{dy}{dx}\right)^2\right\}^{3/2}}{\dfrac{d^2y}{dx^2}} \approx \dfrac{1}{\dfrac{d^2y}{dx^2}}$ for small bending.

Thus $YAK^2\dfrac{d^2y}{dx^2} = W(l - x)$ (1.17)

Integrating $w.\ r.\ t.\ x$

$$YAK^2 \frac{dy}{dx} = W\left(lx - \frac{x^2}{2}\right) + C_1$$

At $x = 0$, $\frac{dy}{dx} = 0 \quad \therefore\ C_1 = 0$

Hence $YAK^2 \frac{dy}{dx} = W\left(lx - \frac{x^2}{2}\right)$ (1.18)

Integrating again $w.\ r.\ t.\ x$

$$YAK^2 y = W\left(l\frac{x^2}{2} - \frac{x^3}{6}\right) + C_2$$

At $x = 0$, $y = 0 \quad \therefore\ C_2 = 0$

$\therefore\ YAK^2 y = W\left(l\frac{x^2}{2} - \frac{x^3}{6}\right)$ (1.19)

At the free end B, $x = l$, $y = \delta$

$$\therefore YAK^2\delta = W\left(\frac{l^3}{2} - \frac{l^3}{6}\right) = Wl^3\frac{3-1}{6} = \frac{Wl^3}{3}$$

or, $\delta = \frac{Wl^3}{3YAK^2}$ (1.20)

Since $A = bd$ and $k^2 = \frac{d^2}{12}$

$\delta = \frac{4Wl^3}{Ybd^3}$ (1.21)

When the beam is not weightless, let it have a weight ω per unit length. Equation (1.17) will be modified by taking into account the moment due to the weight $\omega(l-x)$ acting at the mid point of BC.

Thus $YAK^2 \frac{d^2y}{dx^2} = W(l-x) + \omega\left(\frac{l-x}{2}\right)^2$ (1.22)

Integrating successively twice $w.\ r.\ t.\ x$

$YAK^2 \frac{dy}{dx} = W\left(lx - \frac{x^2}{2}\right) + \frac{\omega}{2}\left\{l^2x - 2l\frac{x^2}{2} + \frac{x^3}{3}\right\}$ (1.23)

and $YAK^2 y = W\left(\frac{lx^2}{2} - \frac{x^3}{6}\right) + \frac{\omega}{2}\left\{\frac{l^2x^2}{2} - \frac{lx^3}{3} + \frac{x^4}{12}\right\}$ (1.24) since, at $x = 0, \dfrac{dy}{dx} = 0, y = 0$

Putting $x = l$ and $y = \delta$ in equation (1.24)

$$YAK^2\delta = W\left\{\frac{l^3}{2} - \frac{l^3}{6}\right\} + \omega\left\{\frac{l^4}{4} - \frac{l^4}{6} + \frac{l^4}{24}\right\}$$

$$= \frac{Wl^3}{3} + \omega l\frac{6-4+1}{24}l^3 = \frac{Wl^3}{3} + \omega l\frac{l^3}{8}$$

$$= \frac{Wl^3}{3} + \frac{W_1 l^3}{8}, \text{ where } W_1 = \omega l = \text{ Total weights of the beam}$$

or, $$\delta = \frac{l^3}{3YAK^2}\left\{W + \frac{3}{8}W_1\right\} \tag{1.25}$$

1.6. A more general method of discussing beam problems

Sign convention used. (a) upword froce + *ve*, (b) clockwise moment + *ve*.

Let us consider an element of a bent beam of length dx on the neutral axis at a distance x from some origin and let the load per unit length be ω. Let F; M; $F + dF$ and $M + dM$ be the shearing forces and bending moments at x and $x + dx$ respectively. Then for equilibrium

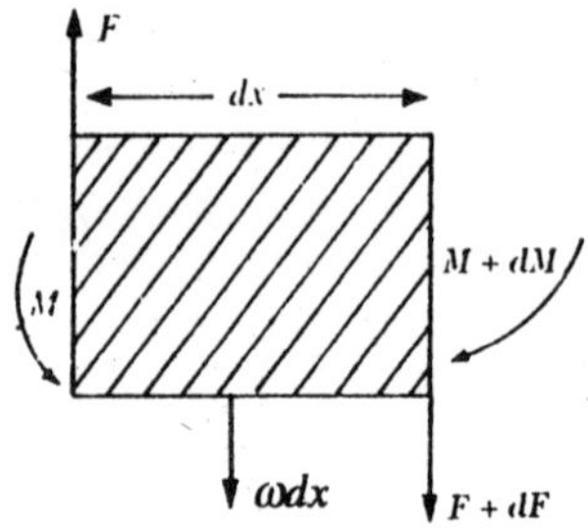

Fig. 1.11

$$F - (F + dF) - \omega dx = 0$$

or, $$\frac{dF}{dx} = -\omega. \tag{1.26}$$

Taking moment over the section at the left extreme,

$$M + dM - M + \omega dx.\frac{dx}{2} + (F + dF)dx = 0$$

or, $$\frac{dM}{dx} = -F \tag{1.27}$$

neglecting small quantities of second order.

Combining (1.26) and (1.27)

$$\omega = -\frac{dF}{dx} = \frac{d^2M}{dx^2} \tag{1.28}$$

Now for small deformation $\dfrac{1}{R} = \dfrac{d^2y}{dx^2}$

$$\therefore M = YAK^2\frac{d^2y}{dx^2} \tag{1.29}$$

Hence $\omega = YAK^2 \dfrac{d^4y}{dx^4}$ (1.30)

and $YAK^2 \dfrac{d^3y}{dx^3} = -F$ (1.31)

Applications :

(a) Weightless beam clamped at one end and loaded at the other :

(i) Variation of shearing force and bending moment :

R = reaction at the clamped end.

M_0 = Couple the wall exerts on the rod.

Considering the equilibrium of the whole beam (Fig. 1.12).

$R - W = 0$ or, $R = W$ (upwards)

and $M_0 + Wl = 0$ or, $M_0 = -Wl$ (anticlockwise)

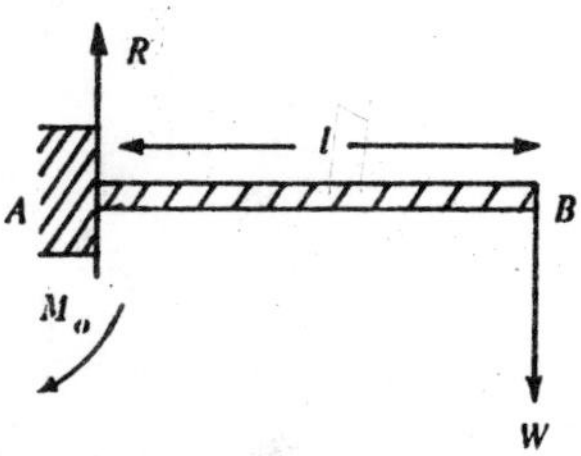

Fig. 1.12

Considering the equilibrium of the portion $AK = x$ (Fig. 1.13)

$(0 < x < l)$, *let F* be the shearing force and M the bending moment at K. We have for equilibrium.

$W + F = 0$

or, $F = -W \leftarrow$ downwards

$M - Wl - Fx = 0$, taking moment about A

or, $M = W(l - x) \leftarrow$ clockwise

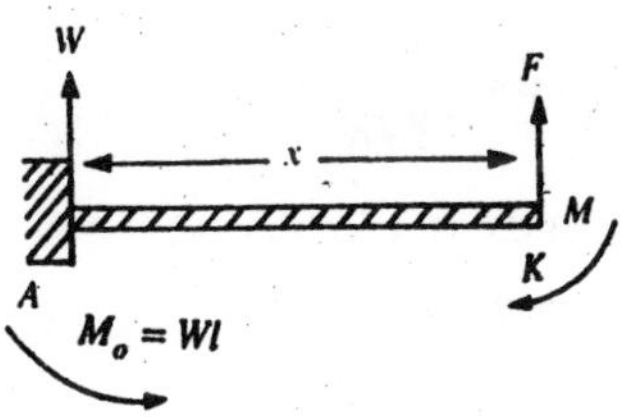

Fig. 1.13

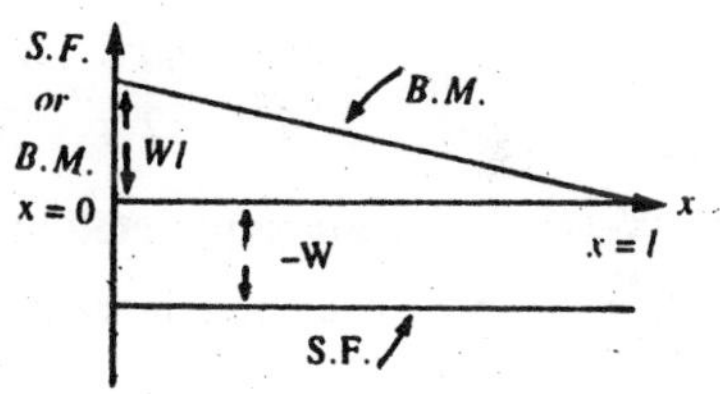

Fig. 1.14

Variation of shearing force and bending moment are shown graphically in Fig. 1.14.

(ii) Calculation of the depression at the free end :

In this case $F = -W$ and $\omega = 0$

Choosing the clamped end A as the origin, x-axis along the undeflected position of the beam and y-axis vertically downwards (Fig. 1.15), we have from equation (1.30)

$$YAK^2 \frac{d^4 y}{dx^4} = 0. \quad \text{... (a)}$$

Integrating *w. r. t. x*

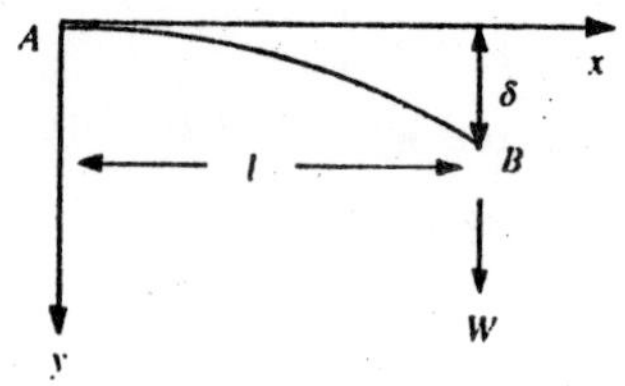

Fig. 1.15

$$YAK^2 \frac{d^3 y}{dx^3} = \text{Constant} = -W \text{ from (1.31)}$$

Integrating successively three times *w. r. t. x*

$$YAK^2 \frac{d^2 y}{dx^2} = -Wx + A \quad \text{... (b)}$$

$$YAK^2 \frac{dy}{dx} = -\frac{Wx^2}{2} + Ax + B \quad \text{... (c)}$$

$$YAK^2 y = -\frac{Wx^3}{6} + \frac{Ax^2}{2} + Bx + C \quad \text{... (d)}$$

At $x = 0$, $YAK^2 \frac{d^2 y}{dx^2} = Wl \quad \therefore A = Wl$

At $x = 0$, $y = 0$ and $\frac{dy}{dx} = 0 \quad \therefore B = C = 0$ Substituting in ...(d)

$$YAK^2 y = -\frac{Wx^3}{6} + \frac{Wlx^2}{2}$$

At B, $x = l$, $y = \delta$

Hence $YAK^2 \delta = -\frac{Wl^3}{6} + \frac{Wl^3}{2} = \frac{Wl^3}{3} \quad \therefore \delta = \frac{Wl^3}{3YAK^2}$

(b) Weightless beam supported at the ends and loaded at the middle :

(i) Variation of shearing force and bending moment :

It is evident that the reactions on the points of support are each $\frac{W}{2}$.

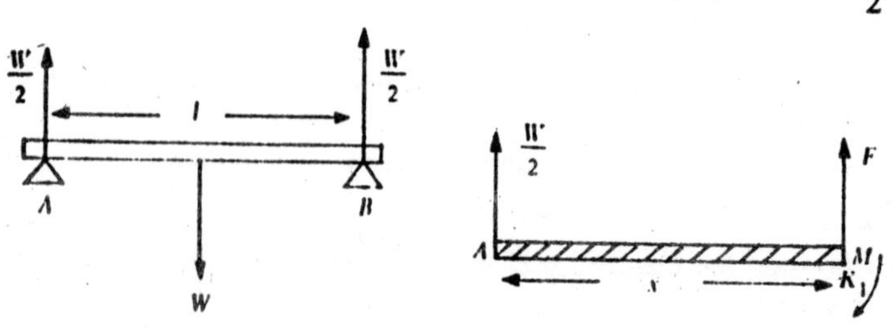

Fig. 1.16

Considering a point K_1 at a distance x from A $\left(0 < x < \frac{l}{2}\right)$, let F be the shearing force and M the bending moment at K_1.

For equilibrium, we have

$$\frac{W}{2} + F = 0 \text{ or, } F = -\frac{W}{2} \text{ (Downwards)}$$

Taking moment about A (Fig. 1.16)

$$M - Fx = 0$$

or, $M = Fx = -\frac{W}{2}x$ (anticlockwise)

Similarly if K_2 is a section at a distance x from A, where $\frac{l}{2} < x < l$, then we have for equilibrium (Fig. 1.17).

$$\frac{W}{2} + F - W = 0 \quad \text{or,} \quad F = \frac{W}{2} \text{ (upwards)}$$

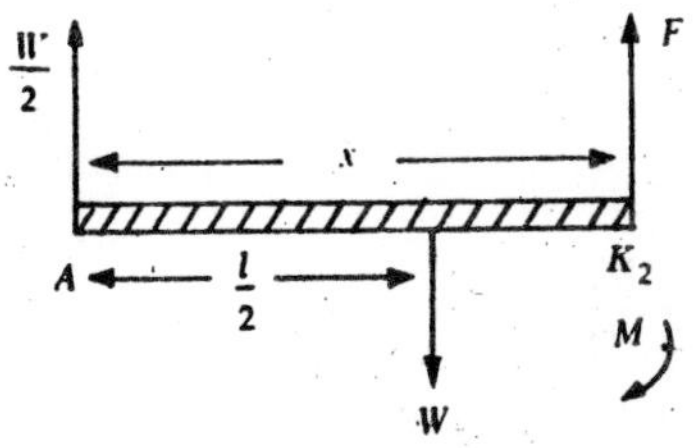

Fig. 1.17

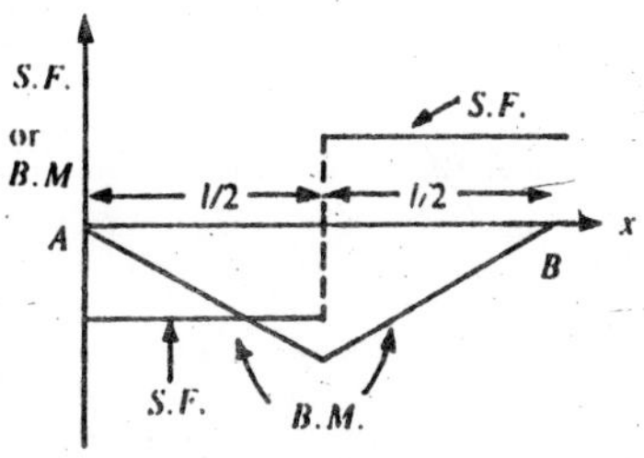

Fig. 1.18

Taking moment over the section at A, $M + \frac{Wl}{2} - Fx = 0$

$$\therefore M = -\frac{Wl}{2} + \frac{Wx}{2} = \frac{W}{2}(x - l) \text{ (Anticlockwise)}$$

Variations of shearing force and bending moment are shown graphically in Fig. 1.18.

(ii) Calculation of the depression at the mid point :

The point A is chosen as the origin. x-axis is taken along the undeflected position of the beam and y-axis vertically downwards.

Let us confine our attention to the portion AC of the beam. From equation (1.30)

$$YAK^2 \frac{d^4 y}{dx^4} = 0. \qquad \text{... (a)}$$

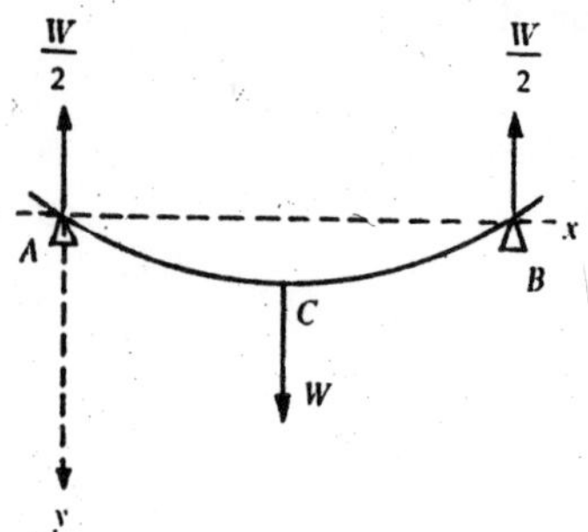

Fig. 1.19

Integrating *w. r. t. x*

$$YAK^2 \frac{d^3 y}{dx^3} = \text{Constant} = -\frac{W}{2} \qquad \text{... (b)}$$

Integrating successively thrice *w. r. t. x*

$$YAK^2 \frac{d^2 y}{dx^2} = -\frac{W}{2}x + A \qquad \text{... (c)}$$

$$YAK^2 \frac{dy}{dx} = -\frac{W}{2}\frac{x^2}{2} + Ax + B \qquad \text{... (d)}$$

$$YAK^2 y = -\frac{W}{2}\frac{x^3}{6} + \frac{Ax^2}{2} + Bx + C \qquad \text{... (e)}$$

At $x = 0$, $y = 0$. $\therefore$ $C = 0$

At $x = 0$, $YAK^2 \frac{d^2 x}{dx^2} = 0 \quad \therefore A = 0$

At $x = l/2$, $\frac{dy}{dx} = 0 \quad \therefore B = \frac{Wl^2}{16}$

Substituting in (e) $YAK^2 y = -\frac{Wx^3}{12} + \frac{Wl^2}{16}x$ ....

At the mid point $x = \frac{l}{2}$, $y = \delta$

Hence $YAK^2 \delta = -\frac{W}{12}\cdot\frac{l^3}{8} + \frac{Wl^2}{16}\cdot\frac{l}{2}$

$$\text{or, } \delta = \frac{Wl^3}{32\ YAK^2}\left(1 - \frac{1}{3}\right) = \frac{Wl^3}{48\ YAK^2}$$

N.B. : By measuring $(W = Mg)$, l, $(A = bd)$ and $K^2 = \frac{d^2}{12}$, Y can be found out.

(c) Weightless beam clamped at the ends and carrying a central load (Encastre beam) :

(i) Variation of bending moment and shearing force :

From symmetry the reacting forces at the clamped end are each equal to $\frac{W}{2}$ upwards (Fig. 1.20).

Let us consider the shearing force and bending moment at a section at K_1, where $AK_1 = x\left(0 < x < \frac{l}{2}\right)$

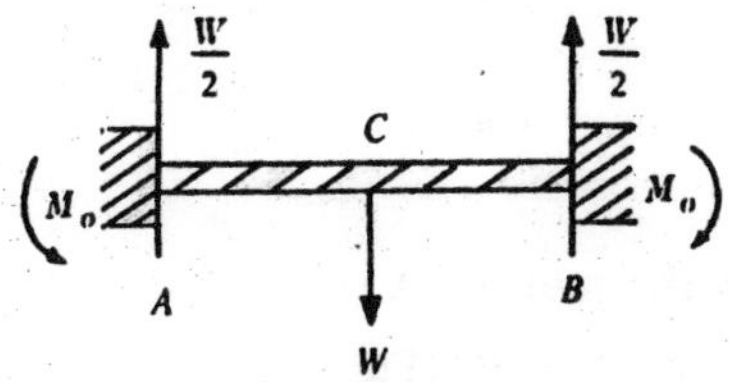

Fig. 1.20

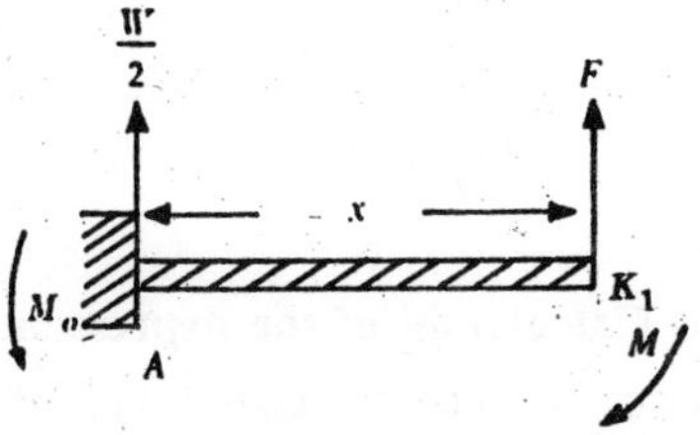

Fig. 1.21

For equilibrium,

$$\frac{W}{2} + F = 0 \quad \text{(Fig. 1.21)}$$

or, $F = -\frac{W}{2}$ (downwards)

Taking moment about A,

$$M - Fx - M_0 = 0$$

or, $M = M_0 + Fx = M_0 - \frac{Wx}{2} = \frac{Wl}{8} - \frac{Wx}{2}$ (from the next section)

$$= \frac{W}{2}\left(\frac{l}{4} - x\right)$$

Similarly if K_2 be a section at a distance x from $A\left(\frac{l}{2} < x < l\right)$ (Fig. 1.22), we have for equilibrium $\frac{W}{2} + F - W = 0$

or, $F = \frac{W}{2}$ (upwards)

Taking moment about A,

$$M + \frac{Wl}{2} - Fx - M_0 = 0$$

$$\text{or, } M = M_0 + Fx - \frac{Wl}{2} = \frac{Wl}{8} + \frac{Wx}{2} - \frac{Wl}{2} = -W\left(\frac{3l}{8} - \frac{x}{2}\right)$$

Variations of shearing force and bending moment are shown **graphically** in Fig. 1.23.

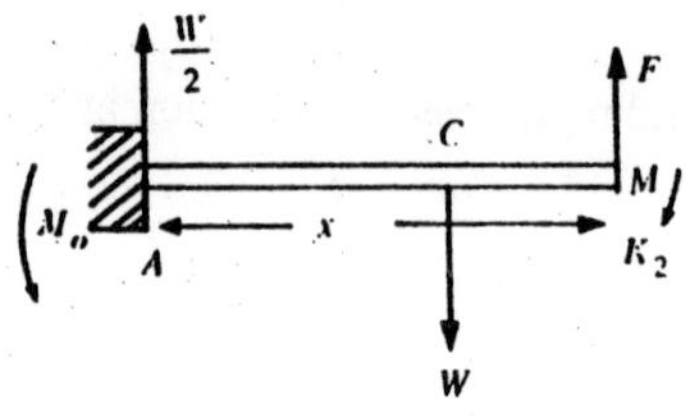

Fig. 1.22

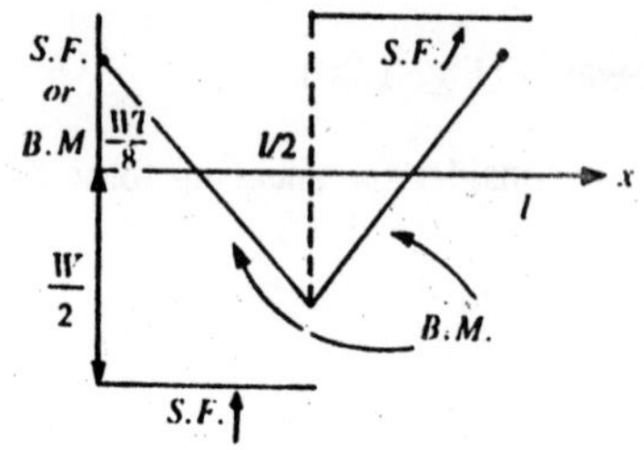

Fig. 1.23

(ii) Calculation of the depression at the mid point :

Let us consider the equilibrium of the portion AC (Fig. 1.24).

Since the beam is weightless we have from equation (1.30)

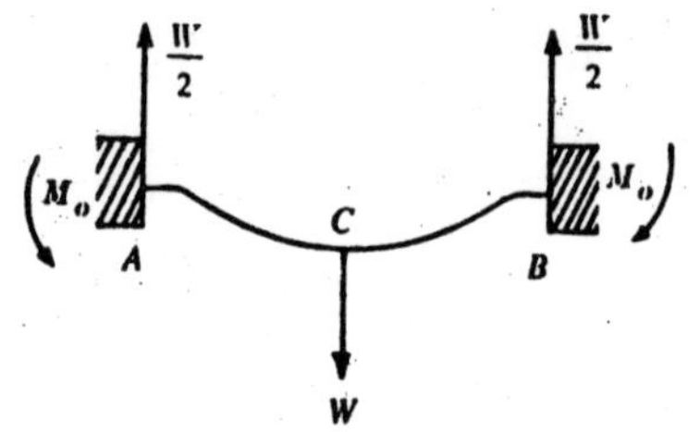

Fig. 1.24

$$YAK^2 \frac{d^4y}{dx^4} = 0. \quad \text{... (a)}$$

Integrating w. r. t. x

$$YAK^2 \frac{d^3y}{dx^3} = \text{Constant} = -\frac{W}{2} \quad \text{... (b)}$$

Integrating successively three times w. r. t. x

$$YAK^2 \frac{d^2y}{dx^2} = -\frac{W}{2}x + A \quad \text{.... (c)}$$

$$YAK^2 \frac{dy}{dx} = -\frac{W}{2}\frac{x^2}{2} + Ax + B \quad \text{.... (d)}$$

$$YAK^2 y = -\frac{W}{2}\frac{x^3}{6} + \frac{Ax^2}{2} + Bx + c \quad \text{.... (e)}$$

At $x = 0$, $y = 0$ and $\frac{dy}{dx} = 0$

$\therefore$ C = 0, B = 0

At $x=\frac{l}{2}$, $\frac{dy}{dx}=0$ $\quad\therefore\ A=\frac{W}{4}\cdot\frac{l}{2}=\frac{Wl}{8}$

$$\left[\text{ From (c) at } x=0, YAK^2\frac{d^2y}{dx^2}=M_0=\frac{Wl}{8}\right]$$

Finally substituting in (c)

$$YAK^2y=-\frac{Wx^3}{12}+\frac{Wlx^2}{16}$$

At C, $x=\frac{l}{2}$, $y=\delta$

$$\therefore YAK^2\delta=-\frac{Wl^3}{96}+\frac{Wl^3}{64}=Wl^3\frac{3-2}{192}=\frac{Wl^3}{192}$$

$$\therefore\ \delta=\frac{Wl^3}{192\,YAK^2}.$$

(d) To show that in the case of a simple cantilever the depression at any point C when the load is placed at B is the same as the depression at B when the load is placed at C.

If the load is placed at B, the depression at C is given by

$$y=\frac{W}{YAK^2}\left(\frac{lx^2}{2}-\frac{x^3}{6}\right)$$ from equation (1.19).

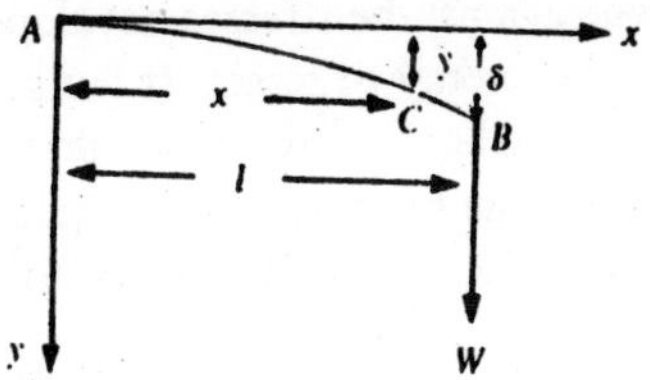

Fig. 1.25

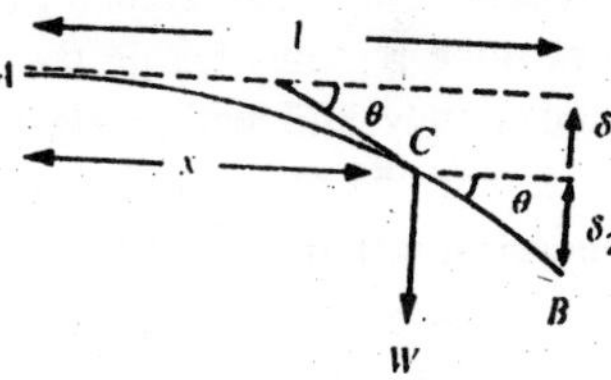

Fig. 1.26

When the load W is placed at C, the situation willl be as shown in Fig. 1.26. In this case the depression at $B=\delta_1+\delta_2$

From equation (1.20) $\delta_1=\frac{Wx^3}{3YAK^2}$

Putting $l = x$ in equation (1.18) we have

$$\tan\theta = \left(\frac{dy}{dx}\right)_{\substack{x=x\\l=x}} = \frac{W}{YAK^2}\left(x^2 - \frac{x^2}{2}\right) = \frac{Wx^2}{2YAK^2} = \frac{\delta_2}{l-x},$$

since the portion CB is straight.

Hence $\delta_2 = \dfrac{Wx^2}{2YAK^2}(1-x)$

Total deflection at $B = \delta_1 + \delta_2 = \dfrac{Wx^3}{3YAK^2} + \dfrac{Wx^2(l-x)}{2YAK^2}$

$$= \frac{W}{YAK^2}\left\{\frac{x^3}{3} + \frac{1}{2}x^2(l-x)\right\}$$

$$= \frac{W}{YAK^2}\left\{\frac{lx^2}{2} + \left(\frac{1}{3} - \frac{1}{2}\right)x^3\right\}$$

$$= \frac{W}{YAK^2}\left\{\frac{lx^2}{2} - \frac{x^3}{6}\right\} = y$$

1.7. Strain Energy

When a body is strained, stress causes a relative displacement of the different parts of the body. Hence work is done against the stresses developed. This work done remains within the body as the energy of strain. If the body is strained slowly and within elastic limit, the whole of the work done is recoverable and the process is reversible. But if the body is strained quickly or beyond elastic limit, a part of the work done is lost as heat and can not be recovered. This process is irreversible.

We now find out expressions for energy in the case of different types of strains.

(i) Energy of longitudinal strain :

Let a wire of length L and area of cross-section α be subjected to a longitudinal stress so that it undergoes a change of length l. At any intermediate stage let it have a change in length x. Instanteneous longitudinal strain produced $= \dfrac{x}{L}$. If Y be the Young's modulus of the material, the longitudinal stress $= Y\frac{x}{L}$ The stretching force $= \frac{Yx}{L}\alpha$. If the wire now suffers an additional change in length dx, the consequent work done $= \frac{Yx\alpha}{L}dx$.

Hence the total work done = energy of longitudinal strain

$$= \frac{Y\alpha}{L}\int_0^l x\,dx = \frac{Y\alpha l^2}{2L}$$

$$= \frac{1}{2}\left(Y\frac{l}{L}\right)\left(\frac{l}{L}\right)L\alpha = \frac{1}{2} \text{ stress} \times \text{strain} \times \text{volume}$$

or work done/volume $= \frac{1}{2}$ stress $\times$ strain.

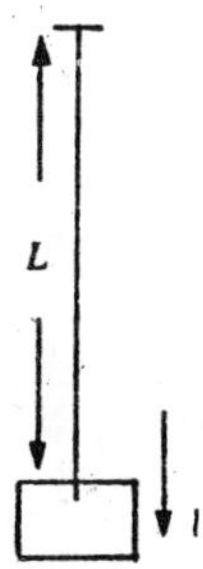

Fig. 1.27

(ii) **Energy of volume strain :**

Let a body of volume V be subjected to a uniform compressive stress so that it undergoes a change in volume by v_1. At any intermediate stage if v be the change in volume, the instanteneous volume strain produced is $\frac{v}{V}$. The corresponding volume stress $= K\frac{v}{V}$, K being the bulk modulus of the material. If body now suffers an additional change in volume δv, work done $= K\frac{v}{V}\delta v$. Hence the total work done or the

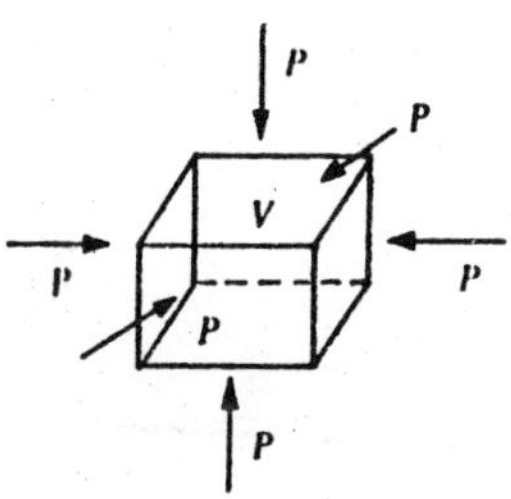

Fig. 1.28

energy of volume strain $= \frac{K}{V}\int_0^{v_1} v\,dv = \frac{K}{V}\frac{v_1^2}{2}$

$$= \frac{1}{2}\left(K\frac{v_1}{V}\right)\left(\frac{v_1}{V}\right)V = \frac{1}{2} \text{ stress} \times \text{strain} \times \text{volume}$$

So work done / volume $= \frac{1}{2}$ stress $\times$ strain.

(iii) **Energy of shearing strain :**

Let us consider a cube of length l and face area α subjected to the action of a tangential couple (Fig. 1.29). n = modulus of rigidity, θ = shearing strain. The tangential force $f = n\theta\alpha$. Work done for a small additional shear $d\theta$ is $fld\theta = \alpha n\theta l d\theta$.

Hence the total work done $= \alpha nl \int_0^\theta \theta d\theta = \frac{1}{2}\alpha nl\theta^2$

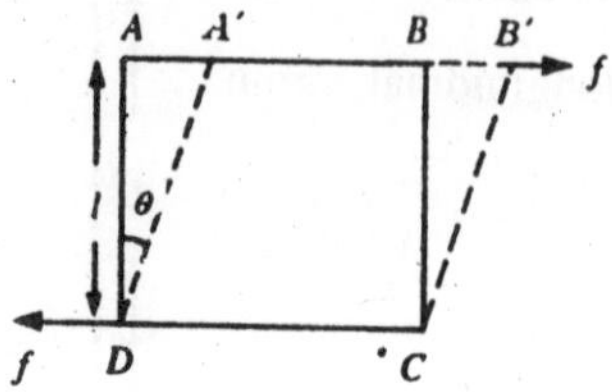

Fig. 1.29

$= \frac{1}{2}(n\theta)(\theta)l\alpha$

$= \frac{1}{2}$ stress × strain × volume

Energy / Volume $= \frac{1}{2}$ stress × strain.

(iv) **Energy of a bent beam :**

Let us consider an element of length dl of a bent beam. Let a filament of cross-section α be situated at a distance z from the neutral surface (Fig. 1.30). The work done in stretching the filament by an amount

e = force × distance = stress × area × distance

$= Y \times$ strain × area × distance $= Y\alpha\frac{e}{dl}de.$

Total work done for the filament is $\int_0^e Y\alpha\frac{e}{dl}de = \frac{1}{2}Y\alpha\frac{e^2}{dl}$

But $\frac{dl}{R} = \frac{e}{z}$ where R = radius of curvature of the neutral surface.

$\therefore\ e^2 = \frac{dl^2}{R^2}Z^2$

$\therefore$ Work done $= \frac{1}{2}Y\alpha\, dl\frac{Z^2}{R^2}$

Hence the total energy of the whole cross-section of the beam of length dl

$= \frac{1}{2}\frac{Y}{R^2}dl\sum \alpha Z^2 = \frac{1}{2}\frac{YAK^2 dl}{R^2}$

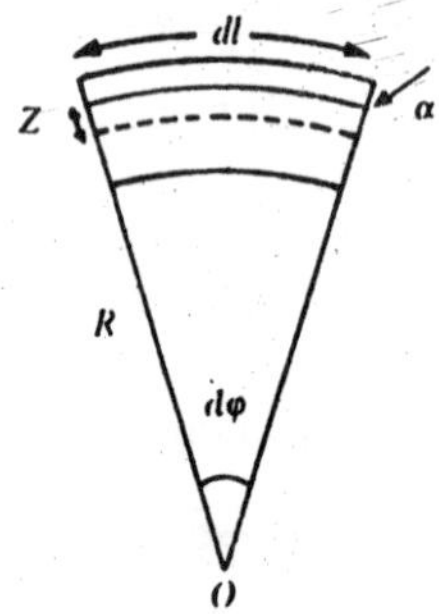

Fig. 1.30

Considering the whole of the beam the total

energy $= \int_0^l \frac{1}{2}\frac{YAK^2}{R^2}dl = \frac{1}{2}\frac{YAK^2}{R}\cdot\frac{l}{R}$

$= \frac{1}{2}C\varphi,$ where C is the internal bending moment and φ is the total bending of the beam (i.e. angle between the tangents at the extremeties of the beam).

Problems :

1. To show theoretically that σ lies between $\frac{1}{2}$ and -1

From the relation $Y = 3k(1-2\sigma)$

$$1-2\sigma = \frac{Y}{3k} = +ve > 0$$

or, $1 > 2\sigma$

or, $\sigma < \frac{1}{2}$

Again from the relation $Y = 2n(1+\sigma)$

$$1+\alpha = \frac{Y}{2n} = +ve > 0$$

or, $\sigma > -1$

Combining $-1 < \sigma < \frac{1}{2}$

2. A suspension thread of length $2l$ consists of a wire of length l and radius r rigidly joined to a second wire of equal length, of radius $2r$ and similar material. The top is clamped, while the lower end is twisted about the axis. Find (a) the torsional rigidity of the complete suspension thread, if the rigidity modulus is η and (b) the ratio of angles of twist between the ends of the two wires.

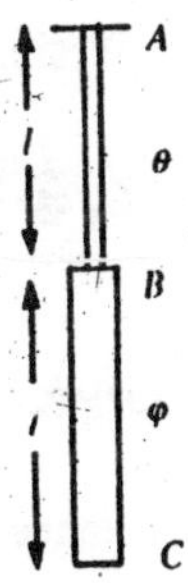

Fig. 1.31

Let θ and φ be relative twists between A and B and between B and C respectively.

At B the internal couple developed from the two sides must be same.

Hence $\frac{\pi\eta\theta r^4}{2l} = \frac{\pi\eta\varphi(2r)^4}{2l} = \tau$

$$\therefore \theta : \varphi = \frac{(2)^4}{1} = 16 : 1$$

Total twist between A and $C = \theta + \varphi = 16\varphi + \varphi = 17\varphi$

Hence the torsional rigidity of the complete system

$$= \frac{\tau}{17\varphi} = \frac{\pi\eta\varphi 16r^4}{17\varphi \times 2l} = \frac{8\pi\eta r^4}{17}.$$

3. Prove that if a number of rods are joined end to end the torsional rigidity of the combination is given by the same formula as that for a number of resistances in parallel. Hence or otherwise find the torsional rigidity of a tapering wire of length l, rigidity modulus η and end radii a and b.

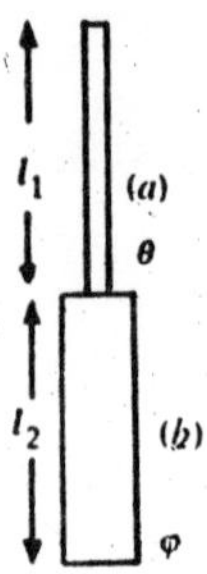

Fig. 1.32

Let us consider the case of two wires of length l_1, l_2 and radii a, and b. Let θ and φ be the relative twists between the ends of the two wires (Fig. 1.32).

At the common section,

$$\frac{\pi\eta\theta a^4}{2l_1} = \frac{\pi\eta\varphi b^4}{2l_2}$$

$$\text{or, } \theta = \left(\frac{b}{a}\right)^4 \frac{l_1}{l_2}\varphi$$

Total twist between the extremities of the combined wire

$$= \theta + \varphi = \varphi\left\{\left(\frac{b}{a}\right)^4 \frac{l_1}{l_2} + 1\right\} = \varphi\,\frac{l_1 b^4 + l_2 a^4}{l_2 a^4}$$

Hence the equivalent torsional rigidity

$$C = \frac{\tau}{\varphi\left(\dfrac{l_1 b^4 + l_2 a^4}{l_2 a^4}\right)} = \frac{\pi\eta\varphi b^4/2l_2}{\varphi\left(\dfrac{l_1 b^4 + l_2 a^4}{l_2 a^4}\right)} = \frac{\pi\eta a^4 b^4}{2(l_1 b^4 + l_2 a^4)}$$

Torsional rigidities of the individual wires are given by

$$C_1 = \frac{\pi\eta a^4}{2l_1} \text{ and } C_2 = \frac{\pi\eta b^4}{2l_2}$$

$$\text{Now } \frac{1}{C_1} + \frac{1}{C_2} = \frac{2l_1}{\pi\eta a^4} + \frac{2l_2}{\pi\eta b^4} = \frac{2}{\pi\eta}\left(\frac{l_1 b^4 + l_2 a^4}{a^4 b^4}\right) = \frac{1}{C}$$

Hence the result.

Let us consider a length δx as shown in Fig. 1.33. If y be the mean radius of cross-section at a distance x from the zero end of the wire, it can be shown that

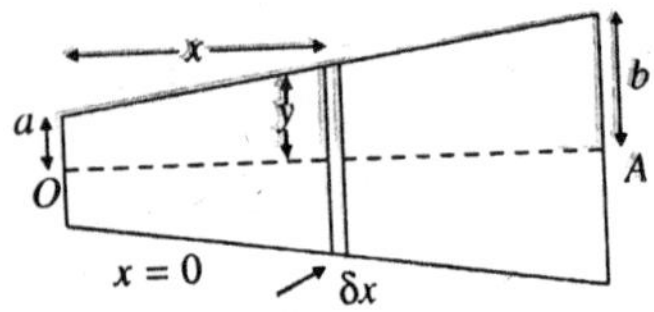

Fig. 1.33

$$y = \left(\frac{b-a}{l}\right)x + a.$$

If $\delta\theta$ be the relative twist between the two ends of the element, we have the moment developed $= \dfrac{\pi\eta y^4 \delta\theta}{2\delta x}$.

Hence the torsional rigidity $= \delta C = \dfrac{\pi\eta}{2}\left\{\dfrac{b-a}{l}x + a\right\}^4 \dfrac{1}{\delta x}$.

The Total Torsional rigidity being C.

$$\frac{1}{C} = \int_{x=0}^{l} \frac{2}{\pi\eta} \frac{dx}{\left(\dfrac{b-a}{l}x + a\right)^4}$$

$$= -\frac{2}{3}\frac{1}{\pi\eta}\cdot\frac{1}{b-a}\left\{\frac{1}{b^2} - \frac{1}{a^2}\right\}$$

$$\text{or, } C = \frac{3}{2}\frac{\pi\eta a^3 b^3}{l}\cdot\frac{1}{a^2 + ab + b^2}$$

4. A long thin rod has a length 100 cm and an elliptic cross section 2 cm wide and 0·5 cm deep. When clamped at the upper end and supporting at the lower end a body of moment of inertia $5 \cdot 21 \times 10^5$ c.g.s. units, the periodic time of torsional oscillations is 0·313 secs. When supported horizontally on knife edges at the ends and carrying a load of 150 gm at the mid point, the sag produced is 6·12 mm. Find the value of each of the elastic constants.

For torsional oscillation $\tau = I\dfrac{d^2\theta}{dt^2} = C\theta$

or, $\dfrac{d^2\theta}{dt^2} = \dfrac{C}{I}$ $\qquad$ C = torsional rigidity

solving $T = 2\pi\sqrt{\dfrac{I}{C}}$

For a beam of elliptic section it can be shown that $C = \dfrac{n\pi}{l}\dfrac{a^3b^3}{a^2+b^2}$,

a and b being semi major and minor axes.

Substituting $T = 2\pi\sqrt{\dfrac{Il(a^2+b^2)}{n\pi a^3 b^3}}$

or, $T^2 = 4\pi^2 \dfrac{Il(a^2+b^2)}{n\pi a^3 b^3}$

$$\therefore n = \frac{4\pi l\left(a^2+b^2\right)I}{a^3b^3T^2} = \frac{4\times 3{\cdot}14\times 100\left(1^2+0{\cdot}25^2\right)\times 10^5\times 5{\cdot}21}{1^3(0{\cdot}25)^3(0{\cdot}313)^2}$$

$$= 4{\cdot}54\times 10^{11} \text{ dynes / cm}^2$$

In the second case $Y = \dfrac{Wl^3}{48AK^2\delta}$

In this case $A = \pi\times 1\times{\cdot}25$

and $k^2 = \dfrac{({\cdot}25)^2}{4}$

Hence $Y = \dfrac{450\times 980\times(100)^3\times 4}{48\times 3{\cdot}14\times 1\times{\cdot}25\times({\cdot}25)^2\times 0{\cdot}612} = 12{\cdot}24\times 10^{11}$ dynes / cm²

Now $Y = 2n(1+\sigma)$

or, $1+\alpha = \dfrac{Y}{2n}$

or, $\sigma = \dfrac{Y}{2n} - 1 = \dfrac{12\cdot 24\times 10^{11}}{2\times 4\cdot 45\times 10^{11}} - 1 = 0\cdot 375$

Also $Y = 3k(1-2\sigma)$

or, $K = \dfrac{Y}{3(1-2\sigma)} = \dfrac{12{\cdot}24\times 10^{11}}{3(1-2\times{\cdot}375)}$

$$= \frac{12{\cdot}24\times 10^{11}}{3(1-{\cdot}750)} = 16{\cdot}32\times 10^{11} \text{ dynes / cm}^2$$

5. A light cantilever beam is clamped horizontally and carries a load of 3 kg at the free end, such as to produce only a slight amount of bending. When depressed and released the beam vibrates with a periodic time of 0·30 secs. The projective length of the beam is 50 cm, its width is 2·5 cm and depth 0·5 cm. Find the value of the Young's modulus of the beam. If the mass of the beam is 500 gm what percentage of error in the value of Young's modulus is made by neglecting to take the mass into account.

Let y be the instanteneous deflection of the loaded end. If F be the instanteneous restoring force developed, we have

$$y = -\frac{4Fl^3}{Ybd^3}$$

But $F = M\dfrac{d^2y}{dt^2}$

$$\therefore\ y = -\frac{4Ml^3}{Ybd^3}\frac{d^2y}{dt^2}$$

or, $\dfrac{d^2y}{dt^2} = -\dfrac{Ybd^3}{4Ml^3}y$

which shows the motion to be simple harmonie having a time period

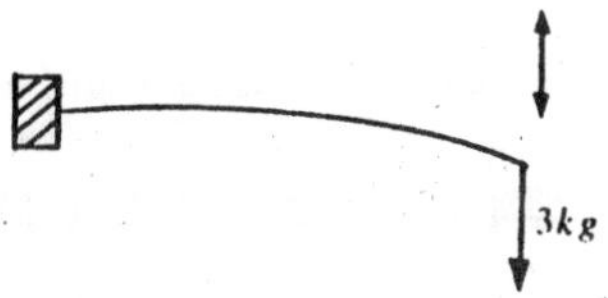

Fig. 1.34

$$T = 2\pi\sqrt{\frac{4Ml^3}{Ybd^3}}$$

or, $T^2 = 4\pi^2.\dfrac{4Ml^3}{Ybd^3}$

$$\therefore Y = \frac{16\pi^2Ml^3}{T^2bd^3} = \frac{16\times 9{\cdot}87\times 3000\times 50^3}{(0{\cdot}30)^2\times 2{\cdot}5\times(\cdot 5)^3} = 2{\cdot}105\times 10^{12}\ \text{dynes / cm}^2$$

When the mass of the vibrating beam is taken into account

$$Y = \frac{16\pi^2\left(M + \frac{33}{140}M_1\right)l^3}{T^2bd^3}.$$

Hence the percentage error in neglecting the mass of the beam

$$= \frac{\frac{33}{140}M_0\times 100}{M} = \frac{33\times 500\times 100}{140\times 3000} = 3{\cdot}92\%$$

Detailed theory of the above problem :

From equation (1.19), deflection at a point at a distance x from the fixed end is given by $YAK^2y = F\left(\dfrac{lx^2}{2} - \dfrac{x^3}{6}\right)$

At $x = l$, the instanteneous deflection is given by

$$YAK^2\delta = F\left(\frac{l^3}{2} - \frac{l^3}{6}\right) = F\frac{l^3}{3}$$

so that $y = \left\{\dfrac{3}{2}\dfrac{x^2}{l^2} - \dfrac{1}{2}\dfrac{x^3}{l^3}\right\}\delta$

Instanteneous velocity at the point is given by

$$\dot{y}=\left\{\frac{3}{2}\frac{x^2}{l^2}-\frac{1}{2}\frac{x^3}{l^3}\right\}\dot{\delta}$$

If m be the mass per unit length of the cantilever, its kinetic energy is given by

$$=\int_0^l \frac{1}{2}m\left\{\frac{3}{2}\frac{x^2}{l^2}-\frac{1}{2}\frac{x^3}{l^3}\right\}\dot{\delta}^2\,dx=\frac{33}{280}ml\,\dot{\delta}^2$$

The kinetic energy of the mass M carried at the end is $\frac{1}{2}M\,\dot{\delta}^2$

The potential energy of the beam at any position is given by

$$\int_0^{\delta} F d\delta=\int_0^{\delta}\frac{3yAK^2}{Rl^3}\delta d\delta=\frac{3YAK^2}{2Rl^3}\delta^2$$

Hence the total energy $=\left(\frac{1}{2}M+\frac{33}{280}ml\right)\dot{\delta}^2+\frac{3}{2}\frac{YAK^2}{Rl^3}\delta^2$ = Constant

Differentiating $w.r.t.t.$, we have

$$\left(\frac{1}{2}M+\frac{33}{280}M_1\right)\ddot{\delta}+\frac{3}{2}\frac{YAK^2}{Rl^3}\delta=0.$$

The equation of motion shows it to be simple harmonic with a period given by $T=2\pi\sqrt{\frac{\left(M+\frac{33}{140}M_1\right)l^3}{3YAK^2}}$.

6. A uniform beam $ABCD$ is supported horizontally at B and C, where $AB=\frac{1}{4}BC=CD=l$. If two equal masses m, m are suspended at A and D, the bar rises by a height h in the middle. Deduce an expression for h in terms of the relevent quantities.

The bar $ABCD$ is placed horizontally upon two knife edges at B and C. On suspending weights mg, mg at A and D, the bar bends. The bent position is as shown in the adjacent diagram (Fig. 1.35). Evidently the reaction at each knife edge is mg. The couple acting at each end is given by $\tau = mgl$, where $l = AB = CD$.

The bar bends into a circular arc of radius r (say). Let the circle be completed. EF is the elevation h. An important property of the circle is

$$EF\times FG = BF^2$$

or, $h(2r-h)=(2l)^2 \quad \because \quad BF=\frac{1}{2}BC=2l$

Neglecting small quantities of higher order $h=\frac{2l^2}{r}$

when the bar is bent, a restoring torque $\frac{YAK^2}{r}$ is developed. This is equal and opposite to the couple τ,

Thus $\frac{YAK^2}{r}=\tau=mgl$

or, $r=\frac{YAK^2}{mgl}$

Substituting for r, we have

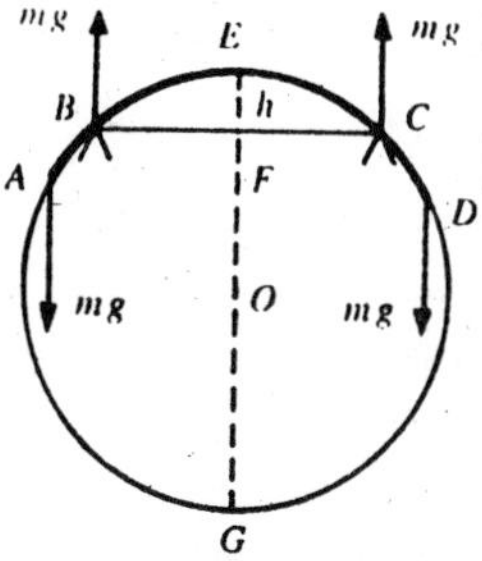

Fig. 1.35

$h=\frac{2mgl^2}{YAk^2}$ which is the required expression.

7. For a weightless beam supported at the ends, calculate the maximum load that can be placed at the middle :

Let a light beam of length l, breadth b and thickness d be supported at its ends. Let us calculate the maximum value of the load W which can be placed at the middle point so that no filament is strained beyond the permissible value. Let the maximum longitudinal stress that can be applied without causing permanent deformation is T dynes/cm^2.

If we consider an element of length dl of the beam to be bent to a curvature $\frac{1}{R}$, its logitudinal strain will be given by $\frac{e}{dl}=\frac{z}{R}$, where e = its elongation and z its distance from the neutral surface. But we have the relation at any point x, $\left(0<x<\frac{l}{2}\right)$.

$$\frac{YAK^2}{R}=\frac{Wx}{2}$$

or, $$\frac{YAK^2}{z}\frac{e}{dl}=\frac{Wx}{2}$$

$$\therefore \quad \text{strain} = \frac{e}{dl}=\frac{Wxz}{2YAK^2}=\frac{6Wxz}{Ybd^3}$$

This shows that the maximum strain will occur at $x = l/2$ and $z = \frac{d}{2}$

$$\therefore \text{ Maximum strain } = \frac{3Wl}{2Ybd^2} = \frac{T}{Y}$$

$$\text{or, } W = \frac{2bd^2T}{3l}.$$

EXERCISE

1. Define elasticity and elastic limit. State Hooke's law and define the different elastic constants.

2. A wire suspended vertically is subjected to the action of a pulling force at the lower end. Is the strain produced in a pure longitudinal one ? Define Poisson's ratio in this connection.

3. A particle of mass m hangs at the end of a light elastic string of unstretched length l and modulus of elasticity λ. Prove that in the position in which the string is stretched by an amount x, the combined gravitational and elastic potential energy is $\frac{\lambda x^2}{2l} + mgx$ + constant. By considering the stationary value of this expression, determine the position of equilibrium of the particle $\left[x = \frac{mgl}{\lambda}\right]$.

4. Two cylindrical shafts have the same length and mass and are made of the same material. One is solid, while the other, which is hollow, has an internal radius half the external radius. Compare their torsional rigidities and the maximum strains produced by equal twisting torques.

[3 : 5, 1 : 1·44]

5. The end of a given strip cantilever depresses 10 mm under a certain load. Calculate the depression under the same load for another cantilever of same material, 2 times in length, 2 times in width and 3 times in thickness.

[1·48 mm]

6. The rectangular cross-section of a long rod has sides in the ratio 1 : 2. Calculate the ratio of the depression under the same load when (i) the smaller side is vertical and (ii) the longer side is vertical. $[\delta_1 : \delta_2 = 4 : 1]$

7. A light cantilever beam 1 m long, of circular cross-section (r = 1 cm) is clamped at one end and carries a load of 2 kg at the free end which produces a very small bending of the beam. The free end is depressed and then released so that it begins to vibrate. Calculate the time period of vibration neglecting damping; $Y = 2\times10^{11}\ N/m^2$. [T = 0·1294 sec.]

8. Describe the determination in a straight wire of circular cross-section when (i) it is slightly bent with terminal couples and when and (ii) it is twisted by axial couples.

How can the Poisson's ratio be measured from the results obtained ?

9. The upper end of a vertical uniform wire of length l cm and of radius a cm is fixed and a horizontal couple of moment G dyne cm is applied at the lower end. Deduce an expression for the angle of twist of the lower end, the modulus of rigidity of the material of the wire being n dynes cm^{-2}.

Describe a method for determining the modulus of rigidity of the wire.

10. Define the three principal moduli of elasticity, and Poisson's ratio. Show that if Y, K and n be respectively, Young's modulus, bulk modulus and modulus of rigidity, then $Y = \dfrac{9nk}{3k+n}$.

A hollow cylindrical cast iron column is 10 inches in external diameter and 8 inches in internal diameter, the length being 10 ft. How much will it contract under a load of 60 tons, Y for cast iron = 8000 ton wt. per inch^2.

[0·032 inch]

11. A, horizontal bar of weight W is suspended by two parallel threads (vertical) each of length l and radius r. The rigidity of the wires is n and they are at a distance d apart. If the upper end of each wire is twisted about its own axis through an angle θ, find the angle through which the bar is deflected.

$$\left[4\pi nr^4\theta \Big/ \left(Wd^2 + 4\pi nr^4\right)\right]$$

12. A rod of uniform cross-section is supported horizontally at three points, one at each end and one in the middle at the same level. Show that the greatest sags occur very nearly at one fifth of the length from each end.

Hints : Let P be any point in AB at a distance x from A and ω the weight per unit length of the beam.

We have $YAk^2D^4y = \omega$

Integrating with respect to x

$YAK^2D^3y = \omega x + A = -F = \omega x - F_o$

Integrating successively three times with respect to x

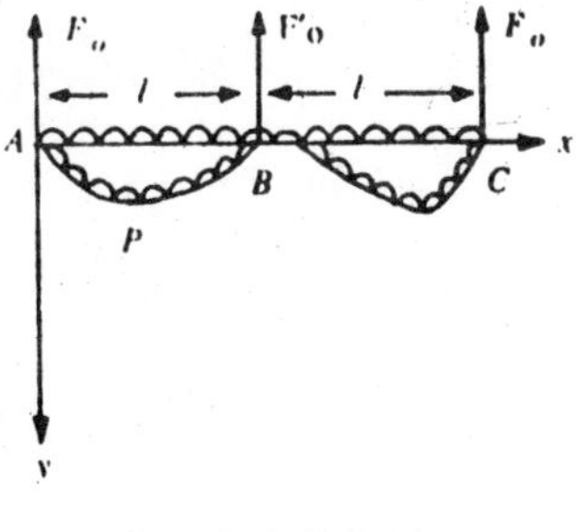

Fig. 1.36

$$YAK^2D^2y = \frac{\omega x^2}{2} - F_o.x + B$$

Since $YAk^2D^2y = 0$, at $x = 0$, $\therefore B = 0$

$$YAK^2Dy = \frac{\omega x^3}{6} - F_0\frac{x^2}{2} + c$$

Since $Dy = 0$ at $x = l$,

$$C = \frac{F_0l^2}{2} - \frac{\omega l^3}{6}$$

$$\therefore\ YAK^2Dy = \frac{\omega x^3}{6} - \frac{F_0x^2}{2} + \frac{F_0l^2}{2} - \frac{\omega l^3}{6} \qquad \text{.....(A)}$$

Lastly, $YAK^2y = \frac{\omega x^4}{24} - \frac{F_0x^3}{6} + \frac{F_0l^2}{2}x \frac{-\omega l^3}{6}x$, the constant of integration vanishing, since at $x = 0$, $y = 0$

Also since $y = 0$ at $x = l$

$$0 = \frac{\omega l^4}{24} - \frac{F_0l^3}{6} + \frac{F_0l^3}{2} - \frac{\omega l^4}{6}$$

or, $F_0l^3\left(\frac{1}{2} - \frac{1}{6}\right) = \omega l^4\left(\frac{1}{6} - \frac{1}{24}\right)$

or, $F_0l^3\frac{3-1}{6} = \omega l^4\frac{4-1}{24}$

or, $\frac{F_0l^3}{3} = \frac{\omega l^4}{8}$ $\quad\therefore\ F_0 = \frac{3}{8}\omega l$

Substituting for F_0 in (A), we have for maximum sag $Dy = 0$

$$\therefore\ \frac{\omega x^3}{6} - \frac{\omega l^3}{6} + \frac{3}{8}\frac{\omega l}{2}\left(l^2 - x^2\right) = 0$$

$$\left(x^3 - l^3\right) + \frac{9l}{8}\left(l^2 - x^2\right) = 0$$

or, $(x - l)\left\{x^2 + lx + l^2 - \frac{9}{8}l^2 - \frac{9}{8}lx\right\} = 0$

$\therefore\ x = l$ and $8x^2 + 8lx + 8l^2 - 9l^2 - 9lx = 0$

or, $8x^2 - lx - l^2 = 0$

$$\therefore x = \frac{l \pm \sqrt{l^2 + 32l^2}}{16} = \frac{l \pm \sqrt{33}l}{16} = \frac{l}{16}\left(1 + \sqrt{33}\right)$$

(Taking this + *ve* value) = ·42l = ·21 (2l) Proved.

13. A body is under stress along three perpendicular axes OX, OY, OZ such that there is no linear strain along OY or OZ. It is then stressed along the same axes so that the linear strains along OY and OZ are equal and there is no strain along OX. Compare the ratio of stress to strain along OX in the first case with that along OY or OZ in the second case.

Hints : Let T_x, T_y, T_z be the stresses along OX, OY, OZ in the first case. As there is no strain along Y & Z directions.

$$\frac{T_y}{Y}-\sigma\left(\frac{T_x}{Y}+\frac{T_z}{Y}\right)$$

$$=\frac{T_z}{Y}-\sigma\left(\frac{T_x}{Y}-\frac{T_y}{Y}\right)=0 \qquad \text{.... (1)}$$

In the second case let T_x', T_y' and T_z' be the corresponding stresses. From the conditions given $\frac{T'_x}{Y}-\sigma\frac{T'_y}{Y}-\frac{\sigma T'_z}{Y}=0$

$$\text{or,}\quad T_{x'}=\sigma\left(T_{y'}+T_{z'}\right) \qquad \text{.... (2)}$$

$$\text{Also}\quad \frac{T_y'}{Y}-\sigma\left(\frac{T_x'}{Y}+\frac{T_z'}{Y}\right)=\frac{T_z'}{Y}-\sigma\left(\frac{T_x'}{Y}+\frac{T_y'}{Y}\right)$$

$$\text{or,}\quad T_y'-\sigma T'_z=T'_z-\sigma T'_y$$

$$\text{or,}\quad T_y'(1+\sigma)=T_z'(1+\sigma)$$

$$\text{or,}\quad T_y'=T_z' \qquad \text{.... (3)}$$

From (2) and (3) $T_x'=2\sigma T_y'=2\sigma T_z'$ .... (4)

Ratio of stress to strain along y direction in the second case

$$=\frac{T_y'}{\frac{T_y'}{Y}-\sigma\left(\frac{T_x'}{Y}+\frac{T_z''}{Y}\right)}=\frac{YT_y}{T_y'-\sigma\left\{2\sigma T_y'+T_y'\right\}}$$

$$=\frac{Y}{1-2\sigma^2-\sigma}=\frac{y}{(1+\sigma)(1-2\sigma)} \qquad \text{.... (5)}$$

Again from (1)

$$T_y-\sigma T_z=T_z-\sigma T_y$$

$$\text{or,}\quad T_y(1+\sigma)=T_z(1+\sigma)$$

or, $T_y = T_z$ and $T_x = \frac{T_y}{\sigma} - T_y = T_y\left(\frac{1}{\sigma} - 1\right)$

In the first case the strain along x direction $= \frac{T_x}{Y} - \sigma\left(\frac{T_y}{Y} + \frac{T_z}{Y}\right)$

$\therefore$ Ratio stress by strain $= \dfrac{T_x}{\frac{T_x}{Y} - \sigma\left(\frac{T_y}{Y} + \frac{T_z}{Y}\right)}$

$$= \frac{Y.T_y\left(\frac{1}{\sigma} - 1\right)}{T_y\left(\frac{1}{\sigma} - 1\right) - \sigma 2 T_y}$$

$$= \frac{Y\frac{1-\sigma}{\sigma}}{\frac{1-\sigma}{\sigma} - 2\sigma} = \frac{Y(1-\sigma)}{1 - \sigma - 2\sigma^2} = \frac{Y(1-\sigma)}{(1-2\sigma)(1+\sigma)} \qquad \text{...... (6)}$$

From (5) and (6)

Required Ratio $= \dfrac{Y(1-\sigma)}{(1-2\sigma)(1+\sigma)} \times \dfrac{(1+\sigma)(1-2\sigma)}{Y} = (1-\sigma) : 1$

14. A uniform beam of span l is fixed horizontally at each end. Two equal loads, W, are placed at equal distance h from the ends of the beam. Neglecting the weight of the beam, prove that the greatest deflection of the beam is

$$\frac{Wh^2}{24\, YAK^2}(3l - 4h).$$

15. A light beam is clamped horizontally at one end. Derive an expression for the lowering of the mid point, when it is loaded at the other end. What would be the lowering if the same beam were supported horizontally on knife edges at its ends and the same load is suspended at the middle.

16. A horizontal beam rigidly fixed at one end is of weight ω per unit length. The vertical shearing force at any point at a distance x from a convenient origin is F and the bending moment is M. Show that $\frac{dF}{dx} = \omega$, $\frac{dM}{dx} = -F$.

The bending being assumed to be small, show that $YAK^2 \frac{d^4 y}{dx^4} = \omega$, symbols having usual meanings.

17. Derive from first principles, an expression for the depression at the mid point of a beam of uniform mass per unit length loaded at the centre and supported at the ends on knife edges at the same level.

In the case of a beam of negligible mass, find the period of small oscillations when the load is slightly displaced vertically from the position of equilibrium and then released.

18. A circular bar, supported horizontally on horizontal knife edges 40 cm apart has a length of 60 cm and carries a load of 10 kg suspended from each end. The radius of the bar is 4 mm and when loads are applied the central point rises by 4·88 mm. Find the value of the Young's modulus of the bar.

$$\left[20\cdot0\times10^{11}\ dynes\ cm^{2}\right]$$

19. One end of a 1·0 meter rod is clamped rigidly. When a 0·5 kg load is suspended at the other end, it is depressed by 0·5 cm. Calculate the depression at 40 cm distance from the fixed end when a 5 kg load is suspended at (i) free end, (ii) 40 cm from the free end.

20. A uniform metal disc of diameter 0·1 m and mass 1·2 kg is fixed to the lower end of a torsion wire of length 1 m and diameter $1\cdot44\times10^{-3}$ m, the upper end of which is fixed. The time period of torsional oscillations is 1·98s. Calculate the modulus of rigidity of the material of the wire.

Hints : $T = 2\pi\sqrt{\frac{I}{C}}$ and $I = \frac{1}{2}MR^2$

$$C = \frac{\pi n r^4}{2l}$$

$$\therefore T = 2\pi\sqrt{\frac{MR^2 l}{\pi n r^4}} \quad \text{or,} \quad T^2 = \frac{4\pi^2 MR^2 l}{\pi n r^4}$$

$$\text{or,} \quad n = \frac{4\pi MR^2 l}{r^4 T^2}$$

$$= \frac{4\pi\times1\cdot2\times(0\cdot05)^2\times1}{(0\cdot72\times10^{-3})^4\times(1\cdot98)^2} = 3\cdot57\times10^{10}\ N/m^2$$

21. A cantilever beam mounted horizontally has a negligible weight. Show that the depression at any point P due to a load applied at a second point Q is the same as the depression at Q produced by a similar load at P.

22. A cylindrical rod of iron, AB, radius 0·50 cm and length 25 cm, is clamped at A, so that the axis is horizontal. Calculate the deflections of the

free end of the rod when a force of 3 kg. wt. is applied horizontally, (a) at (B), (b) the point midway between A and B, Young's modulus of the material of the beam being $2 \cdot 1 \times 10^{12}$ dynes/cm². $\left[\text{(a) } 0 \cdot 467 \, cm, \ \text{(b) } 0 \cdot 145 \, cm \right]$

23. A steel wire of diameter 1 cm and of length 20 cm is twisted by axial couples at the ends, if the modulus of rigidity is 8×10^{11} dynes/cm², calculate the potential energy of twist if the relative angle of twist between the ends is 1 degree.

24. A hollow metal tube $1 \cdot 22$ cm in internal and $1 \cdot 62$ cm in external diameter, whose length is 100 cm is supported at the ends and loaded with a load of 10 kg in the middle. If the sag at the middle is $0 \cdot 48$ cm, calculate the Young's modulus of the material of the wire.

$$\left[18 \cdot 6 \times 10^{11} \ dynes/cm^2 \right]$$

25. A beam of rectangular cross-section is supported at the ends and loaded at the middle. By what per cent the depression at the mid point will change if the beam is clamped at the ends and loaded at the middle with the same load ?

26. Distinguish between—

(a) flexural rigidity,
(b) modulus of rigidity,
(c) torsional rigidity.

CHAPTER II

SURFACE TENSION

Origin of surface tension and surface energy—their interrelations; Angle of contact; Excess pressure on a curved surface (liquid); Shape of liquid drops; Capillarity; Saturated vapour pressure over a curved surface (Experimental determination of surface tension not required); Effect of temperature on surface tension.

2.1. Fundamental definition and relations

Surface tension :

The surface of a liquid behaves as if like a stretched membrane and has a tendency to contract. To explain this behaviour we imagine the existence of a force acting tangential to the liquid surface. This force per unit length is called surface tension. The origin of this tendency to contract can be explained from the stand point of moleculer theory of matter.

Surface energy :

The existence of a force acting tangential to a liquid surface implies the existence of some energy associated with the force, because when a surface

(a) (b)

Fig. 2.1

is produced in presence of this force some work must be done. This work done remains within the surface as its potential energy. This work done per unit area is known as surface energy density.

A very simple experiment to demonstrate this fundamental property of the liquid can be described as follows.

A thread is tied loosely across a circular loop of wire Fig. 2.1(a). The loop is dipped into a soap solution and taken out. A film of soap solution will adhere to the surface of the loop and the thread will lie loosely as before. The lower part of the film is then touched by finger when it will break. The thread will at once be forced to occupy the arc of a circle as in Fig. 2.1(b). The upper part of the film tends to contract and thus pulls the thread which is forced to occupy the arc of a circle.

2.2. Relation between surface tension (*S*) and surface energy density (*E*)

To deduce the relation between surface tension (*S*) and surface energy density (*E*), let us consider a rectangular framework *ABCD* of dimension *x*-*y* of which the arms *AB*, *BC* and *CD* are fixed while the arm *AD* can slide along *BA* and *CD*. Let the framework be dipped into soap solution and taken out so that a film of the solution adheres to it. Let the arm *AD* be displaced to $A'D'$ so that $AA' = \delta y$. The work done for this displacement is $2Sx\delta y$, where x = AD (The factor 2 is due to the two surfaces of the film (Fig. 2.2)).

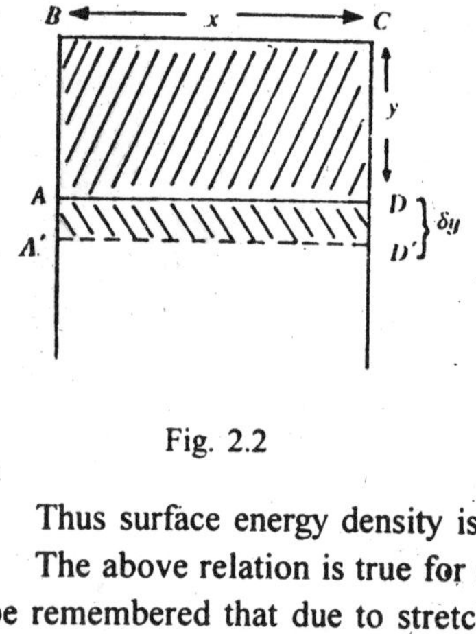

Fig. 2.2

The new area produced $= 2x\delta y$.
Hence the work done per unit area or the

surface energy density $= \dfrac{2Sx\delta y}{2x\delta y} = S$.

Thus surface energy density is numerically equal to the surface tension.

The above relation is true for an adiabatic stretching of the film. It is to be remembered that due to stretching the film gets cooled. During an isothermal stretching an equivalent amount of heat enters the film. Thus $E = S+h$, where S is the mechanical work done per unit area and is known as the free surface energy density and h is the heat absorbed per unit area.

To Find h let us subject the film to a Carnot cycle of operation. Let the surface tensions at absolute temperatures T and $T-\delta T$ be S and $S-\delta S$ respectively. Then since the surface tension is independent of area, the work done when the film is stretched isothermally from a condition represented by D to that represented by C, is $S\ \delta A$, and the heat taken in $h\delta A$.

From C to B there is adiabatic extension, the temperature changing from T to $T-\delta T$. From B to A, let the surface contract isothemally, so that when it contracts adiabatically, it finally arrives back at the original state, as represented by D. The cycle is clearly reversible and so from results of thermodynamics.

$$\frac{\text{Network done during the cycle}}{\text{Heat absorbed at higher temp}} = \frac{\delta T}{T}$$

or, $$\frac{\left(S - \frac{\delta S}{\delta T}\delta T\right)\delta A - S\delta A}{h\delta A} = \frac{\delta T}{T}$$

$$\therefore h = -T\frac{\delta S}{\delta T}$$

Hence $E = S - T\dfrac{\delta S}{\delta T}$ (2.1)

Since $\dfrac{\delta S}{\delta T}$ is always $-ve$, $E > S$

and $E = S$ at $T = 0$.

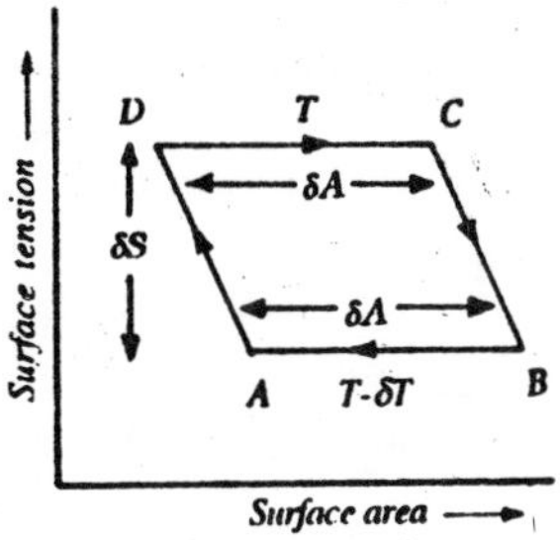

Fig. 2.3

Problems treated by energy considerations :

(a) Shape of small and large drops :

Small drops of liquid on a solid surface not wetted by the liquid appear to assume a spherical surface. For a liquid drop of surface tension S, mass m and surface area A, the total potential energy $= mgh + SA$, where h is the height of the c.g. For a small drop h *is* small and hence the second term is the predominant one. A dynamical system will tend to assume such a form in which it will have minimum P.E. Of all the solid figures having the same volume a sphere has the least area and hence drop will assume a spherical form. For a large drop h is no longer small and hence it will assume such a form in which the sum $mgh + SA$ is minimum *i.e.* its shape will be somewhat flattened.

(b) The method of energy consideration may be employed to determine whether or not a liquid will spread when placed over the surface of another liquid of higher specific gravity.

Let a drop of a liquid of kind B is placed over the surface of a liquid of kind A. Let S_A, S_B be the surface tensions of A and B in contact with air and S_{AB}, the surface tension of A in contact with B. If B spreads over A by an amount δA (of area), the increase in the $P. E.$ of the system is $\delta A\{S_B + S_{AB} - S_A\} = \delta E$ say.

If $S_A > S_B + S_{AB}$, δE is –ve and the spreading involves a decrease of P.E. and it will spread. If on the other hand $S_A \angle S_B + S_{AB}$, the spreading is associated with an increase in energy and so B will not spread over A.

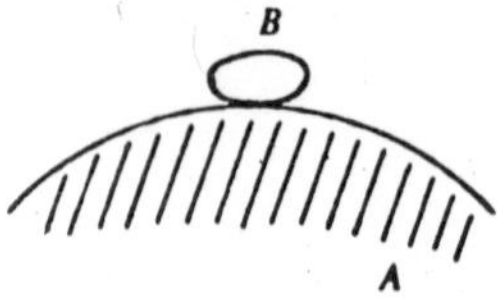

Fig. 2.4

If the second possibility is true, then a triangle can be constructed with its sides proportional, to S_B, S_{AB} and S_A. Such a triangle is called Newmann's triangle. Experimental values of S_B, S_{AB} and S_A, show that no such triangle can be constructed. Hence the conclusion is that B will spread over A. [For more accurate discussion see problems.]

2.3. Liquid in contact with a solid

Angle of contact : When a glass plate is plunged into a liquid, it is found that the surface of the liquid near its place of contact with the solid assumes a curved shape.

In case of ordinary liquids the surface is concave; but with mercury the surface is convex (Fig. 2.5).

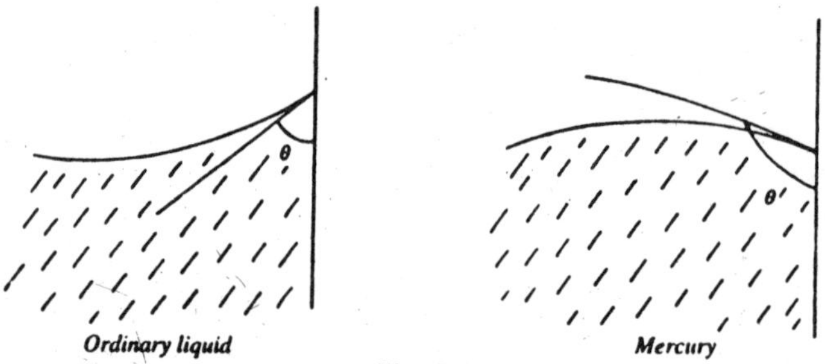

Fig. 2.5

The angle within the liquid between the solid plate and the tangent to the liquid surface is called the angle of contact between the liquid and the solid. With ordinary liquids glass has acute angle of contact, while with mercury it has an obtuse angle of contact.

The cause of this curvature of the liquid surface can be explained if we keep in mind that at any point the surface of the liquid will be perpendicular to the resultant force acting at that point. For a molecule of the liquid near its place of contact with the solid the forces acting are three fold.

These are—(1) The weight of the molecule acting vertically downwords along PM; (2) The force of adheson (resultant force of attraction of the molecules of the solid on the particular liquid molecule) acting perpendicular to the solid plate along *PN*; (3) The force of cohesion (resultant force of attraction on the particular molecule by the rest of the liquid molecules) acting along *PR* (somewhere between the solid plate and the tangent to the liquid surface) (Fig. 2.6).

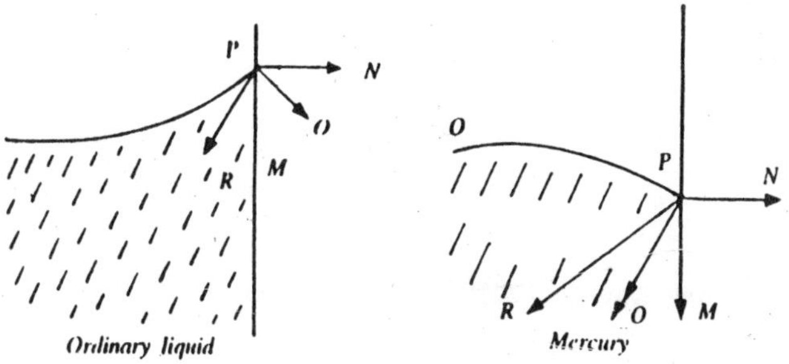

Fig. 2.6

In the case of ordinary liquids adhesive force is much greater than the cohesive force. Hence the resultant (PO) will be inclined more towards the adhesive force. Reverse is the case with mercury. This explains the concavity of the surface in the case of ordinary liquids and convexity in the case of mercury. At a point far removed from the solid plate in both the cases, the adhesive force is negligible. Both the weight of the molecule and the Cohesive force act vertically downwards. The resultant is therefore vertical and hence the surface is horizontal.

2.4. Excess pressure on the concave side of a curved liquid surface

If we consider a curved membrane, it will be in equilibrium only if an excess pressure acts on the concave side to counterbalance the surface tension effect. The magnitude of the excess pressure can be found in the following way.

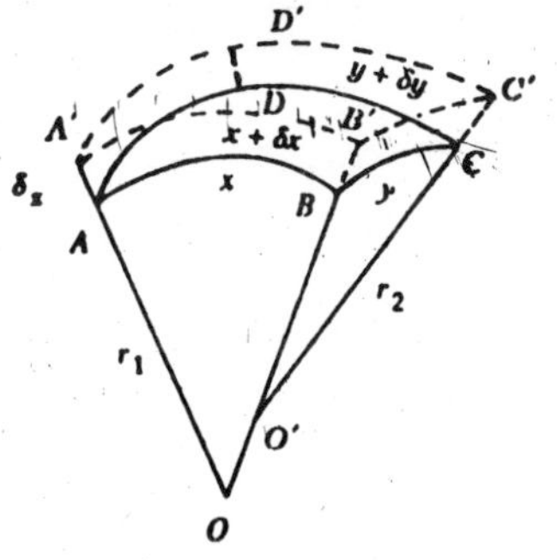

Fig. 2.7

Let *ABCD* be an element of the curved liquid surface having principal radii of curvature r_1 and r_2 (Principal radii of curvature are radii of curvature in two planes in which it has maximum and minimum radii). Let *P* be the required excess pressure and the surface area *xy* be given a normal virtual displacement δz so that it occupies the position *A'B'C'D'* from *ABCD* (Fig. 2.7). The work done by the exess force is $Pxy\delta z$.

Since the increase in area $=(x+\delta x)(y+\delta y)-xy=x\delta y+y\delta z$, neglecting small quantity of higher order, the increase in surface energy $=2(x\delta y+y\delta x)S$ (considering the two surfaces of the element).

Now from the similarity of the triangles ABO and $A'B'O$,

$$\frac{x+\delta x}{r_1+\delta z}=\frac{x}{r_1}$$

or, $(x+\delta x)r_1 = x(r_1+\delta z)$

or, $\delta x = \dfrac{x\delta z}{r_1}$.

In a similar way $\delta y = \dfrac{y\delta z}{r_2}$.

Hence the increase in surface energy $=2xy\delta z\left(\dfrac{1}{r_1}+\dfrac{1}{r_2}\right)S$

Equating the work done by the excess force to the increase in surface energy

$$Pxy\delta z = 2xy\delta z\left(\frac{1}{r_1}+\frac{1}{r_2}\right)S$$

or, $$P = 2S\left(\frac{1}{r_1}+\frac{1}{r_2}\right) \qquad (2.2)$$

The above relation is valid for a liquid film. In the case of a liquid drop

$$P = S\left(\frac{1}{r_1}+\frac{1}{r_2}\right) \qquad (2.3)$$

If the points O and O' do not lie on the same side as in Fig. 2.8.

$$P = S\left(\frac{1}{r_1}-\frac{1}{r_2}\right) \qquad (2.4)$$

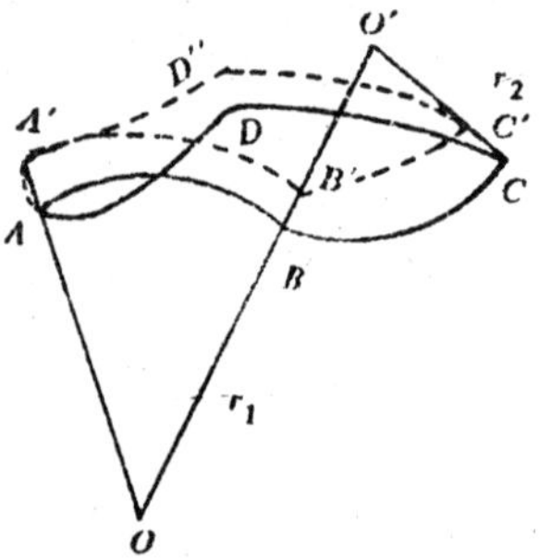

Fig. 2.8

For a spherical bubble blown within a liquid or a spherical drop we have from equation (2.3)

$$P = \frac{2S}{r} \qquad (2.5)$$

Since $r_1 = r_2$

For a spherical bubble blown in air, we have from (2.2) $$P = \frac{4S}{r} \qquad (2.6)$$

2.5. The Capillary Curve

If a perfectly clean glass plate dips vertically in a liquid, then close to the plate, the surface of the liquid assumes a definite **shape**. A section of the surface of the liquid by a vertical plane, perpendicular to the glass plate gives the form assumed by the liquid at the place of contact and is termed the capillary curve and the phenomenon is called capillarity. Let the glass plate or wall be normal to the plane of the diagram (Fig. 2.9), which represents the section the plate and the liquid surface. Of the two principal radii of curvature of the liquid surface at the point of contact, one will be infinite and the other will be the radius of curvature of the curve APQ at P.

Taking x axis horizontal and the axis of y vertical, we obtain,

$$\frac{S}{r} = g\rho . y = \pm S\frac{d\psi}{ds} = \pm S\frac{d\psi}{dy}\frac{dy}{ds}$$

$$= -S\sin\psi\frac{d\psi}{dy}$$ $-ve$. –ve sign is chosen because ψ decreases as y increases.

Putting $\frac{S}{\rho g} = B^2$ and integrating,

$$-B^2\int_{\pi}^{\psi}\sin\psi d\psi = \int_0^y y dy$$

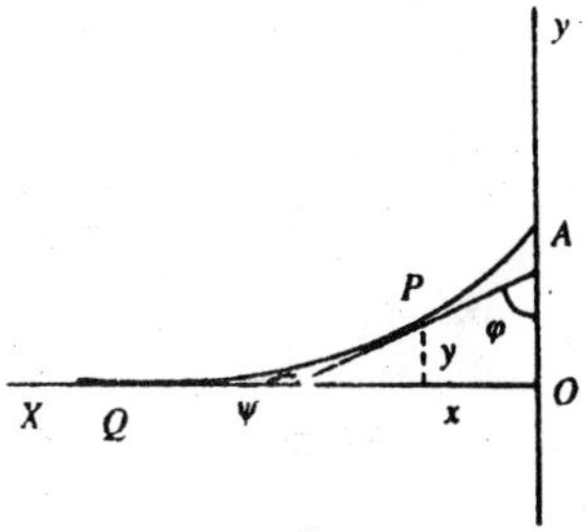

Fig. 2.9

or, $y^2 = 2B^2(1+\cos\psi)$ ...(2.7)

which is the intrinsic equation of the capillary curve.

At $y = y_{max}, (180^\circ - \psi) + \theta = 90^\circ$

$$\therefore \psi = 90^\circ + \theta$$

$$\therefore y^2_{max} = 2B^2(1-\sin\theta) \quad (2.8)$$

where θ = angle of contact.

Equation (2.8) may be used to measure B and hence S, by measuring y_{max} and knowing θ.

2.6. Shape of large drops

Let us consider a drop which is large enough, so that the upper surface is horizontal and the diameter is large enough in comparison with its thickness. A section of such a drop placed on a solid plate not wetted by it is shown in Fig. 2.10.

At any point $P(x, y)$ on the drop the tangent drawn makes an angle ψ with x-axis.

If P_0 be the external pressure, the pressure at P from inside is $P_0 + y\rho g$. If the principal radii of curvature at P are r and R in and perpendicular to the plane of the diagram respectively, then the inside pressure at P

$$= p = S\left(\frac{1}{r}+\frac{1}{R}\right)+P_o,$$ and since R in large compared to r, we have

$$y\rho g = \frac{S}{r} = S\frac{d\psi}{ds} = S\frac{d\psi}{dy}\frac{dy}{ds}$$

$$= S\sin\psi\frac{d\psi}{dy}$$

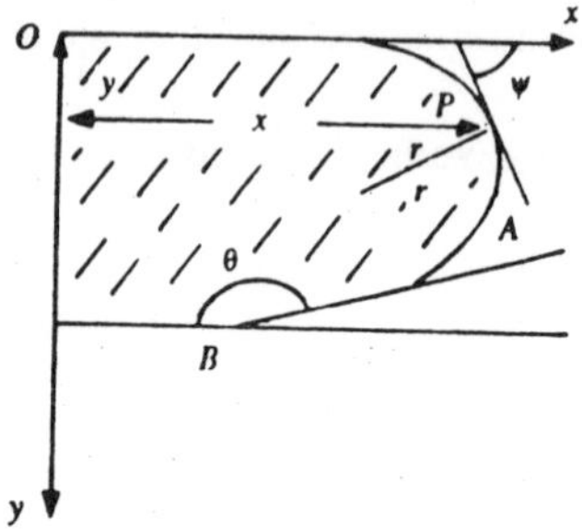

Fig. 2.10

$$\text{or, } g\rho\int_0^y y\,dy = S\int_0^\psi \sin\psi\,d\psi$$

$$\text{or, } \frac{g\rho y^2}{2} = S(1-\cos\psi) \qquad (2.9)$$

At the point where the tangent is vertical, $y = h$ and $\cos\psi = 0$.

$$\text{Hence } \frac{g\rho h^2}{2} = S \qquad (2.10)$$

At the point of contact B, the angle ψ becomes θ, the angle of contact and $y = H$ = thickness of the drop.

$$\text{Thus } \rho g\frac{H^2}{2} = S(1-\cos\theta) \qquad (2.11)$$

Equation (2.10) can be used to determine S, by measuring h. If S be known equation (2.11) can be used to determine θ by measuring H.

2.7. Rise of a liquid in a capillary tube

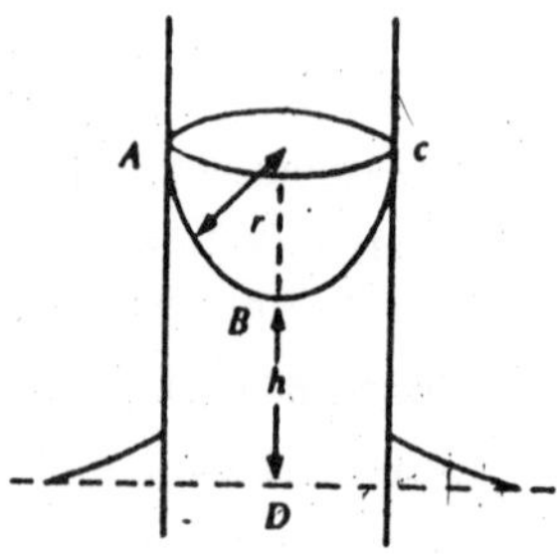

Fig. 2.11

At the first stage let us assume that the tube is narrow and the angle of contact is zero (Fig. 2.11). In this case surface of the liquid within the capillary tube will be part of a sphere of radius, say r.

The pressure over the curved surface is atmospheric at all points At B, a point just below the surface, the pressure is less than the atmospheric by $\frac{2S}{r}$. At D, a point below B and lying in the same horizontal plane as the surface of the liquid outside, the pressure is also atmospheric. Now the difference in pressure between the points B and D is equal to the pressure exerted by a column of liquid of height $BD = h$(Say).

If ρ be the density of the liquid, we have $\frac{2S}{r} = h\rho g$

or, $S = \frac{rh\rho g}{2S}$ (2.12)

In the second stage let us assume that the angle of contact between the liquid and the material of the tube is φ (Fig. 2.12).

Let R be the radius of curvature of the liquid surface at its lowest point.

If the tube is capillary, R is constant at all points on the liquid surface. Hence, if P be the atmospheric pressure, pressure at B is

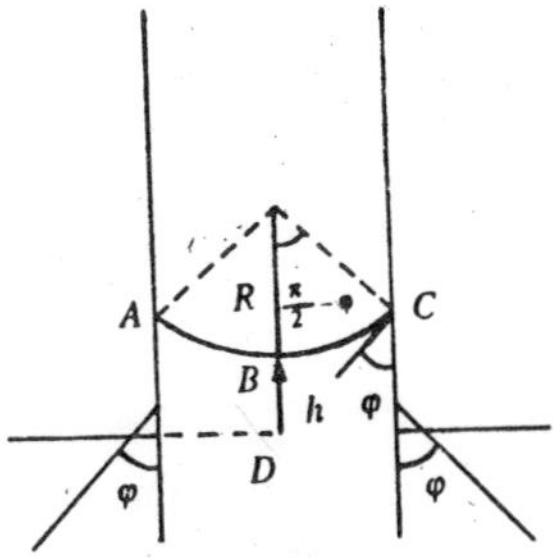

Fig. 2.12

$$P - \frac{2S}{r} = P - h\rho g$$

$$\therefore \quad \frac{2S}{R} = h\rho g \qquad (2.12(a))$$

But $R = r\cos\varphi$; hence $\frac{2S\cos\varphi}{r} = h\rho g$ (2.13)

Since the value of φ is very uncertain and varies with contamination, equation (2.13) is not very suitable for measurement of surface tension.

If we assume that the tube is perfectly cylindrical we may equate the resultant force $2\pi rS$ acting upwards along the circumference of a circle of radius r, to the weight of the liquid raised.

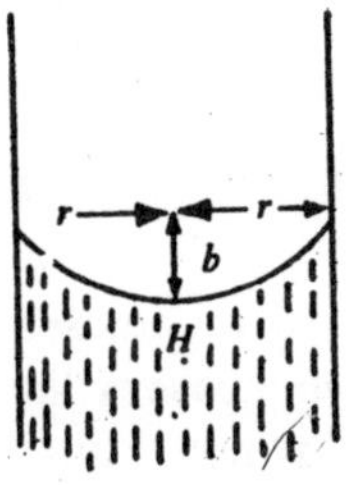

Fig. 2.13

Thus $2\pi rs = \pi r^2 h\rho g + \frac{1}{3}\pi r^2 b\rho g$ (Fig. 2.13) (2.14)

assuming the section of the meniscus to be an ellipse of semi-axes r and b. It can be shown that the radius of curvature at A is given by $R = \frac{r^2}{b}$.

Putting $\frac{S}{g\rho} = a^2$ in (2.12 (a)), we have

$$2a^2 = Rh = \frac{r^2 h}{b} \quad \text{or,} \quad b = \frac{r^2 h}{2a^2} \text{ using equation (2.14)}$$

$$2a^2 = rh\left(1+\frac{r^2}{6a^2}\right)$$

or, $12a^4 - 6a^2rh - r^2h = 0$, which is a quadratic in a^2, giving

$$a^2 = \frac{6rh \pm 6rh\left(1+\frac{4}{3}\frac{r}{h}\right)^{\frac{1}{2}}}{24}$$

Taking the $+ve$ sign, since in the limit $\frac{r}{h} \to 0$

We must have $2a^2 = rh$

$$\therefore \quad \frac{S}{\rho g} = a^2 = \frac{1}{2}rh\left\{1+\frac{1}{3}\frac{r}{h}-\frac{1}{9}\frac{r^2}{h^2}+\ldots\right\} \tag{2.15}$$

which is a close approximation of the correct result.

To a first approximation $S = \frac{rh\rho g}{2}$. This relation can be used to determine S by measuring r and h in the case of a capillary tube.

For a given liquid S and ρ are constants; hence the product rh = Const. This is known as Jurin's law.

2.8. Rise of liquid between vertical (parallel) plates

Let us calculate the rise of a liquid between two parallel plates as shown in Fig. 2.14. The plates are separated by a distance d apart. If we assume the section AC to be cylindrical of diameter d, the pressure at B is less than the atmospheric by $\frac{S}{r}$ or $\frac{2S}{d}$. If h be the amount of rise of the liquid $\frac{2S}{d} = h\rho g$.

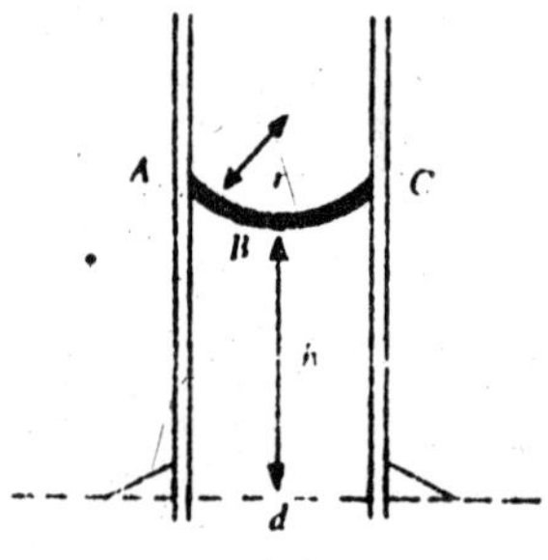

Fig. 2.14

2.9. Saturated vapour pressure over a curved liquid surface

In the following discussion we proceed to prove that the saturated vapour pressure in equilibrium with a convex liquid surface is greater than that over a plane surface. On the other hand saturated vapour pressure over a concave liquid surface is less than that over a plane surface.

To determine the equilibrium between a vapour and a liquid surface let us consider a closed vessel A (Fig. 2.15) containing a liquid and its vapour only and let a capillary tube of material not wetted by the liquid to be dipped vertically in the liquid. The convex meniscus may be considered as hemispherical and is depressed to a depth h below the free surface in A. Let P be the saturation pressure above the plane surface and p, the equilibrium pressure above a curved surface of radius r. The pressure difference $p-P$ equals the weight of a column of vapour of height h, and hence $p-P=\int g\sigma dh$, where σ is the vapour density. The pressure at B is P, while that at C, at the same horizontal level as the meniscus, is $P+g\rho h$, where ρ is the density of the liquid.

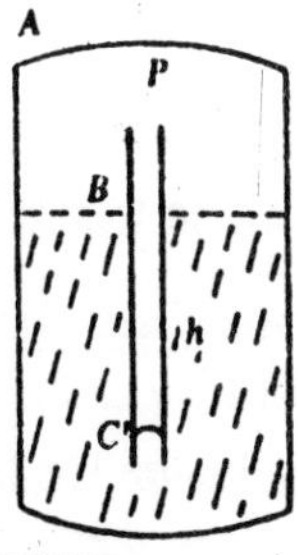

Fig. 2.15

Thus $P + g\rho h - p = \dfrac{2S}{r}$ (2.16)

or, $P - p = \dfrac{2S}{r} - g\rho h$

or, $p - P = h\rho g - \dfrac{2S}{r}$ (2.17)

where r is the radius of the capillary tube.

Thus $\int g\sigma dh = h\rho g - \dfrac{2S}{r}$

or, $\dfrac{2S}{r} = \int g(\rho-\sigma)dh$

Assuming the saturated vapour to obey the gas laws we have,

$P = \sigma RT$

and $dp = g\sigma dh$

Hence $\dfrac{2S}{r} = \int_P^p \dfrac{\rho-\sigma}{\sigma} dp = \int_P^p \dfrac{\rho}{\sigma} dp$

Since σ is small compared to ρ.

Thus $\dfrac{2S}{r} = \rho RT \int_P^p \dfrac{dp}{p} = \rho RT \log \dfrac{p}{P}$

or, $\log \dfrac{P}{p} = \dfrac{2S}{\rho RrT}$ (2.18)

The above equation gives the ratio of the vapour pressures p and P, for equilibrium over curved surface of radius r and a plane surface, respectively. For a given value of p *i.e.*, for a given degree of supersaturation drops of radius less than r will continue to evaporate, while those of greater size will continue to grow.

The above discussion also hold in case of boiling. A pure liquid may be raised to a temperature, much in excess of its normal boiling part, if it is free from all impurities which provide nuclei for the bubbles of vapour to form. When this formation occurs, however the growth of the bubble is very rapid, since the surface tension is increasing insufficient to counteract the excess pressure inside. In this case boiling by bumping occurs. In the normal process of boiling, the bubble contains some gas, as well as vapour, and is in equilibrium when pressure p_2 due to the gas, equals the necessary excess pressure inside over the that outside the bubble. If p be the external and p_1, the vapour pressure inside the bubble,

$$p_1 + p_2 = p + \frac{2S}{r}.$$

2.10. Effect of temperature on surface tension

The surface tension of a liquid decreases with the rise of temperature.

A formula for this temperature variation is $S_T = A\left(1 - \frac{T}{T_c}\right)^n$,

where S_T = Surface tension at absolute temperature T,

A = constant and n has a value in the neighbourhood of 1.21.

Another formula due to Eotvos corrected by Ramsay and Shields is

$$S(M\vartheta x)^{\frac{3}{2}} = K(T_c - T - \delta),$$

where S = Surface tension at absolute temp. T, T_c = critical temp. M = Molecular ωt of the liquid, ϑ = Specific volume of the liquid, K is 2.2 and δ varies from 6 to 8 for different liquid, agrees fairly well with experiment. According to the above formula $S = 0$, when $T = T_c - \delta$, *i.e.* surface tension should vanish 6 to 8 degrees below the critical temperature.

2.11. Molecular theory of surface tension

A fundamental property of a liquid surface is its tendency to contract to explain which surface tension forces have been conceived.

This tendency to contract can be explained in the light of molecular theory of matter In this aspect we shall be guided by the light thrown by Laplace. According to Laplace, the molecules of a liquid are attracting each other with a force which decreases so rapidly with the distance between the molecules that it may be supposed to be vanishingly small if the distance between the molecules exceeds a certain limiting value known as the range of molecular attraction C (Fig. 2.16). With a particular molecule O as centre and radius equal to the range of molecular attraction, if a sphere is imagined to be drawn, it is called the sphere of attraction of that molecule. There will be no attraction between this molecule and those outside the sphere. Confining our attention to Fig. 2.17, let us consider a molecule A well below the liquid surface so that its sphere of attraction lies wholly embeded within the liquid. The resultant force of attraction on this molecule will be zero as it will be subjected to finite attractive forces from all directions which will mutually cancel each other.

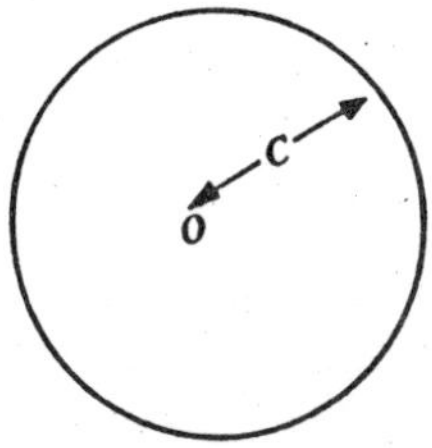

Fig. 2.16

But a molecule B at a distance less than the range of molecular attraction below the surface will be under a different condition. Since a part of its sphere of attraction will be outside the liquid surface and hence devoid of any attracting force, it will be subjected to a net downward force. Hence to bring a molecule on the surface work must be done against this force which will remain within the molecule as its potential energy. For this reason molecules on or near the surface will possess extra potential energy. But a dynamical system will tend to assume such a form in which it can have minimum potential energy. Hence the surface tends to contract so as have as smaller number of molecules on it as is possible. This endows the liquid surface with the properties of a stretched elastic membrane. It can be proved from theoretical considerations that the layer taking part in surface tension is about one molecule thick. Below this depth there is no existence of surface tension.

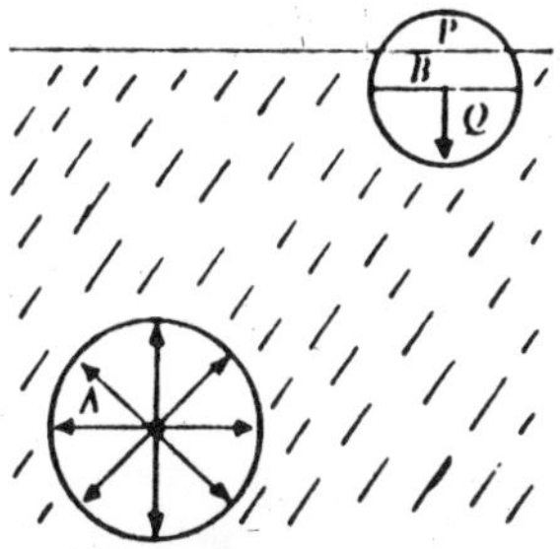

Fig. 2.17

2.12. Appendix. Gist of some important methods of determination of surface tension

(a) Capillary elevation method :

From equation (2.15) we have to a first approvment

$S=\frac{rh\rho g}{2}$. Hence measuring r, the radius of the capillary tube at the point of contact with the liquid and the elevation of the liquid surface above the general level by a travelling microscope, S can be determined.

(b) Maximum bubble pressure (Jaeger's) method :

A sketch of the experimental arrangement is shown below (Fig. 2.18).

The experiment essentially consists in measuring the maximum pressure necessary to produce an air bubble in the experimental liquid. A vertical capillary tube is dipped in the experimental liquid contained in the vessel A. The capillary tube is connected to a manometer C containing kerosene oil or any suitable liquid and also to the Woulf's bottle D fitted to a dropping funnel B. Water or any suitable liquid is allowed to trickle down into D. A difference of pressure will be resistered by the manometer and a bubble will begin to be formed at the end of the capillary tube if the apparatus is air tight. The experiment consists in noting the difference h in the levels of the manometeric liquid when the bubbles break off at the rate of about one in every ten seconds. The depth x of the end of the capillary tube inside the experimental liquid is measured by a mounted pointer and a travelling microscope M. The same microscope is used to measure the radius r of the capillary tube. Finally, the surface tension S of the liquid is determined from the relation $\frac{2S}{r} = g(\rho h - x\sigma)$, where ρ and σ are the densities of the manometric and experimental liquids respectively. More correctly, $2S = f(r)g(\rho h - xa)$, where $f(r)$ is an unknown function of r which can be found out by calibration.

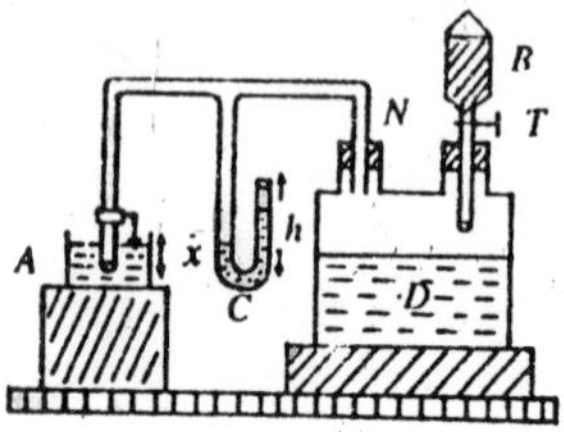

Fig. 2.18

(c) Waves on a liquid surface :

A wave is a continuous transfer of a particular state from particle to particle of the medium without any material transference of the medium. When a wave propagates through a medium the characteristics of the medium which may influence it are (i) the depth and the boundary conditions which impose limitations at the walls and the base of the containing vessel, (b) gravitation, since the changed profile or contour of the surface involves work against gravity, (c) surface tension, because the pressure over a curved surface is defferent from that under a plane surface, (d) viscocity, which is a dissipative energy. Under some circumstances one or more of these factors become negligible compared to others. For instance with waves of long wavelength or gravity waves as they are called, the curvature of the surface is small and surface tension effect may be neglected compared with gravity control. Opposite is the case with short waves or ripples, where surface tension is the main controlling factor. Moreover if the medium is regarded as a perfect fluid, so that viscosity may be neglected.

(i) Velocity of gravity waves on a liquid :

In this particular type of wave, the surface of the liquid is traversed by transverse vertical vibrations controlled by gravity. For simplicity let us assume that every drop of liquid in or near the surface describes a circular path in a vertical plane. This is very nearly the state of affairs for waves of small amplitude. Fig. 2.19 represents the section of a liquid surface traversed by such waves, in a vertical plane parallel to the direction of motion. Let C be the velocity of the waves in a horizontal direction and let us assume that every drop of liquid describes a circle of radius r in an anticlockwise direction with a time period τ. For a praticle x on the crest the instanteneous horizontal velocity is given by

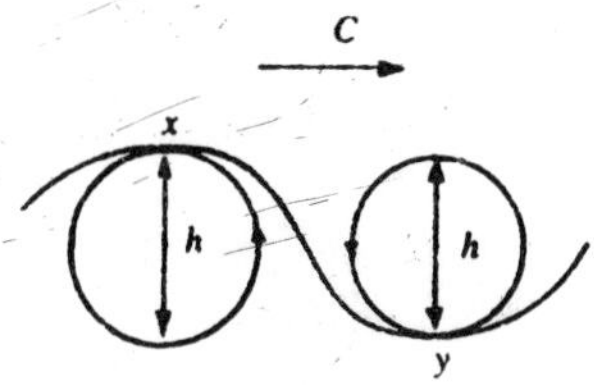

Fig. 2.19

$$q_1 = c - 2\pi\frac{r}{\tau} \tag{2.19}$$

Similarly for a point y in the trough this horizontal velocity

$$q_2 = c + 2\pi\frac{r}{\tau} \tag{2.20}$$

By Bernoulli's theorem

$$q_2^2 = q_1^2 + 2gh = q_1^2 + 4gr \tag{2.21}$$

From (2.19) and (2.20).

$$q_2^2 - q_1^2 = \frac{8\pi Cr}{\tau} \tag{2.22}$$

so that $4gr = \dfrac{8\pi Cr}{\tau}$

$$\therefore\ C = \frac{g\tau}{2\pi} = \frac{g\lambda}{2\pi C} \quad \because \lambda = C\tau$$

$$\text{or, } C = \sqrt{\frac{g\lambda}{2\pi}} \tag{2.23}$$

(ii) Effect of surface tension on the gravity waves :

In order to allow for the effect of surface tension on gravity waves, it is to be noted that whenever there is a curved surface, there is an excess pressure directed from the concave side to the convex side equal to $S\left(\dfrac{1}{R_1}+\dfrac{1}{R_2}\right)$. Let us consider a simple harmonic wave travelling over a liquid, the displacement equation of the wave being $y = a\,\mathrm{Sin}\frac{2\pi}{\lambda}(ct - x)$, symbols having their usual meanings.

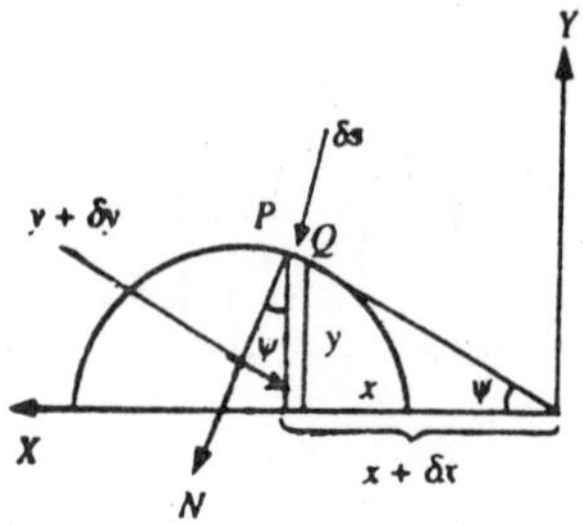

Fig. 2.20

If we assume the wave surface to be cylindrical, then of the two radii of curvaturer one will be (say R_1) infinity and

$$R_2 \approx \frac{1}{\frac{d^2y}{dx^2}}, \text{ since } \frac{dy}{dx} \text{ is small compared to}$$

unity in the case of ripples.

$$\text{Again } \frac{d^2y}{dx^2} = \frac{4\pi^2 y}{\lambda^2}, \text{ since } y = a\,\mathrm{Sin}\frac{2\pi}{\lambda}(ct - x)$$

Thus at a point such as $P(x + \delta x,\ y + \delta y)$ as shown in figure (2.20) surface tension causes an excess of pressure directed along NP of magnitude $\frac{S}{R_2}$ or along PN of magnitude $\frac{-S}{R_2}$, that is of magnitude $\frac{4\pi^2 Sy}{\lambda^2}$. The excess force on the element PQ whose profile curve is ds and thickness unity measured perpendicularly to the plane of the figure is $\frac{4\pi^2 Syds}{\lambda^2}$ along PN or,

$$\frac{4\pi^2 Syds}{\lambda^2}\cos\psi = \frac{4\pi^2 Sy}{\lambda^2} ds\frac{dx}{ds} = \frac{4\pi^2 Sydx}{\lambda^2} \text{ measured vertically downwards.}$$

The weight of the column standing on dx is $(dx \times 1)y\rho g = y\rho g dx$. The total downward force on the element is therefore $g\rho y dx + \frac{4\pi^2 Sydx}{\lambda^2}$

$$= y\rho dx\left\{g + \frac{4\pi^2 S}{\lambda^2\rho}\right\}.$$

The effect of surface tension is as it were to change g to $g + \frac{4\pi^2 S}{\lambda^2\rho}$. Now the velocity of the gravity waves on deep liquids has been found to be $C = \sqrt{\frac{g\lambda}{2\pi}}$. Hence the velocity of the waves controlled both by gravity and surface tension is $C = \sqrt{\frac{\lambda}{2\pi}\left(g + \frac{4\pi^2 S}{\lambda^2\rho}\right)}$

or, $C^2 = \frac{g\lambda}{2\pi} + \frac{2\pi S}{\lambda\rho}$. (2.24)

For sufficiently large values of λ the first term is large compared with the second and the wave velocity is given by $C = \sqrt{\frac{g\lambda}{2\pi}}$. (2.25)

These are gravity waves.

If on the other hand λ is very small, the second term predominates and waves are mainly governed by surface tension. In this case $C = \sqrt{\frac{2\pi S}{\rho\lambda}}$ (2.26)

These are called ripples. It will be noticed that as the wave-length diminishes from infinity to zero, the wave velocity as given by (2.24) falls to a minimum and then increases. The minimum value occurs at the wave length λm, where

$$2C\frac{dc}{d\lambda} = \frac{g}{2\pi} - \frac{2\pi S}{\lambda^2\rho} = 0$$

or, $\lambda m^2 = \frac{4\pi^2 S}{\rho g}$. (2.27)

Lord Kelvin has suggested that waves of wavelength less than

$\lambda m = 2\pi\sqrt{\frac{S}{\rho g}}$ (2.28)

may be called ripples.

(d) Ripple method of measurement of surface tension :

From equation (2.24), the velocity of the waves controlled both by gravity and surface tension is given by $C^2 = \frac{g\lambda}{2\pi} + \frac{2\pi S}{\rho\lambda}$. The method used by *R.C.* Brown is discussed briefly here. The liquid of which S is to be determined is contained in a shallow brass tank whose edges have been previously waxed (Fig. 2.21).

The ripples are produced by the dipper D rigidly attached to reed R_2 which can be energised by passing alternating current through the loud speaker unit P_2. The length of the reed is so adjusted that its natural frequency coincided with that of the a.c. supply. The wavelength of the ripples is measured by stroboscopic arrangement. Light from a lamp S is focussed by the lens L_1 on a horizontal slit S_1. By means of a concave mirror M_1, light from S_1 is focussed at S_2, a point on the first focal plane of the lens L_2. In this way a beam of parallel light is made incident on the plane mirror M_2 inclined at such angle to the horizontal that a parallel beam of light falls on the disturbed surface of the liquid. Now the concave mirror M_1 is attached to a reed R_1 tuned to the frequency of the a.c. supply and energised by the current through a second laud speaker P_1 in series with P_2. Hence for every ripple produced by D i.e. during one complete vibration of R_2, the beam of light from M_1 passes twice a cross S_2 so that the intermittent illumination of water surface has a frequency twice that of the dipper. The telescope T enables the intermittent light reflected from the water surface to be examined. The image will consist of a series of straight lines parallel to the edge of the dipper. The distance between the successive lines when doubled gives the wavelength

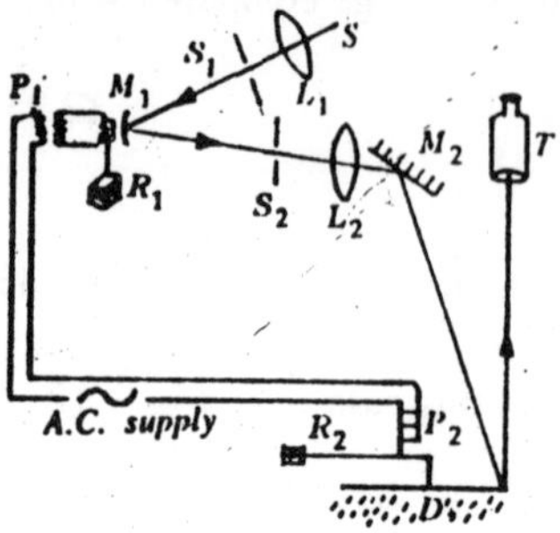

Fig. 2.21

λ and consequently S can be determined from the relation $n^2\lambda^2 = \dfrac{g\lambda}{2\pi} + \dfrac{2\pi S}{\rho\lambda}$

$$\text{or, } S = \frac{\lambda^3 n^2 \rho}{2\pi} - \frac{g\lambda^2\rho}{4\pi^2}.$$

Problems :

(a) Floating needle :

A sewing needle carefully placed on the surface of greese free water floats. This can be explained in the light of surface tension effects.

As shown in Fig. (2.22) which shows a section by a vertical plane, the forces acting on the needle are—

(i) The surface tension A and B.

(ii) The weight of the needle.

(iii) The water pressure acting on AFB equal to the weight of a column of water whose cross-section is $DCAFBD$. These forces are in equilibrium. Let θ be the angle of contact between the needle and water, m = mass per unit length of the needle, z is the height of the axis of the needle above the free surface of water, r = radius of the needle, $2\varphi = \angle BOA$ and ρ = density of water.

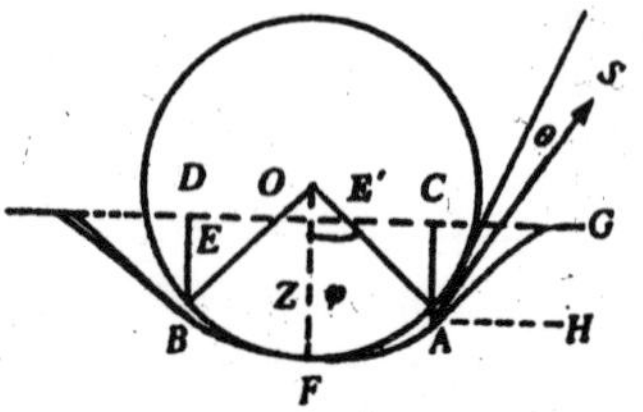

Fig. 2.22

[Lemma : $\angle CAS = 90° - \varphi + \theta = 90° - (\varphi - \theta)$

$$\text{Area } OAB = \int_0^{2\varphi} \frac{1}{2} r^2 d\varphi = r^2\varphi$$

Area $ACE' =$ Area $DBE = \frac{1}{2}(r\cos\varphi - z)\ \ (r\sin\varphi - z\tan\varphi)$

$= \frac{1}{2}\left\{r^2 \text{ Sin }\varphi\cos\varphi - 2rz\sin\varphi + z^2\tan\varphi\right\}$

Area $OEE' = 2.\frac{1}{2}z.z\tan\varphi = z^2\tan\varphi$

$\therefore$ Area $DCAFBD$ = Area OAB + 2 Area BDE – Area OEE'

$= r^2\varphi + r^2\sin\varphi\cos\varphi - 2rz\sin\varphi + Z^2\tan\varphi - Z^2\tan\varphi$

Hence for vertical equilibrium

$2S\sin(\varphi-\theta) + g\rho\left\{r^2\varphi + r^2\sin\varphi\cos\varphi - 2rz\sin\varphi\right\} = \text{mg}$

Similarly for horizontal equilibrium, the horizontal component of surface tension at A together with the horizontal pressure at GA due to water is equal to the surface tension at G.

$\therefore S\cos(\varphi-\theta) = \frac{1}{2}(r\cos\varphi - z)(r\cos\varphi - z)\rho g + S$

or, $S\left\{1 - 2\sin^2\frac{\varphi-\theta}{2}\right\} = \frac{1}{2}(r\ \cos\varphi - \text{z})^2\rho g + S$

or, $4S\text{ Sin}^2\frac{\varphi-\theta}{2} + g\rho(r\cos\varphi - z)^2 = 0$

(b) A spherical soap bubble of 5 cm raidus contracts until its radius is 2 cms. Calculate the volume at N. T. P. of dry air expelled from the bubble. Room temperature $=20°\text{C}$ Atmospheric pressure $=760$ mm of mercury Density of mercury $=13{\cdot}6$ gm/cc. vapour pressure of soap solution at 20°C = 17·4 mm and surface tension of soap solution = 20 dynes/cm. **[C.U. (Hons.)]**

Solution :

Pressure of dry air in the former case

$= \left\{\frac{4\times 20}{5} + 76\times 13{\cdot}6\times 981\ -\ 1{\cdot}74\times 13{\cdot}6\times 981\right\}$

$= 16 + 1{\cdot}013\times 10^6 - 2{\cdot}321\times 10^4$

$= \{{\cdot}0016\times 101{\cdot}3 - 2{\cdot}321\}\times 10^4$

$= 98{\cdot}9806\times 10^4$ dynes/cm^2

$= P_1$ say

Pressure of dry air in the latter case

$$=\left\{\frac{4\times 20}{2}+1\cdot013\times10^{6}-2\cdot321\times10^{4}\right\}$$

$$=\{\cdot004+101\cdot3-2\cdot321\}\times10^{4}=98\cdot983\times10^{4}\text{dynes/cm}^{2}=P_2 \text{ say}$$

Let M_1 and M_2 be the masses of dry air present in the two cases and R the gas constant for unit mass

Then, $\dfrac{P_1\times\frac{4}{3}\pi\times5^3}{T}=M_1R$ and

$$\frac{P_2\times\frac{4}{3}\pi\times2^3}{T}=M_2R$$

$$\therefore M_1R-M_2R=MR=\frac{4\pi}{3T}\{P_1\times5^3-P_2 2^3\}$$

M is the mass of air expelled at 760 mm and 20°C. If v be its volume at N.T.P.

$$MR=\frac{76\times981\times13\cdot6\times V}{273}=\frac{\frac{4}{3}\pi\{P_1\times5^3-P_2\times2^3\}}{20+273}$$

$$\therefore V=\frac{4\times3\cdot14\times273}{3\times293\times1\cdot013\times10^6}\{98\cdot9806\times125-98\cdot983\times8\}\times10^4$$

$$=446\cdot2 \text{ c.c.}$$

(c) A soap film of thickness 0·0001 cm is at the temperature of melting ice. Calculate the fall in temperature due to stretching it adiabatically until its area is doubled. Given $\dfrac{dS}{dT}=\cdot15$, sp. heat = sp. gr = 1.

From the relation $E=S-T\dfrac{\partial S}{\partial T}$, we have

$$T\frac{\partial S}{\partial T}=\text{heat absorbed per unit area}=\frac{ms\theta J}{2A}$$

$$\therefore \theta=T\frac{\partial S}{\partial T}\frac{2A}{msJ}=\frac{273\times\cdot15\times2A}{AtsJ}$$

$$=\frac{273\times\cdot15\times2}{4\cdot2\times10^7\times\cdot0001}=0\cdot0195^\circ\text{ C}$$

(d) Two light flexible threads AC and BD hanging from a fixed horizontal bar AB support a rod CD of mass m gms, such that $ABCD$ is a rectangle. When a soap film is stretched on $ABDC$, the surface tension of the film causes each

of the threads to take the form of arc of a circle of radius r and the distance between the mid points of the threads reduces to x cm. Neglecting the weight of the film, derive expressions for the tensions of the threads and the surface tension of the film in terms of r and x. [C.U. (Hons.)]

Solution :

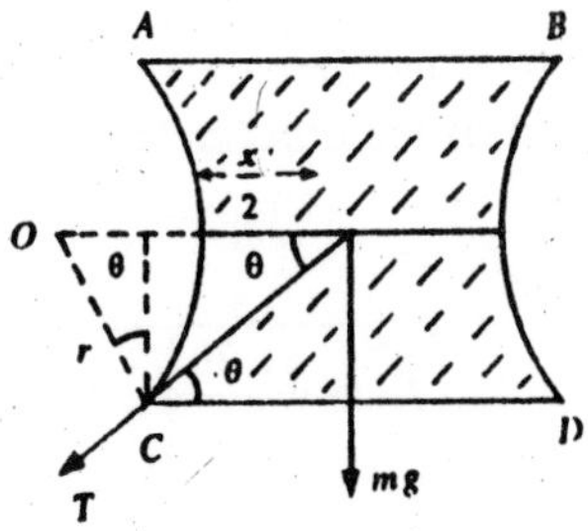

Fig. 2.23

The fact is shown in Fig. 2.23.

With reference to this figure

$$\text{Sin } \theta = \frac{r}{r + \frac{x}{2}} \quad \text{... (1)}$$

Considering the horizontal equilibrium of the thread AC,

$$2T \text{ Cos } \theta = 2\{2.r \text{ Cos } \theta\}S$$

$$\therefore T = 2rS \quad \text{.... (2)}$$

For vertical equilibrium of the rod CD,

$$mg = 2T \text{ Sin } \theta + 2S \left\{2\left(\frac{x}{2} + r - r \text{ Sin } \theta\right)\right\}$$

$$= 2 \times 2rS \text{ Sin } \theta + 2xS + 4rS - 4rS \text{ Sin } \theta$$

$$= 2S\,(x + 2r)$$

$$\left.\begin{aligned} \therefore\ S &= \frac{mg}{2(x+2r)} \\ T &= \frac{mgr}{x+2r}. \end{aligned}\right\} \quad \text{.... (3)}$$

(e) Two spherical soap bubbles of diameter 10 cm and 6 cm respectively are formed, one at each end of a narrow horizontal glass tube. What is the pressure difference between the ends of the tube, if the surface tension of soap solution be 30 dynes/cm? What happens when the bubbles are put into pressure communication with each other through the tube. [C.U. (Hons.)]

Solution :

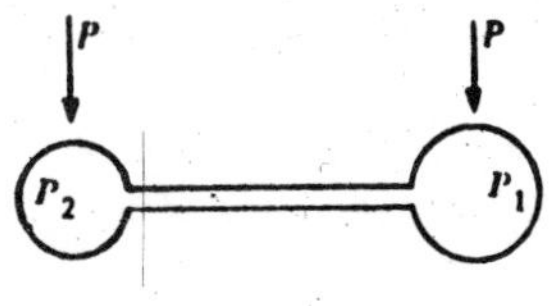

Fig. 2.24

$P_1 = P + \frac{4S}{r_1}$; $P_2 = P + \frac{4S}{r_2}$, where P is the atm. pressure

$$\therefore P_1 - P_2 = 4S\left(\frac{1}{r_1} - \frac{1}{r_2}\right)$$

$$= 4 \times 30 \left\{ \frac{1}{3} - \frac{1}{5} \right\}$$

$$= 4 \times 30 \times \frac{5-3}{16} = 16 \text{ dynes/cm}^2.$$

When in communication the radius of the smaller drop will increase and that of the other drop will decrease until they assume a common pressure and hence common radius r can be found out if the external pressure P is known

$$\left\{ P + \frac{4 \times 30}{3} \right\} \frac{4}{3} \pi 3^3 + \left\{ P + \frac{4 \times 30}{5} \right\} \frac{4}{3} \pi 5^3$$

$$= \left\{ P + \frac{4 \times 30}{r} \right\} \frac{4}{3} \pi r^3 \times 2$$

(f) Show that during a capillary rise half the energy is last in overcoming the dissipative forces (viscosity etc.). [C.U. (Hons. '87)]

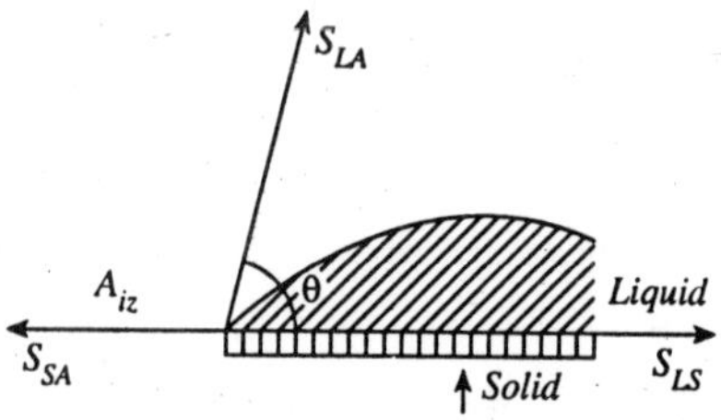

Fig. 2.25

Let us first consider the interface of a liquid, solid and air. Let S_{LA}, S_{SA} and S_{LS} be the surface tensions of the liquid in contact with air, solid in contact with air and liquid in contact with solid respectively. If θ be the angle of contact between the liquid and the solid, we have (Fig. 2.25), for equilibrium $S_{SA} = S_{LS} + S_{LA} \cos\theta$ or $S_{LA} \cos\theta = S_{SA} - S_{LS}$. Now let us consider the rise of the liquid through a height h in a capillary tube.

Let r be the radius of the capillary tube and h, the height through which the liquid rises in the tube (Fig. 2.26).

Loss of free surface energy

$$= 2\pi rh(S_{SA} - S_{LS})$$

$$= 2\pi rh\, S_{LA} \cos\theta = 2\pi rh \frac{rh\rho g}{2}$$

$$= \pi r^2 h^2 \rho g$$

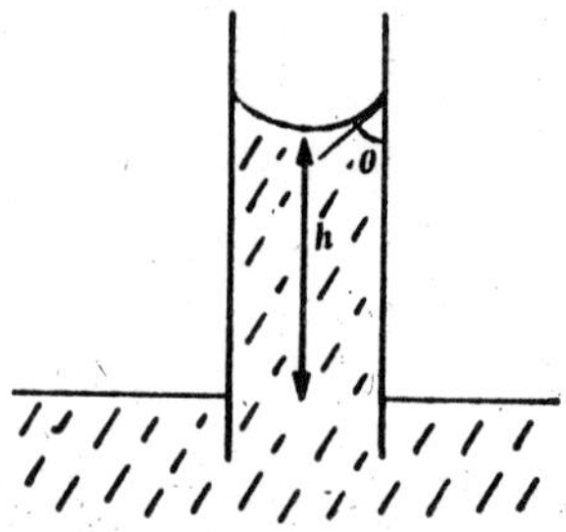

Fig. 2.26

[Area previously covered by air is replaced by the liquid.]

Gain of gravitational $P.E. = \pi r^2 h\rho g \cdot \frac{h}{2}$

Hence the energy lost in overcoming dissipative forces is $\pi r^2 h^2 \rho g - \pi r^2 h^2 \rho g/_2 = \pi r^2 h^2 \rho g/_2$

(g) One gm of mercury is enclosed between two horizontal glass plates so that the liquid spreads in a circle of radius 5 cm. Find the downward force on the upper plate if the angle of contact is 140°, surface tension of mercury is 440 dynes/cm and its density is 13·6 gm/c.c. **[C.U. (Hons.) '89]**

Solution :

Ans. Let d be the separation between the plates. As is evident from Fig. 2.27,

$$\frac{d/2}{r} = \cos(180° - \theta)$$

$$= -\cos\theta$$

$$\therefore r = -\frac{d}{2\cos\theta}$$

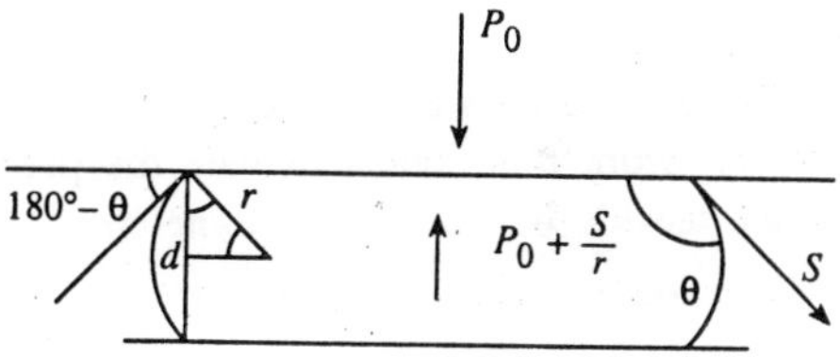

Fig. 2.27

Downward force on the upper plate = F

$$= P_0\pi R^2 + 2\pi RS\sin(180° - \theta) - \left(P + \frac{S}{r}\right)\pi R^2$$

$$= 2\pi RS\sin\theta - \frac{S}{r}\pi R^2$$

$$= 2\pi RS\sin\theta + \frac{2\pi R^2 S\cos\theta}{d}$$

Again $\pi(5)^2 d \times 13{\cdot}6 = 1$ or, $d = \dfrac{1}{3{\cdot}14 \times 25 \times 13{\cdot}6}$

Substituting, $F = 2\pi \times 5 \times 440 \sin 40^o - 2\pi \times 25 \times 440 \cos 40^o \times 3{\cdot}14 \times 25 \times 13{\cdot}6$ dynes

N.B. : In the case of liquid having an acute angle of contact we have as shown in Fig. 2.28.

$$\frac{d}{2r} = \cos\theta \quad \text{or,} \quad r = \frac{d}{2\cos\theta}$$

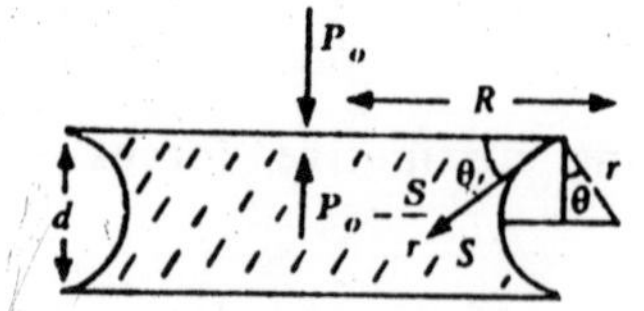

Fig. 2.28

Downward force on the upper plate

$$= P_0\pi R^2 + 2\pi R \sin\theta - \left(P_0 - \frac{S}{r}\right)\pi R^2$$

$$= 2\pi RS \sin\theta + \frac{2S\pi R^2 \cos\theta}{d}$$

(h) A soap bubble of surface tension 25 dynes/cm has a radius of 5 cm. Find the work done in blowing the bubble to the radius of 10 cm.

Ans. Work done = Surface tension × increase in area

$$= 2 \times 25 \times 4\pi\left\{10^2 - 5^2\right\} = 25 \times 4 \times 3 \cdot 14 \times 75 \times 2 = 4 \cdot 71 \times 10^4 \text{ erg}$$

(Considering two surfaces)

(i) A drop of water 0·5 cm in radius is split into 1000 equal droplets. Find the work done and calculate the pressure inside one of the droplets, surface tension of water is 72 dyne/cm.

Solution :

Volume of the original drop $= \frac{4}{3}\pi(\cdot 5)^3$

If r be the radius of the droplets,

$$\frac{4}{3}\pi r^3 \times 1000 = \frac{4}{3}\pi(\cdot 5)^3$$

or, $r = 0 \cdot 05$ cm.

Increase in surface area due to the formation of the droplets

$$= 4\pi(\cdot 05)^2 \times 1000 - 4\pi \times (\cdot 5)^2$$

$$= 4\pi \times 2 \cdot 5 - 4\pi \times \cdot 25$$

$$= 4\pi \times 2 \cdot 25 = 9\pi$$

Hence the required work done $= 9\pi \times S$

$$= 9 \times 3 \cdot 14 \times 72 = 2035 \text{ ergs}$$

Pressure inside the one of droplets

$$= P_0 + \frac{2S}{r} = P_0 + \frac{2 \times 72}{0 \cdot 05}$$

$$= P_0 + 2880 \text{ dynes/cm}^2,$$

P_0 being the atomic pressure.

(j) If a number of little droplets of water, each of radius r coalesce to form a single drop of radius R, show that the rise in temperature will be gv by $\frac{3S}{J}\left(\frac{1}{r}-\frac{1}{R}\right)$, where S is the surface tension of water and J the mechanical equivalent of heat.

Solution :

When a number of droplets coalesce to form a single drop, the surface area decreases and hence the surface energy also decreases. The decrease in surface energy is equal to the energy of the decreased surface. Let us suppose that n drops coalesce. Then

surface area of n droplets $= n4\pi r^2$

surface area of the single drop $= 4\pi R^2$

$\therefore$ decrease in surface area $= n4\pi r^2 - 4\pi R^2$

$$= 4\pi\left(nr^2 - R^2\right)$$

Energy released $= 4\pi\left(nr^2 - R^2\right)S$

If θ be the rise in temperature

$$4\pi\left(nr^2 - R^2\right)S = \frac{4}{3}\pi R^3 \times 1 \times 1 \times \theta . J$$

Since density of water $= 1$

Sp. heat of water $= 1$

$$\therefore \ \theta = \frac{3\left(nr^2 - R^2\right)S}{JR^3} = \frac{3S}{J}\left\{\frac{nr^2}{R^3} - \frac{1}{R}\right\}$$

But $\frac{4}{3}\pi r^3 n = \frac{4}{3}\pi R^3$ or, $R^3 = nr^3$

$$\therefore \theta = \frac{3S}{J}\left\{\frac{1}{r} - \frac{1}{R}\right\}$$

(k) If the rate of change of surface energy E, of a liquid with temperature is proportional to the absolute temperature T, show that $\frac{dE}{dt}+\frac{dS}{dt}$ is constant and that the surface tension S is a quadratic function of the temperature.

Solution :

We have the relation

$$E = S - T\frac{dS}{dt} \qquad \text{.... (1)}$$

According to problem

$$\frac{dE}{dT} = KT \quad \text{.... (2)}$$

From (1) $\frac{dE}{dT} = \frac{dS}{dT} - \frac{dS}{dT} - T\frac{d^2S}{dT^2} = KT$

or, $\frac{d^2S}{dT^2} = -K$

Integrating twice *w. r. t. T*

$$\frac{dS}{dT} = -\mathrm{KT} + \mathrm{K}_1$$

and $\mathrm{S} = -\frac{\mathrm{KT}^2}{2} + \mathrm{K}_1\mathrm{T} + \mathrm{K}_2$ which is a quadratic function in *T*

Also $\frac{dE}{dT} + \frac{dS}{dT} = KT - KT + K_1 = K_1 =$ constant.

(l) The *surface* tension of mercury is 520 dynes/cm the angle of contact with glass is 140°, the density of mercury is 13·6 gm/c.c. Find the maximum length of a mercury thread which can be supported by surface tension in a vertical capillary tube of radius 0·010 cm.

Solution :

If *l* be the maximum length,

thus $\pi r^2 l \rho g = 2\pi S r \cos\theta$

or, $l = \frac{2\pi r S\cos\theta}{\pi r^2 \rho g} = \frac{2\times 3{\cdot}14\times 520\times \cos 140^\circ \times {\cdot}01}{3{\cdot}14\times 13{\cdot}6\times 980\times({\cdot}01)^2}$

$$= \frac{2\times 3\cdot 14\times 520\times 0\cdot 766\times\cdot 01}{13\cdot 6\times 980} = 1\cdot 87\,\mathrm{cm}\times 10^{-3} = 0\cdot 00187\,\mathrm{cm}.$$

(m) If two soap bubbles of radii *a* and *b* colesce to form a single bubble of radius *c*, show that the surface tension *S* of the liquid is given by $S = \frac{B\left(C^3 - a^3 - b^3\right)}{4\left(a^2 + b^2 - c^2\right)}$, Where *B* is the external pressure.

Solution :

Using the relation $P_1V_1 + P_2V_2 = PV$

$$\left(B + \frac{4S}{a}\right)\frac{4}{3}\pi a^3 + \left(B + \frac{4S}{b}\right)\frac{4}{3}\pi b^3 = \left(B + \frac{4S}{c}\right)\frac{4}{3}\pi c^3$$

or, $B\left(a^3+b^3-c^3\right)=4S\left(c^2-a^2-b^2\right)$

$$\therefore\ S=\frac{B\left(a^3+b^3-c^3\right)}{4\left(c^2-a^2-b^2\right)}=\frac{B\left(c^3-a^3-b^3\right)}{4\left(a^2+b^2-c^2\right)}$$

EXERCISE

1. Define surface tension and surface energy. Deduce the relation between them.

2. A glass tube of circular cross-section is closed at one end. This rod is weighted and the tube floats vertically in water, heavy end down. How far below the water surface is the end of the tube? Given outer radius of the tube 0·14 cm; mass of weighted tube 0·2 gm; surface tension of water 73 dynes/cm and $g = 980$ cm/sec².

Solution :

Let l be the length of the tube inside water.

The forces acting on the tube are :

(i) Upthrust of water $=\pi r^2 l\times 1\times 980$

$$=3\cdot 14\times(0\cdot 14)^2 l\times 980=60\cdot 368l\ \text{dynes}$$

(ii) Weight of the system acting downwards

$= 0\cdot 2 \times 980 = 196$ dynes.

(iii) Force of surface tension acting downward

$=2\pi rS=2\times 3\cdot 14\times 0\cdot 14\times 73=64\cdot 24$ dynes

Since the tube is in equilibrium

$60\cdot 368l=196+64\cdot 24=260\cdot 24$

$$\therefore l=\frac{260\cdot 24}{60\cdot 368}=4\cdot 31\,\text{cm}$$

3. Show that the excess pressure on the concave side of a liquid surface is $S\left(\frac{1}{r_1}+\frac{1}{r_2}\right)$, where r_1 and r_2 are the principal radii of curvatures of the surface.

4. A soap bubble of radius a is blown on a tube with a solution whose surface tension is S. The atmospheric pressure is P. The tube on which bubble is blown is connected to a vessel of volume V filled with air at atmospheric pressure. Show that bubble will shrink until the radius is

$$\frac{1}{2}r \text{ if } 2\pi r^2 S + \frac{1}{11}\pi r^3 P = \frac{4SV}{r}.$$

5. A large air bubble is collected between the lower side of a concave glass surface of radius 50 cm and the upper surface of some ether. The following observations are made : radius of the circle of contact of the air bubble and glass plate 5 cm, depth of ether surface below the centre of glass surface 0·565 cm; height of the widest horizontal section of bubble above bottom point 0·225 cm. If the density of ether is 0·714 gm/c.c. and g = 981, find the surface tension of air ether surface and the angle of contact between ether and glass.

[17·7 dynes/cm 10·5°]

6. A vertical *u*-tube contains a liquid. One end of this tube is open to the atmosphere while a soap film is formed across the end of the other limb. A side tube is attached to the *u*-tube so that, by blowing, the soap film may be distended. Show that for different bubbles the product of the radius of the bubble and the difference in height of the liquid levels in the *u*-tube is constant. If the liquid in the *u*-tube is water and the above constant is 0·123 cm², calculate the value of the surface tension of the soap.

Solution :

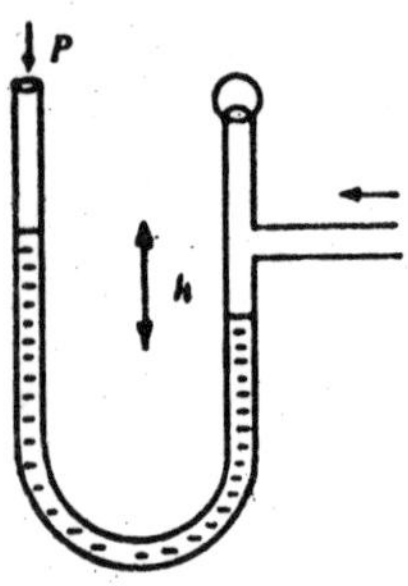

Fig. 2.29

Let P be the atmospheric pressure, h the difference in liquid levels and r the radius of the bubble.

Excess pressure inside the bubble

$$= P + h\rho g - P = \frac{4S}{r}$$

$$\text{or,} \quad hr = \frac{4S}{\rho g} = \text{Constants}$$

For water $\rho = 1$ gm/C.C.

$$\therefore S = \frac{\text{Const.} \times g}{4} = \frac{0 \cdot 123 \times 980}{4}$$

$= 30·135$ dynes/cm.

7. A capillary tube of 0·8 mm bore stands vertically in a wide vessel containing a liquid of surface tension 30 dynes/cm. The liquid wets the tube and has a density 0·8 gm/c.c. Calculate the rise of the liquid in the tube. What would be the rise if the capillary is inclined at 45° with the vertical.

Solution :

Assuming the angle of contact to be zero, the rise h is given by

$$S = \frac{rh\rho g}{2} \quad \text{to a first approximation}$$

or, $h = \frac{25}{r\rho g} = \frac{2 \times 30}{0.04 \times 0.8 \times 980} = 1 \cdot 89\,\text{cm}$

when the tube is inclined at 45° with the vertical,

$h^1 = 1 \cdot 9 \text{ Cos } 45° = 1 \cdot 9 \times \sqrt{2} = 2 \cdot 7\,\text{cm}.$

8. A soap bubble of surface tension 30 dynes/cm is slowly enlarged from a radius of 2 cm to a radius of 20 cm. Calculate the amount of work necessary for the enlargement and explain why this is less than if the increase were made at a faster rate.

Hints : Work necessary = Increase in area × surface tension

$= 2 \times 4 \times 3 \cdot 14(400 - 4) \times 30 = 2 \cdot 98 \times 10^5$ ergs

when the area is increased at a faster rate no heat will enter the film and the change will be adiabatic.

9. A Soap solution of surface tension S is used to form a film between a horizontal rod of length l and a length of weightless inextensible thread attached to each end of this rod. A weight W is attached to the mid point of the thread. Prove that (a) the shape of each half of the thread is circular and (b) tension of the thread is equal to

$\frac{Sl}{\cos\theta_2 - \sin\theta_1}$ θ_1 is the angle which the tangent to the thread at end makes with the horizontal and $2\theta_2$ is the angle between the tangents to the thread at the point where W is attached.

Solution :

Fig. 2.30 explains this problem.

To start with let us consider the equilibrium of the element PQ of length δs of one part of the thread. Let the normals at P and Q meet at C, where $\angle PCQ = \delta\theta$, and let r be the radius of curvature at P (Fig. 2.31).

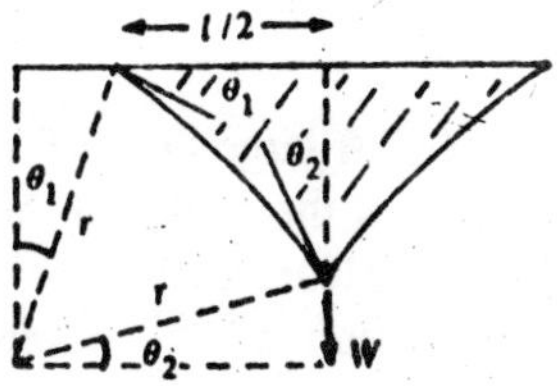

Fig. 2.30

If T_1 and T_2 are the tensions at P and Q, then neglecting the weight of the element,

$T_1 \text{Cos}\frac{\delta\theta}{2} - T_2 \text{Cos}\frac{\delta\theta}{2} = 0$ for equilibriun along PQ and $2S\delta s = (T_1 + T_2)$

$\text{Cos}\left(90° - \frac{\delta\theta}{2}\right)$ for equilibrium normal to PQ.

Solving $T_1 = T_2 = T$ say i.e. **the tension** in the thread is constant, and

$$2S\delta s = 2T\sin\frac{\delta\theta}{2} = T\delta\theta$$

$$\therefore \frac{ds}{d\theta} = Lt\frac{\delta s}{\delta\theta} = \frac{T}{2S} = \text{Constant}$$

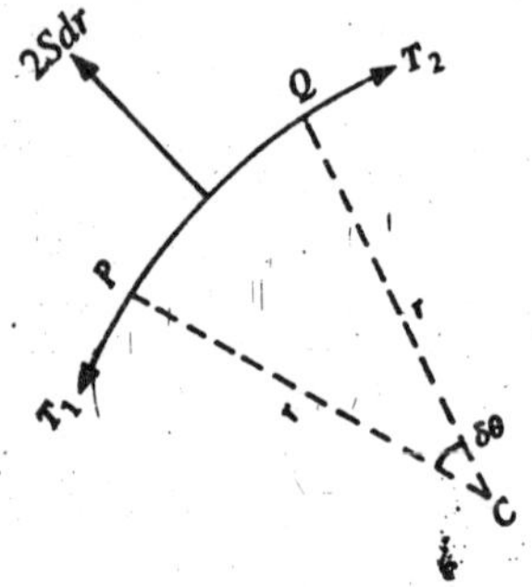

Fig. 2.31

In other words the radius of curvature is constant, or the thread takes the form of a circle of raius r, where $r = PC$. Hence

$$T = 2Sr$$

For the second part of the problem we have from Fig. 2.30, $r\cos\theta_2 - r\sin\theta_1 = \frac{l}{2}$

$$\text{or, } 2r = \frac{l}{\cos\theta_2 - \sin\theta_1}$$

$$\text{Hence } T = \frac{ls}{\cos\theta_2 - \sin\theta_1}.$$

10. A drop of water of radius r is formed at the lower end of a vertical capillary tube. The height of water in the tube above the lowest point of the drop is h_1, when the drop and lower end of the tube are immersed in a beaker of water at such a depth that the level in the capillary tube is unaltered this level is at a heigt h_2 above the level of the beaker.

Show that

$$\frac{2S}{g\rho} = r(h_1 - h_2) - \frac{r^2}{3},$$ where S is the surface tension and ρ is the density.

11. A large drop of a liquid lies on a horizontal surface in air. Take the origion at the centre of the upper surface and show that at any point P in a vertical plane section the ordinate y and the slope ψ of the tangent at the intersection of the plane and the surface are related by the equation

$$\frac{g\rho y^2}{2S} = 1 - \cos\psi,$$ where ρ is density of the liquid and S the surface tension.

Assume the curvature in horizontal section through P is very small.

Describe and explain how this formula can be applied to determine the surface tension and the angle of contact in certain cases.

12. A narrow tube of internal radius r is in the form of inverted u and at the top point air can be forced in through a tube connected by a T joint. If the vertical limbs stand at depths h_1 and h_2 in two liquids of densities ρ_1 and ρ_2 and surface tensions S_1 and S_2, find the condition that the bubble should be formed in the liquid of density ρ_1 and show that as the depth h_1 is increased bubbling will be suddenly changed over to the other liquid.

$$\left[(h_2\rho_2 - h_1\rho_1)gr > 2(S_2 - S_1)\right]$$

13. Calculate the minimum aqueous vapour pressure required to prevent evaporation of a drop of water of radius 10^{-4} mm at 0°C. Density of saturated aqueous vapour at 0°C $= 4.8\times10^{-6}$ gm/c.c., surface tension of water at $0°C = 75$ dyne/cm. saturated aqueous vap. pressure at 0°C over a plane surface $= 6\times10^3$ dynes /cm^2.

Solution :

Hints : Let P = saturated vap. pressure over a plane surface,

P_1 = saturated vap. pressure over a drop of radius r to prevent its evaporation.

Then $log_e \frac{P_1}{P} = \frac{2S}{\rho RTr}$, ρ = density of the liquid.

The density σ of the saturated vapour is given by

$$P = \sigma RT \text{ or, } RT = \frac{P}{\sigma}$$

$$\therefore \log_e \frac{P_1}{P} = \frac{P_1}{P} = \frac{2S\sigma}{P\rho r}$$

$$\text{or, } 2\cdot3\left(\log_{10} P_1 - \log_{10} 6\times10^3\right) = \frac{2\times76\times4\cdot8\times10^{-6}}{1\times6\times10^3\times10^{-5}}$$

$$\therefore P_1 = 6074 \text{ dynes}/\text{cm}^2.$$

14. What are gravity waves? How do they differ from ripples? Derive an expression for the velocity of gravity waves. How will the expression be modified if the waves are also controlled by surface tension? Calculate the velocity of the slowest moving ripples.

15. (a) Explain why small drops of liquid are spherical while larger drops are somewhat flattened.

(b) Derive an expression for the excess pressure on the concave side of a curved liquid surface.

16. A thin circular rubber band when unstretched has a radius of 3·8 cm. It is placed on a soap film and the film within the ring is thus broken. The new radius of the ring is found to be 3·9 cm. The ring is then cut at a place, and it is found that a force of 0·28 gm ωt is required to increase its natural length by 1 cm. Calculate the value of the surface tension of the soap solution from which the film was formed.

Hints for solution :

As in problem no. 9, if T be the tension on the rubber band,

$$T = 2Sr, \quad S = \text{ surface tension},$$

r = radius of the ring. This tension causes an increase in length of $2\pi(\cdot 1)$.

Hence unit increase of length corresponds to a tension of

$$\frac{2Sr}{2\pi(\cdot 1)} = \cdot 28 \times 980$$

$$\text{or, } S = \frac{\cdot 2\pi \times \cdot 28 \times 980}{2 \times 3 \cdot 8} = \frac{2 \times 3 \cdot 14 \times \cdot 28 \times 980}{2 \times 3 \cdot 8}$$

$$= 22{\cdot}67 \text{ dynes/cm.}$$

17. A glass plate of length 10 cm, breadth 1·54 cm and thickness 0·20 cm weighs 8·2 gm in air. It it is held vertically with the long side horizontal and the lower half immersed in water what will be its apparent weight? Surface tension of water 73 dynes/cm.

Hints for solution :

The forces acting on the plate are :

(a) Its weight

(b) upthrust of water

(c) downward force due to surface tension acting along the edges immersed in water.

Now upthrust = ωt. of the displaced water $\dfrac{10 \times 1 \cdot 54 \times \cdot 20 \times 1 \times 980}{2}$

$= 1{\cdot}54 \times 1 \times 980$ dynes = $1{\cdot}54$ gm. ωt.

Force due to surface tension acting downwards along a distance equal to twice the sum of the length and thickness $= 2\{10 + \cdot 2\} = 20 \cdot 4$ cm.

Total downward force due to surface tension = $20{\cdot}4 \times 73$ dynes

$= \frac{20 \cdot 4 \times 73}{980}$ gm wt. $= 1{\cdot}518$ gm ωt

Hence the net upward thrust on the plate $= 1 \cdot 54 - 1 \cdot 518 = 0 \cdot 022$ gm ωt.

18. A conical glass tube 12 cm long, 0·12 cm in diameter at one end and 0·04 cm at the other is fixed vertically with its wider end just touching the surface of water, for which the surface tension is 72 dyne cm^{-1}. To what height will the water rise in the tube?

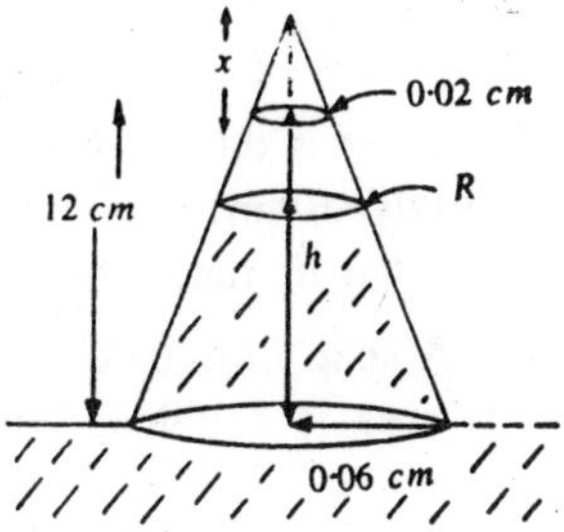

Fig. 2.32

Hints for solution :

Let h be the required height and R the radius of the cone at this place (Fig. 2.32).

Now $\frac{x}{0{\cdot}02} = \frac{x+12}{0{\cdot}06}$ from similar triangles.

Solving $x = 6$ cm.

We have $\frac{2S}{R} = hPg$

$$\text{or,} \quad h = \frac{2S}{R\rho g} \qquad \text{... (1)}$$

If r be the radius (0·06 cm) at the lowest point and l the total height (18 cm) of the cone.

Then $\frac{R}{r} = \frac{l-h}{l}$ whence $R = \left(\frac{l-h}{h}\right)r$

Substituting for R in (1), we have

$$h = \frac{2Sl}{(l-h)r\rho g}$$

$$\text{or,} \quad h(l-h) = \frac{2lS}{r\rho g}$$

$$\text{or,} \quad h^2 - lh + \frac{2lS}{r\rho g} = 0$$

So that $h = \dfrac{l \pm \sqrt{l^2 - 4.\frac{2lS}{r\rho g}}}{2}$

$$= \frac{l}{2} - \sqrt{\frac{l^2}{4} - \frac{2lS}{r\rho g}}$$ (taking the – *ve* sign only, +*ve* sign would give inadmissible root)

$$= 9 - \sqrt{81 - \frac{2 \times 18 \times 72}{0{\cdot}06 \times 1 \times 980}} = 9 - \sqrt{81 - 44{\cdot}08}$$

$$= 9 - 6{\cdot}075 = 2{\cdot}925 \text{ cm.}$$

19. What is the work done in blowing a soap bubble of radius 10 cm? (S.T.= 30 dynes/cm). What additional work will be performed in further blowing it, so that its radius becomes 15 cm? [7·5 × 10^4 ergs, 9·42 × 10^4 ergs]

20. A soap film, 0·001 mm thick and at 0°C is stretched adiabatically, until its area is doubled. What is the resulting fall in temperature, if we assume that the sp. heat and density area both unity, that $\frac{ds}{dt}$ at 0°C is –0·15 dyne per cm per 1°C and that $J = 4{\cdot}2 \times 10^7$ ergs per calorie? [0·0195°C]

21. A vertical *u*-tube containing mercury has one limb of diameter 5 cm and the other of diameter 0·1 cm. Calculate the difference in levels of mercury column in the two limbs. Surface tension for mercury is 550 dynes/cm, its density is 13·5 gms/c.c. and angle of contact between the walls and the liquid is 140°. [1·04 cm]

22. Two circular glass plates 4 cm diameter are separated by a film of water 0·1 mm thick. Assuming that the surface tension of water is 73 dynes/cm, find the normal pull in gm. ω*t* which must be exerted in order to separate the plates. [374 gm ω*t*]

Chapter III

VISCOSITY

Viscosity : Stream line and turbulent flow, critical velocity and Reynold's number; Poiseuille's equation for incompressible fluids with corrections (Experimental determination of coefficient of viscosity of fluid not required); Stokes' law; effect of temperature on viscosity.

3.1. Fundamental Definitions

Experiments show that whenever there is a relative motion among the various layers of a fluid, motion of one layer will be resisted by the other. Faster moving layers tend to accelerate the slower moving ones, while the slower moving layers tend to retard the faster moving ones. This property of the fluids which opposes the relative motion among its various parts is called viscosity. This force thus called into play is known as the viscous force and is due to internal friction among the molecules of the fluid.

A fluid may posses two characteristic types of motion—(1) Stream line and (2) turbulent. If the path followed by a particular fluid particle (Fig. 3.1) coincides with the direction of motion of the fluid, the motion will be called streamline.

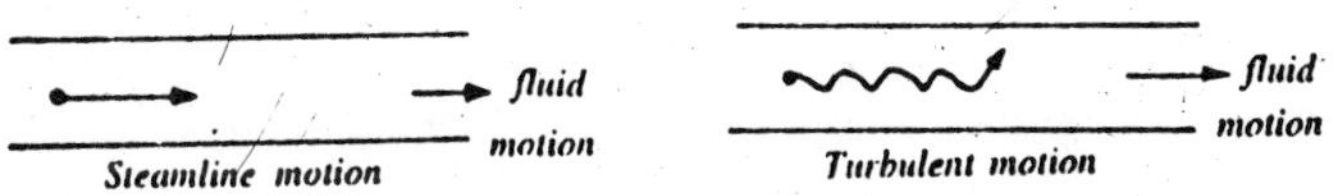

Fig. 3.1

If it does not, the motion will be called turbulent.

It is found that whether the motion will be streamline or turbulent will depend upon the velocity of flow of the fluid. Motion remains streamline if the velocity of flow is less than certain value known as critical velocity v_c. The critical velocity depends on the density ρ of the liquid, its viscosity η and the lateral dimension r of the channel through which the flow takes place. In case of flow of a liquid through a capillary tube the critical velocity is given by $v_c = \frac{k\eta}{\rho r}$, where η represents the coefficient of viscosity of the liquid, r

the radius of the tube through which the flow takes place and K is a constant, having a value nearly equal to 1000 for narrow tubes and is known as Reynold's number.

3.2. Newtonian flow of fluids and coefficient of viscosity

Let us consider the stream line motion of a fluid over a solid plane AB (Fig. 3.2). The layer of the fluid just in contact with AB will be at rest and as we proceed higher up along x direction, velocities of the respective layers go on increasing. Let v and $v + \delta v$ be the velocities at distances x and $x + \delta x$ from the plane. The velocity gradiant is $\frac{\partial v}{\partial x}$. Newton's assumption is that the tangential viscous stress acting on any layer CD moving with a velocity v due to the adjascent layer will be proportional to the velocity gradiant i.e. $\frac{F}{A} \propto \frac{\partial v}{\partial x}$

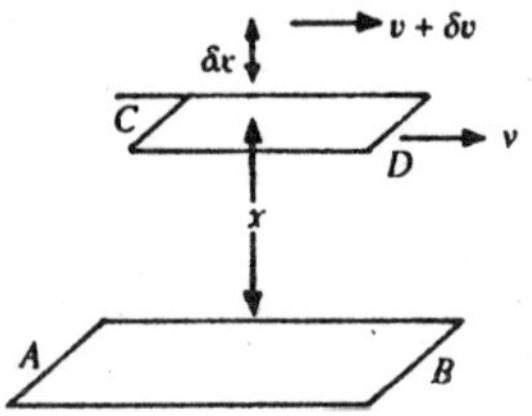

Fig. 3.2

$$\therefore F = A\frac{\partial v}{\partial x}\eta \tag{3.1}$$

where η is a constant known as the coefficient of viscosity of the fluid. It is to be noted that Newton's assumption is valid for small velocities, and fluids obeying equation (3.1) are called Newtonian fluids. A formal definition of this coefficient of viscosity may be given as viscous force per unit area per unit vel. grad. c.g.s. unit is gm/cm/sec.

3.3 Steady flow of an incompressible fluid through a horizontal capillary tube

Let us confine our attention to the stream line flow of an incompressible fluid through a capillary tube of length l and radius a under a pressure difference P applied between its ends. In this consideration it is assumed

(i) that the flow is everywhere parallel to the axis of the tube.

(ii) that the flow is stream line.

(iii) that the liquid in contact with the wall of the tube is at rest.

To fulfil these requirements, it is needed that the pressure gradiant along the axis of the tube should be constant. Choosing x axis along the axis of the tube we must have $\frac{\partial P}{\partial x}$ = constant = a, say.

Integrating $p = ax + b$

If $p = p_1$ at $x = 0$ and $p = p_2$ at $x = l$

$$b = p_1 \text{ and } p_2 = al + p_1 \quad \text{or, } a = \frac{P_2 - P_1}{l}$$

i.e. $\frac{\partial p}{\partial x} = \frac{p_2 - p_1}{l} = \frac{-p}{l}$ (3.2)

To proceed further let us consider the forces acting on an elementary liquid shell of radii r and $r + dr$ along the axis of the tube and of length dx (Fig. 3.3). The tube is considered to be horizontal so that gravity does not play any part.

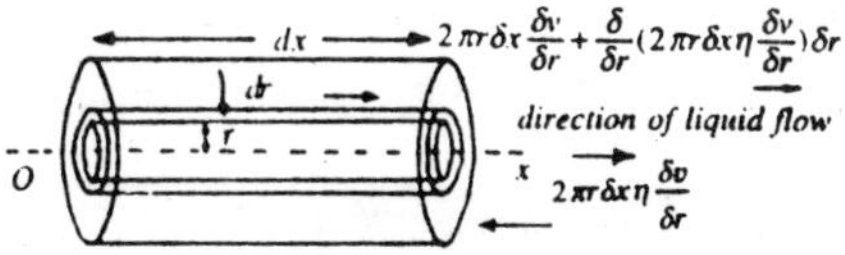

Fig. 3.3

The liquid nearer to the axis of this element exerts a force on the right of magnitude $2\pi r\ \delta x\ \eta\left(-\frac{\partial v}{\partial r}\right)$ i.e. a force of magnitude $2\ \pi r\ \delta x\ \eta\frac{\partial v}{\partial r}$ acting towards the left. The liquid outside the shell exerts on it a force $2\pi r\ \delta x\ \eta\frac{\partial v}{\partial r} + \frac{\partial}{\partial r}\left\{2\pi r\ \delta x\ \eta\ \frac{\partial v}{\partial r}\right\}\delta r$. directed towards right. These are forces due to viscosity. On the otherhand, forces on the element due to fluid pressure are $p 2\pi r\delta r$ directed towards right and $2\pi r\delta r\left(p + \frac{\partial p}{\partial x}\ \delta x\right)$ directed towards left. Since the liquid is in the steady state of motion, the net force acting on the shell is zero. Hence

$$2\pi r\ \delta x\ \eta\ \frac{\partial v}{\partial r} - 2\pi r\ \delta x\ \eta\frac{\partial v}{\partial r} - \frac{\partial}{\partial r}\left\{2\pi r\ \delta x\ \eta\frac{\partial v}{\partial r}\right\}\delta r$$
$$-p\ 2\pi r\ \delta r + 2\ \pi r\ \delta r\left(p + \frac{\partial p}{\partial x}\ \delta x\right) = 0$$

or, $$\frac{\partial}{\partial r}\left(2\pi r\ \delta x\ \eta\ \frac{\partial v}{\partial x}\right)\delta r = 2\pi r\ \delta r\ \frac{\partial p}{\partial x}\ \partial x$$

or, $$\frac{\partial p}{\partial x} = \frac{\eta}{r}\ \frac{\partial}{\partial r}\left(r\ \frac{\partial v}{\partial r}\right)$$

or, $$-\ \frac{pr}{l\eta} = \frac{\partial}{\partial r}\left(r\frac{\partial v}{\partial r}\right)$$

Integrating successively twice w. r. t. r.

$$v = -\frac{pr^2}{4l\eta} + A\log_e^r + B,\ A \text{ and } B \text{ are constants of integration}$$

Since v is not infinite at $r = 0$, $A = 0$

Again $v = 0$ at $r = a$, $B = \frac{Pa^2}{4l\eta}$

Substituting $v = \frac{P}{4l\eta}(a^2 - r^2)$ (3.3)

The volume of the liquid passing per unit time through the annular space between two co-axial circles of radii r and $r+\delta r$ is $2\pi r dr v$.

Hence the volume passing through the tube in unit time is given by

$$V = \int_0^a 2\pi r\, dr\, v = \frac{2\pi p}{4l\eta}\int_0^a (a^2 - r^2)\, r dr$$

$$= \frac{\pi a^4 p}{8l\eta} \tag{3.4}$$

The above relation is known as Poiseuille's equation.

3.4. Corrections to Poiseuille's Formula

In Poiseuille's equation corrections due to two factors are necessary. In the first place the pressure difference P is partly used to give $K.E.$ to the liquid and not wholly in overcoming the viscous effect. Secondly, it has been assumed that there are no accelerations along the axis of the tube. This condition is not fulfilled near the inlet end of the tube. In fact accelerations do not decrease to zero value until appreciable length of the tube has been traversed. To correct for the second error the expression for l is to be increased to l + 1.64a.

To calculate the correction due to the first factor, let us assume that P_l is the effective pressure difference which is used to counteract the viscous effect.

In one second work done against viscosity is P_lV, while the $K.E.$ given to the liquid is $\int_0^a \frac{1}{2} 2\pi r\, dr\, \rho.\, v.\, v^2 = \pi\rho\left(\frac{P}{4l\eta}\right)^3 \int_0^a r(a^2 - r^2)^3\, dr$

$$\left[\because v = \frac{P}{4l\eta}(a^2 - r^2)\right]$$

$$= \pi\rho\left(\frac{P_1}{4l\eta}\right)^3 \frac{a^8}{8} = \left(\frac{P_1 \pi a^4}{8l\eta}\right)^3 \frac{\rho}{\pi^2 a^4}$$

Thus the total loss of energy is

$P_1V + \frac{V^3\rho}{\pi^2 a^4}$ and this must be equal to PV. Hence

$$P_1 = P - \frac{V^2\rho}{\pi^2 a^4}.$$

Experimental verification and correction by Edser shows that

$P_1 = P - \frac{V^2\rho}{\pi^2 a^4} \cdot K$, where K is a constant whose value depends upon the form of the apparatus and although its value is nearly unity, for more accurate work it should be found out by calibration.

When both the corrections are applied in Poiseuille's formula, it becomes

$$V=\pi\left(P-\frac{V^2\rho k}{\pi^2 a^4}\right)\frac{a^4}{8\eta(l+1.64a)} \tag{3.5}$$

3.5. Stokes' Law

Deduction by the method of dimensions :

A spherical body moving through a viscous medium is subjected to a force of resistance given by $F = 6\pi\eta rv$, where r = radius, v = velocity of the body and η = coefficient of viscosity of the medium. The above expression is known as Stokes's Law. A form of deduction of the above expression can be given from the method of dimensions (Discussed later).

The force F experienced by a spherical body of radius r moving with a velocity v through a medium having a coefficient of viscosity η depends on r, v and η. Thus we may write $F = k r^{\alpha} v^{\beta} \eta^{\gamma}$, where α, β and γ are constants and K is a non-dimensional factor. Equating the dimensions on both sides

$MLT^{-2} = L^{\alpha}(LT^{-1})^{\beta}(ML^{-1}T^{-1})^{\gamma}$ [For dimension of a physical quantity see later.]

Equating the powers of M, L and T on both sides

$$\gamma = 1,$$

$$\alpha+\beta-\gamma=1$$

and $-\beta-\gamma=-2$

Solving $\alpha=\beta=\gamma=1$

Hence $F = k r v \eta$

Correspondence with experiment shows that the value 6π fits to K.

$$\text{Thus } F = 6\pi\eta r v. \tag{3.6}$$

Equation (3.6) is valid for a liquid having infinite stretch and infinite depth. For a liquid contained in cylinder of radius R which is filled with a height h, the corrected expression due to Ladenberg is

$$F = 6\pi\eta rv\left(1+2.4\frac{r}{R}\right)\left(1+3.3\frac{r}{h}\right). \tag{3.7}$$

3.6. Effect of temperature on viscosity

Viscosity of a liquid decreases with the rise of temperature. The empirical relation $\eta_t = \dfrac{\eta_0}{1+\alpha t+\beta t^2}$, where α and β are constants is in approximate agreement with experiment. The formula $\eta v^{\frac{1}{3}} = Ae^{\frac{c}{vT}}$ where v is the specific volume of the liquid and A and C are constants agree closely with experiment except for water and certain tertiary alcohols.

The property of viscosity of a liquid is explained as fugitive elasticity. Liquids are assumed to sustain and exert a certain amount of shearing stress which breaks away after a regular interval of time. This time is called the time of relaxation of the medium.

On the otherhand, the viscosity of gas increases with the rise of temperature. Viscosity of a gas is explained as due to transport of momentum from layer to layer when a velocity gradiant exists between them. Sutherland's formula regarding the variation of viscosity of a gas with temperature is

$\eta = \dfrac{A\sqrt{T}}{1+\dfrac{b}{T}}$, where A and b are constants.

3.7. Appendix

(a) Experimental determination of the coefficient of viscosity of a liquid by capillary flow :

Experimental arrangement is shown in Fig. 3.4. Liquid from a constant level tank is allowed to flow through a horizontal capillarv tube AB.

The flow is adjusted to be stream line so that the liquid trickles out in drops and is collected in a beaker C. The liquid is collected for a given time t and its volume is measured (v') by a measuring cylinder. Dividing the volume collected (v') by the time of collection (t), the volume V flowing per second is determined as $V = \dfrac{v'}{t}$.

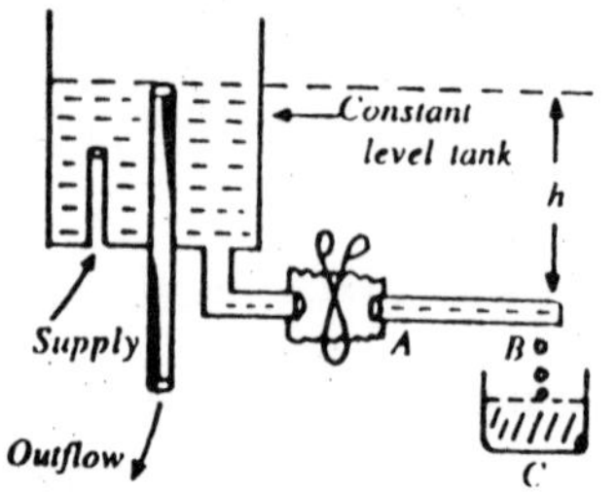

Fig. 3.4

The difference (h) between the heights of the free surface of the liquid in the reservoir and the free end of the capillary when multiplied by the density ρ of the liquid and the acceleration due to gravity g, gives, the pressure difference P between the ends of the capillary tube. At the end the flow is stopped, the capillary tube is dried and a thread of mercury is sucked within it. The mean length l' of the thread for different positions within the tube is measured by a travelling microscope. The mass m of the thread is measured by weighing. Finally, the mean radius r of the capillary tube is calculated from the relation $m = \pi r^2 l'\rho'$, where ρ' is the density of mercury. The coefficient of viscosity of the liquid is found out from the relation.

$V = \dfrac{\pi p r^4}{8l\eta}$ or, $\eta = \dfrac{\pi h\rho g r^4}{8lV}$, l being measured by a scale.

(b) Measurement of the viscosity of very viscous liquids by application of Stokes' law :

The theory of the method lies in that fact that when a spherical body of radius r moves with a velocity v through a viscous medium having a coefficient of viscosity η, the force of resistance encountered by it is given by Stokes' Law as $F = 6\pi\eta rv$. If the body falls vertically through the medium, the accelerating force on it is its effective weight $\frac{4}{3}\pi r^3(\rho-\sigma)g$, where ρ is the density of the body and σ that of the liquid. In falling the body will attain a steady (terminal) velocity when the retarding force equals the accelerating force. At this stage $6\pi\eta rv = \frac{4}{3}\pi r^3(\rho-\sigma)g$

$$\text{or, } \eta = \frac{2}{9}\frac{r^2(\rho-\sigma)}{v}g$$

or, making Ladenberg correction

$$\eta = \frac{2}{9}\frac{r^2(\rho-\sigma)g}{v\left(1+2\cdot4\frac{r}{R}\right)\left(1+3\cdot3\frac{r}{h}\right)} \quad (3.8)$$

Experimental arrangement consists of a tall glass jar containing the required liquid. At the lower part of the jar two scratch marks are given at a measured distance d apart. A small sphere of low density alloy is chosen and its mean diameter is measured by a travelling microscope. Half this diameter gives r. The density ρ of the material of the sphere and σ that of the experimental liquid are measured by a specific gravity bottle. The sphere is then allowed to fall from rest from the surface of the surface of the liquid. It is timed when seen cross the scratch marks. If t be this time, the velocity (steady) of the sphere is given by $v = \frac{d}{t}$. The radius R of the jar and the height h of the liquid being measured, η is found out term equation (3.8).

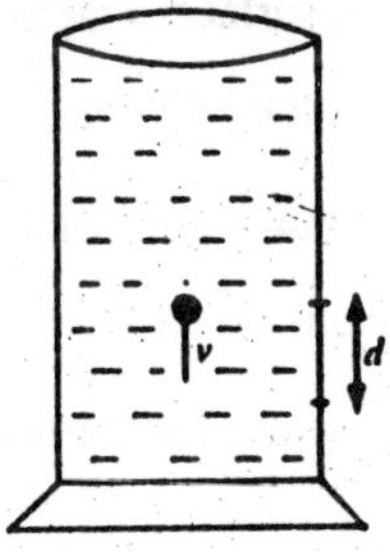

Fig. 3.5

(c) Torque on a cylinder placed in a rotating fluid :

If a cylinder is suspended inside a coaxial hollow cylinder which is made to rotate about the common axis and the space between the two is filled with a viscous fluid, the first cylinder is subjected to a torque whose magnitude can be found out in the following way.

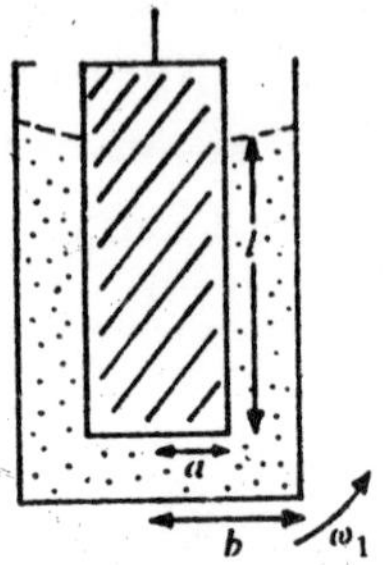

Fig. 3.6

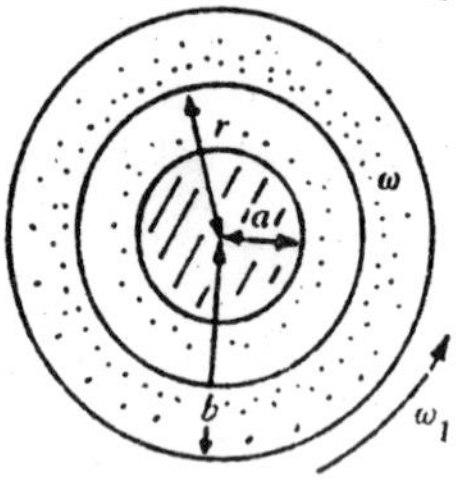

Fig. 3.7

Let the inner and the outer cylinders have radii a and b (Fig. 3.6 and Fig. 3.7) and the fluid covers a length l of the inner cylinder.

If the outer cylinder rotates with a constant angular veocity ω_1, the fluid layer in contact with it will also rotate with the same angular velocity, but the fluid layer in contact with the inner cylinder will be at rest. If we confine our attention to the space between the two cylinders it will be evident that different fluid cylinders will rotate with different angular velocities. Let a fluid cylinder of radius r $(a < r < b)$ be rotating with an angular velocity ω. Linear velocity on its surface $= \omega r$. The velocity gradient $= \frac{d}{dr}(\omega r) = \omega + r\frac{d\omega}{dr}$. The first part ω of the velocity gradient is utilised to prevent the relative slipping among the various rotating fluid cylinders and the second part $r\frac{d\omega}{dr}$ is used to produce viscous effect. The viscous torque τ_1 acting over the side of this cylinder is given by $\tau_1 = 2\pi r l \eta r \frac{d\omega}{dr}.r$

$$= 2\pi l \eta r^3 \frac{d\omega}{dr}.$$

Since the liquid between this boundary and the inner cylinder is in the steady state of motion, τ_1 will also be the torque acting on the inner cylinder.

Thus $\tau_1 \frac{dr}{r^3} = 2\pi l \eta \, d\omega$

Integrating both sides

$$\tau_1 \int_b^a \frac{dr}{r^3} = 2\pi l \eta \int_{\omega_1}^0 d\omega$$

or, $$\tau_1 \left(\frac{1}{b^2} - \frac{1}{a^2}\right) = 4\pi l \eta \, (-\omega_1)$$

or, $$\tau_1 = \frac{4\pi\eta \, l\omega_1 \, a^2 b^2}{b^2 - a^2} \tag{3.9}$$

τ_1, however gives the torque on the sides of the suspended cylinder. There is also a torque τ_2, due to viscosity acting over the bottom surface. The magnitude of the torque will depend on the values of a, b, ω_1 and the distance d between the bases of the cylinders. If this distance remains fixed while the length of the inner cylinder acted upon by the fluid varies τ_2 remains the same. Thus $\tau_2 = f(a, b, d, \omega_1)$.

If l_1 and l_2 be the length of the suspended cylinder immersed in the liquid in two cases, then $\tau = \tau_1 + \tau_2 = \dfrac{4\pi\eta\omega_1 a^2 b^2 l_1}{b^2 - a^2} + f(a, b, d, \omega_1)$

and $$\tau' = \tau'_1 + \tau_2 = \frac{4\pi\eta\omega_1 a^2 b^2 l_2}{b^2 - a^2} + f(a, b, d, \omega_1)$$

Whence $$\tau - \tau' = \frac{4\pi\eta\omega_1 a^2 b^2}{b^2 - a^2} (l_1 - l_2) \tag{3.10}$$

Equation 3.10 may be used to determine the coefficient of viscosity of a very viscous liquid. The experimental arrangement is shown in the Fig. 3.8.

The solid cylinder is suspended by a suspension fibre with a plane mirror M attached to it. The hollow cylinder is rotated by an electric motor and the space between the two is filled with a viscous fluid whose coefficient of viscosity is to be measured.

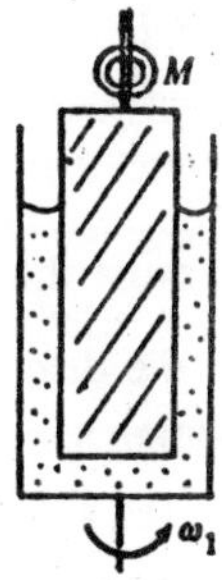

Fig. 3.8

If θ_1 and θ_2 be the angles of rotation of the suspended cylinder (measured by lamp and scale arrangement), then from equation (3.10)

$$\frac{4\pi\eta\omega_1 a^2 b^2}{b^2 - a^2}(l_1 - l_2) = c(\theta_1 - \theta_2) \tag{3.11}$$

where c is the torsional rigidity of this suspension fibre and can be determined from $$T = 2\pi\sqrt{\frac{I}{c}} \tag{3.12}$$

I being the moment of inertia of the suspended cylinder about the axis of suspension (determined as $I = \frac{1}{2}Mr^2$) and T is the time period of oscillation of the cylinder when it excutes torsional oscillations with the outer cylinder removed. ω_1 is determined from the relation $\omega_1 = \frac{2\pi}{t_1}$, t_l being the time period of rotation of the motor. a and b are measured by slide callipers and l_l measured by a scale attached to the suspended cylinder. From equation (3.11) η can be found out.

(d) Torque acting on a disc suspended on a rotating fluid :

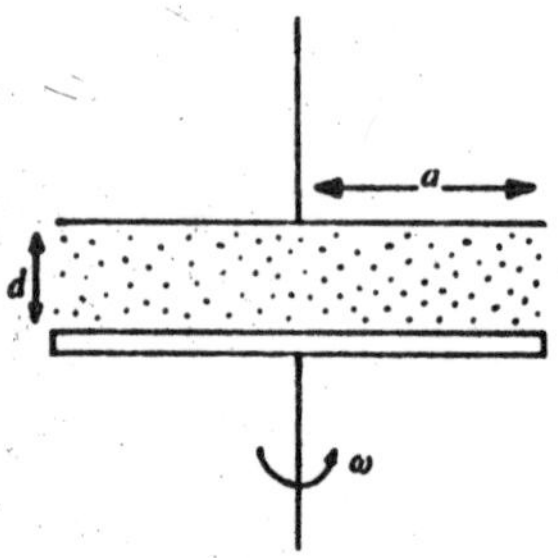

Fig. 3.9

Let us consider two discs mounted coaxially and let the lower one be rotated with a constant angular velocity ω (Fig. 3.9).

Let the space between the discs to be filled with a viscous fluid of coefficient of viscosity η. The vertical velocity gradiant at a distance r from the common axis will be on the average $\frac{\omega r}{d}$, where d is the separation between the discs. The torque $d\tau$ acting on an annular ring of radius r and thickness dr of the suspended disc, is $d\tau = 2\pi r\, dr \frac{\omega r}{d} \eta r$.

The total torque acting on the whole suspended disc is given by

$$\tau = \int_0^a \frac{2\pi\eta\omega}{d} r^3 dr = \frac{\pi\eta\omega a^4}{2d} \qquad (3.13)$$

where a is the radius of the upper disc. The relation is only approximate, because the average velocity gradient $\frac{\omega r}{d}$ is not necessarily that at the surface of the disc and because of the edge effect the distribution of stream lines is complex.

(e) Flow of a compressible fluid through a capillary tube :

In the case of an incompressible fluid density is independent of pressure and hence it can be assumed that the volume crossing any section of the tube is constant. This will not be true in the case of compressible fluids. For these it is the mass of the fluid which cross any section in a given time which is constant. If V be the volume passing any section per sec. and ρ the density of the fluid at that place, then

$$\rho V = \text{constant}$$

and since $\rho \propto p$

$$PV = \text{constant.}$$

If we consider an element of the tube of length dx having a difference of pressure dp at its ends, the volume V leaving the element per sec. is given

by $V = -\frac{\pi a^4}{8\eta}\frac{dp}{dx}$. (3.14)

The negative sign being used since P decreases as x increases. If P_l is the pressure at the inlet end V_1 is the volume entering the tube per second, then we have

$$P_1V_1 = PV = -\frac{\pi a^4}{8\eta} p \frac{dp}{dx}$$

so that $\int_0^l P_1V_1\, dx = -\frac{\pi a^4}{8\eta}\int_{p_1}^{p_2} p\, dp$

where P_2 is the pressure at the outlet end of the tube.

Thus $P_1V_1 = P_2V_2 = \frac{\left(P_1^2 - P_2^2\right)\pi a^4}{16l\eta}$

is the equation of flow of the compressible fluid through a capillary tube also known as Meyer's formula.

(f) On Non-Newtonian flow :

There are a number liquids which show that the viscous force between adjescent layers is not proportional to the velocity gradient. On the other hand, these forces are some functions of the velocity gradient. Different colloids and pastes belong to this class. In these cases the volume of the liquid discharged per second through a capillary tube is not proportional to the pressure difference applied. The observed viscosity changes with pressure in the capillary tube or angular velocity in a viscometer. Fluids of this type are called non-Newtonian fluids. An efficient lubricant has in general a high viscosity at zero shear but exihibit very small viscosity when the moving parts gain speed. Thus a lubricant shows the property of decrease of viscosity with the increase of shear. On the otherhand, some pastes like wet-sand exihibit the inverse property where the viscosity increases with the increase of shear.

Problems :

1. A glass tube AB of length 20 cm is mounted horizontally with the end A connected to a water reservoir giving a constant pressure head of 5 cms, the end B being open to the atmosphere. The tube has a bore of diameter 5 mm along one half of its length from A, and of diameter 2·5 mm along the other half. Calculate the pressure difference between B and the mid point of the tube. **[C.U. (Hons.)]**

Solution :

Let P be the atmospheric pressure in cms of water. P_1 = Pressure at the mid-point C in cms of water.

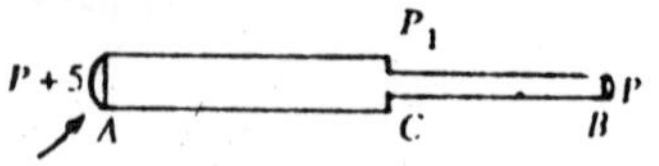

Fig. 3.10

Considering the viscous flow through the composite tube

$$\frac{\pi\{(P_1+5)-P_1\}\left(\frac{5}{2}\right)^4}{8\times 10\times\eta} = \frac{\pi\{P_1-P\}\left(\frac{25}{2}\right)^4}{8\times 10\times\eta}$$

$$\text{or, } (P_1-P)\left\{(25)^4+(5)^4\right\} = 5\times(5)^4$$

$$\text{or, } P_1 - P = \frac{5\times(5)^4}{(25)^4+(5)^4} = 4.706 \text{ cm of water}$$

2. Two vessles of equal cross-section α, are joined near their bases by a horizontal narrow tube of length l and internal radius r. Initially the liquid surfaces are at heights $3h$ and h respectively above the capillary tube. Cal·ulate the time taken for the difference to become h if the coefficient of viscosity is η and the density is ρ. The flow is assumed to be slow.

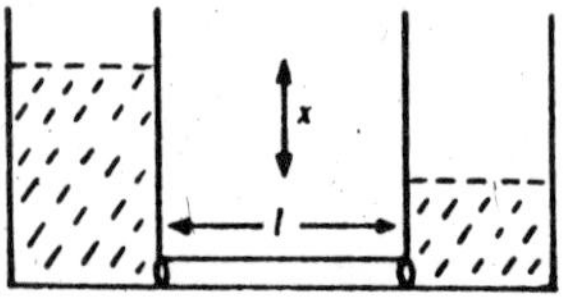

Fig. 3.11

Solution :

Let x be the difference in levels at any instant t. In subsequent time dt let the difference change by dx. Volume of liquid that flows from one vessel to another $= \alpha\,\frac{dx}{2}$.

Hence the instanteneous rate of flow

$$= -\frac{\alpha}{2}\frac{dx}{dt} = \frac{\pi x\rho g r^4}{8l\eta}$$

$\because$ x decreases with the increase of t

$$\text{or, } -\frac{dx}{x} = \frac{\pi\rho g r^4}{4l\eta\alpha}dt \text{ Integrating, } \int_h^{2h}\frac{dx}{x} = \frac{\pi\rho g r^4}{4l\eta\alpha}\int_0^t dt$$

$$\text{or, } \log_e 2h - \log_e h = \frac{\pi\rho g r^4}{4l\eta\alpha} t$$

$$\therefore\ t = \frac{4l\eta\alpha \log_e 2}{\pi\rho g r^4}.$$

3. A thin coire of radius b is placed coaxially in a narrow tube of length l and radius a. Find the volume of the liquid which flows per second through the annular space between the coire and the tube when a pressure difference P is maintained between the ends of the tube, if η is the coefficient of viscosity of the liquid and steady flow conditions are maintained.

[C.U. (Hons.) 96]

Solution :

If v be the fluid velocity at a distance r from the common axis we have as in the deduction of Poiseuille's equation.

$$v = -\frac{P}{4l\eta} r^2 + A \log_e r + B \quad \text{.... (i)}$$

Fig. 3.12

$v = 0$ at $r = a$ and at $r = b$

$$\text{Hence } 0 = -\frac{P}{4l\eta} a^2 + A \log_e a + B$$

$$\text{and } 0 = -\frac{P}{4l\eta} b^2 + A \log_e b + B$$

$$\text{Solving } A = \frac{P\left(a^2 - b^2\right)}{4l\eta} \left[\frac{1}{\log_e\left(\frac{a}{b}\right)}\right]$$

$$\text{and } B = \frac{Pv^2}{4l\eta} - \frac{P(a^2 - b^2)}{4l\eta} \left\{\frac{\log_e b}{\log_e\left(\frac{a}{b}\right)}\right\}$$

$$\therefore v = \frac{P}{4l\eta}\left\{\left(b^2 - r^2\right) + \frac{a^2 - b^2}{\log_e \frac{a}{b}} \log_e \frac{r}{b}\right\}$$

Volume of the liquid flowing out per sec.

$$= \int_b^a v\,2\pi r dr = \frac{\pi p}{2l\eta}\int_b^a \left\{(b^2 - r^2) + \frac{a^2 - b^2}{\log_e \frac{a}{b}}\log_e\left(\frac{r}{b}\right)\right\} r\,dr$$

$$= \frac{\pi P}{2l\eta}\left\{\left[b^2\frac{r^2}{2} - \frac{r^4}{4}\right]_b^a + \frac{a^2 - b^2}{\log_e \frac{a}{b}}\int_b^a (\log_e r - \log_e b)\, r\,dr\right\}$$

$$= \frac{\pi p}{2l\eta}\left\{\left(\frac{a^2b^2}{2} - \frac{a^4}{4} - \frac{b^4}{2} + \frac{b^4}{4}\right) + \frac{a^2 - b^2}{\log_e \frac{a}{b}}\left[\log_e r.\frac{r^2}{2} - \frac{r^2}{4} - \log_e b\frac{r^2}{2}\right]_b^a\right\}$$

$$= \frac{\pi P}{2l\eta}\left\{\frac{a^2b^2}{2} - \frac{a^4}{4} - \frac{b^4}{4}\right.$$

$$\left. + \frac{a^2 - b^2}{\log_e \frac{a}{b}}\left[\log_e a.\frac{a^2}{2} - \log_e b.\frac{b^2}{2} - \frac{a^2}{4} + \frac{b^2}{4} - \log_e b\left(\frac{a^2 - b^2}{2}\right)\right]\right\}$$

$$= \frac{\pi P}{8l\eta}\left\{2a^2b^2 - a^4 - b^4 + \frac{4(a^2 - b^2)}{\log_e \frac{a}{b}}\left[\log_e \frac{a}{b}\cdot\frac{a^2}{2} - \frac{a^2}{4} + \frac{b^2}{4}\right]\right\}$$

$$= \frac{\pi P}{8l\eta}\left\{2a^2b^2 - a^4 - b^4 + 2\left(a^2 - b^2\right)a^2 - \frac{(a^2 - b^2)^2}{\log_e \frac{a}{b}}\right\}$$

$$= \frac{\pi P}{8l\eta}\left\{a^4 - b^4 - \frac{(a^2 - b^2)^2}{\log_e \frac{a}{b}}\right\}$$

4. Two circular horizontal discs of radius 5 cm and distance apart 1 mm are separated by a layer of oil of viscosity 1·01 c.g.s. units. The upper disc is fixed while the lower one is revolved in vertical frictionless bearings by the tangential pull of a flexible string the tension in which is 6000 dynes, which unwraps from a cylinder, radius 2 cm, coaxial with the disc and fastened rigidly to it. Find approximately the time taken by the rotating system to make one complete revolution.

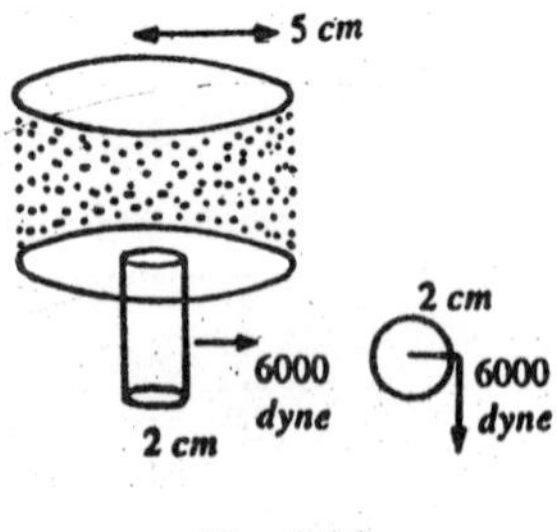

Fig. 3.13

Solution :

For rotation with constant angular velocity, the viscous torque = applied torque.

$$\therefore \frac{\pi\eta\omega a^4}{2d} = 6000 \times 2$$

$$\text{or, } \frac{\pi\eta 2\pi}{2dT} a^4 = 12000$$

$$\therefore T = \frac{9{\cdot}86 \times 1{\cdot}01 \times (5)^4}{0{\cdot}1 \times 12000} = 5{\cdot}19 \text{ secs.}$$

5. A cylindrical vessel of constant cross-sectional area A is filled with a liquid to a depth h above an orifice in its sides. A horizontal capillary tube of length l and mean internal radius a is attached to the orifice. Show that the volume of the liquid in the vessel and above the side orifice will be halved in a time $A\frac{8\eta l}{\pi a^4 g\rho}\log_e 2$, where η is the viscosity of the liquid, ρ is the density and g is the intensity of gravity.

Solution :

Let x be the height of the liquid above the capillary tube any instant of time t.

If the height decreases by dx in subsequent time dt, the volume of the liquid flowing out per sec $= -A\frac{dx}{dt} = \frac{\pi x\rho g a^4}{8l\eta}$

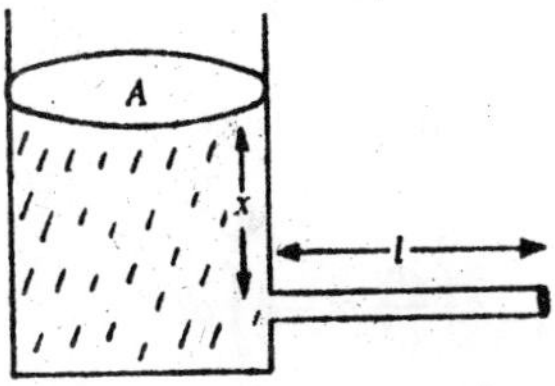

Fig. 3.14

$$\text{or, } -\int_h^{h/2} \frac{dx}{x} = \int_0^t \frac{\pi\rho g a^4}{8l\eta A} dt$$

$$\text{or, } \log_e h - \log_e \frac{h}{2} = \frac{\pi\rho g a^4}{8l\eta A} t = \log_e 2 \qquad \therefore t = \frac{8lA\eta}{\pi\rho g a^4}\log_e 2$$

6. A particle moves in a medium which offers a resistance proportional to the cube of the velocity; no other force act on the particle. During a time τ the velocity of this particle diminishes from v_1 to v_2. Show that in this time τ the particle traverses a distance $\frac{2v_1v_2\tau}{v_1+v_2}$.

Solution :

It is evident that

$$\frac{dv}{dt} = -kv^3$$

or, $\frac{dv}{ds}\,\frac{ds}{dt} = -kv^3$

or, $\frac{dv}{ds}v = -kv^3$

or, $\int_{v_1}^{v_2} \frac{dv}{v^2} = \int_0^s -k\,ds$

or, $\left[-\frac{1}{v}\right]_{v_1}^{v_2} = -ks$

or, $\frac{1}{v_1} - \frac{1}{v_2} = -ks$

or, $\frac{v_2 - v_1}{v_1 v_2} = -ks$

$$\therefore \quad s = \frac{v_1 - v_2}{kv_1 v_2} \qquad (1)$$

Again $\frac{dv}{dt} = -kv^3$

or, $\int_{v_1}^{v_2} \frac{dv}{v^3} = -k\int_c^{\tau} dt$

or, $\left[\frac{1}{2v^2}\right]_{v_1}^{v_2} = K\tau = \frac{1}{2v_2^2} - \frac{1}{2v_1^2}$

$$\therefore k = \frac{1}{2\tau}\left(\frac{v_1^2 - v_2^2}{v_1^2 v_2^2}\right) \qquad (2)$$

From (1) and (2)

$$s = \frac{v_1 - v_2}{v_1 v_2}\,\frac{2\tau v_1^2 v_2^2}{v_1^2 - v_2^2}$$

$$= \frac{2\tau v_1 v_2}{v_1 + v_2}$$

7. A cylindrical vessel of radius 0·05 meter has at its bottom a horizontal capillary tube of length 0·2 meter and internal radius 0·4 mm. The vessel is completely filled with water and then the water is allowed to flow out of the capillary. How long will it take for the water level to fall to half its initial height in the vessel. The viscosity of water is 0·001 MKS units, $g = 9{\cdot}8$ meter /sec^2; density of water = 1×10^3 kg/m^3 $\log_e 2 = 0{\cdot}693$.

Solution :

Let x be the height of water in the vessel at any instant t. The pressure difference across the ends of the tube of that instant is $x\rho g$, when ρ is the density of water. If the height of water falls through dx is a small time dt, then the volume of water flowed through the capillary tube is Adx, where A is the cross-section of that vessel. The instanteneous rate of flow of water is $-\frac{Adx}{dt}$, the $-ve$ sign is because x decreases with this increase of t

$$\text{Thus } \frac{\pi p a^4}{8l\eta} = \frac{\pi x \rho g a^4}{8l\eta} = -\frac{Adx}{dt}$$

$$\text{or, } dt = \frac{-8lA\eta}{\pi\rho g a^4}\frac{dx}{x}$$

$$\text{Integrating } \int_0^t dt = -\frac{8lA\eta}{\pi\rho g a^4}\int_h^{\frac{h}{2}}\frac{dx}{x}$$

$$\text{or, } t = \frac{-8lA\eta}{\pi\rho g a^4}\left\{\log_e\frac{h}{2} - \log_e h\right\}$$

$$= \frac{8lA\eta}{\pi\rho g a^4}\log_e 2$$

$$= \frac{8\times0{\cdot}2\times0{\cdot}001\times3{\cdot}14\times(0{\cdot}05)^2}{3{\cdot}14(1\times10^3)\times9{\cdot}8\times(0{\cdot}4\times10^{-3})^4}\times 0{\cdot}693$$

$$= 1{\cdot}1 \times 10^4 \text{ sec.}$$

8. A soap bubble of radius r_1 cm is blown at the end of a capillary tube of length l cm and internal radius a cm. Find the time taken by the bubble to be reduced to a radius of r_2 cm, given the viscosity of air at the temperature in η c.g.s. units.

Solution :

Let r = radius of the bubble of any instant t and P_0 = atmospheric pressure.

$$\text{Then } \left(P_0 + \frac{4S}{r}\right) 4\pi r^2 \frac{dr}{dt} = \left\{\left(P_0 + \frac{4S}{r}\right)^2 - P_0^2\right\} \frac{\pi a^4}{16l\eta}$$

$$= \left(2P_0 + \frac{4S}{r}\right) \frac{4S}{r} \frac{\pi a^4}{16l\eta}$$

$$\text{or, } -P_0 4\pi r^2 \frac{dr}{dt} = 2P_0 \cdot \frac{4S}{r} \frac{\pi a^4}{16l\eta} \quad \text{Since } \frac{4S}{r} << P_0$$

$$\text{or, } -r^2 \frac{dr}{dt} = \frac{Sa^4}{8l\eta}$$

$$\text{Whence } t = \frac{1}{4} \frac{8l\eta}{Sa^4}\left(r_1^4 - r_2^4\right)$$

$$= \frac{2l\eta}{Sa^4}\left(r_1^4 - r_2^4\right)$$

9. A plate of area 100 sq. cm is placed on the upper surface of caster oil 2 mm thick. Taking the coefficient of viscosity to be 15·5 paise, calculate the horizontal force necessary to move the plate with a velocity of 3 cm/sec.

Solution :

The viscous force acting is given by $F = \eta A \frac{dv}{dy} = \eta A \frac{v}{y}$ approximately

$$= 15.5 \times 100 \times \frac{3}{0.2}$$

$$= 2{\cdot}3025 \times 10^5 \text{ dynes.}$$

10. An oil drop of density 0·95 gm/c.c. and radius 10^{-4} cm and carrying a charge of q e.s.u. is driven vertically upwards through air of density 0·0013 gm/c.c. by an electric field of 2000 volt/cm with a constant velocity of 0·036 cm/sec. Calculate the magnitude of q being given that the viscosity of air is 180×10^{-6} gm cm^{-1} sec^{-1}.

Solution :

For constancy of speed the effective upward force must be balanced by the viscous resistance.

Hence $Eq - \frac{4}{3}\pi r^3(\rho - \sigma)g = 6\pi\eta r v$, where E intensity of the electric field, r = radius of the drop, ρ = density of oil, σ = density of air, v = constant velocity of the drop.

$$q = \frac{6\pi\eta r v + \frac{4}{3}\pi r^3(\rho - \sigma)g}{E}$$

$$= \frac{6 \times 3{\cdot}14 \times 180 \times 10^{-6} \times 10^{-4} \times 0{\cdot}036 + \frac{4}{3} \times 3{\cdot}14 (10^{-4})^3 (0{\cdot}95 \times 980 - 0{\cdot}0013)}{\frac{2000}{300}}$$

$$= \frac{122{\cdot}08 \times 10^{-10} + 389{\cdot}8 \times 10^{-11}}{6{\cdot}66}$$

$$= \frac{(122{\cdot}08 + 38{\cdot}98) \times 10^{-10}}{6{\cdot}66}$$

$$= 24{\cdot}18 \times 10^{-10} \text{ e.s.u.}$$

EXERCISE

1. Define viscosity of a fluid. What are stream line and turbulent motions? How are they related to the velocity of flow ? What is Reynold's No. ?

2. Describe the viewpoints from which the viscosity of liquids and gases can be explained.

3. Describe the theory of the capillary tube method of determining the coefficient of viscosity of a liquid. What are the assumptions on which the deduction of the theory is based ?

4. A soap bubble of radius 4 cm and surface tension 30 dynes per cm is blown at the end of a tube of length 10 cm and internal radius 0·100 cm. If the viscosity of air is $1{\cdot}85 \times 10^{-4}$ c.g.s. units, find the time taken by the bubble to be reduced to a radius of 2 cm. [4 min 55·6 sec.]

5. Deduce an expression for the volume of the liquid flowing out of a capillary tube in terms of the pressure difference applied, the radius of the tube and the coefficient of viscosity of the liquid. What is the name of the expression ? What are the factors for which correction is to be applied to this expression.

6. Find theoretically an expression for resistance force acting on a spherical body moving through a viscous medium. What this expression is called ?

7. An air bubble of radius 1 cm rises through a long column of liquid of density $1{\cdot}47 \times 10^3$ kg/m^3 with a steady velocity of 0·21 cm/sec. Find the viscosity of the liquid. Take $g = 9{\cdot}8$ m/sec^2 and neglect the density of air. [$1{\cdot}52 \times 10^2$ kg/m sec.]

8. Water having coefficient of viscosity 0·01 gm/cm. sec. is escaping from a tank through a horizontal capillary tube 50 cm long and 0·012 cm in radius. What is the rate of flow of water from the tank when the level in it is 50 cm above that of water in the capillary tube. How long will it take for the water level to fall a further 10 cm if the cross-sectional area of the tank is everywhere 100 cm^2? [$7{\cdot}98 \times 10^{-4}$ cm/sec ; $7{\cdot}41 \times 10^6$ sec.]

9. A glass bulb of volume 500 c.c. has a capillary tube of length 40 cm and radius 0·020 cm leading from it. The bulb is filled with hydrogen at an initial pressure of 86 cm of mercury of density 13·6 gm/c.c. It is found that if the volume of the gas remaining in the vessel is kept constant, the pressure falls to 80 cm of mercury in 25·4 sec. If the height of the barometer is 76 cm and $g = 981$ cm/sec^2, find the viscosity of hydrogen.

Hints for solution :

Let v = volume of the vessel and p the instanteneous pressure within it. The gas of volume v increases to $v + dv$ in time $t + dt$; hence

$$p\frac{dv}{dt} = \frac{\pi a^4}{16l\eta}(p^2 - P^2),$$ P being the atmospheric pressure.

The gas occupying a volume v at pressue p increases in volume $v + dv$ at pressure $p + dp$, from which, at constant temperature

$$pdv = -vdp$$

$$\therefore -\frac{vdp}{dt} = k(p^2 - P^2) \text{ where } k = \frac{\pi a^4}{16l\eta}$$

$$\text{or, } dt = -\frac{v}{k}\frac{dp}{p^2 - P^2}$$

$$= \frac{v}{2kP}\left\{\frac{1}{P+p} + \frac{1}{P-p}\right\}dp$$

$$\text{Integrating, } t = \frac{v}{2kP}\int_{p_1}^{p_2}\left\{\frac{1}{P+p} + \frac{1}{P-p}\right\}dp$$

$$= \frac{v}{2kP}\left[\log_e(P+p) - \log_e(P-p)\right]_{p_1}^{p_2}$$

$$= \frac{v}{2kP}\left[\log_e\frac{P+p}{P-p}\right]_{p_1}^{p_2} = \frac{v}{2kP}\left\{\log_e\frac{P+p_2}{P-p_2} - \log_e\frac{P+p_1}{P-p_1}\right\}$$

$$= \frac{v}{2kP}\log_e\frac{(P+p_2)(P-p_1)}{(P+p_1)(P-p_2)}$$

$$= \frac{8vl\eta}{\pi a^4 P}\log_e\frac{(P+p_2)(p_1-P)}{(P+p_1)(p_2-P)}$$

$$\text{or, } \eta = \frac{\pi a^4 Pt}{8vl} + \log_e \frac{(P+P_2)(p_1-P)}{(P+p_1)(p_2-P)}$$

$$= \frac{3{\cdot}14 \times (0{\cdot}02)^4\, 76 \times 13{\cdot}6 \times 981 \times 25{\cdot}4}{8 \times 40 \times 500} + \log_e \left\{ \frac{(76+80)\times 10}{(76+86)\times 4} \right\}$$

$$= 8{\cdot}086 \times 10^{-5} + 0{\cdot}8785 = 9{\cdot}204 \times 10^{-5} \text{ c.g.s. units.}$$

10. Explain the statement that viscosity of water is 0·0101 poise at 20°C. Using the method of dimensional analysis, show how the resistance to motion of a sphere moving through a fluid varies with the velocity. What happens when the velocity becomes large.

11. Deduce the velocity profile for streamline flow of a liquid through a capillary tube of circular cross-section. Deduce the fraction of liquid which flows through the section upto a distance $\frac{a}{2}$ from the axis, where a is the radius of the capillary tube.

Hints : Show that the ratio between the volume v' and $v = \frac{7}{16}$.

$$\frac{v'}{v} = \frac{\int_0^{a/2} v\, 2\pi r\, dr}{\int_0^{a} v 2\pi r\, dr}$$ v = velocity at a distance r from the axis

12. Water at 20°C is flowing through a capillary tube of length 1 meter and diameter 2 mm under a constant pressure head of 10 cm of water. Calculate the velocity of water at a point (i) on the axis, (ii) adjascent to the wall of the tube, (iii) at 0·5 mm distance from the axis of the tube. Coefficient of viscosity of water at 20°C is 0·01 Poise.

[(i) 24·5 cm/sec, (ii) 0, (iii) 375 cm/sec.]

13. A narrow tube of radius a and length l is connected horizontally to a tank of uniform horizontal cross-section A. Water in the tank has initially a depth H and after an interval t it falls to a depth h. The viscosity of water being η, determine the relation between h and t assuming the flow to be stream line.

14. A particle of mass m, initially at rest, is acted on by a constant force fm and by a resisting force $mk\dot{x}$ where $\dot{x}$ is the velocity of the particle at a distance x from the origin. Derive expressions for (a) the terminal velocity acquired, (b) the velocity at a time t after the application of the force, (c) the distance travelled by the particle in this time.

15. A cylindrical vessel is maintained full of a liquid to a depth of 20 cm and has protuding from it three similar horizontal capillary tube, each 45 cm long, fixed at heights 0, 5 and 10 cm respectively, from the base, show that the rate of supply to the cylinder is the same as would be necessary for a single outflow tube of length 20 cm and similar radius protuding horizontally at the bottom of the cylinder.

16. A plate of metal 10^{-2} m^2 in area rests on a layer of castor oil 2×10^{-3} m thick, whose coefficient viscosity is 1·55 N.S./m^2. Calculate the horizontal force required to move the plate with a uniform speed of 3×10^{-2} m / sec.

From Newton's formula

$$F = \eta A \frac{dv}{dx}$$

$$\text{Hence } \frac{dv}{dx} = \frac{3\times10^{-2}}{2\times10^{-3}} = 15$$

$$\therefore F = 1{\cdot}55\times10^{-2}\times15$$
$$= 0{\cdot}2325 \text{ N.}$$

17. A cylinder 20 cm deep is filled up with muddy water and water is throughly stirred up. Assuming that the suspended particles are spherical and have a density 2·6 gm/c.c. Find the radius of the largest particle remaining in suspension after the settling process was allowed to continue for 24 hours. Viscosity of water may be taken to be 0·016 c.g.s. units.

[C.U. (Hons.) '95]

Solution (Hints) :

The required particle, are those who will traverse 20 cm in 24 hours. Thus

$$v = \frac{20}{24\times60\times60} \text{ cm/sec.}$$

Again for steady fall

$$6\pi\eta rv = \frac{4}{3}\pi r^3\rho g$$

$$\text{or, } r^2 = \frac{9}{2}\frac{\eta v}{g\rho} = \frac{9\times0{\cdot}016\times20}{2\times24\times60\times60\times981\times2{\cdot}6} = 6{\cdot}53\times10^{-9}$$

$$\therefore r = 8{\cdot}08\times10^{-5} \text{ cm.}$$

CHAPTER IV

MECHANICS OF IDEAL FLUIDS

Mechanics of ideal Fluids : Euler's equation of motion : Equation of continuity : Bernoulli's equation—its applications.

4.1.

The problem of the flow of liquids and gases is not very simple. They possess inertia and therefore *K.E.* when in motion. They are also subject to pressure and possess *P.E.* on account of their position. Besides these there may arise forces due to viscosity if there is a velocity gradient among the different parts. Lastly, density may vary under different pressure. For our present purpose viscosity is neglected and liquids are taken to be incompressible. Such fluids are called perfect fluids.

4.2. Equation of Continuity

Let us consider a surface S to be entirely inside a fluid. If ρ be the density of the fluid and $d\tau$ an element of volume. The total mass of the fluid inside S is given by

$$M = \iiint_\tau \rho(x, y, z, t)\, d\tau. \qquad (4.1)$$

Differentiating *w.r.t.* time remembering that (x, y, z) are variables of integration, we obtain

$$\frac{dM}{dt} = \iiint_\tau \frac{\partial \rho}{\partial t} d\tau. \qquad (4.2)$$

Now there are three ways in which the mass of the fluid inside S can change—(1) fluid may be entering or leaving the surface. The contribution to this effect is $\iint_S \vec{v}\rho . d\vec{S}$, $\vec{v}$ being the velocity at a point where the density is ρ.

[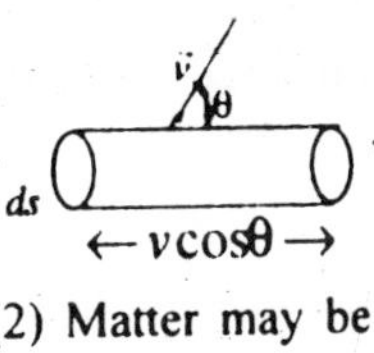

, volume passing through dS per sec. $= v \cos\theta\, dS = \vec{v} . \vec{dS}$]

(2) Matter may be created (source) or (3) matter may be destroyed (sink).

Let ψ (x, y, z, t) be the amount of matter created or destroyed per unit volume per second. For a source $\psi > 0$ and for a sink $\psi < 0$. The net gain

of the fluid is therefore $\iiint_\tau \psi d\tau - \iint_S \vec{v\rho}.\vec{dS}$

Hence $\frac{dM}{dt} = \iiint_\tau \frac{\partial \rho}{\partial t} d\tau = \iiint_\tau \psi\, d\tau - \iint_S \vec{v\rho}.d\vec{S}$

Applying Gauss's divergence theorem

$$\iiint_\tau \frac{\partial \rho}{\partial t} d\tau = \iiint_\tau \psi\, d\tau - \iiint_\tau \left(\vec{\nabla}.\vec{v\rho}\right) d\tau \qquad (4.3)$$

Since the element of volume $d\tau$ is quite arbitrary, $\frac{\partial \rho}{\partial t} = \psi + \vec{\nabla}.(\vec{v\rho})$ (4.3a)

This is the equation of continuity. For no source or sink equation (4.3a) reduces to $\frac{\partial \rho}{\partial t} + \vec{\nabla}.\left(\vec{v\rho}\right) = 0.$ (4.4)

Furthermore, if the fluid is incompressible (liquid) $\frac{\partial \rho}{\partial t} = 0$ and equation

(4.4) becomes $\vec{\nabla}\,\vec{v} = 0.$ (4.5)

If the motion is irrotational, then $\vec{\nabla} \times \vec{v} = 0.$

This requires $\vec{v} = \vec{\nabla}\varphi$, φ being the scalar function known as the velocity potential. Hence the equation of continuity for no source and sink and having irrotational motion is given by $\nabla^2\varphi = 0.$ (4.6)

4.3. Equation of motion of a perfect fluid—Bernoulli's equation

Let us consider the motion of a perfect fluid inside a simply connected region of volume τ and area S. The forces acting on the volume are—

(1) The external force (gravity etc.) say $\vec{f}$ per unit mass.

(2) Thrust on the surface due to fluid pressure $= -p\vec{dS}$, since $\vec{dS}$ points outwards and p is the magnitude of the pressure. The total force acting on τ is given by $\vec{F} = \iiint_\tau \rho \vec{f}\, d\tau - \iint_s p\vec{dS}$

$$= \iiint \left(\rho\vec{f} - \vec{\nabla}p\right) d\tau$$

[If $\vec{p} = \hat{a}p$ where $\hat{a}$ is unit vector in the direction of $\vec{p}$,

Then $\iint_S \vec{p}.\vec{ds} = \iint_S p\,\hat{a}.\vec{ds}$

$$= \iiint_\tau \vec{\nabla}.\hat{a}\,p d\tau = \hat{a}.\iiint_\tau \vec{\nabla}\,p d\tau$$

or, $\hat{a}.\iint_S p\vec{ds} = \hat{a}.\iiint_\tau \vec{\nabla}\,p d\tau$

or, $\iint_S p\vec{ds} = \iiint_\tau \vec{\nabla}\,p d\tau$]

The linear momentum of τ is $\vec{L} = \iiint_\tau \rho\vec{v}\,d\tau$

The time rate of change of linear momentum is $\frac{d\vec{L}}{dt} = \frac{d}{dt}\iiint_\tau \rho\vec{v}\,d\tau$

$$= \iiint_\tau \rho\frac{d\vec{v}}{dt}d\tau + \iiint_\tau \vec{v}\frac{d}{dt}(\rho d\tau)$$

Though the volume $d\tau$ change with time, $(\rho\ d\tau)$ = mass of the volume $d\tau$ remains constant throughout the motion, so that $\frac{d}{dt}(\rho\,d\tau) = 0$.

Since $\vec{F} = \frac{d\vec{L}}{dt}$, we obtain $\iiint_\tau \left(\rho\vec{f} - \vec{\nabla}p\right)d\tau = \iiint_\tau \rho\frac{d\vec{v}}{dt}d\tau$ (4.7)

This equation is true for all τ's, so that $\rho\vec{f} - \vec{\nabla}\,p = \rho\frac{d\vec{v}}{dt}$

or, $\frac{d\vec{v}}{dt} = \vec{f} - \frac{1}{\rho}\vec{\nabla}\,p$ (4.8)

This is Euler's hydrodynamical equation of motion. Since $\vec{v}$ is a function of (x, y, z, t)

$$\vec{dv} = \frac{\partial\vec{v}}{\partial x}dx + \frac{\partial\vec{v}}{\partial y}dy + \frac{\partial\vec{v}}{\partial z}dz + \frac{\partial\vec{v}}{\partial t}dt$$

$$= \left(\vec{dr}.\vec{\nabla}\right)\vec{v} + \frac{\partial\vec{v}}{\partial t}dt$$

$$\because \vec{dr} = \hat{i}\,dx + \hat{j}\,\partial y + \hat{k}\,dz$$

$$\therefore \frac{d\vec{v}}{dt} = \left(\frac{d\vec{r}}{dt}.\vec{\nabla}\right)\vec{v} + \frac{\partial \vec{v}}{\partial t}$$

$$= \left(\vec{v}.\vec{\nabla}\right)\vec{v} + \frac{\partial \vec{v}}{\partial t}$$

Hence equation (4.8) reduces to

$$\frac{\partial \vec{v}}{\partial t} + \left(\vec{v}.\vec{\nabla}\right)\vec{v} = \vec{f} - \frac{1}{\rho}\vec{\nabla} p \tag{4.9}$$

If the external force field is conservative $\vec{f} = -\vec{\nabla}\Omega$, Ω being the potential per unit mass from which the external force field is derivable.

So that $$\frac{\partial \vec{v}}{\partial t} + \left(\vec{v}.\vec{\nabla}\right)\vec{v} = -\vec{\nabla}\left(\Omega + \frac{p}{\rho}\right) \tag{4.10}$$

This equation is valid for an incompressible fluid, since ρ is constant in this case.

[Now $\vec{u}\times\left(\vec{\nabla}\times\vec{v}\right) = \vec{\nabla}_v\left(\vec{u}.\vec{v}\right) - \left(\vec{u}.\vec{\nabla}\right)\vec{v}$

$\vec{\nabla}_v$ signifies the operation on $\vec{v}$, $\vec{u}$ remaining constant

Again $\vec{v}\times\left(\vec{\nabla}\times\vec{u}\right) = \vec{\nabla}_u\left(\vec{v}.\vec{u}\right) - \left(\vec{v}.\vec{\nabla}\right)\vec{u}$

Adding, $\vec{u}\times\left(\vec{\nabla}\times\vec{v}\right) + \vec{v}\times\left(\vec{\nabla}\times\vec{u}\right)$

$$= \vec{\nabla}_v\left(\vec{u}.\vec{v}\right) + \vec{\nabla}_u\left(\vec{v}.\vec{u}\right) - \left(\vec{u}.\vec{\nabla}\right)\vec{v} - \left(\vec{v}.\vec{\nabla}\right)\vec{u}$$

$$= \vec{\nabla}\left(\vec{u}.\vec{v}\right) - \left(\vec{u}.\vec{\nabla}\right)\vec{v} - \left(\vec{v}.\vec{\nabla}\right)\vec{u}$$

Putting $\vec{u} = \vec{v}$

$$2\vec{v}\times\left(\vec{\nabla}\times\vec{v}\right) = \vec{\nabla}(v^2) - 2\left(\vec{v}.\vec{\nabla}\right)\vec{v}$$

If $\vec{v}$ is irrotational, $\vec{\nabla}\times\vec{v} = 0$

$$\therefore \left(\vec{v}.\vec{\nabla}\right)\vec{v} = \frac{1}{2}\vec{\nabla}v^2 \ ]$$

Equation (4.10) yields

$$\frac{\partial \vec{v}}{\partial t} = -\vec{\nabla}\left(\Omega + \frac{p}{\rho} + \frac{1}{2}v^2\right)$$

For steady motion $\frac{\partial \vec{v}}{dt} = 0$

$$\therefore \quad \vec{\nabla}\left(\Omega + \frac{p}{\rho} + \frac{1}{2}v^2\right) = 0$$

Hence $\Omega + \frac{p}{\rho} + \frac{1}{2}v^2$ = constant (4.11)

along a stream line for the steady motion of an incompressible fluid flowing irrotationally under a conservative system of forces.

If gravity is the only external force acting on the moving fluid, $\Omega = gh$

$\therefore \quad \frac{p}{\rho} + \frac{1}{2}v^2 + gh$ = constant (4.12)

along a stream line. gh is the potential energy per unit mass, $\frac{1}{2}v^2$ = $K.E.$ per unit mass and $\frac{p}{\rho}$ is called the pressure energy i.e. the energy to be imparted to the fluid so that it is just at the point of motion without acquiring any velocity. The formal statement of Bernoulli's theorem may be made as— the sum of $P.E.$, $K.E.$ and pressure energy is constant at any point on a stream line for steady flow if an incompressible fluid moving under a conservative system of forces.

If the flow takes at a constant height gh = constant. In this case equation (4.12) reduces to $\frac{p}{\rho} + \frac{1}{2}v^2$ = constant

or, $\quad p + \frac{1}{2}\rho v^2$ = constant (4.13)

p is called the static pressure and $\frac{1}{2}\rho v^2$, the dynamic pressure and the sum of the two will be constant.

4.4. Efflux of fluids : Torricelli's theorem

- In the case of a liquid flowing **from a reservoir through a small orifice** in the base of the containing vessel, it is found that after leaving the orifice, the cross-section of the jet contracts to a minimum value after which it increases. This minimum cross-section is called the 'contracted vein' and the ratio of its area to that of the orifice is termed the 'coefficient of contraction'.

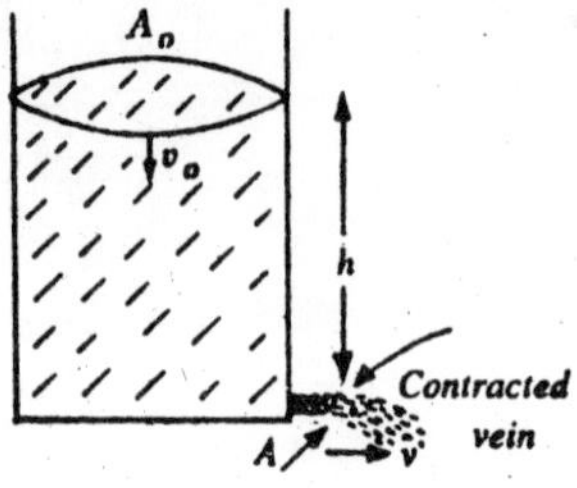

Fig. 4.1

Let A_0 be area of the free surface of the liquid and v_0 its velocity. If h be depth of the orifice below the free surface—the head of the liquid being maintained constant (Fig. 4.1), we have from Bernoulli's theorem

$$\frac{P}{\rho}+\frac{1}{2}v_0^2+gh_0=\frac{P}{\rho}+\frac{1}{2}v^2+gh^1,$$

where P is the atmospheric pressure and h_0 and h^1 the heights of the free surface and the contracted vein respectively.

or, $v^2 = v_0^2 + 2g(h_0 - h^1) = v_0^2 + 2gh$

For incompressible fluids (liquids) $v_0 A_0 = vA$

Substituting above, $v^2 = \dfrac{v^2 A^2}{A_0^2} + 2gh$

or, $v^2\left(\dfrac{A_0^2 - A^2}{A_0^2}\right) = 2gh$ or, $v^2 = \dfrac{A_0^2}{A_0^2 - A^2} 2gh$

If the orifice is small $A_0^2 >> A^2$.

Hence $v = \sqrt{2gh}$. (4.14)

Thus when the liquid reaches the contracted vein the velocity acquired is the same as if it fell directly to the orifice. This is Torricelli's theorem.

Let A be the area of the orifice and C the coefficient of contraction. The volume of the issuing from the vessel in unit time is $CA\sqrt{2gh}$.

To find the time in which a vessel of any form will be emptied through a small orifice, let A_0 be the area of the free surface, then $A_0 dx = CA\sqrt{2gx}\,dt$, where A is the area of the orifice and C the coefficient of contraction.

Integrating, $\displaystyle\int_0^{x_0} \frac{dx}{\sqrt{2gx}} = \int_0^{t_0} \frac{CA}{A_0}\,dt$, where x_0 = height of the liquid and t_0 the time to empty the vessel.

4.5. Flow through a constricted tube (Venturimeter)

When a liquid flows through a tube of varying cross-section, the velocity varies along the tube and the pressure also varies.

Let v_1, v_2 and v_3 be the velocities at three places (Fig. 4.2) at which the pressures are p_1, p_2 and p_3 respectively.

Applying Bernoulli's theorem

$$\frac{p_1}{\rho}+\frac{1}{2}v_1^2+h_1g=\frac{p_2}{\rho}+\frac{1}{2}v_2^2+h_2g$$

$$=\frac{p_3}{\rho}+\frac{1}{2}v_3^2+h_3g$$

If the tube is kept horizontal $h_1 = h_2 = h_3$

Hence

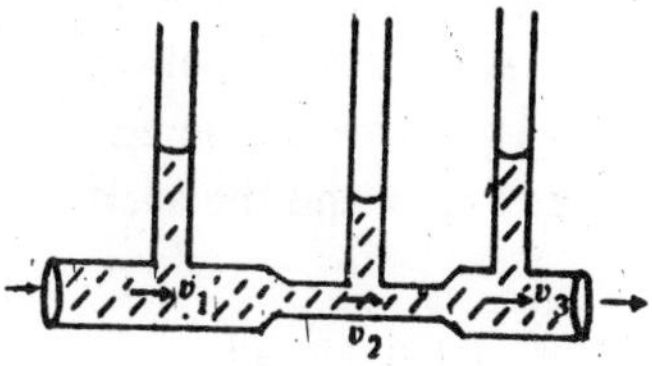

Fig. 4.2

$$\frac{p_1}{\rho}+\frac{1}{2}v_1^2=\frac{p_2}{\rho}+\frac{1}{2}v_2^2=\frac{p_3}{\rho}+\frac{1}{2}v_3^2$$

since the volume of the liquid passing per sec. through any section is constant.

$\alpha_1 v_1=\alpha_2 v_2=\alpha_3 v_3$, where α_1, α_2 and α_3 are the cross-sections of the tube at the three places respectively.

Since $\alpha_3>\alpha_2$

$$v_2>v_3$$

Hence $p_3>p_2$

The distribution of pressure can be proved by attaching three vertical tubes at 1, 2 and 3.

The height of liquid at 3 will be higher than at 2.

The principle has been used to measure the amount of flow in pipes.

The arrangement is known as **Venturimeter.** Considering the places 1 and 2, we have

$$\frac{1}{2}v_2^2=\frac{1}{2}v_1^2+\frac{p_1-p_2}{\rho}.$$

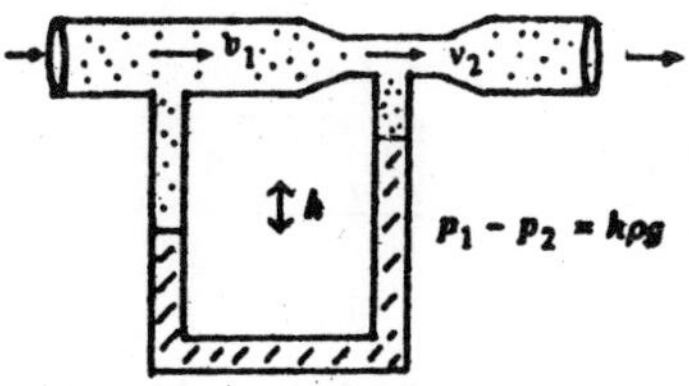

Fig. 4.3

Since $\alpha_1 v_1=\alpha_2 v_2$ and $\rho = 1$ gm/l c.c. in the case of water, we have

$$\frac{1}{2}\frac{\alpha_1^2}{\alpha_2^2}v_1^2-\frac{1}{2}v_1^2=p_1-p_2$$

$$\text{or, } v_1=\alpha_2\sqrt{\frac{2(p_1-p_2)}{\alpha_1^2-\alpha_2^2}}$$

The amount of flow through the pipe $=\alpha_1 v_1=\alpha_1\alpha_2\sqrt{\dfrac{2(p_1-p_2)}{\alpha_1^2-\alpha_2^2}}$ (4.15)

Thus equation (4.15) enables us to measure the amount of flow in terms of $p_1 - p_2$ and α_1, α_2 $(p_1 - p_2)$ can be measured with the help of the appliance shown in Fig. 4.3.

4.6. Pitot Tube (Application of Bernoulli's Principle)

This is another arrangement like venturimeter to measure the amount of flow in pipes. The principle of the method can be explained with the help of Fig. 4.4. It consists of two vertical tubes with small apertures at their lower ends. The plane of the aperture in one tube PQ is parallel to the direction of water flow and the aperture in the other tube RS is perpendicular to this direction. The rise of liquid in the tube PQ measures the pressure at Q. Since water is stopped at S, the velocity of flow is zero there. Water rises to a higher level in RS. If h be the difference of level of water in the two tubes, we have

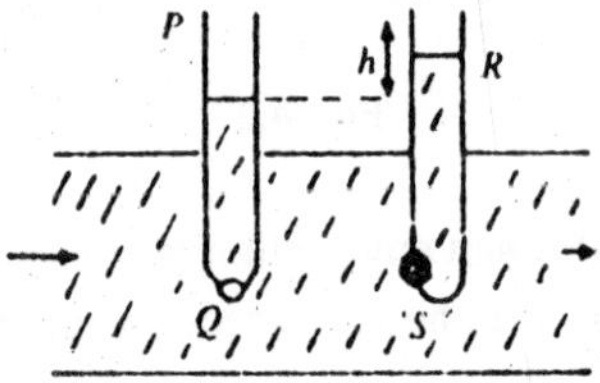

Fig. 4.4

$$\frac{1}{2}v^2 = hg, \quad \text{or, } v^2 = 2gh$$

$$\therefore \ v = \sqrt{2gh}.$$

This multiplied by the cross-section of pipe where the tubes are attached gives the volume flowing per second.

4.7. Other applications of Bernoulli's Principle are Filter pumps, Atomisers etc.

(a) The **Principle of Atomiser** is discussed below :

To blow air through the nozzle D (Fig. 4.5), the rubber baloon B is compressed. Air flows past A with high speed. The pressure at A falls below the atmospheric pressure. Liquid from the reservior C rises up. Caught by the high velocity jet of air, the liquid is broken into small drops which moves with a high velocity and comes out of the nozzle.

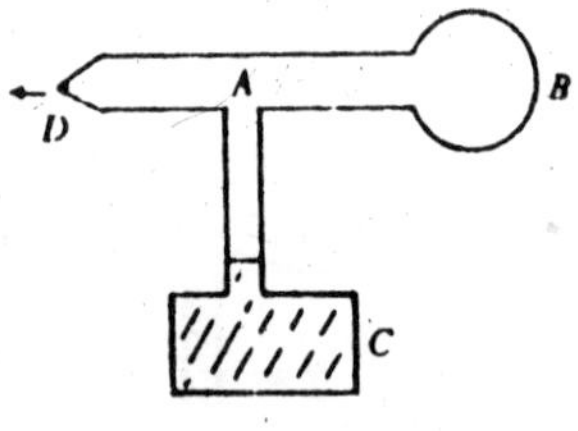

Fig. 4.5

(b) **The filter pump :**

It is an appliance based on Bernoulli's principle and is used to produce moderately low pressure inside a vessel (Fig. 4.6).

This is effected by a stream of water flowing out of narrow orifice B connected to a tap. Due to high velocity near the orifice, the pressure in this region falls. Due to this fall in pressure air enters the region by the side tube connected to the vessel to be evacuated.

The air entering is entrapped by the current of water and is taken away.

Problems :

1. Liquid flows in streamline motion through a horizontal tapering tube of circular section, the section at the exit end having half the inlet area. If the tube length is L, inlet area A, input velocity v, pressure at entrance P, density of the liquid ρ, find (a) the velocity u at a distance x from the entrance, (b) pressure at the same point, (c) the pressure at the exit end; the area at x being $A\left(1-\frac{x}{2L}\right)$.

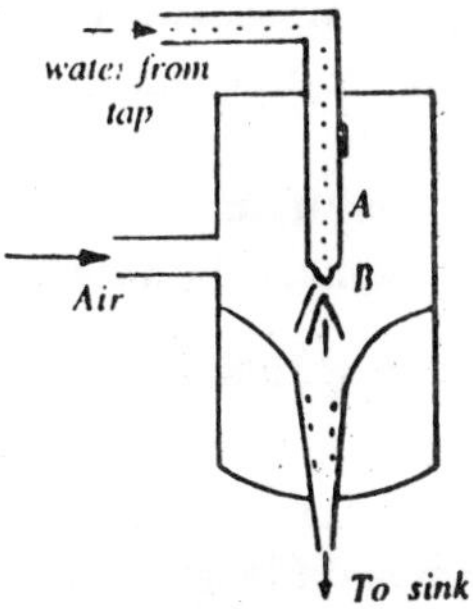

Fig. 4.6

Solution (Ref. Fig. 4.7) :

(a) Since volume/sec = constant

$$vA = uA\left(1-\frac{x}{2L}\right)$$

$$= uA\frac{2L-x}{2L} \quad \text{or,} \quad u = \frac{v2L}{2L-x}$$

(b) Let p be the required pressure at the point where the velocity is u. By Bernoulli's theorem

$$\frac{P}{\rho}+\frac{1}{2}v^2 = \frac{p}{\rho}+\frac{1}{2}u^2$$

$$\text{or,} \quad p = P+\frac{1}{2}\left(v^2-u^2\right)\rho$$

$$= P+\frac{1}{2}v^2\left\{1-\frac{4L^2}{(2L-x)^2}\right\}\rho$$

$$\text{or,} \quad p = P-\frac{\rho v^2 x(4L-x)}{(2L-x)^2}$$

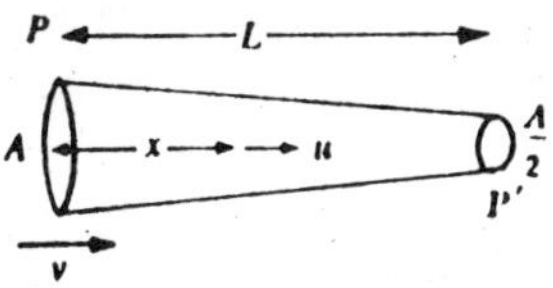

Fig. 4.7

(c) Let P' be the pressure and v' the velocity at the exit end.

We have $\frac{1}{2}v^2+\frac{P}{\rho} = \frac{1}{2}v'^2+\frac{P'}{p}$

But $vA = v'\frac{A}{2}$

or, $v' = 2v$

Substituting $p' = P + \frac{1}{2}\left(v^2 - v'^2\right)\rho$

$$= P - \frac{3v^2}{2}\rho$$

2. Water is maintained at a height of 20 cm in vessel which has a small circular hole radius 1 mm. Find the rate of flow if the coefficient of contraction is ·52.

Solution :

Volume flowing sec. = vel × Area

$$= \sqrt{2gh}\, cA = \sqrt{2\times 981\times 20}\times{\cdot}52\times 3{\cdot}41\times({\cdot}1)^2 = 3{\cdot}23 \text{ cm}^3/\text{sec.}$$

3. A venturimeter is connected to two points in a main where the radii are 20 cm and 15 cm respectively and the levels of water columns in the two limbs differ by 10 cm. How much water flows through the pipe per hour ?

Solution :

By equation (4.15) amount of water flowing per hour

$$= \alpha_1\alpha_2\sqrt{\frac{2(p_1 - p_2)}{\alpha_1^2 - \alpha_2^2}}\times 60\times 60$$

$$= \left(\pi r_1^2\right)\left(\pi r_2^2\right)\sqrt{\frac{2\times 10\times 1\times 981}{\left(\pi r_1^2\right)^2 - \left(\pi r_2^2\right)^2}}\times 60\times 60$$

$$= r_1^2\, r_2^2\sqrt{\frac{2\times 10\times 981}{r_1^4 - r_2^4}}\times 60\times 60$$

$$= (20)^2\,(15)^2\sqrt{\frac{2\times 10\times 981}{(20)^4 - (15)^4}}\times 60\times 60$$

$$= 400\times 225\sqrt{\frac{2\times 10\times 981}{160000 - 225\times 225}}\times 3600$$

$$= \sqrt{\frac{2\times 10\times 981}{109375}}\times 324\times 10^6 = 433\times 10^6 \text{ c.c.}$$

$43{\cdot}3\times 10^4$ lits.

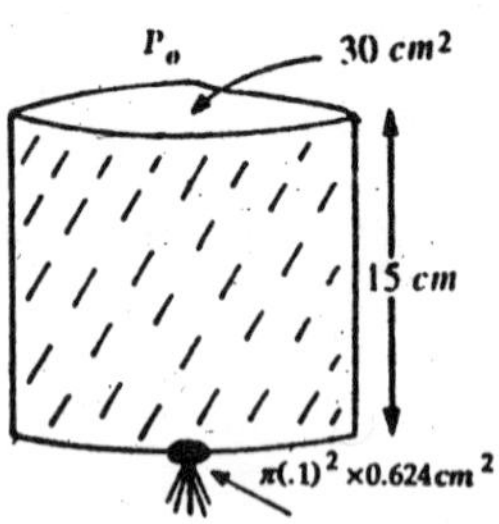

Fig. 4.8

4. Water is maintained at a height of 15 cm in a vessel which has a small circular hole in its thin horizontal base. The radius of the hole is 1 mm and the area of the free surface is 30 sq. cm.

Find the rate at which the liquid leaves the vessel if the coefficient of contraction is 0·624.

Solution :

From Torricelli's theorem the velocity of efflux $= \sqrt{2gh}$

Hence the volume of water flowing out per sec. $= \sqrt{2gh}\ c.A$

$$= \sqrt{2 \times 981 \times 15} \times 0{\cdot}624 \times 3{\cdot}14 \times ({\cdot}1)^2$$

$$= 3{\cdot}36 \text{ c.c.}$$

5. If a cylindrical vessel of cross-sectional area 10 cm² is filled upto height 20 cm, above an orifice of area 0·01 cm² at its side, calculate the time required for the free surface to fall down to 10 cm [Coefficient of contraction being 0·62].

Solution :

Let v be the velocity of the liquid when the free surface is a height h above the orifice,

From Torricelli's theorem

$$0{\cdot}62 \times 0{\cdot}01\sqrt{2gh}\,dt = -10dh$$

$$\text{or, } \int_0^{t_0} dt = \int_{h_1}^{\frac{h_1}{2}} -\frac{dh}{\sqrt{h}} \times \frac{10}{0{\cdot}62 \times 0{\cdot}01\sqrt{2g}}$$

$$\text{or, } t_0 = \frac{1000 \times 2}{{\cdot}62\ \sqrt{2 \times 981}} \left[\sqrt{h}\right]_{10}^{20} = 72{\cdot}8\left\{\sqrt{20} - \sqrt{10}\right\}$$

$$= 72{\cdot}8 \times 1{\cdot}31 = 95{\cdot}36 \text{ secs.}$$

6. If the diameters of a pipe are 10 cm and 6 cm at the points where a venturimeter is connected and the pressures at the points are found to differ by 5 cms of water column. Find the volume of water flowing through the pipe per second.

Solution :

Volume passing per sec.

$$= \alpha_1\alpha_2\sqrt{\frac{2(p_1 - p_2)}{\alpha_1^2 - \alpha_2^2}}$$

$$= 25\pi \times 9\pi \sqrt{\frac{2\times5\times1\times981}{(25\pi)^2-(9\pi)^2}}$$

$$= 225\pi^2 \sqrt{\frac{9810}{34\pi\times16\pi}}$$

$$= 225\pi \sqrt{\frac{9810}{34\times16}} = 3002 \text{ c.c.}$$

Appendix :

4.8. Efflux of Gases

For steady motion of a compressible fluid (density depending on pressure), Bernoulli's equation may be written as $\int \frac{dp}{\rho} + \frac{1}{2}v^2 + \Omega =$ constant.

Let p_1 and ρ_1 be the pressure and density respectively of a gas when it is contained in a vessel. Let the gas issue through an orifice into the open space, the pressure and density of the gas when it issues being P_2 and ρ_2 respectively.

Assuming that adiabatic conditions prevail, $\frac{p}{\rho^\gamma}$ = constant, where γ is the ratio of the two specific heats of the gas.

Hence in the absence of external forces $\frac{1}{2}v^2 = \int_{p_1}^{p_2} \frac{dp}{\rho}$

$$\text{or, } v^2 = \frac{2\gamma}{\gamma-1}\left\{1-\left(\frac{p_2}{p_1}\right)^{\frac{\gamma-1}{\gamma}}\right\} \tag{4.16}$$

Equation (4.16) shows that an increase in velocity is attended by a decrease in pressure. A vacuum cleaner works on this principle.

CHAPTER V

UNITS AND DIMENSIONS

Units and Dimensions : c.g.s. and S.I. units. Dimensional analysis : Principle of homogeneity of dimensions and its applications.

5.1. Units and Dimensions

To measure a physical quantity we must choose a standard and the quantity is to be expressed in terms of the standard. This standard is called the unit of the physical quantity. There are thousands of physical quantities and to avoid complications fundamental units are chosen to be three in number. These are units of length (L), mass (M) and time (T). Units of all other physical quantities are fundamental units raised to different powers. They are called derived units. The two commonly used system of units are the c.g.s. and S.I. units.

c.g.s. (centimeter—gramme—second) unit :

In this system the unit of length is centimeter, that of mass, the gramme and that of time, the second. The centimeter is one hundredth part of a metre, which is the distance, at a temperature 0°C, between two lines in a platinum-irridium bar, preserved at the International Bureau of Metric Weights and Measures near Paris.

The gramme is one thousandth part of a lump of platinum-irridium called a kilogramme, made by Borda, in accordance with a decree of the French Republic also preserved at the above place.

The second is $\frac{1}{86400}$ part of the mean solar day which is the average time between two successive transits of the sun across the meridian.

Most widely used in the scientific world is 'Systeme International de units' or briefly S.I. unit. The fundamental units in this system are Meter, Kilogramme, Second as usual. But besides these there are units like Candela · light, Ampeare in electricity and Kelvin in heat. Candela refers to the .surement of luminous intensity, ampeare to current strength and Kelvin to the measurement of temperature.

5.2. Dimensions

Dimension of a physical quantity is an expression which shows how and which of the fundamental units enter into these expression for the unit of the physical quantity. Dimensions of several familiar physical quantities are calculated below :

$$[v] = [\text{Velocity}] = \frac{L}{T} = LT^{-1}$$

$$[f] = [\text{Acceleration}] = \frac{LT^{-1}}{T} = LT^{-2}$$

$$[F] = [\text{Force}] = [m][f] = MLT^{-2}$$

$$[\text{Pressure}] = \frac{[\text{Force}]}{[\text{Area}]} = \frac{MLT^{-2}}{L^2} = ML^{-1}T^{-2}$$

$$[\text{Moment of inertia}] = [I] = ML^2$$

$$[\text{Gravitational constant}] = [G] = \frac{[F][r^2]}{[m_1][m_2]}$$

$$= \frac{MLT^{-2}L^2}{M^2} = M^{-1}L^3T^{-2}$$

$$[\text{Surface tension}] = [S] = \left[\frac{F}{l}\right] = \frac{MLT^{-2}}{L} = MT^{-2}$$

$$[\text{Coefficient of viscosity}] = [\eta] = \left[\frac{F}{A\frac{dv}{dx}}\right]$$

$$= \frac{MLT^{-2}}{L^2\,\frac{LT^{-1}}{L}} = ML^{-1}T^{-1}$$

$$[\text{Modulus of elasticity}] = \frac{[\text{Stress}]}{[\text{Strain}]} = \frac{\left[\frac{F}{A}\right]}{1}$$

$$= \frac{MLT^{-2}}{L^2} = ML^{-1}T^{-2}$$

$$[\text{Poisson ratio } (\sigma)] = \frac{[\text{Leteral Strain}]}{[\text{Longitudinal Strain}]} = 1$$

$$[\text{Density}] = [\rho] = \frac{[M]}{[V]} = \frac{M}{L^3} = ML^{-3}$$

$$[\text{Work}] = [W] = [F \times x] = MLT^{-2}L = ML^2T^{-2}$$

$$[\text{Power}] = \frac{[W]}{[t]} = \frac{ML^2\,T^{-2}}{T} = ML^{2}\,T^{-3}$$

5.3. Dimensional Analysis (Homogeneity of Dimensions)

In the expression deduced for a given physical quantity the dimensions on both sides must be same. This is known as the principle of homogeneity of dimensions. This priniciple can be used for many important purposes; some of which are the follows :

(i) Conversion of one system of units into another.
(ii) Checking the results already arrived at.
(iii) Deducing a correct relationship among different physical quantities.
(iv) Locating 'the errors in the development of formulae'.

The operation is known as dimensional analysis.

The following short examples will illustrate these uses.

(i) Conversion of one system of units into another :

Example : To convert Newton into dyne

Newton is the unit of force in *M.K.S.* system.

Thus 1 Newton $\doteq MLT^{-2}$ $(M.K.S.)$

$$= 1000\ M\ 100\ LT^{-2}\ \text{c.g.s.}$$

$$= MLT^{-2}\text{c.g.s.} \times 10^5 = 10^5\text{dyne.}$$

(ii) Checking suggested theoretical relations for possibility :

Our knowledge on hydrodynamics suggests that in some cases, the velocity C of the waves on a liquid surface are connected with the wave-length λ and the acceleration due to gravity by the relation

$$C^2 = \frac{g\lambda}{2\pi}.$$

Now $\left[C^2\right] = L^2T^{-2}$

And $[g\lambda] = LT^{-2}L = L^2T^{-2}$

As 2π is non-dimensional constant, the above relation is possible.

(iii) Deducing a correct relationship between different physical quantities :

The basis of this method is also the principle of homogeneity of dimensions. Knowing the factors on which a physical quantity may possibly depend, an expression for it can be obtained in terms of these factors. Equating the dimensions on both sides the uncertainties in the powers of these factors can be found out.

Illustration :

(a) For a liquid having streamline motion through a capillary tube, the volume of the liquid flowing per sec. through the tube depends on the following factors :

(i) the pressure gradient $\left(\frac{P}{l}\right)$ along the axis of the tube.

(ii) the radius r of the tube.

(iii) the coefficient of viscosity η of the liquid.

Thus we may write

$$V = k\left(\frac{P}{l}\right)^{\alpha} r^{\beta}\,\eta^{\gamma},$$

where α, β, γ are constants to be determined and K is a non-dimensional constant.

Equating the dimensions on both sides

$$L^3T^{-1} = \left(\frac{ML^{-1}T^{-2}}{L}\right)^{\alpha} L^{\beta}\left(ML^{-1}T^{-1}\right)^{\gamma}$$

$$= M^{\alpha+\gamma}L^{-2\alpha+\beta-\gamma}T^{-2\alpha-\gamma}$$

Equating the powers of M, L and T on both sides.

$$\alpha+\gamma = 0 \qquad \text{... (i)}$$

$$-2\alpha+\beta-\gamma = 3 \qquad \text{... (ii)}$$

$$-2\alpha-\gamma = -1 \qquad \text{... (iii)}$$

From (i) and (iii) $\alpha = -\gamma$ and $\alpha = 1$, $\gamma = -1$

From (ii) $-2+\beta+1 = 3$

or, $\beta = 4$

Substituting above $V = K\frac{P}{l\eta}r^4$

correspondence with experimental results shows that the value $\frac{\pi}{8}$ fits to K.

Hence $V = \frac{\pi Pr^4}{8l\eta}$.

(b) **Velocity of Transverse waves in a stretched string :**

It is known from experience that the velocity v of these transverse waves depends on the tension T of the string and its material density ρ.

Hence we can write $v = KT^{\alpha}\rho^{\beta}$

Equating the dimensions on both sides

$$LT^{-1} = (MLT^{-2})^{\alpha}\ (ML^{-1})^{\beta} = M^{\alpha+\beta}\ L^{\alpha-\beta}\ T^{-2\alpha}$$

Equaing the power of T on both sides

$$2\alpha = 1, \text{ or, } \alpha = \frac{1}{2}$$

Equating the power of M

$$\alpha + \beta = 0 \quad \text{or, } \beta = -\alpha = -\frac{1}{2}$$

Hence $v = K\sqrt{\frac{T}{p}}$.

(iv) Locating the errors in the development of formula :

It sometimes happens in deriving consequences from physical principles that because of an error in the theoretical considerations, they can be seen to be wrong by reason of dimensional inhomogeneity which has been detected. The successive developmental equations may also be checked dimensionally and the source of error found out.

5.4. Limitations of Dimensional Analysis

The method of dimensions with its decent application has a number of limitations. The most notable of these are the following :

(a) This method gives little or no information as regards the non-dimensional constants involved in different expressions. The values of these constants cannot be determined by the dimensional analysis.

(b) This method is applicable only in the case of those relations which are power functions. Relationships which involves trigonometric and exponential functions cannot be tackeled by this method.

(c) The most outstanding defect of this method is that it fails in the case which involves the dependence on more than three variables. This is because at most three equations can be formed by equating the powers of L, M and T on both sides. These three independent equations enable us to furnish the values of three unknowns.

(d) The last but not the least demerit is that the application of this method is risky without precise knowledge of the variables on which a particular physical quantity may possibly depend.

EXERCISES WITH HINTS FOR SOLUTION

1. Experiment suggests that the velocity of a sphere allowed to fall from rest in a viscous medium varies with time according to the law $\log\left(1 - \frac{kv}{g}\right) = -kt$ where k is a constant for a given sphere and medium. Examine this suggested law for dimensional homogenity and express k in terms of the radius r and the mass M of the sphere and η the viscosity of the medium.

Hints for solution :

$$\log\left(1-\frac{kv}{g}\right) = -kt$$

or, $1-\frac{kv}{g} = e^{-kt}$

Differentiating w.r.t. t

$$-\frac{k}{g}\frac{dv}{dt} = -ke^{-kt}$$

or, $-\frac{f}{g} = -e^{-kt} = -\left(1-\frac{kv}{g}\right)$

or, $f = g - kv$

kv has the dimension of acceleration

Let $K = A\ M^{\alpha} r^{\beta} \eta^{\gamma}$, A being a constant

Now $AM^{\alpha} r^{\beta} \eta^{\gamma} v = f$

or, $M^{\alpha} L^{\beta} (ML^{-1} T^{-1})^{\gamma} LT^{-1} = LT^{-2}$

Equating the powers of M, L and T on both sides

$$\alpha + \gamma = 0$$

$$\beta - \gamma + 1 = 1, \text{ or } \beta - \gamma = 0, \text{ or } \beta = \gamma$$

and $-\gamma - 1 = -2$, or $\gamma = 1 = \beta$

and $\alpha = -1$

Substituting the value of α, β and γ

$$K = AM^{-1} r\eta = \frac{Ar\eta}{M}.$$

2. If the velocity of light C, the Newtonian Constant of Gravitation G and Plank's constant of action h are chosen as the fundamental units, find the dimensions of the ordinary units of mass, length and time in the new system.

Solution :

Let us use the familiar equations $F = \frac{Gm_1 . m_2}{r^2}$

or, $MLT^{-2} = \frac{GM^2}{L^2}$

or, $M = \frac{L^3}{G} T^{-2}$ ... (1)

Also $mC^2 = h\upsilon$

or, $MC^2 = hT^{-1}$

or, $T = \frac{h}{MC^2}$ ... (2)

From (1) and (2)

$$M=\frac{L^3}{G}\frac{M^2C^4}{h^2} \quad \text{... (3)}$$

From the uncertainty principle

$$MLT^{-1}L=h$$

or, $T=\frac{ML^2}{h}=\frac{h}{MC^2}$

$$\text{or, } L^2=\frac{h^2}{M^2C^2} \quad \text{... (4)}$$

From (3) and (4)

$$M=\frac{M^2C^4}{h^2G}\frac{h^3}{M^3C^3}=\frac{hC}{GM}$$

$$\text{or, } M^2=hC\,G^{-1} \quad \text{... (5)}$$

Substituting in (4)

$$L^2=\frac{h^2}{C^2\,Ch\,G^{-1}}=C^{-3}\,h\,G \quad \text{... (6)}$$

$$\text{From (2) } T^2=\frac{h^2}{M^2C^4}=\frac{h^2}{C^4ChG^{-1}}=hGC^{-5} \quad \text{... (7)}$$

3. In the new unitary system suggested in the previous problem c, h and G are all of unit value. Find the values of the gram, centimeter and second in terms of the new units of mass, length and time respectively, given that in $c.g.s.$ units $C = 3\times10^{10}$, $G = 6{\cdot}7\times10^{-6}$, $h = 6{\cdot}6\times10^{-27}$

Solution :

As shown in the previous problem

$$M^2=\frac{hC}{G}$$

$$\text{or, } M=\sqrt{\frac{Ch}{G}}=\sqrt{\frac{3\times10^{10}\times6{\cdot}6\times10^{-27}}{6{\cdot}7\times10^{-8}}}$$

$$=\sqrt{\frac{3\times6{\cdot}6\times10^{-8}}{67}}$$

when c, h and G have unit values

$$\text{(gram)} = M = \sqrt{\frac{67}{3\times 6\cdot 6\times 10^{-8}}} = 1\cdot 84\times 10^{4}$$

$$\text{Similarly } L = \sqrt{\frac{hG}{c^3}} = \sqrt{\frac{6\cdot 6\times 10^{-27}\times 6\cdot 7\times 10^{-8}}{\left(3\times 10^{10}\right)^3}}$$

$$= \sqrt{\frac{6\cdot 6\times 67\times 10^{-65}}{27}}$$

When c, h and G have unit values

$$\text{centimeter} = \sqrt{\frac{27}{6\cdot 6\times 67}\times 10^{65}} = \sqrt{\frac{270}{6\cdot 6\times 6\cdot 7}}\times 10^{32} = 2\cdot 47\times 10^{32}$$

$$\text{Lastly } T = \sqrt{\frac{hG}{c^5}} = \sqrt{\frac{6\cdot 6\times 10^{-27}\times 6\cdot 7\times 10^{-8}}{(3\times 10^{10})^5}}$$

When c, h and G have unit values

$$\text{second} = \sqrt{\frac{3^5\times 10^{50}}{6\cdot 6\times 10^{-27}\times 6\cdot 7\times 10^{-8}}} = \sqrt{\frac{(3)^5}{66\times 6\cdot 7}}\times 10^{43} = 7\cdot 41\times 10^{42}$$

4. If the fundamental units are the velocity of light in vacuum, the acceleration due to gravity at Greenwich and the density of mercury at 0°C, find the units of mass length and time.

C = 3 × 10^{10} cm/sec.; g Greenwich = 981 cm/sec.^{2}

density of mercury =13·6 gm /c.c.

Hints for Solution :

Unit of velocity $(LT^{-1}) = 3\times 10^{10}$ cm/sec. ... (i)

Unit of acceleration (LT^{-2}) = 981 cm/sec^2 ... (ii)

Unit of density $(ML^{-3}) = 13\cdot 6$ gm/c.c. ... (iii)

From (i) and (ii)

Unit of time $(T) = \frac{3\times 10^{10}}{981} = 3\cdot 058\times 10^7$ sec.

Substituting in (i), unit of length $(L) = LT^{-1}.\ T$

$$= 3\times 10^{10}\times 3\cdot 058\times 10^7 = 9\cdot 174\times 10^{17}\text{cm}$$

Again from (iii) unit of mass $(M) = ML^{-3}\times L^3$

$$= 13\cdot 6\times (9\cdot 174\times 10^{17})^3 = 1\cdot 051\times 10^{55}$$

5. Find the dimension of velocity and acceleration. Assuming that when a body falls from rest under gravity the velocity v is given by $Kg^p h^q$, where h is the distance fallen through, g is the acceleration due to gravity. K, p and q are constants. Show by consideration of dimensions involved, that $v = k\sqrt{gh}$.

Hints for Solution :

$$[g^p] = (LT^{-2})^p$$
$$[h]^q = L^q$$

Hence $[Kg^p h^q] = L^p T^{-2p} L^q$

$$= L^{p+q} T^{-2p}$$

Thus $LT^{-1} = L^{p+q} T^{-2p}$

Equating the power of L and T on both sides

$$p + q = 1 \text{ and } -2p = -1$$

Solving $p = \frac{1}{2}, \quad q = \frac{1}{2}$

Hence $v = k\sqrt{gh}$

6. The intensity I of a sound wave—energy flux per unit area per unit time —depends upon the amplitude a, the medium density ρ, the frequency f and the velocity ϑ. Find the form of the connection and show that for equal intensities in the same medium, waves must have amplitudes inversely proportional to their frequencies.

Hints for Solution :

Let $I = ka^\alpha \rho^\beta f^\gamma \vartheta^\delta$

Equating the dimensions on both sides

$$\frac{MLT^{-2}L}{L^2 . T} = MT^{-3} = L^\alpha (ML^{-3})^\beta T^{-\gamma} (LT^{-1})^\delta$$

Equating the powers of M, L and T on both sides

$$\beta = 1$$
$$\alpha - 3\beta + \delta = 0$$
$$\gamma + \delta = 3 \quad \text{or, } \delta = 3 - \gamma$$
$$\therefore \alpha = 3 - \delta = 3 - (3 - \gamma) = \gamma$$

Hence $I = \rho\vartheta^3 \varphi\left(\frac{af}{\vartheta}\right)$

7. Assuming that the vibration frequency ν of atoms in a crystal depends only on the atomic mass m, the spacing distance d and compressibility β, find the nature of the relationship.

Solution (Hints) :

Let $v = km^a d^b \beta^c$

Equating the dimensions on both sides

$$T^{-1} = M^a L^b \left(\frac{1}{\frac{MLT^{-2}}{L^2}} \right)^c = M^{a-c} L^{b+c} T^{2c}$$

Equating the powers of M, L and T on both sides

$$a - c = 0 \quad \text{or,} \quad a = c$$

$$b + c = 0 \quad \text{or,} \quad c = -b$$

and $2c = -1$

$$\text{or,} \ c = -\frac{1}{2}$$

$$b = \frac{1}{2}$$

$$a = -\frac{1}{2}$$

Hence, $\hat{v} = k\sqrt{\dfrac{d}{m\beta}}$

8. Assuming that the mass M of the largest stone that can be moved by a flowing river depends on v, the velocity, ρ, the density of water and g, show that M varies with the sixth power of the velocity of flow of the river.

Hints for Solution :

Let $M = k\, v^\alpha \rho^\beta g^\gamma$

Equating the dimensions on both sides

$$M = (LT^{-1})^\alpha \, (ML^{-3})^\beta \, (LT^{-2})^\gamma$$

$$= M^\beta L^{\alpha - 3\beta + \gamma} T^{-\alpha - 2\gamma}$$

Equating the power of M, L and T on both sides

$$\beta = 1,$$

$$\alpha - 3\beta + \gamma = 0 \qquad \text{Solving} \quad \alpha - 3 - \frac{\alpha}{2} = 0$$

$$-\alpha - 2\gamma = 0 \quad \text{or,} \ \alpha = -2\gamma \ \text{or,} \ \frac{\alpha}{2} = 3 \ \therefore \alpha = 6 \ \text{Hence etc.}$$

9. Investigate dimensionally the connection between the reverberation period t in a room, its volume V_1 its surface area A and the velocity of sound v.

Hints for Solution :

Let $t = kV^{\alpha} A^{\beta} v^{\gamma}$

Equating the dimensions on both sides

$T = (L^3)^{\alpha} (L^2)^{\beta} (LT^{-1})^{\gamma}$

Equating the power if L and T on both sides

$$3\alpha + 2\beta + \gamma = 0$$

$$-\gamma = 1 \quad \text{or, } \gamma = -1$$

substituting above

$$3\alpha + 2\beta = 1 \text{ or, } \alpha = \frac{1-2\beta}{3}$$

Hence $t = kV^{\frac{1-2\beta}{3}} A^{\beta} v^{-1}$

$$= \frac{k}{v} V^{\frac{1-2\beta}{3}} A^{\beta}$$

$$= \frac{k}{v}\left(\frac{V^2}{A^3}\right)^{\beta'} V^{\frac{1}{3}} \text{ where } \beta' = -\beta$$

10. The time period of rotation t of small particle of mass m_1 around a heavy nucleus of mass m_2 depends upon these masses, the distance l and the constant of gravitation G.

Hence $t = k\, G^{\alpha} m_1^{\beta} m_2^{\gamma} l^{\delta}$

Again $F = \dfrac{Gm_1m_2}{l^2}$

whence $MLT^{-2} = [G]\, M^2 L^{-2}$

or, $[G] = M^{-1} L^3 T^{-2}$

The dimensional equation reduces to

$T = (M^{-1} L^3 T^{-2})^{\alpha} (M)^{\beta+\gamma} L^{\delta}$

Equating the power of M, L and T on both sides

$$-\alpha + \beta + \gamma = 0$$

$$3\alpha + \delta = 0$$

$$-2\alpha = 1$$

Solving $\alpha = -\frac{1}{2},\ \delta = \frac{3}{2},\ \beta + \gamma = -\frac{1}{2}$

Hence $t = kG^{-\frac{1}{2}} m_1^{\beta+\gamma} m_1^{-\gamma} m_2^{\gamma} l^{\frac{3}{2}}$

$$= K\sqrt{\frac{l^3}{Gm_1}}\left(\frac{m_2}{m_1}\right)^\gamma = K\sqrt{\frac{l^3}{Gm_1}}f_1\left(\frac{m_2}{m_1}\right)$$

11. Assuming that the period of vibration of a timing fork depends upon the length of the prongs, on the density and Young's modulus of the material, find by the method of dimensions a formula for the period of vibration.

Hints for Solution :

Let $t = l^\alpha \rho^\beta y^\gamma$

Equating dimensions on both sides

$$T = L^\alpha (ML^{-3})^\beta \left(\frac{MLT^{-2}}{L^2}\right)^\gamma$$

Equating the powers of M, L and T on both sides

$$\beta + \gamma = 0$$

$$\alpha - 3\beta - \gamma = 0$$

and $-2\gamma = 1$ or, $\gamma = -\frac{1}{2}$

$$\beta = +\frac{1}{2}$$

$$\alpha - \frac{3}{2} + \frac{1}{2} = 0$$

or, $\alpha = 1$

Hence $t = kl\sqrt{\frac{\rho}{y}}$; k being a non-dimensional constant.

12. It has been suggested that for liquids $S^2 \beta^4 = k$, a constant, S being the surface tension and β the compressibility. Show that the quantity k can not be a mere numeric.

Hints for Solution :

$$[S]^2 = \left(\frac{MLT^{-2}}{L}\right)^2 = M^2 T^{-4}$$

$$\text{Again } [\beta]^4 = \left\{\frac{1}{\frac{MLT^{-2}}{L^2}}\right\}^4 = \left(M^{-1} LT^2\right)^4 = M^{-4} L^4 T^8$$

$$[S]^2 \, [\beta]^4 = (M^2 \, T^{-4}) \, (M^{-4} \, L^4 \, T^8)$$

$$= M^{-2} \, L^4 \, T^4$$

Hence k cannot be a mere numeric, but it will have the dimension of $M^{-2} \, L^4 \, T^4$.

13. Assuming that the range R a of projectile thrown at a given angle with the horizontal depends only on the velocity v of projection and the acceleration due to gravity g, find the form of the relation.

Hints for Solution :

Let $R = k \, v^{\alpha} g^{\beta}$

Equating the dimensions on both sides,

$$L = (LT^{-1})^{\alpha} \; (LT^{-2})^{\beta}$$

Equating the powers of L and T on both sides

$$\alpha + \beta = 1$$

and $-\alpha - 2\beta = 0$ or, $\alpha = -2\beta$

Substituting, $-2\beta + \beta = 1$

or, $\beta = -1$

and $\alpha = 2$

Hence $R = \dfrac{kv^2}{g}$

14. Using the method of dimensions obtain expressions for (a) acceleration of a particle moving with uniform speed v in a circle of radius r.

(b) the tension τ in a uniform circular wire of radius r and mass m per unit length, rotating in its own plane with angular velocity ω about an axis passing through its centre and pendicular to its plane.

Hints :

(a) Let $f = kr^{\alpha} v^{\beta}$

Equating dimensions on both sides

$$LT^{-2} = L^{\alpha} (LT^{-1})^{\beta}$$

Equating the powers of L and T on both sides

$$\alpha + \beta = 1$$

$-\beta = -2$ or, $\beta = 2$

$\therefore \; \alpha = -1$

Hence $f = k\frac{v^2}{r}$

(b) Let $\tau = r^{\alpha} m^{\beta} \omega^{\gamma}$

Equating the dimensions on both sides

$MLT^{-2} = L^{\alpha} (ML^{-1})^{\beta} T^{-\gamma}$

Equating the powers of M, L and T on both sides

$$\beta = 1$$

$$\alpha - \beta = 1 \quad \text{or,} \quad \alpha = 2$$

$$-\gamma = -2 \quad \text{or,} \quad \gamma = 2$$

Hence $\tau = kmr^2\omega^2$

15. Under a pressure difference p a viscous liquid is forced at a rate of Vcc./sec. through a circular opening of radius r in a plate. If the liquid density is ρ and the viscosity is η, find how V is related to the other quantities.

Hints for Solution :

Let $V = kp^{\alpha} r^{\beta} \rho^{\gamma} \eta^{\delta}$

Equating the dimensions on both sides

$$L^3 T^{-1} = \left(\frac{ML\,T^{-2}}{L^2}\right)^{\alpha} L^{\beta} (ML^{-3})^{\gamma} (ML^{-1} T^{-1})^{\delta}$$

$$= M^{\alpha+\gamma+\delta} L^{-\alpha+\beta-3\gamma-\delta} T^{-2\alpha-\delta}$$

Equating the powers of M, L and on both sides

$$\alpha + \gamma + \delta = 0 \qquad \text{... (i)}$$

$$-\alpha + \beta - 3\gamma - \delta = 3 \qquad \text{... (ii)}$$

$$-2\alpha - \delta = -1 \qquad \text{.... (iii)}$$

From (i) and (ii)

$$\beta - 2\gamma = 3 \qquad \text{... (iv)}$$

From (i) and (iii)

$$-\alpha + \gamma = -1 \qquad \text{... (v)}$$

Also $\delta = 1 - 2\alpha$

and $\beta = 1 + 2\alpha$

$\gamma = \alpha - 1$

Hence $V = kr^{1+2\alpha} \rho^{\alpha-1} \eta^{1-2\alpha}$

$$= k\frac{\eta r}{\rho}\left(\frac{r^2 p\,\rho}{\eta^2}\right)^{\alpha} = \frac{\eta r}{\rho}\,\varphi\left(\frac{p\,r^2\rho}{\eta^2}\right)$$

16. A drop of a liquid is suspended in another liquid of same specific gravity but with which it is immisible. If the drop is distorted from the spherical shape and released, deduce by the method of dimensions a formula for the period of oscillation, given (i) that the latter depends on the surface tension S, the radius r of the drop and its density ρ.

Hints for Solution :

Let the period of oscillation $t = ks^{\alpha} r^{\beta} \rho^{\gamma}$

Equating the dimensions on both sides

$$T = \left(\frac{MLT^{-2}}{L}\right)^{\alpha} L^{\beta} \left(ML^{-3}\right)^{\gamma}$$

Equating the powers of M, L and T on both sides

$$\alpha + \gamma = 0 \quad \text{... (i)}$$

$$\beta - 3\gamma = 0 \quad \text{... (ii)}$$

$$-2\alpha = 1 \quad \text{... (iii)}$$

From (iii) $\alpha = -\dfrac{1}{2}$

From (i) $\gamma = -\alpha = \dfrac{1}{2}$

From (ii) $\beta = 3\gamma = \dfrac{3}{2}$

Hence $t = KS^{-\frac{1}{2}} r^{\frac{3}{2}} \rho^{\frac{1}{2}} = k\sqrt{\dfrac{r^3 \rho}{S}}$, K being a non-dimensional constant.

17. Assuming that the exess pressure inside a soap bubble depends on the surface tension S of the soap solution and the radius r, derive a relationship between the different quantities.

Hints for Solution :

Let $p = KS^{\alpha} r^{\beta}$

Equating the dimensions on both sides

$$\frac{MLT^{-2}}{L^2} = \left(\frac{MLT^{-2}}{L}\right)^{\alpha} L^{\beta}$$

Equating the power of M and L on both sides

$$\alpha = 1$$

$$\beta = -1$$

Hence $p = k\dfrac{S}{r}$

18. Assuming that the critical velocity in the case of flow through a capillary tube depends upon the viscosity of fluid, its density and the radius of the tube derive a relation between them.

Hints for Solution :

Let the critical velocity

$v_c = K\,\eta^{\alpha}\,\rho^{\beta}\,r^{\gamma}$, symbols having usual meanings,

Equating the dimensions on both sides,

$$LT^{-1} = (ML^{-1}\,T^{-1})^{\alpha}\,(ML^{-3})^{\beta}\,L^{\gamma}$$

Equating the powers of M, L and T on both sides

$$\alpha + \beta = 0 \qquad \text{or, } \alpha = -\beta \qquad \text{... (i)}$$

$$-\alpha - 3\beta + \gamma = 1 \qquad \text{or, } \gamma = -1 \qquad \text{... (ii)}$$

$$-\alpha = -1 \qquad \text{or, } \alpha = 1$$

$$\text{or, } \beta = -1$$

Hence $V_c = \dfrac{k\eta}{\rho r}$, K, being a non-dimensional constant known as Reynold's No.

19. Assuming that the pressure P which a gas exerts on the wall of the container depends on the mass m of a molecule, the average velocity v and the number of molecules per unit volume, show that if the temperature θ is proportional to the mean $K.E.$ of a single molecule, $PM = \rho R\theta$, where M is the molecular weight, ρ is the density and R a constant which is same for all gases.

Hints for Solution :

Let $P = Km^{\alpha} v^{\beta} n^{\gamma}$.

Equating the dimensions on both sides

$$\frac{MLT^{-2}}{L^2} = M^{\alpha}\left(LT^{-1}\right)^{\beta} L^{-3\gamma}$$

Equating the powers of M, L and T on both sides

$\alpha = 1,\ \beta - 3\gamma = -1,\ -\beta = -2$

or, $\beta = 2$ and $\gamma = 1$

Hence $P = K\,mv^2\,n$

Again $\theta \propto \frac{1}{2}mv^2$ or, $mv^2 = \dfrac{2\theta}{k^1}$

$\therefore\ P = K\dfrac{2\theta\, n}{k^1}$

or, $PM = \dfrac{K}{K^1}\,2\theta\, nM = R\theta\rho$

TOPIC INDEX

PART–I

Methods of Mathematical Physics and Mechanics

CHAPTER I

Vector Analysis

CHAPTER II

Ordinary Differential Equations

Chapter III

Equations

Chapter IV

Fourier Series

Chapter V

Matrix

Chapter VI

Mechanics I

CHAPTER VII

Rotational Motion

CHAPTER VIII

Gravitation

CHAPTER IX

Mechanics II

CHAPTER X

Mechanics II

CHAPTER XI

Small Oscillations

PART–II

General Properties of Matter

CHAPTER I

Elasticity

CHAPTER II

Surface Tension

CHAPTER III

Viscosity

CHAPTER IV

Mechanics of Ideal Fluids

CHAPTER V

Units and Dimensions